Applications of X-ray Photoelectron Spectroscopy to Catalytic Studies

From Routine Analysis to Cutting-Edge Surface Characterization

CATALYTIC SCIENCE SERIES

ISSN 1793-1398 (Print)
ISSN 2399-4495 (Online)

Series Editor: Chris Hardacre (*The University of Manchester, UK*)

Catalysis is at the forefront of the chemical industry and is essential to many fields in the chemical sciences. This series explores all aspects of catalysis in authored and edited volumes drawing on expertise from around the globe in a focussed manner. Volumes are accessible by postgraduate students and professionals in academia and industry.

Published

Vol. 21 *Applications of X-ray Photoelectron Spectroscopy to Catalytic Studies: From Routine Analysis to Cutting-Edge Surface Characterization*
edited by Spyridon Zafeiratos

Vol. 20 *Noble-Metal-Free Electrocatalysts for Hydrogen Energy*
edited by Qingsheng Gao and Lichun Yang

Vol. 19 *Iron Catalysis: Design and Applications*
edited by Jose M. Palomo

Vol. 18 *Photoorganocatalysis in Organic Synthesis*
edited by Maurizio Fagnoni, Stefano Protti and Davide Ravelli

Vol. 17 *Hydroprocessing Catalysts and Processes: The Challenges for Biofuels Production*
edited by Bo Zhang and Duncan Seddon

Vol. 16 *Electro-Catalysis at Chemically Modified Solid Surfaces*
by Jacques Simonet

Vol. 15 *Noble Metal Noble Value: Ru-, Rh-, Pd-catalyzed Heterocycle Synthesis*
edited by Xiao-Feng Wu

Vol. 14 *Enantioselective Titanium-Catalysed Transformations*
by Hélène Pellissier

Vol. 13 *Gold Catalysis: An Homogeneous Approach*
edited by F. Dean Toste and Véronique Michelet

Vol. 12 *Catalysis by Ceria and Related Materials (Second Edition)*
edited by A. Trovarelli and P. Fornasiero

More information on this series can be found at http://www.worldscientific.com/series/css

(Continued at end of book)

Catalytic Science Series — Vol. 21
Series Editor: Chris Hardacre

Applications of X-ray Photoelectron Spectroscopy to Catalytic Studies

From Routine Analysis to Cutting-Edge Surface Characterization

edited by
Spyridon Zafeiratos
CNRS & University of Strasbourg, France

World Scientific
NEW JERSEY • LONDON • SINGAPORE • BEIJING • SHANGHAI • HONG KONG • TAIPEI • CHENNAI • TOKYO

Published by

World Scientific Publishing Europe Ltd.

57 Shelton Street, Covent Garden, London WC2H 9HE

Head office: 5 Toh Tuck Link, Singapore 596224

USA office: 27 Warren Street, Suite 401-402, Hackensack, NJ 07601

Library of Congress Cataloging-in-Publication Data
Names: Zafeiratos, Spyridon, editor.
Title: Applications of x-ray photoelectron spectroscopy to catalytic studies : from routine analysis to cutting-edge surface characterization / edited by Spyridon Zafeiratos, CNRS & University of Strasbourg, France.
Description: New Jersey : World Scientific, [2023] | Series: Catalytic science series, 1793-1398 ; vol. 21 | Includes bibliographical references and index.
Identifiers: LCCN 2022041380 | ISBN 9781800613287 (hardcover) | ISBN 9781800613294 (ebook for institutions) | ISBN 9781800613300 (ebook for individuals)
Subjects: LCSH: Catalysts--Analysis. | X-ray photoelectron spectroscopy. | Catalysis.
Classification: LCC QD505 .A63 2023 | DDC 541/.395--dc23/eng20221121
LC record available at https://lccn.loc.gov/2022041380

British Library Cataloguing-in-Publication Data
A catalogue record for this book is available from the British Library.

For any available supplementary material, please visit
https://www.worldscientific.com/worldscibooks/10.1142/Q0392#t=suppl

Desk Editors: Sanjay Varadharajan/Adam Binnie/Shi Ying Koe

Typeset by Stallion Press
Email: enquiries@stallionpress.com

To Ioanna

https://doi.org/10.1142/9781800613294_fmatter

Foreword

The activation of water, carbon oxides and nitrogen with renewable energy and through interfacial catalysis is *the* critical challenge of the time to realistically counteract climate change. Chemical energy conversion is the technology allowing bulk transport and storage of renewable electricity. Interfacial catalysis will become the fundament of global energy markets replacing their fossil counterparts. The challenge to keep the chemical industry with its multitudes of products and processes working also when oil and gas are no longer available is another task for re-designing product chains using interfacial catalysts.

This pivotal relevance of a seemingly well-established technology can only be met when we expand its foundations into a rigorous understanding of processes and materials involved. Fortunately, the last two decades saw the advent of a multitude of *operando* spectro-microscopies. A cumulated result of their application was the evolution of our concept of static catalytic materials into the now being accepted "dynamical" picture. Catalysts are functional materials that adopt their phase inventory with little kinetic barrier according to the chemical potential of the reactant surrounding. In this way, tailored active masses are formed from the pre-catalysts synthesized such as to allow under the chosen process conditions the desired geometric and electronic structures to result from their restructuring.

This is, however, a property of many materials. Catalytic materials exhibit under working conditions in addition to restructuring chemical dynamics. The active phase resulting from restructuring is metastable and tries to undergo a phase transition. This requires first the formation of nuclei of the stable phase. In this state of an attempt to the phase transition, the material is in its highest potential energy form and thus can provide the required high energy sites for catalytic transformation. In the process of the reaction, the nuclei lose their high energy and return to the metastable phase allowing the final liberation of the products from the reaction between the reagents. The design task is thus to cater for the structure of the active phase for a facile formation of nuclei of the stable phase without allowing them to grow into the phase itself. Catalysts are materials in a state of frustrated phase transition.

It is clear that the detailed knowledge of the chemistry of the active phase is a crucial pre-requisite if one wishes to design the frustrated phase transition. This knowledge is to a substantial part delivered by X-ray photoelectron spectroscopy (XPS) and its ancillary techniques ultraviolet photoelectron spectroscopy (UPS), Auger electron spectroscopy (AES), and resonant photoemission spectroscopy (ResPES) applied under controlled conditions (*in situ*) or under working conditions (*operando*). The present book is thus also from the methodical perspective timely and gives a state-of-the-art overview of the XPS analysis of catalytic materials. *Operando* experiments are not only conducted at "near ambient pressure" but are also measured under simultaneous registration of the catalytic performance. This allows for the definition of steady state being pre-requisite to assign the structure observed to the relevant form of the catalyst. In addition, *operando* experiments can be performed in periodically modified conditions of chemical potential. Such modulation of the chemical potential offers the chance for assigning spectral features following the pattern of the potential status of an intermediate discerning them from irrelevant or poisonous species.

A significant number of precautions are necessary in order to perform such an XPS analysis on catalytic materials. These are documented and exemplified in a comprehensive form likely for the first time in respective chapters of this book. The precautions are needed from preparing the samples over designing the experiment and reach to the correct alignment of spectral features to the binding energy scale. Assignments to the correct number of species and translating this information to chemical structures using computational spectroscopy are additional tasks. Catalysts are frequently characterized by high chemical and structural inhomogeneity (e.g., a metal nanoparticle in core-shell structure with a perimeter of a reduced defective support oxide on a matrix of a photochemically active support material). The various *levels of complexity* should ideally be considered during *catalyst* characterization. Only if all of these multiple challenges are met correctly and verified against instrumental artifacts such as differential charging, contamination from the reaction cell and beam damage effects, then one can begin to derive the desired information about the chemical structure of a working catalyst.

The present expert compendium guides the catalysis scientists through many aspects of meaningful XPS analysis of interfacial catalysts. It is unfortunate that in the last two decades, a substantial number of publications derived erroneous chemical information about catalysts from incorrectly performed and falsely analyzed XPS studies. This observation alone is a strong motivation for reading and observing the recommendation within this book. It presents a wealth of examples for the correct application of the techniques and gives useful information in all chapters on the physical basis of the best practice recommendations. The book does not intend and is not a handbook on the physics of photoemission. Such books exist and should be consulted by those readers who are not familiar with the basics of photoemission. This understanding is pre-requisite for being able to follow the comprehensive collection of practical tips and conceptual recommendations for conducting studies on interfacial catalysts.

It is noted that meaningful *operando* studies still require access to synchrotron radiation experiments. As these are scarce, it is highly advisable to prepare campaigns at the beamline with extensive pre-studies of pristine and used catalyst materials conducted in laboratory environments. The different excitation conditions in beam focus, lateral resolution, irradiation intensity and kinetic energy will lead in many cases to different results than those obtained at the synchrotron where maximum surface sensitivity and optimal resolution can be obtained. The comparison of the results from both families of experiments is, however, highly valuable. It not only gives hints on the lateral and depth distribution of phase formation being frequently inhomogeneous in performance catalysts. In addition, it allows important conclusions about the chemical processes leading to the formation of the active phase and the unwanted stable phase characterizing the deactivated state of the system. It is thus clearly not a waste of resources if catalysts are studied in the laboratory XPS as well as at the synchrotron. In this way, this book also gives essential hints although the specifics of both families of experiments are not broadly discussed in some of the chapters.

When reading through the book and recognizing the multiple hurdles towards meaningful XPS experiments on catalysts, the reader may consider to revert to other *operando* methods using hard X-rays which are in many aspects easier to apply. This is also reflected in the scientific literature where multiple more experiments with non-XPS methods applying X-rays are found. These tend to deliver structural information complementing the essentially electronic information offered by XPS. Hard X-ray methods are transmission techniques with limited surface sensitivity on a given catalyst material unless the catalyst is atomically flat (model crystals) allowing to apply grazing incidence geometries. XPS has its strongest plus in being highly chemically specific to light elements such as C, N, O, and S where hard X-rays are little sensitive. These elements constitute the ligands of the metal centers to which hard X-rays are highly sensitive. XPS, in addition, offers unprecedented direct and indirect chemical information for metal centers not only when analyzing the "chemical" shift information, which has many pitfalls in it, but also when interpreting the hidden, but unambiguous, information encoded in satellite structures. In addition, the reagents of catalysis being converted consist by large also of these light elements offering the chance for XPS to address the nature of reactants, intermediates, spectators and poisons of surface reactions in direct observation.

These arguments strongly suggest that catalyst studies using XPS are highly valuable even if their execution requires the observation of many details exemplified and enumerated in this book. The organization of the material in this book is such that it can serve as a "checklist" for how to prepare an experimental campaign. With this information at hand, the preparation, execution and analysis of XPS experiments on interfacial catalysts are facilitated substantially.

Comprehensive understanding of a working catalyst requires in any case both structural and electronic information. The XPS family of methods provides the toolbox for uncovering the surface electronic structure resulting from the chemical constitution of the working system plus their adsorbates. Its comparison to information from scattering needs to be weighted with the different material compartments (lateral and in depth) from which it stems. The methodical resolution for observing the restructuring of the pre-catalyst into its active form is substantially different in XPS than in scattering techniques due to the differing requirements for structural ordering. Microscopy with photoelectrons is an excellent method to observe kinetically controlled transformation on a meso-scale and assign inhomogeneities in morphology to different phase formations (often defective variants of the matrix phase). The reactive dynamics is not specifically detected unless one conducts experiments intended to freeze the dynamics. To observe this set of phenomena, one needs high time resolution offered today by (X-ray) laser-based methods that are not covered in this book.

Interfacial catalysts are functional materials exhibiting a whole spectrum of properties acting together on the reaction of interest at different scales of space and time. XPS in the various forms described in the present expert compendium can deliver a substantial and indispensable fraction of the information required to create an evidence-based insight into the mode of operation of this class of materials. The diversity of depth and style of presentation of the various aspects of XPS with interfacial catalysts renders this book an inspiring companion useful not only as a study book but also at the spectrometer where instant advice may be required. May this book find a broad readership and minimize the occurrence of wrong information from XPS analysis of catalytic materials.

Robert Schlögl
Fritz Haber Institute of the Max Planck Society,
Department of Inorganic Chemistry,
Faradayweg 4–6, 14195 Berlin, Germany
acsek@fhi-berlin.mpg.de
Summer 2022

https://doi.org/10.1142/9781800613294_fmatter

Preface

In 1956, the group of Kai Siegbahn at Uppsala University reported about the first high-resolution spectrometer to observe low-energy electrons formed upon irradiating a solid surface with X-rays. Soon, it was realized that the technique, which now is known as X-ray photoelectron spectroscopy (XPS), can provide precise information about the composition and electronic structure of matter, with high potential for catalytic research. In the next decades, instruments for recording X-ray photoelectron spectra were rapidly developed and installed in several laboratories around the world. In the early years of XPS applications, the technique was mainly carried out by specialized surface physicists and chemists. Today, XPS has become a mature technique and automated XPS facilities can be found in industry and in universities all over the world. This transformed XPS from an advanced characterization method for dedicated research to a rather standard analysis technique for surface characterization.

The catalyst's surface state is probably the most prominent factor that influences catalytic performance. It is therefore without surprise that XPS has become an indispensable tool in studies of solid catalysts. It has been directly used to investigate issues such as the surface composition of the active catalyst and the reaction and deactivation mechanisms. Several excellent essays exist, where one can find a detailed description of the theoretical and practical aspects of X-ray photoelectron spectroscopy. The objective of the this volume is to provide a comprehensive overview of the current status and future perspectives of X-ray photoelectron spectroscopy dedicated to catalytic applications including thermal catalysis, electrocatalysis, and photo(electro)catalysis. The book contains 13 chapters, starting with a brief historical outline and the necessary introduction to the background of the technique, including basic phenomena and instrumentation aspects. The second part of the book focuses on the presentation of long-established applications of the technique such as XPS studies of model catalysts. In the last part, we refer to relatively recent developments of this method for cutting-edge surface characterization mainly using synchrotron X-ray radiation.

The book is intended primarily for researchers and graduate students but can be equally used by catalysis industry professionals. The aim is to give a complete spectrum of various ways the technique is used in catalysis research and provide material scientists and engineers with a valuable reference source. I hope that this book will help researchers to gain a broader understanding and better visualize how X-ray photoelectron spectroscopy can be employed to solve specific problematics. In addition, I expect that professionals will find particularly helpful the practical aspects described in the book and will be provided with a concrete reference source of the state-of-the-art techniques in their technological/business field.

As the editor, I am particularly thankful to the 23 contributors of this book spanning different backgrounds and three different continents. All of them are renowned academics and experts in their chosen fields, which I believe makes the content of the book sufficiently comprehensive as well as of the highest scientific standards. I would like to thank in particular Professor R. Schlögl for writing the foreword of the book and for the four wonderful years I spent in Fritz Haber Institute. Finally, I would like to take the opportunity to express my sincere gratitude to Spyros Ladas and Stella Kennou, my PhD mentors, for introducing me to the exciting world of photoelectron spectroscopy, ευχαριστώ.

Spyridon Zafeiratos
Strasbourg
March 2023

https://doi.org/10.1142/9781800613294_fmatter

About the Editor

Spyridon Zafeiratos is a CNRS Research Director affiliated with the Institute of Chemistry and Processes for Energy, Environment, and Health (ICPEES), which is a joint research unit between the French National Center for Scientific Research (CNRS) and the University of Strasbourg. After his Ph.D. in surface chemistry at Patras University, he joined the Inorganic Chemistry department at the Fritz Haber Institute in Berlin, to work on the application of ambient pressure XPS to heterogeneous catalysis. He participates in national committees for photoelectron spectroscopy and hydrogen research, while he is currently visiting professor at Shanghai University. His current research interests focus on the atomic level understanding of functional nanomaterials with applications in the fields of heterogeneous catalysis and high-temperature solid oxide electrochemical cells.

https://doi.org/10.1142/9781800613294_fmatter

About the Contributors

Peter Amann holds a doctorate in physics from the University of Innsbruck, Austria, and a degree in mechanical engineering from the Austrian Chamber of Commerce. During his post-doctoral period, he contributed to building up the Research Centre for Energy at the Vorarlberg University of Applied Sciences. He then became a researcher at the University of Stockholm, where he led the development of a new high-pressure X-ray photoelectron spectroscopy system and concentrated on research on catalyst materials using synchrotron-based technologies. Currently, Peter is a product manager at Scienta Omicron being responsible for hard X-ray and ambient pressure photoemission spectroscopy. He has contributed to the fields of magnetism, low-dimensional physics, surface science, instrument development, batteries, and catalysis.

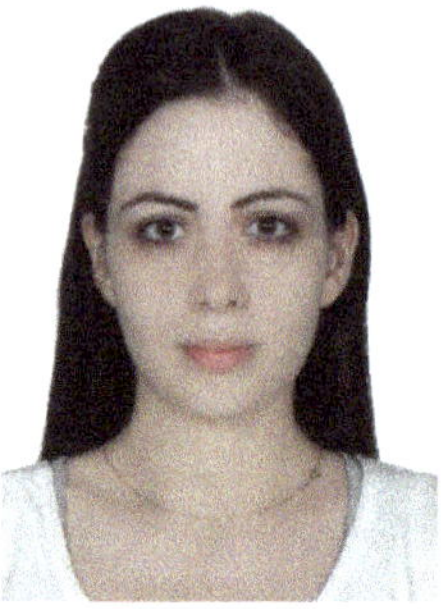

Pinar Aydogan Gokturk is currently a post-doctoral fellow in the Advanced Light Source at the Lawrence Berkeley National Laboratory. She is also a member of Center for Materials for Water and Energy Systems (MWET). Previously, she received her PhD degree in Chemistry from Bilkent University (2018), Turkey, where she also obtained her B.Sc. degree in Chemistry and Minor degree in Molecular Biology and Genetics (2013). Her research focuses on the understanding and control of solid/gas and solid/liquid interfaces for water desalination/purification and electrochemical systems under *in situ* and *operando* conditions.

Emilia A. Carbonio studied Chemistry at the Faculty of Chemical Sciences, National University of Córdoba in Argentina. She obtained her Ph.D. at the São Carlos Institute of Chemistry, São Paulo University in Brazil, where she worked on the synthesis of supported nanoparticles for the electrochemical oxygen reduction reaction. Later she did a post-doctoral stage at the Department of Applied Physics of Gleb Wataghin Institute of Physics, State University of Campinas in Brazil, where she worked with X-ray photoelectron spectroscopy and scanning tunneling microscopy studying decorated surfaces as model systems for materials of interest in (electro)catalysis. She is now a researcher at the BESSY II synchrotron of Helmholtz Zentrum Berlin für Materialen und Energie and at Fritz Haber Institute of the Max Planck Society in Germany, where she works with *in situ* and *operando* X-ray photoelectron spectroscopy and X-ray absorption spectroscopy applied to the study of (electro)catalytic materials at the solid-gas and solid-liquid interfaces.

Ethan J. Crumlin is a Career Staff Scientist at the Advanced Light Source (ALS) and in the Chemical Sciences Division at Lawrence Berkeley National Laboratory (LBNL) in Berkeley California. He received his B.S., M.S., and Ph.D. in Mechanical Engineering from the Massachusetts Institute of Technology in 2005, 2007, and 2012, respectively. His research group focuses on the utilization and development of ambient pressure X-ray photoelectron spectroscopy (APXPS) to study chemical and electrochemical reactions at the solid/gas, solid/liquid, and solid/solid interfaces for catalysts, batteries, corrosion, and electrochemical chemical transformations under *in situ* and *operando* conditions. He has been recognized for his scientific achievements receiving the International Solid State Ionics Young Scientist Award, the American Ceramics Society Ross Coffin Purdy Award, the Department of Energy Early Career Research Award, the LBNL Director's Award for Exceptional Early Scientific Career Achievement, and the International Society of Electrochemistry-Elsevier Prize for Experimental Electrochemistry.

Neal Fairley has worked in academic and commercial environments for over 30 years. His first academic publications were in computational physics and quantum mechanics which provided a natural path to the application of computers to data analysis and software tools for research. Most of his career has been spent in industrial environments initially in computer-aided design and subsequently in computer systems for mass spectrometry and surface-sensitive electron spectroscopy. In 1999, Dr. Fairley released the first version of the commercial software CasaXPS and as a consequence has had the privilege to support academic research and commercial applications of data analysis for XPS, AES, SIMS, Dynamic SIMS, ToF-MS, and other scientific software solutions.

Vincent Fernandez is a professor of physical and material chemistry at the Jean Rouxel Institute of Materials. He is an expert in X-ray photoelectron spectroscopy and Auger electron spectroscopy, including X-ray photoelectron microscopy and imaging. Fernandez received his Ph.D. working in the Crystal Growth Mechanisms Research Center in collaboration with the Department of Controlled Fusion Research. He did his post-doctoral studies at the Fritz Haber Institute of the Max Planck Society and at the European Synchrotron Radiation Facility. Currently, he has co-authored over 65 peer-reviewed publications and holds an *h* factor of 27.

Luca Gregoratti received his Ph.D. in 2003 at King's College, London. Since 2001, he has held a permanent research position at Elettra Sincrotrone Trieste where his activity has been mainly focused in the field of photoemission spectromicroscopy and its application in various domains of materials science. Presently, he is the responsible senior scientist for the ESCA Microscopy beamline, and since 2011, for nine years, he was also the group coordinator of the microscopy-diffraction beamlines of the Elettra Laboratory. He is a co-author of more than 200 articles in peer-reviewed scientific journals and is the co-inventor of two patents.

Saulius Kaciulis received his Ph.D. degree at the University of Vilnius (Lithuania) in 1983. At present time, he is Director of Research at the Institute for the Study of Nanostructured Materials (ISMN), which belongs to the National Research Council (CNR) of Italy. He is a member of the international advisory board of European conference ECASIA, member of the committee of Italian Association of Metallurgy (AIM), and member of editorial boards of the journals Surface and Interface Analysis (Wiley), Nanomaterials (MDPI), and Coatings (MDPI). He is a co-author of over 230 scientific articles and two books. Current research interests of Prof. S. Kaciulis fall in the field of the surface analysis and depth profiling of innovative solid-state materials, including thin films and heterostructures, anti-wear and corrosion-resistant coatings, biocompatible materials, and nanostructured and 2D materials.

Maya Kiskinova graduated in Chemistry in 1972 and received Ph.D. in 1977 and Sc. D. Habilitation in 1989 in Physical Chemistry. In the period 1977–1990, she held research positions at the Bulgarian Academy of Science, visiting scientist at NIST (USA), IGF-KFA (Germany), and the University of Pittsburgh (USA). Since 1990, she is a senior scientist at Elettra-Sincrotrone Trieste, where she has coordinated research activities focused on exploring properties and transient states of matter at sub-micrometer length scales and the development of synchrotron-based microscopy methods. She obtained Italian citizenship for scientific merits in 2002 and Distinguished Humboldt Research Grant in 2005 for studies in catalysis. She is the author or co-author of over 350 articles, 16 reviews, four book chapters, one book, and two U.S. Patents. She gave over 100 invited, keynote, and plenary lectures at International Congresses, Conferences, Symposia, and Workshops.

Andreas Klein is professor in the Department of Materials and Earth Sciences at Technische Universität Darmstadt, Germany, and leading the research group Electronic Structure of Materials. He has an education in physics and is specialized in the electronic structure of materials and their interfaces. During his career, he has worked on layered chalcogenide semiconductors and thin-film solar cells. His current emphasis is on non-metallic oxides, particularly transparent conductive oxides and dielectric and ferroelectric perovskites. He is using photoelectron spectroscopy with *in situ* sample preparation of thin films by magnetron sputtering and electrical transport measurements in dependence on temperature and oxygen partial pressure. Of special interest is how defects affect electronic properties. He has authored more than 250 publications in scientific journals and a number of feature articles and book chapters.

Matthew R. Linford is a professor of chemistry at Brigham Young University. His research focuses on thin film and material synthesis, material characterization, and the statistical analysis of data. He is an expert in X-ray photoelectron spectroscopy, XPS, focusing on XPS peak fitting. Linford is currently interested in making materials for sample preparation/chromatography, quantifying the number of functional groups at surfaces, the chemical vapor deposition of diamond, and data analysis through techniques like multivariate curve resolution and pattern recognition entropy. Linford graduated from Stanford in 1996 with a Ph.D. in chemistry and an M.S. in materials science, did a post-doc at the Max Planck Institute of Colloids and Interfaces in Germany, and then spent three years in industry before coming to BYU. Linford enjoys teaching general chemistry and graduate courses in his areas of specialty. By Google Scholar, his h-index is 49, his i10-index is 161, and he has 11,830 citations.

Andrea Locatelli graduated in Physics at the University of Trieste in 1994. He continued his studies in the group of Prof. Sir. David A. King in University of Cambridge (GB) and obtained a Ph.D. in Physical Chemistry in 1999. Since 2000, he is a scientist at Elettra-Sincrotrone Trieste and currently is the coordinator of the Spectroscopy, Photoemission, and Dynamics group. His main scientific interests are in the field of surface chemistry and materials sciences, which he addresses using synchrotron-based spectroscopy and microscopy. His research activity mainly focuses on the study of surfaces, interfaces, and ultra-thin films, tackling topics related to chemical, electronic, and magnetic properties of complex and low-dimensional systems. He is a co-author of more than 200 papers published in international refereed journals and gave more than 60 presentations at international conferences, workshops, as well as in various universities and research centers.

George H. Major is a graduate student working with Dr. Matthew Linford at Brigham Young University. His research focuses on the development of novel materials for sample preparation, advanced instrumentation for deposition, and quantifying reproducibility problems within the X-ray photoelectron spectroscopy literature. Major graduated with a degree in neuroscience and psychology from Brigham Young University in 2017, which he uses for a unique perspective to material science problems. According to Google Scholar, he currently has an h-index of 8, an i10-index of 8, and 351 citations.

David J. Morgan obtained his Ph.D. in Surface Science from Cardiff University in 2002, under the supervision of Prof. Philip Davies, followed by post-doctoral positions under the Prof Graham Hutchings FRS studying model catalysts by high-pressure XPS and TAP reactor studies. He is currently the surface analysis manager within the Cardiff Catalysis Institute and School of Chemistry at Cardiff University and also serves as technical manager of the EPSRC national facility for photoelectron spectroscopy, "HarwellXPS", located at the Research Complex at Harwell. In 2021, he was awarded the Vickerman prize by the UK Surface Analysis Forum (UKSAF) and serves as a member of UKSAF, the British Vacuum Council (BVC) and is a member of the British Standards Institute (BSI) as part of the ISO/TC 201 committee for standardization in the field of surface chemical analysis.

David N. Mueller received his Ph.D. in Physical Chemistry from RWTH Aachen University, Germany, in the group of Prof. Manfred Martin under the mentorship of Prof. Roger A. De Souza on the topic of mixed conducting perovskites for oxygen separation membranes. Here, he got his first exposure to *in situ* X-ray studies with home-built equipment at European synchrotron facilities. In 2012, he moved to Stanford University, CA, as a post-doctoral scholar in the group of William C. Chueh, expanding the use of *in situ* and *operando* X-ray spectroscopies on mixed conducting oxides for energy conversion devices, mainly employing facilities of the Advanced Light Source, Berkeley, CA. Since 2014, he is a staff scientist at the Peter Gruenberg Institute (PGI-6) at Research Centre Juelich, Germany, where his main interests are the development and utilization of X-ray spectroscopic and microscopic techniques to assess fundamental physicochemical concepts and processes in materials science.

Kenichi Ozawa received his B.Sc. from Shizuoka University (1992) and an M.Sc. (1994) and a Ph.D. (1998) from Tokyo Institute of Technology. He had worked as an Assistant Professor at the Chemistry Department of Tokyo Institute of Technology since 1996 and became an Associate Professor at the Institute of Materials Structure Science, High Energy Accelerator Research

Organization in 2021. He has more than 20 years of experience in photoelectron spectroscopy utilizing synchrotron radiation. His research interests encompass many aspects of physics and chemistry of metals and semiconductors, including surface electronic structure, energy-level alignment at heterojunctions, ultrafast carrier dynamics, and surface chemical reactions.

Lars G. M. Pettersson obtained his Ph.D. in Theoretical Physics specializing in quantum chemistry from Stockholm University in 1984. He spent two years as post-doc in California with IBM, San Jose, and NASA Ames, Moffett Field, before returning to Stockholm University where he is now Professor in Theoretical Chemical Physics. His research interests cover chemical bonding, heterogeneous catalysis, X-ray spectroscopies, and the properties and structure of liquid water. He has authored or coauthored more than 350 scientific publications.

Gabriel L. S. Rodrigues is from the city of Belo Horizonte in Brazil where he got his Ph.D. of Science in Chemistry from the Federal University of Minas Gerais (UFMG). His research focus is on molecular modeling, where he has contributed to theoretical studies in different areas of chemistry, such as metal-ligand bonds in inorganic complexes with biological interest, and environmental and surface chemistry. He has been a post-doctoral researcher at Stockholm University working with simulations of X-ray spectroscopy applied to surface catalysis and, presently, he is a post-doctoral fellow at KTH Royal Institute of Technology in Sweden. His current research interests are in strongly correlated systems and the computational implementation of multiconfiguration density functional theory methods.

Günther Rupprechter received his Ph.D. in Physical Chemistry from the University of Innsbruck. After post-doctoral work at the University of California at Berkeley, the Lawrence Berkeley National Laboratory, and the Fritz Haber Institute of the Max Planck Society in Berlin, he accepted a Full Professorship in Surface and Interface Chemistry at Technische Universität Wien in 2005. His research emphasis is on heterogeneous catalysis, particularly *in situ* (*operando*) spectroscopy/microscopy on model and technological catalysts, applied to studies of the mechanisms and kinetics of processes relevant to energy and environment. He received the Jochen Block Award of the German Catalysis Society for "the application of surface science methods to heterogeneous catalysis". He is corresponding member of the Austrian Academy of Sciences and "Renowned Overseas Professor" at the Shanghai University of Engineering Science.

Detre Teschner is a staff scientist at the Max Planck Institute for Chemical Energy Conversion, Mülheim an der Ruhr, Germany, and located mostly in Berlin at the Fritz Haber Institute and the Helmholtz-Zentrum Berlin. He obtained his PhD from the University of Veszprém and the former Institute of Isotopes in Budapest, Hungary, in 2001. He then moved to Berlin, Germany, joining the group of Prof. Robert Schlögl as a post-doctoral researcher with a strong focus on *in situ* characterization of solid catalysts. His current research interest includes heterogeneous, electro-, and photocatalysis.

Mikael Valter obtained his Ph.D. at the Chalmers University of Technology in 2020 on the topic of modeling electrocatalysis of glycerol and methanol on transition-metal surfaces. He is currently a post-doctoral researcher at Stockholm University, working with machine-learning methods to connect theory, such as electronic structure calculations and microkinetic modeling, with experimental data in heterogeneous catalysis.

Yifan Ye is a Staff Scientist at National Synchrotron Radiation Laboratory (NSRL), University of Science and Technology of China (USTC). He received B.S. and Ph.D. degrees from the Department of Chemical Physics and NSRL at USTC in 2011 and 2016, respectively. After several years of working in the Lawrence Berkeley National Laboratory, he returned to USTC in January 2021. His research interests are centered on the field of synchrotron radiation techniques and their applications. On the basis of synchrotron radiation facility, he has developed various *in situ* and *operando* synchrotron-based X-ray spectroscopy techniques to conduct fundamental researches in the fields of heterogeneous catalysis and rechargeable batteries.

Dmitry Y. Zemlyanov received his Ph.D. in Physics and Mathematics from the Novosibirsk State University, Russia, in 1995. He is currently a Senior Research Scientist — Surface Science Application at the Birck Nanotechnology Center at Purdue University. He held research and teaching positions in premier scientific organizations in Russia, Ireland, and Germany prior to joining Purdue and published over 175 journal papers in the areas of surface science, pharmacy, catalysis, nanotechnology, and materials science.

https://doi.org/10.1142/9781800613294_fmatter

Contents

https://doi.org/10.1142/9781800613294_fmatter

List of Abbreviations

Acronym	Definition
2PB	Double phase boundary
3PB	Triple phase boundary
AAS	Atomic absorption spectroscopy
AdC	Adventitious carbon
AES	Atomic emission spectroscopy
AES	Auger electron spectroscopy
AP	Ambient pressure
APXPS	Ambient pressure X-ray photoelectron spectroscopy
BE	Binding energy
BSCF	$(Ba,Sr)(Co,Fe)O_{3-\delta}$
CE	Counter electrode
DFT	Density functional theory
DOS	Density of states
EAL	Electron attenuation length
EDL	Electrical double layers
EDX	Energy dispersive X-ray
ESCA	Electron spectroscopy for chemical analysis
EXAFS	Extended X-ray absorption fine structure
FTIR	Fourier transform infrared spectroscopy
FWHM	Full width at half maximum
GC	Gas chromatography
GDC	Gadolinium doped ceria, $Gd_xCe_{1-x}O_{2-\delta}$

HOPG	Highly oriented pyrolytic graphite
IL	Ionic liquid
IMFP	Inelastic mean free path
IR	Infrared
LEED	Low energy electron diffraction
LEIS	Low energy ion scattering
LSC	$(La,Sr)CoO_{3-\delta}$
LSC_{113}	Perovskite type LSC
LSC_{214}	K_2NiF_4 type LSC
LSCF	$(La,Sr)(Co,Fe)O_{3-\delta}$
LSCrF	$La_{0.75}Sr_{0.25}Cr_{0.9}Fe_{0.1}O_{3-\delta}$
LSF	$(La,Sr)FeO_{3-\delta}$
LSF55	$La_{0.5}Sr_{0.5}FeO_{3-\delta}$
MB	Molecular beams
MIEC	Mixed ionic electronic conductor
MKS	Mass flow controllers
MS	Mass spectroscopy
NAP-XPS	Near ambient pressure XPS
OER	Oxygen evolution reaction
PE	Pass energy
PEM	Polymer electrolyte membrane
PEMFC	Polymer electrolyte membrane fuel cell
PLD	Pulsed laser deposition
PM-IRAS	Polarization modulation infrared reflection-adsorption spectroscopy
PZC	Potential of zero charge
RIXS	Resonant inelastic X-ray scattering
SFG	Sum frequency generation
SOEC	Solid oxide electrolyzer cell
SOFC	Solid oxide fuel cell
SPEM	Scanning photoelectron microscopy
SSEC	Solid state electrochemical cell
STEM	Scanning transmission electron microscopy
STF	$Sr(Ti,Fe)O_{3-\delta}$
STM	Scanning tunneling microscopy
STO	$SrTiO_3$

STXM	Scanning transmission X-ray microscope
TEM	Transmission electron microscopy
TPD	Temperature-programmed desorption
TUW	Technische universität wien
UHV	Ultra-high vacuum
UPS	Ultraviolet photoelectron spectroscopy
WE	Working electrode
XPS	X-ray photoelectron spectroscopy
YSZ	Yttria stabilized zirconia; $Y_xZr_{1-x}O_{2-x/2}$

https://doi.org/10.1142/9781800613294_0001

Chapter 1

X-ray Photoelectron Spectroscopy in Catalysis: Impact and Historical Background

Spyridon Zafeiratos

Institute of Chemistry and Processes for Energy, Environment and Health (ICPEES), University of Strasbourg, UMR 7515 CNRS-UdS, 25 Rue Becquerel, Strasbourg 67087, France

spiros.zafeiratos@unistra.fr

Abstract

This chapter discuss first a short historical background of X-ray photoelectron spectroscopy, starting from the discovery of the photoelectric effect in 1887 till the modern ambient pressure instruments. Particular focus is given to the developments that find applications in catalysis research. With the aid of scientific databases, the fast-growing impact of XPS in this field is demonstrated. Finally, this chapter finishes with a short overview, including the reasoning behind the selection of the different topics and their organization into chapters.

1.1 A Brief Historical Background

The history of electron spectroscopy began almost 140 years ago when H. Hertz observed that light could cause an electrical current to flow from a sample [1]. Some years later, A. Einstein gave the theoretical explanation of what is today known as the photoelectric effect [2]. After that, and for about 50 years, spectroscopic methods based on the photoelectric effect were exploited in several laboratories in Europe and in the United States, but without a big impact among the scientific community. Notably, one has to refer to the early experimental work of H. Robinson in the laboratory of E. Rutherford at Manchester University. In a paper published in 1923, we can find an account of a method which resembles very much the description of modern XPS apparatuses [3]. I cite from this work: *Briefly, the method consisted in exposing a narrow strip of metal to the beam of X-rays and using a uniform magnetic field to bend the stream of ejected electrons onto a photographic plate, the experiment being performed in a high vacuum* [3]. One should not omit here the description of the Auger effect by P. Auger [4] almost two years after the work of Robinson. A very detailed review on the historical development of modern photoelectron spectroscopy has been given by J. G. Jenkin *et al.* in a paper published in 1977 [5].

The X-ray photoelectron spectroscopy as we know it today starts in the mid-1950s with the construction of the first high-resolution spectrometer at Uppsala University (Sweden) [6], and the concerted efforts of Kai Siegbahn and his coworkers, also mentioned in several chapters. The newly established technique was benefited by the revolutionary technological progress in the ultra-high vacuum technology, with the utilization of ion pumps, demountable stainless steel chambers, and ionization gauges for pressure measurement. In its first years, the interest of the technique was mainly to study the chemical structure of atoms, molecules, and solids [7] and was not yet encompassed with the concept of surface science. This might be because the average analysis depth of XPS (or Electron Spectroscopy for Chemical Analysis (ESCA) as referred by K. Siegbahn) was initially considered in the order of 10 nm, therefore the contribution of the outer one or two monolayers as compared to the overall signal was regarded minor [7]. A review article co-authored by C. Fadley in 1971 was the first report that I managed to find, where the potential applications of XPS in catalysis research were highlighted [8]. One year after, in 1972, also appears the first XPS study of an industrial hydrogenation catalyst composed of rhodium on charcoal, performed on a Hewlett-Packard spectrometer, which was, in addition, the first commercial XPS instrument in the market [9]. The conclusions of this study, preformed 50 years ago, summarized the correlation between catalytic performance and surface state, still remaining today in the focus of the XPS investigations of catalytic materials. I cite from this work: *The spectra reveal the presence of multiple rhodium species on the catalyst surface. A direct correlation exists between the catalytic activity and the oxide to metal ratio* [9].

In the early 1970s also appears the first demonstration showing that photoelectron spectroscopies can be applied for the investigation of solid surfaces [10] and adsorbates [11] that acted as simplified "model" catalysts. Soon, it became evident that XPS is an extreme surface analysis technique, and the provided information concerns a depth of a few nanometers. After that, photoemission was systematically employed to analyze the surface of solid samples and in a few cases, also liquids and gases. Gradually, XPS became one of the most powerful tools for studying the composition and electronic structure of matter. In the intervening decades, there has been much progress, and new modes of measurement and more precise theoretical interpretation methodologies continue to be developed. Many of these advancements are being discussed in the chapters of this book.

Another revolutionary technological development that evolved almost in parallel with XPS was the establishment of the synchrotron radiation technology [12]. Originally built to serve the research of high energy physics for the discovery of new particles, synchrotrons were rapidly evolved around the production of electromagnetic radiation. With the development of the second- and third-generation facilities, synchrotron radiation became an indispensable research tool for a wide range of scientists, including physicists, biologists, chemists, and materials scientists. The first synchrotron XPS experiment reports the Au 4f spectrum of a gold foil [13] and was done in 1974. It is interesting to mention that X-rays with an energy of 8 keV were used as excitation energy, which makes this particular experiment also the first photoemission experiment in the hard X-ray regime.

Sabatier, Haber, and Langmuir, the pioneers in the field of heterogeneous catalysis, noticed very early the important role of surfaces in catalytic performance. Several years later, the evolution of surface analysis techniques provided the researchers in the field with a valuable tool to get an atomic-level understanding of catalytic surfaces [14,15], recognized by the 2007 Nobel Prize in chemistry awarded to Ertl [16]. Auger electron spectroscopy and low energy electron diffraction (LEED) were the two major surface analysis techniques for surface investigation in the early days of surface science, and studies were mainly performed on well-defined single crystal surfaces [14,17]. However, progressively X-ray and UV photoelectron spectroscopies were employed to characterize catalytic surfaces and their interaction with gas molecules [18,19]. A nice demonstration of XPS studies on model/simplified catalysts and their contribution to understand key catalytic phenomena is the focus of Chapter 7 in this book. The timeline diagram in Figure 1.1 displays in chronological order a list of events, which, according to the editor, marked significant milestones in the evolution of XPS and its application in catalytic studies.

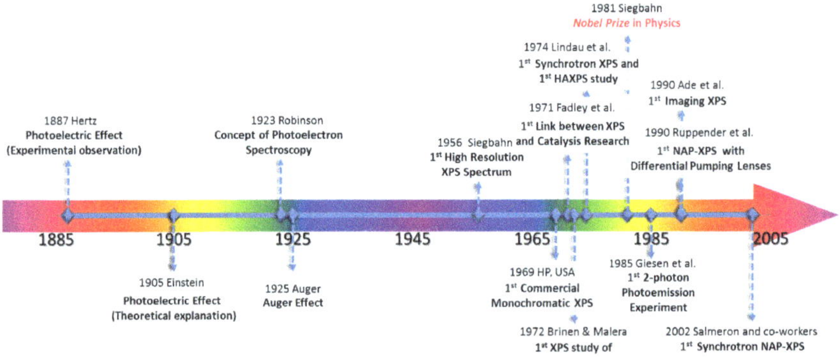

Fig. 1.1. Timeline of the key elements and milestones that contributed to the evolution of X-ray photoelectron spectroscopy, with particular focus on catalytic studies.

1.2 Contribution of XPS in Catalytic Studies

In the early years of photoemission spectroscopy, most of the XPS spectrometers were custom-built, according to the specifications of each individual research group. Owing to their complexity, XPS apparatuses were mainly operated by specialists with inside knowledge on the hardware and good theoretical background on the technique. With time, several companies have emerged offering complete analytical systems, often combining a handful of surface analysis methods. Today, the modern XPS apparatuses have a certain degree of automation and they are relatively easy to operate even for non-experts in the field. Therefore, over the last two decades, XPS has become a standard tool for surface characterization, and catalytic research, both in academia and in industry, has largely benefited from that. This is clearly reflected in the exponential increase in the number of publications utilizing XPS in catalysis research over the last 30 years (Figure 1.2(a)). If one would like to give a more precise account regarding the implication of XPS in catalytic research, then the representation in Figure 1.2(b) is more appropriate. It is evident that the proportion of articles that utilize XPS in catalysis increases constantly within the last 20 years and today about 10% of all the articles published in catalysis area (including heterogeneous, homogeneous, and theoretical catalysis) contain XPS results.

The characterization of heterogeneous catalysts covers several features of the material, notably its structure, morphology, composition, elemental analysis, and chemical state. A variety of photoelectron spectroscopy methods exist today. The various photoemission methods can be distinguished according to the energy of the incoming photons (hard or soft X-rays and UV radiation), the type of electron analysis (energy, spin, and emission angle), and the experimental conditions (vacuum or gas environment and lateral resolution). In its variant versions, XPS can provide direct or indirect information for all the critical features of catalysts described above. In this book, we deal only with the variations of photoelectron spectroscopy that are relevant to catalysis research.

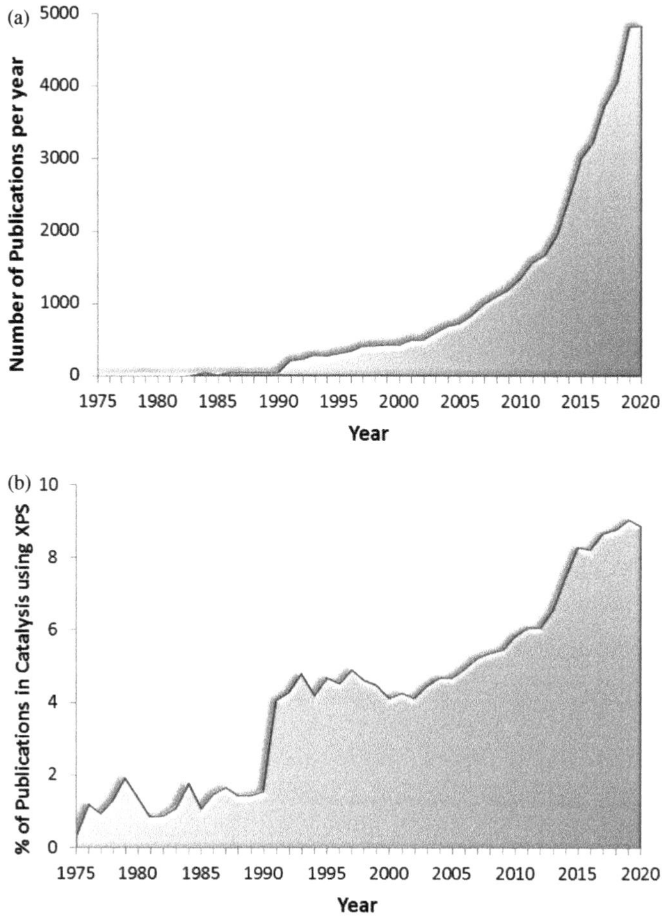

Fig. 1.2. Evolution of the research articles that utilize XPS to characterize catalytic materials published in the period 1975–2020: (a) the number of publications per year and (b) the percentage of the articles presented in (a) normalized to the total number of published articles in catalysis. Details of the bibliographic research: key words "catalysis" OR "catalyst" AND "XPS" OR "ESCA" OR "photoelectron spectroscopy". Web of Science™ (all databases). Please note that the key words "catalysis" and "catalysts" include articles from heterogeneous catalysis and also from homogeneous and theoretical catalysis. Since XPS studies are mainly applied to solid materials, it is reasonable to expect that the % ratio in Figure 1.2(b) should be considerably higher in heterogeneous catalysis studies.

Different acronyms are used in the literature to describe each method, such as HAXPS for hard X-ray photoelectron spectroscopy or NAPXPS for near-ambient pressure X-ray photoelectron spectroscopy. In order to be consistent, we have tried to use the same acronym for each XPS variant throughout this book, however we are aware that there is no common consensus in the literature and some methods can be referred to with different acronyms elsewhere. For instance, X-ray photoelectron spectroscopy, apart from the most commonly used XPS acronym, is also referred to as photoelectron or photoemission spectroscopy (PES) and Electron Spectroscopy for Chemical Analysis (ESCA). The use of these acronyms has evolved over the years, for example, in early publications, the (near-)ambient pressure XPS (NAPXPS) was referred to as "*in situ* XPS" or as "high-pressure XPS" [20].

In reality, the impressive progress made in the utilization of XPS in catalysis is definitely not without danger. As reported recently by a committee of XPS experts that analyzed 409 articles published in three high-quality journals over a period of six months, in one out of three papers that report XPS data, there are major errors in the analysis that could lead to erroneous conclusions [21]. This is a grave issue that has wider implications for the reliability of scientific research, justifying, in part, the purpose of this book. One should be fully aware not only of the capabilities but also of the limits of the XPS technique, and the authors have tried to highlight them in several cases.

1.3 Organization of the Book

The book contains in total 13 chapters and can be divided into three parts. The first part (Chapters 1–4) is mainly addressed to occasional XPS users that would like to have a fast assessment of the various aspects of the technique. After a short historical background (Chapter 1), Chapter 2 introduces readers to the fundamentals of the technique. More precisely, it describes the basic phenomena and instrumentation aspects of XPS. This part also contains many useful hints on how to process spectra (Chapters 2–4) and how to distinguish spectral features from artifacts that may arise during measurements. In addition, I hope that the readers will appreciate practical information regarding the handling and preparation of samples for XPS analysis, provided in Chapters 2 and 4.

The second part of the book (Chapters 5–8) deals with some particular topics which, however, might be essential for catalysis researchers who would like to deepen into the technique. For example, a very critical aspect, that is a source of misinterpretation of XPS results, is the electrostatic charging that might appear in some samples. This issue is extremely important in the case of catalysts since often they are electronic insulators or metals supported on wide band-gap oxides. Taking into account this specificity for XPS in catalysis, an entire chapter (Chapter 5) is dedicated to describe the background of charging accumulation and to discuss the methods to measure insulating catalytic materials. One should refer here to Chapter 6, which is dedicated to computational studies, and a critical report of investigations on simplified catalytic systems (Chapter 7), the so-called model catalysts. The second part of the book is completed with Chapter 8 which describes how photoelectron spectroscopy can be applied to determine the energy level alignment of photocatalysts.

The third, and last part of the book, surveys relatively recently established photoemission techniques and demonstrates their potential in the fields of photo-, electro-, and thermal catalysis. These exciting developments take advantage of the characteristics of synchrotron X-rays, namely high brightness, broad energy spectrum, high collimation, and pulsed time structure, for cutting-edge surface characterization. In particular, Chapter 9 provides a detailed account of time-resolved XPS and how it can be used to get insights into photocatalytic phenomena. In Chapters 10 and 12, researchers from Fritz Haber Institute (FHI) in Berlin and Lawrence Berkeley National Laboratory (LBNL) in Berkeley, the two laboratories that initially developed Near-Ambient Pressure XPS (NAPXPS), discuss its application to studies of heterogeneous catalysts and solid/liquid interfaces, respectively. NAPXPS is also the focus of Chapter 13 that gives a detailed account of *operando* studies of mixed ionic electronic conducting materials for solid-state electrochemical applications.

Finally, the power of imaging XPS to explore the properties of planar model (electro)catalysts is demonstrated in Chapter 11.

In the chapters, one can find useful summaries, references to key papers, and outlines of the historical developments in the field. Almost all chapters contain enlightening schematic illustrations, which facilitate comprehension of complex concepts and can be used as a teaching aid. The book intentionally omitted complex calculations, extensive technical details, and deep theoretical analysis and describes the essential (according to the authors' opinion) and practical aspects, without being oversimplistic. Some concepts, for example, the Auger parameter or electrostatic sample charging, are mentioned in more than one chapter. I am aware of these repetitions; however, I preferred to retain them, in order to maintain the integrity of each chapter. The book should equally appeal to graduate students, both early stage and experienced researchers as well as XPS users from the industry. I hope that, in addition to the practical information, they will also discover new approaches and with them the potential of photoemission methods to give fine information regarding catalysts and catalytic processes.

References

[1] Hertz H. Ueber einen Einfluss des ultravioletten Lichtes auf die electrische Entladung. Ann Der Phys Und Chemie. 1887;267:983–1000. Doi:10.1002/andp.18872670827.

[2] Einstein A. Über einen die Erzeugung und Verwandlung des Lichtes betreffenden heuristischen Gesichtspunkt. Ann Phys. 1905;322:132–148. Doi:10.1002/andp.19053220607.

[3] Robinson H, Rutherford E. The secondary corpuscular rays produced by homogeneous X-rays. Proc R Soc London Ser A, Contain Pap a Math Phys Character. 1923;104:455–479. Doi:10.1098/rspa.1923.0121.

[4] Auger P. Sur les rayons secondaires produits dans un gaz par des rayons X. C R Acad Sc. 1925;180:65.

[5] Jenkin JG, Leckey RCG, Liesegang J. The development of X-ray photoelectron spectroscopy: 1900–1960. J Electron Spectros Relat Phenomena. 1977;12:1–35. Doi:10.1016/0368-2048(77)85065-2.

[6] Siegbahn K, Edvarson K. β-Ray spectroscopy in the precision range of 1: 105. Nucl Phys. 1956;1:137–159. Doi:10.1016/S0029-5582(56) 80022-9.

[7] Siegbahn K, Nordling C, Fahlman A, Nordberg R, Hamrin K, Hedman J, Johansson G, Bergmark T, Karlsson S-E, Lindgren I, Lindberg BJ. ESCA – Atomic, molecular and solid state structure studied by means of electron spectroscopy. Nova acta Regiae Societatis Scientiarum Upsaliensis. ser. 4. vol. 20. 1967.

[8] Delgass WN, Hughes TR, Fadley CS. X-ray photoelectron spectroscopy: A tool for research in Catalysis. Catal Rev. 1971;4:179–220. Doi:10.1080/01614947108075489.

[9] Brinen JS, Melera A. Electron spectroscopy for chemical analysis (ESCA) studies on catalysts. Rhodium on charcoal. J Phys Chem. 1972;76:2525–2526. Doi:10.1021/j100662a007.

[10] Brundle CR, Roberts MW. Surface sensitivity of esca for submonolayer quantities of mercury adsorbed on a gold substrate. Chem Phys Lett. 1973;18:380–381. Doi:10.1016/0009-2614(73)80195-2.

[11] Madey TE, Yates JT, Erickson NE. ESCA study of fractional monolayer quantities of chemisorbed gases on tungsten. Chem Phys Lett. 1973;19:487–492. Doi:10.1016/0009-2614(73)85132-2.

[12] Cramer SP. *X-Ray Spectroscopy with Synchrotron Radiation: Fundamentals and Applications* (1st ed.). Springer Cham: Springer Nature Switzerland, 2020. Doi:10.1007/978-3-030-28551-7.

[13] Lindau I, Pianetta P, Doniach S, Spicer WE. X-ray photoemission spectroscopy. Nature. 1974;250:214–215. Doi:10.1038/250214a0.

[14] Conrad H, Ertl G, Latta EE. Adsorption of hydrogen on palladium single crystal surfaces. Surf Sci. 1974;41:435–446. Doi:10.1016/0039-6028(74)90060-0.

[15] Christmann K, Schober O, Ertl G, Neumann M. Adsorption of hydrogen on nickel single crystal surfaces. J Chem Phys. 1974;60: 4528–4540. Doi:10.1063/1.1680935.

[16] Wandelt K, Campbell CT. Special issue of surface science dedicated to Prof. Dr. Dr. h.c. mult. Gerhard Ertl, Nobel-Laureate in chemistry 2007. Surf Sci. 2009;603:vii. Doi:10.1016/S0039-6028(09)00333-1.

[17] Joyner RW, Lang B, Somorjai GA. Low pressure studies of dehydrocyclization of *n*-heptane on platinum crystal surfaces using mass

spectrometry, auger electron spectroscopy, and low energy electron diffraction. J Catal. 1972;27:405–415. Doi:10.1016/0021-9517(72)90177-7.

[18] Lo WJ, Somorjai GA. Temperature-dependent surface structure, composition, and electronic properties of the clean $SrTiO_3$(111) crystal face: Low-energy-electron diffraction, Auger-electron spectroscopy, electron energy loss, and ultraviolet-photoelectron spectr. Phys Rev B. 1978;17:4942–4950. Doi:10.1103/PhysRevB.17.4942.

[19] Ferrer S, Somorjai GA. UPS and XPS studies of the chemisorption of O_2, H_2 and H_2O on reduced and stoichiometric $SrTiO_3$(111) surfaces; The effects of illumination. Surf Sci. 1980;94:41–56. Doi:10.1016/0039-6028(80)90155-7.

[20] Zhong L, Chen D, Zafeiratos S. A mini review of in situ near-ambient pressure XPS studies on non-noble, late transition metal catalysts. Catal Sci Technol. 2019;9:3851–3867. Doi:10.1039/c9cy00632j.

[21] Major GH, Avval TG, Moeini B, Pinto G, Shah D, Jain V, Carver V, Skinner W, Gengenbach TR, Easton CD, Herrera-Gomez A, Nunney TS, Baer DR, Linford MR. Assessment of the frequency and nature of erroneous X-ray photoelectron spectroscopy analyses in the scientific literature. J Vac Sci Technol A. 2020;38:61204. Doi:10.1116/6.0000685.

https://doi.org/10.1142/9781800613294_0002

Chapter 2

Practical Aspects of XPS: From Sample Preparation to Spectra Interpretation

Dmitry Y. Zemlyanov

Birck Nanotechnology Center, Purdue University,
1205 West State Street, West Lafayette IN 47907-2057, USA
dimazemlyanov@purdue.edu

Abstract

Nowadays, X-ray photoelectron spectroscopy (XPS) is one of the most widely used surface characterization techniques in materials science and chemistry, spreading to pharmacy, biochemistry, biology, nanotechnology, etc. The basic principle of XPS is simple: the sample under investigation is irradiated by an X-ray beam and the photoelectrons generated are collected by an energy analyzer. The binding energy of the photoelectrons correlates with the energy of an electron level providing information about the element composition and chemical state of elements. The intensity of the photoelectrons is proportional to the number of atoms, "emitters", and thus, the quantitative data can be extracted. However, despite the apparent simplicity of the technique, in many publications, the XPS data are misinterpreted or are reported in a limited or not complete way. The goal of this chapter is to provide good practice guidance for users with different education levels and experiences to apply XPS appropriately.

2.1 Introduction

In 1981, Kai Siegbahn was awarded the Noble Prize in Physics *for his contribution to the development of high-resolution electron spectroscopy* [1]. By this time, X-ray photoelectron spectroscopy (XPS) was already a mature technique, and nowadays, it is one of the most widely used surface characterization techniques in materials science and chemistry, spreading further to pharmacy [2–7], biochemistry [8–13], biology [14–16], etc. The basic principle of XPS is fairly simple: the sample under investigation is irradiated by an X-ray beam and the photoelectrons generated are collected by an energy analyzer. The kinetic energy, E_{KE}, of the ejected photoelectrons is related to the energy of the incident X-ray photon, hv, the electron binding energy, E_{BE}, measured with respect to the Fermi level, and the work function of the spectrometer, φ_{sp}:

$$E_{\mathrm{KE}} = hv - E_{\mathrm{BE}} - \varphi_{\mathrm{sp}}. \tag{2.1}$$

The binding energy correlates with the energy of an electron level, ε_i. Figure 2.1 shows the energy diagram of the photoemission process.

For a conductor, the Fermi level of the sample and the spectrometer ground are connected. Therefore, the work function of the spectrometer, which is typically in the range of ~4 eV, appears in Eq. (2.1) (not the work function of the surface). At the entry to an electron energy analyzer, the photoelectron either accelerates (if the work function of a sample, $\varphi > \varphi_{\mathrm{sp}}$) or decelerates (if $\varphi < \varphi_{\mathrm{sp}}$). The kinetic energy of the photoelectrons measured by an electron energy analyzer does not depend on the work function of a sample and this makes the binding energy calculated in Eq. (2.1) a good descriptor of the electron level.

Since every element has a unique electron level structure, the element's presence can be identified [17–19]. The photoemission peak intensities (actually the peak areas) could deliver information about the element's composition of a sample surface [20]. The small shift of the electron levels, typically a few eV, due to a chemical bond with heteroatoms, referred as to a chemical shift, provides data about the chemical states of the elements of interest.

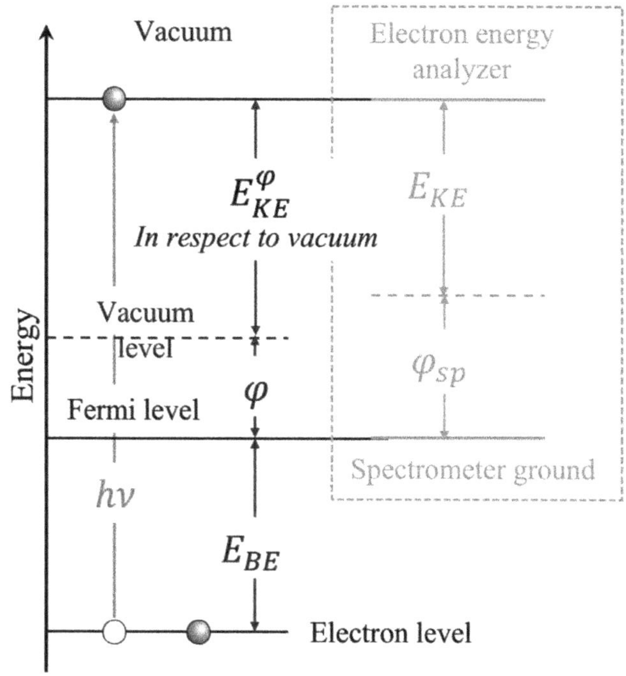

Fig. 2.1. Cartoon representation of the photoemission process.

However, despite the apparent simplicity of the technique, in many publications, the XPS data are misinterpreted or are reported in a limited or incomplete way. The goal of this chapter is to provide good practice guidance for users with different education levels and experiences to apply XPS appropriately.

2.1.1 *Practical aspects of the XPS analysis*

There are several questions that should be raised before starting the XPS experiment. The main questions are as follows:

- *What kind of information do I want to get using XPS and can XPS provide the information I need?*
- *How should a sample be handled for XPS analysis?*
- *What measurement conditions should be used to achieve the goal?*

And finally, after the XPS data acquisition, the important questions are as follows:

- *How can the XPS data be analyzed?*
- *What should be reported in a publication?*

Many factors can contribute to the successful and dependable application of XPS but more or less clear answers to these questions greatly increase the chance of the accomplishment. Also, many steps involving setting up and verifying the instrument performance should be performed by qualified personnel or by the vendor of an equipment, otherwise the data reliability could be compromised.

2.2 Practical Guidelines for the XPS Analysis

As a widely used reliable technique, the XPS received significant attention in the literature. Many features of the techniques are covered in the detailed textbooks [21,22] and in the useful web pages [17,18], and good practical advice can be found in Ref. [23] and references therein. In the following, we discuss the important XPS aspects, which a researcher should know before, during, and after an XPS experiment.

2.2.1 *Sample–XPS compatibility*

XPS instruments are designed to achieve ultra-high vacuum (UHV) conditions to keep the surface from being contaminated during analysis and to avoid the photoelectron being scattered by gas. However, if the surface cleanliness is not a concern (especially when a sample was transferred from air), at least 10^{-7} mbar pressure is required to operate the X-ray gun and the detector (channeltrons or channelplates) and to permit the photoelectrons traveling without scattering. Since traditional XPS measurements are performed under high vacuum conditions, the sample of interest should be vacuum-compatible. Dedicated XPS instruments, such as (near-) ambient pressure XPS (see Chapter 10), should be exploited for samples with high vapor pressure or to study *in situ* chemical reaction [24–31]. In the design of XPS experiments, it is a good practice to check the vapor pressure of the material of interest before introducing it into the vacuum chamber. Sample freezing may be needed (cryo-XPS) if the sample is not vacuum-compatible due to high vapor pressure, especially when dealing with biological samples [14–16,32]. The sample temperature may increase during the XPS measurement, for instance, due to the use of a charge neutralizer source and a non-monochromatic X-ray gun (as the filament is in the proximity of the sample surface). The possible heating during the measurement (up to 50°C above the ambient temperature) should be considered to estimate the vapor pressure. For "unknown" materials, the rule of thumb is that a sample is claimed to be "XPS-vacuum-compatible" if it can be investigated by secondary electron microscopy (SEM). It is interesting that a liquid sample might have low vapor pressure and it can be compatible with XPS, for instant, gallium or $AgNO_3$ [33,34]. Some solids, such as, for instance, iodine, sublimate at room temperature and cannot be investigated under vacuum conditions.

X-ray and thermal degradation of a sample are the common issues. Inorganic samples are typically much less prone to X-ray damage, while organic surfaces suffer from X-ray degradation much more. Color and/or morphology changes should be closely monitored under X-ray exposure. A good practice is to compare a few fast scans of the characteristic core level. Use of monochromatic Al Kα radiation in a laboratory-based XPS allows the mitigation of the sample damage because the X-ray monochromator cuts (i) Bremsstrahlung radiation, which contains a lot of low-energy photons having high absorption cross-section and (ii) the thermal radiation from the filament of an X-ray gun, which otherwise might cause sample heating. The synchrotron radiation has too high intensity, and this makes the XPS analysis of organic samples challenging for synchrotron-based instruments.

2.2.2 *The information provided by XPS*

2.2.2.1 *Surface sensitivity*

Since electrons are detected in XPS, the information depth is limited by the electron attenuation length (EAL). In the past, instead of EAL, the inelastic mean free path (IMFP) was used [35,36]. Around 95% of the photoelectrons reach an energy analyzer from the depth equal to three EAL. NIST database 86 is a good resource helping to calculate EAL for different materials in a simple and an intuitive way [37–40]. For Al Kα (1486.6 eV) and Mg Kα (1253.6 eV), which are the most used X-ray excitations in a laboratory-based XPS, the information depth is in the range of 4–10 nm depending on the binding energy of the photoelectrons. The larger information depth was achieved with a higher energy X-ray source, such as Ag Lα (2984.3 eV) or synchrotron radiation [41–45]. For real element bulk composition, the sample should be investigated by energy dispersive X-ray (EDX), atomic absorption spectroscopy (AAS), and atomic emission spectroscopy (AES). Another possibility to access the element composition at greater depth is the ion sputtering with consecutive XPS analysis (see, for instance, Refs. [33,46,47]); this approach allows us to get element composition at μm depth. However, the ion beam, which is usually argon or other noble gas ions with the kinetic energy of 0.5–5 keV, is highly destructive and the information about the chemical states will be lost. Recently, big carbon molecules, such as fullerene, coronene, and Ar clusters, were successfully used for the low destructive depth profiling of organic samples [48–53].

2.2.2.2 *Element composition and detection limit*

XPS is a quantitate technique, which can detect all elements, except for hydrogen and helium, in the near-surface region of a sample by measuring the binding energies of photoelectrons emitted during X-ray excitation. The intensity of the photoemission peaks (or more precisely the peak area) is proportional to the element concentration within the information depth. However, the absolute intensities of the photoemission peaks from different samples cannot be compared directly because the absolute intensity is affected by many parameters. A few good practical quantification approaches are proposed by C. Fadley [20]. In most publications, the element composition, $C_i[atomic\ \%]$, is calculated using the assumption of homogeneously mixed elements within the information depth. The calculations are based on the area of the *i*th photoemission peaks, A_i, normalized to a relative sensitivity factor (RSF_i), the instrument function, F_i^{inst}, which depends on the kinetic energy of the photoelectrons, E_{KE}^i, and corrected to the electron attenuation length, EAL_i, which is also specific to the kinetic energy of the photoelectrons and sample nature. The normalized area can be expressed as follows:

$$A_i^{\text{norm}} = \frac{A_i}{RSF_i\, F_i^{\text{inst}}\, EAL_i}. \tag{2.2}$$

Then, $C_i[atomic\ \%]$ is

$$C_i\left[atomic\ \%\right] = \frac{A_i^{\text{norm}}}{A_{\text{total}}^{\text{norm}}}. \tag{2.3}$$

Here, $A_{\text{total}}^{\text{norm}} = \Sigma_{j=1}^{n} A_j^{\text{norm}}$ is the total normalized area for all elements of interest. The electron attenuation length, EAL_i, in Eq. (2.2) can be estimated as $\approx (E_{\text{KE}}^i)$, where α is an empirical parameter ranging from 0.4–1.0.

The calculation of sample element composition based on the assumption of homogeneously mixed elements within the information depth (Eq. (2.3)) is a standard built-in function for many XPS software (for instance, CasaXPS [54]). Both the survey spectrum and the high-resolution spectra can be used for quantification. One characteristic peak (usually the most intense or with the highest RSF and with the smallest width) should be taken for each element present in the sample. It is important that in the case of high-resolution spectra, the core-level regions of choice are collected under identical conditions: the spectra obtained with the different pass energy of an energy analyzer or with the different X-ray gun power cannot be simply combined for the quantification.

More sophisticated quantification approaches can help calculate the surface coverage of adsorbed species or thin film thickness (the thickness should be less than the information depth) [20,55–60]. The simulation of electron spectra for surface analysis (SESSA) is advanced and it covers a number of different models [61].

The nominal sensitivity of the XPS technique is commonly assumed to be approximately 0.1 atomic %. However, the practical element sensitivity is determined by several factors, such as RSF, which might vary two orders of magnitude depending on the specific element, depth distribution, possible overlapping of photoemission peaks, sample material density, etc. Relatively simple charts were reported to help determine the XPS detection limit for different elements [62–64].

2.2.2.3 *Chemical shift*

The identification of the chemical state of the element, or elsewhere the chemical state analysis, is one of the most famous features of XPS. The original name of the technique was electron spectroscopy for chemical analysis (ESCA), which fully reflects this capability. The chemical shift of core electron levels was convincingly demonstrated for several elements in numerous different free atom and molecule compounds (gasses) by Siegbahn *et al.* [21]. This is quite a remarkable fact because the core-level electrons do not involve the chemical bond much. The chemical shift originates from the interaction of core-level electrons with the outer-level valence electrons. The shift is mainly due to electrostatic interaction and its typical value is very roughly equal to the magnitude of the chemical bond energy [21].

A redistribution of the valence electrons in the molecule as compared to the atom, which involves increasing or decreasing the electron density of the valence electrons, results in an almost uniform change of the effective potential experienced by the core-level electrons of the atom. The addition of a uniform perturbation potential to the effective original Hamiltonian does not change the shape of the orbital wave function, and using first-order perturbation theory, the core-level wave function can be considered as "frozen" [65]. The approximately uniform variations of the effective potential inside the electron shell result in that all core levels on the same atom shift by approximately the same amount [21].

The change of the effective potential (in other words, the change of the core-level electron energies) due to a redistribution of valence electrons in the molecule can be divided into two components: (1) one-center contribution associated with the change of the density of the valence electrons on the atom *A* and (2) two-center contribution due to interaction between the core-level electrons and the charge of the other atom in the molecule. The shift of the electron levels, $\Delta\varepsilon_A$, can be written as follows [21]:

$$\Delta\varepsilon_A = \Delta q_A k_{\text{core-valence}} + \sum_{A\neq B} \frac{q_B}{R_{AB}}. \tag{2.4}$$

Here, Δq_A is the charge difference between atom *A* in the molecule of interest and the reference molecule or atom. $k_{\text{core-valene}}$ is the average interaction between core electrons and the valence electrons of atom *A*. The second term in Eq. (2.4) is the interatomic effective potential, i.e., the molecular potential, $V^{\text{molecular potential}}$, and Eq. (2.4) can be rearranged as follows [21]:

$$\Delta\varepsilon_A = \Delta q_A k_{\text{core-valence}} + V^{\text{molecular potential}} + constant. \tag{2.5}$$

The energy of the *i*th electron level, εi, correlates with the binding energy of the photoelectron, E^i_{BE}, measured by XPS. This relation can be understood assuming that the photoemission is a simple one-electron process, and no relaxation follows the emission of the electron. Then, the binding energy of the electron, E_{BE}, is simply the difference between the energy of the initial state in an atom with n electrons, $E_{\text{initial}}(n)$, and the energy of the final state, $E_{\text{final}}(n-1)$, which is a single-charged atom with $n-1$ electrons and free photoelectron:

$$E_{\text{BE}} = E_{\text{final}}(n-1) - E_{\text{initial}}(n). \tag{2.6}$$

Using the approximation of the frozen orbitals [65], the energies of the initial state and the final state are eigenvalues of the wave functions and they can be calculated using the non-relativistic Hartree–Fock (HF) method. The absence of relaxation assumes no electronic rearrangement (neither intra- nor extra-atomic) while the photoelectron is leaving the atom, and therefore the same unmodified set of the wave functions can be used for the original non-ionized atom and for the single-charged atom. The approximate solution is

$$E^i_{\mathrm{BE}} \approx -\varepsilon_i. \tag{2.7}$$

Most useful is that the shift of the binding energy, ΔE^i_{BE}, is proportional to the shift of the electron level, $\Delta\varepsilon_i$,

$$\Delta E^i_{\mathrm{BE}} \approx -\Delta\varepsilon_i. \tag{2.8}$$

This establishes the relation between the shift of the binding energy observed in XPS and the shift of the corresponding core level due to a chemical bond. The binding energy shift is often referred to as a chemical shift because the formation of the chemical bond results in the changing of the binding energies of the photoelectrons emitted by the atom of interest. Equation (2.4) can be simplified even further neglecting the molecular potential:

$$\Delta\varepsilon_A = \Delta q_A k_{\mathrm{core-valence}}, \tag{2.9}$$

and the chemical shift can be described using only the properties of the atom of interest. This approximation was confirmed to work well for free gas molecules [21]. Figure 2.2 shows a schematic of the concepts using the C 1s as an example.

When the chemical bond forms between a carbon atom and more electronegative atom(s), the electron density is transferred from the carbon atom making it positively charged, $\Delta q_C > 0$. As a result, the C 1s level shifts down; the magnitude of the shift depends on the magnitude of the charge transferred, and the binding energy of the C 1s peak increases accordingly. For instance, the binding

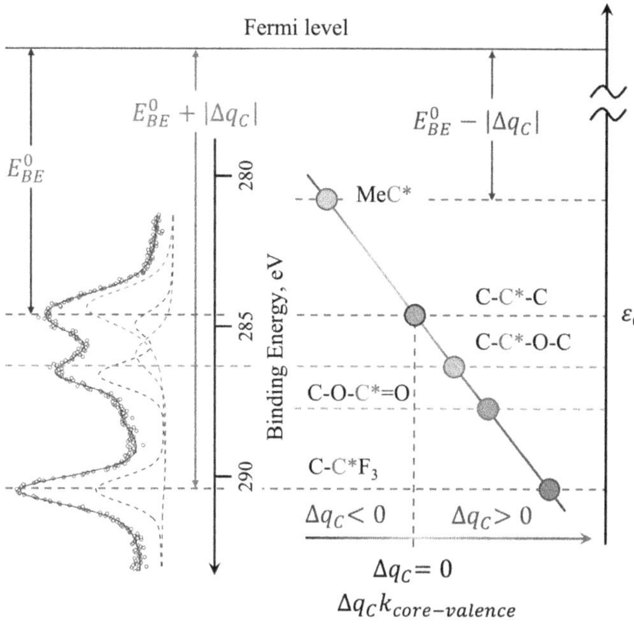

Fig. 2.2. Cartoon representation of the correlation between the typical measured binding energies of C 1s in hydrocarbons and the charge difference on the carbon atom in the different chemical states. The electron-emitting atom is marked with a star (*).

energy shift of the carbon atom from the CF_3 group is bigger than the corresponding value for the atom in C–C*–O–C or in C–O–C*=O functional groups. *Vice versa,* when carbon forms the chemical bond with the less electronegative element, such as a metal, the carbon atom becomes negatively charged, and the binding energy of the C 1s peak decreases with respect to the reference.

The chemical shift is often classified as the initial state effect because the core levels shift before the photoemission process takes place. The change in charge density on the atom of interest explains the shift in the core-level binding energy quite well only for free gasses. The binding energy measured experimentally in various molecules, macromolecules, and compounds can be found, for instance, in Ref. [66]. Recently, density functional theory (DFT) was proved to be helpful for the interpretation of the XPS data from adsorbed species on a metal surface, when the simple approach described above was applied [67,68].

2.2.2.4 *Factors leading to binding energy shift*

A few other factors should be taken into consideration to understand the shift of the binding energy and to compare the experimental results obtained from different materials:

- In the HF method, the binding energy of a given orbital is calculated with respect to the vacuum level, meaning that the photoelectron is released without experiencing any potential. This is not the case for the condensed matter, where the orbital energy changes because of the placing of the atom into the solid matrix. This shift is often referred to as the Madelung potential, V^i_{Madelung} [69].
- Even if the binding energy is corrected to the Madelung potential, the calculated value is different by several eVs from the experimentally observed binding energies. Thus, the C 1s photoemission peaks of hydrocarbons are shifted by approximately +6 eV in the HF calculations [70]. This is due to the relaxation energy, arising from the re-arrangement of the electron density during photoemission. The process of the photoelectron leaving the parent atom is not "sudden" as it is supposed by the approximation of the frozen orbitals, and therefore the electrons from the parent atom and the neighboring atoms start to screen the core-level hole changing the interaction between the leaving photoelectron and the ion left behind. The relaxation can be intra-atomic, R^i_{intra}, when the screening electrons originate from the parent atom or extra-atomic, R^i_{extra}, if neighboring atom electrons participate in the process. In general, this relation leads to a constant shift in the binding energy with respect to the value calculated using the approximation of the frozen orbitals. But in certain circumstances, the intra- and/or extra-atomic relaxation might result in the abnormal shift of the photoemission peak.

For instance, the Ag 3d peaks shift towards lower binding energies for Ag_2O compared to silver metal [71]. Since the relaxation occurs after the ionization of the atom, it is classified as a final state effect.

With the above-discussed contributions, the binding energy from Eq. (2.7) can be written as follows:

$$E^i_{\mathrm{BE}} \approx -\varepsilon_i + V^i_{\mathrm{Madelung}} - R^i_{\mathrm{intra}} - R^i_{\mathrm{extra}}. \tag{2.10}$$

There are other factors resulting in the binding energy shift such as protonation of organics [6,7] or Fermi level shift for semiconductors [72–74], but Eq. (2.10) covers the most common contributions.

The final state effects are not limited by the relaxation and include other phenomena, which can complicate the "simplicity" of the identification of the chemical states with XPS. To understand the XPS spectra of metal oxides, detailed DFT modeling might be required. Thus, for instance, in the case of CeO_2, the satellite structure of the Ce 3d peaks was correctly described using the semi empirical Anderson model, which includes the charge transfer from oxygen atoms to the empty Ce 4f shell [75]. More details about theoretical calculations of XPS spectra can be found in Chapter 6.

2.2.2.5 *X-ray-induced Auger process*

Photoemission is an excitation process, and the ejection of the photoelectrons results in a single-charged atom with the hole at the core level. This excitation can relax in two ways: (1) through Auger process or (2) through X-ray fluorescence. In the former process, Auger electrons are emitted, and they appear in the XPS spectra particularly in the survey spectrum. In the Auger process, an electron from an upper level, often from the valence band region, fills the hole formed during photoemission and the energy released in this transition is adsorbed by another electron from the valence band region; this electron is ejected leaving behind double-ionized atom. There are several possibilities to mix the transitions involving three electrons, therefore the Auger peaks have a complex shape containing several overlapping transitions. Since the kinetic energy of the Auger electron, E_{KE}^{Auger}, is determined by only intrinsic electron levels of the atom including the energy of the core-level hole, E_{BE}^{hole}, which is in fact the binding energy of the photoelectron, the binding energy of the upper-level electron filling the hole, $E_{BE}^{filling}$, and the binding energy of the upper-level, $E_{BE}^{e-Auger}$, from which the Auger electron is emitted,

$$E_{KE}^{Auger} = E_{BE}^{hole} - E_{BE}^{filling} - E_{BE}^{e-Auger}. \quad (2.11)$$

Remarkably, as one can see in Eq. (2.11), the kinetic energy of the Auger electrons does not depend on the excitation energy and the type of excitation. During the planning of the XPS experiment, a user should check that the Auger peaks do not overlap with the region of interest, especially when the photon energy varies at a synchrotron-based XPS or when the anode is switched at laboratory-based XPS.

The Auger process includes two electrons from the valence band region, which is mainly involved in the formation of a chemical bond. This makes the Auger electrons very sensitive to the chemical states of an element. Wagner has introduced the so-called modified Auger parameter, α', to identify the chemical state [76,77]

$$\alpha' = E_{KE}^{ijk} + E_{BE}^{i}. \quad (2.12)$$

Here, E_{BE}^{i} is the binding energy of the photoelectron on the *i*th core level, which corresponds to the energy of the hole. E_{KE}^{ijk} is the kinetic energy of the *ijk*th Auger peak. The modified Auger parameter is a good descriptor of the chemical state of an element [76] as it was demonstrated in Ref. [33] for the identification of the silver compounds.

The modified Auger parameter could be also very useful in the case of non-conducting samples, as discussed in Chapter 5. In particular, the electrostatic charging causes an equal magnitude shift of the binding energy and the kinetic energy, but the sign is different, and, therefore, the charge-associated shift is canceled out in Eq. (2.12). Also, α' could be helpful for the investigation of the chemical state of nanoparticles. The binding energy typically shifts towards higher values by decreasing the size of nanoparticles [78–80] and the oxidation results in the increase of the binding energy as well. The modified Auger parameter could discriminate between these two effects.

2.2.2.6 *Is XPS the right technique for the characterization of a sample?*

Before starting the XPS experiment, the following questions should be raised:

1. Is the surface (topmost of 10 nm) of a sample the subject of interest?
2. Is the element composition of a sample surface the subject of investigation?
3. Should the chemical states of elements in the near-surface region be examined?

The XPS is the right technique if the answer to all these questions is positive and a sample is vacuum-compatible. However, an XPS user should always remember the details discussed above. In particular, (i) the detection limit of XPS is approximately 0.1 atomic %, and the technique might not be suitable for the ppm level detection, such as a dopant in semiconductors. (ii) The chemical shift of the electron core levels provides information about the chemical states of the element of interest. (iii) On the other hand, there are other factors, which can result in the shift of the binding energies as discussed above (see also Chapter 5). (iv) The modified Auger parameter can be very helpful for the identification of the chemical state as well.

2.2.3 *XPS sample handling*

Keeping in mind that XPS is a surface-sensitive technique, the proper handling of a sample for XPS becomes crucial. Carbon deposition on the sample surface might hinder the XPS analysis reducing the actual information depth. Other characterization techniques applied before the XPS analysis might cause cross-contamination on the sample surface. For instance, SEM causes substantial carbon deposition due to the hydrocarbon decomposition by the SEM electron beam. Often, SEM samples have to be coated with metals or carbon to mitigate charging. All these will have a negative effect on the XPS measurements. It is strongly recommended to explore with XPS as freshly prepared samples as possible. Any contact with the examined surface should be avoided: even gloves can leave contamination. No part of a transport container should make contact with the surface for the XPS analysis. The best practice to carry a flat sample (foils, chips, films, tablets, etc.) is shown in Figure 2.3. Details about handling powder samples can be found in Chapter 4.

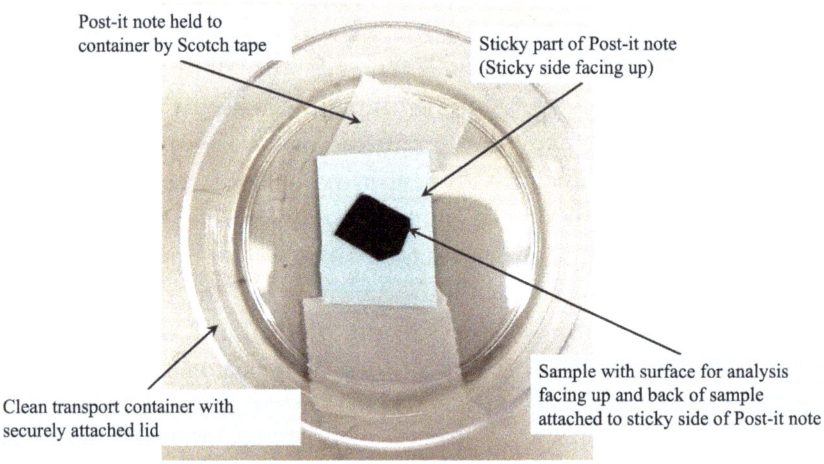

Fig. 2.3. Best practice example to ship/transport a flat sample, like a piece of a silicon wafer, for XPS analysis. A clean transport container should be used with a lid, which can be securely attached with a Scotch tape. Multiple samples can be clearly labeled on a Post-it note, for instance, numerically (1, 2, 3, ...) with the legend included in the package for the sample identification.

A glass vial is the best for powder because it is really hard to take a powder out of a charged plastic vial, especially when dealing with a small amount of sample. Normally, it is recommended to dry powder samples before the XPS analysis, however, slurry might be acceptable in some instruments, but the non-dry (water- or hydrocarbon-containing) slurry might make the analysis difficult by (i) increasing the pumping time, (ii) growing carbon deposit on the surface, (iii) causing a pressure increase in a chamber during the XPS analysis, and/or (iv) requiring sample cooling during the XPS acquisition.

Oxidation is another possible issue emerging even during a fast transfer of a sample through air. This problem can be solved by using either a transfer suitcase (available commercially or self-build) or a glove box attached directly to an XPS instrument or reduction treatment in a reaction cell directly attached to an XPS instrument. Both approaches have their own advantages. A glove box allows us to analyze a sample following a specific treatment performed 1000 miles away as shown in Figure 2.4. It is remarkable that even a short sample transfer within a university campus caused massive oxidation.

Battery research is another example that can be benefited from the use of a glove box. A battery, which received specific treatment or which was aged, can be open inside the glove box and then can be immediately transferred for the XPS analysis keeping the surface chemistry non-intact. On the other hand, the reaction cell attached to an XPS spectrometer allows us to reduce sample mitigating oxidation and carbon contamination. Also, a chemical reaction under realistic conditions can be run in the reaction cell for a catalytic sample [55].

2.2.4 *XPS acquisition conditions*

The acquisition conditions are pretty much determined by answering the following question: "What kind of information do I want to get by using XPS?" A user should decide if high-resolution spectra should be collected or the survey spectrum is sufficient.

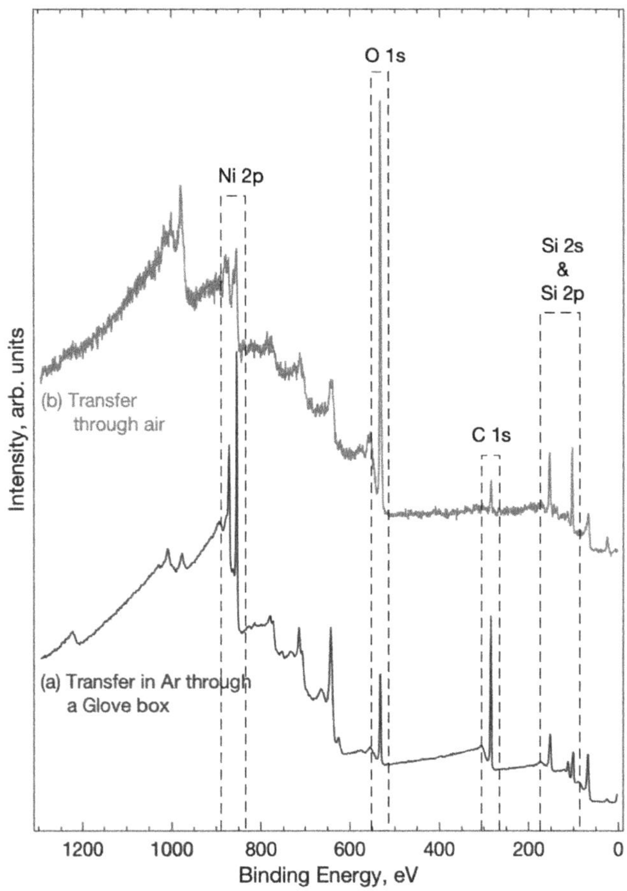

Fig. 2.4. Survey spectra collected from two identical samples. One sample (a) has traveled over see (three days in transit) in a glass vial sealed under argon and then loaded through a glove box without any contact with air; another sample (b) was transferred within a university campus (10 min of walk) in air.

2.2.4.1 *Survey spectrum*

The survey (or wide) spectrum acquires a wide range of binding energies covering almost the full spectrum of X-ray photon excitation. Modern XPS spectrometers always operate the energy analyzer at constant pass energy mode, where the kinetic energy of photoelectrons at the entrance of the analyzer is fixed, and this provides a reliable quantification of the XPS data. The survey spectrum is collected at a high pass energy (PE) typically in the range of 100–160 eV, which delivers high intensity of a signal (greater number of electrons can reach the detector or more correctly speaking, better signal-to-noise ratio) and, therefore, the highest element sensitivity is achieved in the survey spectrum. However, due to the high pass energy used, the energy resolution is poor, and the chemical shift can be hardly detected. In cases where only the elemental composition of a sample surface is required, the survey spectrum is sufficient. In general, it takes 1–15 minutes to collect a survey spectrum. The typical acquisition parameters for a survey spectrum are shown in Table 2.1.

2.2.4.2 *High-resolution spectra*

The high-resolution spectra are acquired at lower PES than those used for the surveys. The intensity of the signal (or signal-to-noise ratio) is lower, and this applies a longer acquisition time to achieve the acceptable signal-to-noise ratio. However, the resolution, usually measured as full width at half maximum (FWHM), of a characteristic peak is much higher than those for a survey spectrum, allowing us to identify the chemical state(s) of the element of interest. The typical parameters for the acquisition of high-resolution spectra are shown in Table 2.1. If the chemical state is the subject of a study, then the high-resolution spectra should be collected. In the case of the photoemission peak overlapping, the high-resolution spectra

Table 2.1. The typical acquisition parameters used for a survey spectrum and high-resolution spectra for a laboratory-based XPS.

	Acquisition parameters				
	Pass energy, eV	Energy step size, eV	Energy range, eV	Resolution, eV	Acquisition time, min
Survey	100–160	0.5–1	~1200	~2–3	1–15
High-resolution spectra	10–40	0.05–0.1	~15	0.4–1.0	30 and up

could be very helpful for the quantification analysis. The low resolution of the survey spectrum does not allow us to separate the photoemission peaks of interest, but those peaks could be distinguished in a high-resolution spectrum. One photoemission peak, which (i) has the highest intensity (in other words, the highest relative sensitivity factor, RSF) and which is the narrowest (the highest resolution or smallest FWHM), should be collected for each element detected in the survey spectrum. To save the acquisition time, some elements can be omitted, but the photoemission peak appropriate for the charge correction must always be collected (see the following section). Ignoring to obtain some binding energy regions might cause (i) the lower accuracy of the data quantification and (ii) possible loss of information required for further data analysis.

2.2.4.3 *Sample sputtering*

As repeatedly said, XPS is a surface-sensitive technique. The μm depth can be accessed if the sample is sputtered by the ion beam of a noble gas. However, sputtering is highly destructive and Ar^+ beam can even cause the reduction of an oxide [81]. Sputtering depth profiling should not be used for organic samples; however, recently, big carbon molecules, such as fullerene and coronene, and non-destructive (or low destructive) ion beams are explored, such as Ar clusters or coronene or fullerene [48–53].

The sputtering rate should be calibrated for each instrument and for each experimental condition [81]. Reference samples with precise thickness of repeating metal layers are available and they can be used for such purposes. The sputtering rate varies depending on atomic weight and ion beam energy [81,82]; these might result in changing the element composition of the sample surface due to preferential sputtering.

2.2.4.4 *XPS error*

"What is the quantification error of the XPS measurement?" is a quite frequent question, which does not have a straightforward answer because several sources can contribute to the experimental error.

The first contribution comes from the measurement of the peak area of a noisy signal after background subtraction. The shape of the background is chosen by the XPS user, and it can affect the peak area significantly. Also, slight variations of the beginning and the end of the background might result in significant changes in the photoemission peak area, especially in the case where a survey spectrum is used for quantification. XPS specialized software (for instance, CasaXPS) allows the XPS user to estimate such errors automatically. For a spectrum with a reasonable signal-to-noise ratio, the peak area changes less than 1% with variations of the beginning and the end of the background.

The second possible contribution is the relative sensitivity factor (RSF). Even though the excitation cross-sections of the different electron levels are carefully calculated and tabulated for different photon energies, RSFs are still theoretically calculated numbers, for which the error is hard to estimate in practical applications. On the other hand, empirical atomic sensitivity factors (ASFs) also were reported by Wagner *et al.* [83] providing a good cross-checking for the theoretical values. But it should be noted that the accuracy of Wagner's empirical ASFs depends on the purity of a reference compound. More discussions about this issue can be found in Chapter 3.

The third source of the XPS quantification error is the quantification model itself. As discussed above, typically, the element composition is calculated using an assumption of homogeneously mixed elements within the information depth. Several assumptions were done to derivate this model and these assumptions introduce an error. The XPS quantification model likely makes the biggest contribution to the overall quantification error.

There are other sources of the quantification error. Organic samples often degrade under an X-ray beam, and this causes changing the element's composition. If non-monochromatic radiation is used, the photoemission peak might contain the contributions due to the satellite peaks or the ghost peaks, which is difficult to subtract. The best practice is probably (i) to use a calibration compound with well-known element composition and (ii) to perform multiple acquisitions at the different spots of a sample to get statistic errors for a particular sample. In general, the accuracy of 5% (the deviation of the calculated atomic concentration of a particular element) is considered to be a very good result.

2.2.5 *Artifacts*

Artifacts and the "strange" peaks can be observed in a photoemission spectrum. Sample charging is one of the common artifacts. The Auger peaks, plasmon excitation, satellite, and ghost peaks can be seen as well.

2.2.5.1 *Sample charging*

Since the illumination of a surface with an X-ray beam results in the flux of the photoelectrons, non-conducting samples get charged during the XPS analysis. This issue is discussed in detail in Chapter 5, however some information is included here for the sake of consistency. Modern XPS instruments are equipped with a charge neutralizer to deal with non-conducting samples. A charge neutralizer typically mitigates non-homogeneous charges allowing to obtain high-resolution spectra. For a non-conducting sample, the spectrometer ground is not connected to the Fermi level of the sample, as in the case of an electron conductor shown in Figure 2.1, therefore Eq. (2.1) cannot be applied. To identify the chemical states in a non-conducting sample, a charge correction (should not be mistaken with energy calibration) should be performed individually for each sample and for each acquisition condition (if it was varied).

The good practice of the charge correction procedure is described by NIST [18]:

- Adventitious carbon (AdC) or internal hydrocarbon or condensed hydrocarbon can be used as the binding energy scale (charge) reference with an assumed C 1s binding energy of 284.8 eV. Typically, carbon can be found in almost any sample, therefore it is the most popular charge reference. The exceptions are samples in which the carbon chemical states are under investigation; in those cases, an alternative charge correction reference is required.

- Some samples may contain argon, for instance, Ar can be implanted during Ar sputtering, then the Ar $2p_{3/2}$ peak at 241.82 eV can be used as the charge reference [84]. However, one should be cautious since the Ar 2p binding energy might vary depending on the material in which the ions were implanted [85].
- A noble metal (typically an inert one, like Au) can be vapor deposited over the investigated sample and the characteristic peak of the noble metal photoelectrons can serve as a reference. But this approach is destructive and the surface under investigation will be altered.

The charge correction, despite the apparent simplicity and the frequent usage, requires creativity and consistency between alike samples. Mistreatment of the charge correction causes misinterpretation of the chemical states.

Powders are typically non-conducting, and a powder sample gets often charged under an X-ray beam. Proper preparation and mounting of powder samples can help mitigate charging. Powders can be pressed as a pellet or they can be evenly distributed on a double-sided sticking tape. A tablet holds shape well and does not cause the powder to spread around, whereas a thick non-conducting material is more prone to charging. It might be even worse due to the static bulk charge built up inside of a tablet, while a material was pressed. The thin layer of a powder sample evenly distributed over a conducting double-sided sticking tape suffers charging much less. More details about the analysis of electrically insulating materials can be found in Chapter 5.

2.2.5.2 *"Other" peaks in XPS spectrum*

In the case of non-monochromatic excitation, the so-called ghost peaks and satellite peaks can appear in a spectrum. Plasmon can be detected as well disregarding the excitation source, monochromatic

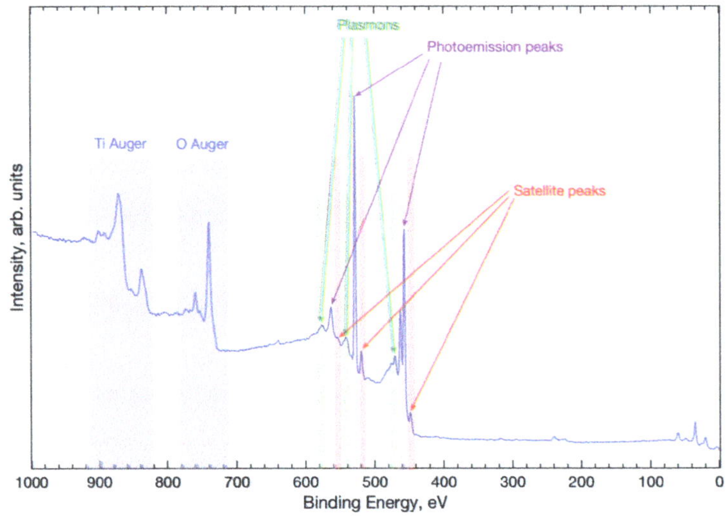

Fig. 2.5. Survey spectra collected from TiO_2 using non-monochromatic Mg Kα radiation (1253.6 eV). The Auger peaks are marked with blue rectangles, the photoemission peaks with gray, the plasmons with green, and the satellite peaks with red.

or non-monochromatic. A survey spectrum of TiO_2 shown in Figure 2.5 was collected using non-monochromatic Mg Kα radiation and demonstrates photoemission and Auger peaks marked by gray and blue rectangles, correspondently, along with "strange" peaks marked with green and red rectangles. The peaks appearing at higher binding energy following a photoemission peak (green rectangles) are plasmon excitations. This phenomenon is not analyzed often. The peaks at lower binding energy preceding a photoemission peak (red rectangles) are the satellites, which are generated by the X-ray photons with different energies due to various electron transitions [19] and these photons originated in the same anode material.

The ghost peaks are caused by the X-ray photons, which are generated by heteroelement material other than the anode material [19]. These could be due to impurities, or, in the case of a dual anode, the ghost peaks appear due to X-ray from an adjacent anode material. It should be underlined that the ghosts and satellites are

Table 2.2. The shift of the ghost peaks due to excitations from heteroelements for a dual Mg/Al anode.

Contaminating radiation	Anode material: Al	Mg	Comments
Mg Kα	+233	–	Because of poor design of a dual anode, the electrons hit over another anode material.
Al Kα	–	–233	Because of poor design of a dual anode, the electrons hit over another anode material.
O Kα	+961.7	+728.7	Bad sign, maybe an anode has a leak, and water appeared on the anode surface.
Cu Lα	+556.9	+323.9	Bad sign too, maybe anode material is gone.

the features of non-monochromatic radiation. The shift of a ghost peak with respect to a "primer" peak is equal to the energy difference between primer radiation and contaminating radiation. The frequently observed shifts for a dual Mg/Al anode are summarized in Table 2.2. Figure 2.6 shows the O 1s ghost peak, which appears in the C 1s region when radiation was used. The sample, $TiO_2(110)$, was characterized by an intensive O 1s peak at 530.5 eV, the O 1s ghost is at 297.5 eV, which corresponds to the energy difference between Mg Kα photons and contaminating and Al Kα [58]. To get a correct fitting of the spectrum, the O 1s ghost peak should be included.

2.2.6 *XPS data analysis*

XPS provides both qualitative and quantitative information. The data analysis depends on what kind of information is needed. The quantitative data can be obtained from the survey spectrum assuming that the photoemission peaks and Auger peaks of different elements do not overlap. Otherwise, the high-resolution spectra should be used to measure the area of the photoemission peak more precisely without the contribution of other peaks. Higher resolution spectra allow us to separate overlapping peaks or in some cases, the curve fitting can help us to determine the contribution of the

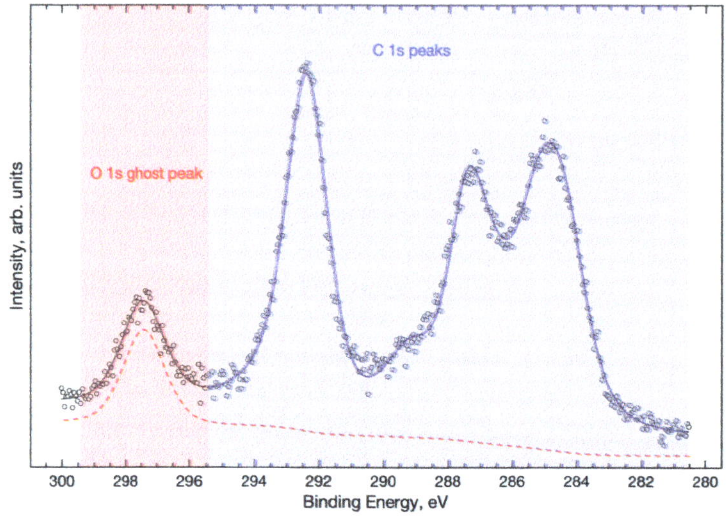

Fig. 2.6. O 1s ghost peak appeared in the C 1s region when the spectrum was collected using non-monochromatic Mg Kα radiation. The ghost peak is due to "contaminating" Al Kα radiation.

different peaks in a complex overlapping spectrum. The curve fitting and shape of the photoemission peak form the subject of Chapter 3.

The qualitative information (or chemical state data) is obtained from the high-resolution spectra. First, the charge correction should be applied using the protocol described above in this chapter. A curve fitting by using a reference peak or a least-squares fit of a parabola to the top of the peak can be used to determine the precise peak position [86]. This might be called a user-independent approach permitting to minimize the uncertainty of a peak position. Following the charge correction, the binding energy of the peak of interest becomes a good descriptor of the chemical states. Reference data of the various chemical states are widely available; for instance, NIST provides a thorough and reliable database in Ref. [18].

Combining both qualitative and quantitative XPS information is the most fruitful approach to the XPS data analysis.

The stoichiometry obtained by quantitative analysis should be consistent with the chemical states and *vice versa*; the chemical state data can predict if any possible contamination leads to an error in the elemental composition. For example, in case of a metal oxide, the metal to oxygen ratio should match the oxidation state of the metal found by the chemical shift. Details about the quantitative analysis of *XPS* spectra in order to determine the amount of the different elements at the surface of a catalyst can be found in Chapter 4.

2.2.7 *Reporting XPS data in publication*

General guidelines are that the data and information reported in a publication should be sufficient to understand and to reproduce the experiment. Elsevier publishing group, for instance, recommends the reviewers of an experiment-based manuscript to first check the experimental/methods section before reading results and discussion. The major flaws are unsound methodology, missing processes known to be influential in the area of reported research, and a conclusion drawn in contradiction to the statistical or qualitative evidence reported in the manuscript. Keeping this in mind, in the experimental session of the manuscript containing XPS data, we would recommend reporting the following information:

- the type of XPS instrument (instrument manufacturer and model) used,
- the acquisition parameters such as radiation source (monochromatic or non-monochromatic, photon energy or which anode-type was used), pass energy, using or not of a charge neutralizer, time under the X-ray beam, and the X-ray gun power if sample degradation was detected,
- the software used for data analysis and curve fitting with a clear description of a background type and a peak shape used,
- the charge correction method (if applicable) and the energy scale calibration,
- quantification parameters such as what set of RSFs were used and what model for the quantification was assumed; the

associated parameters can be provided in the Supporting Information of a manuscript; how the quantification data were corrected for the instrument function and the information depth should be described as well,

- sputtering conditions (if any), such as the type of an ion gun, the kind of ion gas (Ar, Ne, Ar cluster — size, coronene, fullerene, etc.), the energy of the ion, the ion current, the sputtering area (to estimate ion current density), and the sputtering rate,
- description of customized setups (reaction cell, dedicated gas system, glove box, etc.).

The main idea is to make sure that sufficient descriptive elements are presented.

2.3 Concluding Remarks

In the following, we briefly summarize the important aspects discussed in this chapter.

- *Surface sensitivity*: Due to the limited EALs of the photoelectrons through a solid, XPS is a surface-sensitive technique with the information depth in the range of 4–10 nm for typical laboratory-based instruments using Mg Kα or Al Kα radiation. Using other anode materials, for instance, Ag, "tender" Ag Lα X-rays (2984.3 eV) can be produced, and therefore, the XPS information depth can be increased. Synchrotron-based XPS can achieve higher surface sensitivity, if the softer X-ray radiation is used, or higher bulk sensitivity, in the case of tender or hard X-ray radiation. On the other hand, depth profiling with ion sputtering can help us to investigate the element composition in the bulk.
- *Quantification*: XPS is a quantification technique, which provides the element composition in the near-surface region. Typically, the element composition is calculated assuming the homogeneous mixture of elements within the information depth; this simple model works well in many cases. The typical sensitivity of XPS is in the range of 0.1 atomic %. For thin films or adsorbed

layers (non-homogeneous distribution of the elements), more complicated quantification models should be explored.

- *Chemical state analysis*: The binding energy of photoelectrons is a good descriptor of the chemical state of an emitting atom. However, there are quite a few other factors, which might lead to the shift of the binding energy as well; therefore, each case should be carefully examined. The modified Auger parameter could be a very good descriptor of the chemical state too.
- *Acquisition parameters and resolution*: The pass energy of the electron energy analyzer determines the resolution and the signal intensity (or the signal-to-noise ratio). At high PE, the sensitivity is high, but the resolution is low. This is the condition typically used for the acquisition of a survey spectrum. At low PE, the situation is reversed: the sensitivity is low, but the resolution is high. This is suitable for the high-resolution spectra. The resolution is one of the key factors to separate the chemical states, therefore PE should be chosen properly. The resolution also should be considered during the XPS data analysis, especially for the curve fitting procedure.
- *Sample transportation*: Even though it is often neglected, sample transportation is one of the critical parameters contributing to the quality of the XPS data. A sample should be freshly prepared for XPS analysis, and if possible, after preparation, the sample surface should not come in contact with the container or any handling instrument. An example of the sample mounting for the transfer is given in this chapter. Oxygen- and/or moisture-sensitive samples should be handled in a glove box without contact with air.
- *Charging*: A non-conducting sample gets charged under an X-ray beam. The proper charge correction should be applied for every acquisition spot.
- *Artifacts and obstacles*: The X-ray and thermal degradations along with the sample charging were discussed. However, in the case of non-monochromatic excitation, the so-called ghost and satellite peaks can appear in a spectrum. The ghost peaks are caused by the X-ray photons, which are generated by heteroelement

material other than the anode material [19]. These could be impurities or adjacent anode material in the case of a dual anode. The satellite peaks are generated by the X-ray photons with different energies due to different electron transitions, and these photons originated in the same anode material [19]. The ghost and satellite peaks can hinder the data analysis.

References

[1] Siegbahn K. Electron spectroscopy for atoms, molecules and condensed matter, nobel lecture; 1981.

[2] McCarthy CA, Zemlyanov DY, Crean AM, Taylor LS. Comparison of drug release and adsorption under supersaturating conditions for ordered mesoporous silica with indomethacin or indomethacin methyl ester. Mol Pharmaceut. 2020;17:3062–3074.

[3] Que CL, Qi QQ, Zemlyanov DY, Mo HP, Deac A, Zeller M, Indulkar AS, Gao Y, Zhang GGZ, Taylor LS. (2020). Evidence for halogen bonding in amorphous solid dispersions. Cryst Growth Des. 2020;20:3224–3235.

[4] Saboo S, Mugheirbi NA, Zemlyanov DY, Kestur US, Taylor LS. Congruent release of drug and polymer: A “sweet spot” in the dissolution of amorphous solid dispersions. J Control Release. 2019;298: 68–82.

[5] Bhujbal SV, Zemlyanov DY, Cavallaro A, Mangal S, Taylor LS, Zhou QT. Qualitative and quantitative characterization of composition heterogeneity on the surface of spray dried amorphous solid dispersion particles by an advanced surface analysis platform with high surface sensitivity and superior spatial resolution. Mol Pharmaceut. 2018;15:2045–2053.

[6] Song Y, Zemlyanov D, Chen X, Su ZY, Nie HC, Lubach JW, Smith D, Byrn S, Pinal R. Acid-base interactions in amorphous solid dispersions of lumefantrine prepared by spray-drying and hot-melt extrusion using X-ray photoelectron spectroscopy. Int J Pharmaceut. 2016;514:456–464.

[7] Song Y, Zemlyanov D, Chen X, Nie HC, Su ZY, Fang K, Yang XH, Smith D, Byrn S, Lubach JW. Acid-base interactions of polystyrene

sulfonic acid in amorphous solid dispersions using a combined UV/FTIR/XPS/ssNMR study. Mol Pharmaceut. 2016;13:483–492.

[8] Wilson NE, Mutukuri TT, Zemlyanov DY, Taylor LS, Topp EM, Zhou QT. Surface composition and formulation heterogeneity of protein solids produced by spray drying. Pharm Res. 2020;37:14.

[9] Jedlicka SS, Little KM, Nivens DE, Zemlyanov D, Rickus JL. Peptide ormosils as cellular substrates. J Mater Chem. 2007;17:5058–5067.

[10] Jedlicka SS, Rickus JL, Zemlyanov D. Controllable surface expression of bioactive peptides incorporated into a silica thin film matrix. J Phys Chem C. 2010;114:342–344.

[11] Slavin JWJ, Jarori U, Zemlyanov D, Ivanisevic A. Characterization of amino acid adlayers on InAs surfaces using X-ray photoelectron spectroscopy. J Electron Spectrosc. 2009;172:47–53.

[12] Flores-Perez R, Zemlyanov DY, Ivanisevic A. DNA molecules on GaP (100) surfaces: Spectroscopic characterization and biospecificity assessment. Chemphyschem. 2008;9:1528–1530.

[13] Dolash BD, Lahiji RR, Zemlyanov DY, Reifenberger R, Bergstrom DE. DNA-associated single-walled carbon nanotubes as a platform for drug delivery. Nsti Nanotech 2008, Vol 2, Technical Proceedings; 2008. pp. 419–422.

[14] Shchukarev A, Gojkovic Z, Funk C, Ramstedt M. Cryo-XPS analysis reveals surface composition of microalgae. Appl Surf Sci. 2020; 526:146538–146543.

[15] Gojkovic Z, Shchukarev A, Ramstedt M, Funk C. Cryogenic X-ray photoelectron spectroscopy determines surface composition of algal cells and gives insights into their spontaneous sedimentation. Algal Res. 2020;47:101836–101844.

[16] Ramstedt M, Shchukarev A. Analysis of bacterial cell surface chemical composition using cryogenic X-ray photoelectron spectroscopy. Methods Mol Biol. 2016;1440:215–223.

[17] XPS simplified, https://xpssimplified.com/index.php. Thermo Fisher Scientific Inc.

[18] Naumkin AV, Kraut-Vass A, Gaarenstroom SW, Powell CJ. NIST X-ray photoelectron spectroscopy database. In: NIST Standard Reference Database 20 Gaithersburg MD, 20899: National Institute of Standards and Technology; 2000.

[19] Moulder JF, Stickle WF, Sobol PE, Bomben K. *Handbook of X-ray Photoelectron Spectroscopy*. Eden Prairie, USA: Physical Electronics, Inc., 1992.

[20] Fadley CS. Electron spectroscopy: Theory, techniques and applications. In *1 Basic Concepts of X-ray Photoelectron Spectroscopy* (pp. 1–156). London, New York, San Francisco: Academic Press, Published, 1978.

[21] Siegbahn K, Nordling C, Johansson G, Hedman J, Heden PF, Hamrin K, Gelius U, Bergmark T, Werme LO, Manne R, Baer Y. *ESCA Applied to Free Molecules*. Amsterdam: North-Holland Pub. Co., 1969.

[22] Briggs D, Grant JT (Eds.). *Surface Analysis by Auger and X-ray Photoelectron Spectroscopy*. Chichester, Manchester, UK: IM Publications and SurfaceSpectra Ltd., 2003.

[23] Baer DR, Artyushkova K, Brundle CR, Castle JE, Engelhard MH, Gaskell KJ, Grant JT, Haasch RT, Linford MR, Powell CJ, Shard AG, Sherwood PMA, Smentkowski VS. Practical guides for X-ray photoelectron spectroscopy: First steps in planning, conducting, and reporting XPS measurements. J Vac Sci Technol A. 2019;37: 31401–31411.

[24] Gabasch H, Hayek K, Klotzer B, Unterberger W, Kleimenov E, Teschner D, Zafeiratos S, Havecker M, Knop-Gericke A, Schlogl R, Aszalos-Kiss B, Zemlyanov D. Methane oxidation on Pd(111): In situ XPS identification of active phase. J Phys Chem C. 2007;111: 7957–7962.

[25] Zemlyanov D, Klotzer B, Gabasch H, Smeltz A, Ribeiro FH, Zafeiratos S, Teschner D, Schnorch P, Vass E, Havecker M, Knop-Gericke A, Schlogl R. Kinetics of palladium oxidation in the mbar pressure range: Ambient pressure XPS study. Top Catal. 2013;56:885–895.

[26] Axnanda S, Crumlin EJ, Mao B, Rani S, Chang R, Karlsson PG, Edwards MOM, Lundqvist M, Moberg R, Ross P, Hussain Z, Liu Z. Using "Tender" X-ray ambient pressure X-ray photoelectron spectroscopy as a direct probe of solid-liquid interface. Sci Rep. 2015;5: 9788–9799.

[27] Salmeron M, Schlogl R. Ambient pressure photoelectron spectroscopy: A new tool for surface science and nanotechnology. Surf Sci Rep. 2008;63:169–199.

[28] Arrigo R, Hävecker M Fau — Schuster ME, Schuster Me Fau — Ranjan C, Ranjan C Fau — Stotz E, Stotz E Fau — Knop-Gericke A, Knop-Gericke A Fau — Schlögl R, Schlögl R. In situ study of the gas-phase electrolysis of water on platinum by NAP-XPS. Angewandte Chemie. 2013;52:11660–11664.

[29] Butcher DR, Grass Me Fau — Zeng Z, Zeng Z Fau — Aksoy F, Aksoy F Fau — Bluhm H, Bluhm H Fau — Li W-X, Li Wx Fau — Mun BS, Mun Bs Fau — Somorjai GA, Somorjai Ga Fau — Liu Z, Liu Z. In situ oxidation study of Pt(110) and its interaction with CO. J Am Chem Soc. 2011;133:20319–20325.

[30] Stoerzinger KA, Hong WT, Crumlin EJ, Bluhm H, Shao-Horn Y. Insights into electrochemical reactions from ambient pressure photoelectron spectroscopy. Acc Chem Res. 2015;48:2976–2983.

[31] Maibach J, Källquist I, Andersson M, Urpelainen S, Edström K, Rensmo H, Siegbahn H, Hahlin M. Probing a battery electrolyte drop with ambient pressure photoelectron spectroscopy. Nat Commun. 2019;10:3080–3080.

[32] Shchukarev A, Ramstedt M. Cryo-XPS: Probing intact interfaces in nature and life. Surf Interface Anal. 2017;49:349–356.

[33] Zemlyanov DY, Nagy A, Schlogl R. The reaction of silver with NO/O_2. Appl Surf Sci. 1998;133:171–183.

[34] Bilodeau RA, Zemlyanov DY, Kramer RK. Liquid metal switches for environmentally responsive electronics. Adv Mater Interfaces. 2017;4: 1600913–1600921.

[35] Tougaard S. QUASES-IMFP-TPP2M. http://www.quases.com/products/quases-imfp-tpp2m/. QUASES.

[36] Tanuma S, Powell CJ, Penn DR. Calculations of electron inelastic mean free paths 0.5. Data for 14 organic-compounds over the 50–2000 Ev range. Surf Interface Anal. 1994;21:165–176.

[37] Powell CJ, Jablonski A. *NIST Electron Effective-Attenuation-Length Database, Version1.3, SRD 82.* Gaithersburg, MD: National Institute of Standards and Technology, 2011.

[38] Jablonski A, Powell CJ. The electron attenuation length revisited. Surf Sci Rep. 2002;47:35–91.

[39] Jablonski A, Powell CJ. Comparisons of practical attenuation lengths obtained from different algorithms for application in XPS. Surf Sci. 2002;520:78–96.

[40] Powell CJ, Jablonski A. Surface sensitivity of X-ray photoelectron spectroscopy. Nucl Instrum Meth A. 2009;601:54–65.

[41] Fadley CS, Nemsak S. Some future perspectives in soft- and hard-X-ray photoemission. J Electron Spectrosc. 2014;195:409–422.

[42] Chainani A, Yokoya T, Takata Y, Tamasaku K, Taguchi M, Shimojima T, Kamakura N, Horiba K, Tsuda S, Shin S, Miwa D, Nishino Y, Ishikawa T, Yabashi M, Kobayashi K, Namatame H, Taniguchi M, Takada K, Sasaki T, Sakurai H, Takayama-Muromachi E. Hard-X-ray photoelectron spectroscopy of $Na_xCoO_2 \cdot yH_2O$. Nucl Instrum Meth A. 2005;547:163–168.

[43] Kobayashi K, Yabashi M, Takata Y, Tokushima T, Shin S, Tamasaku K, Miwa D, Ishikawa T, Nohira H, Hattori T, Sugita Y, Nakatsuka O, Sakai A, Zaima S. High resolution-high energy X-ray photoelectron spectroscopy using third-generation synchrotron radiation source, and its application to Si-high k insulator systems. Appl Phys Lett. 2003;83: 1005–1007.

[44] Dallera C, Duo L, Braicovich L, Panaccione G, Paolicelli G, Cowie B, Zegenhagen J. Looking 100 A deep into spatially inhomogeneous dilute systems with hard X-ray photoemission. Appl Phys Lett. 2004; 85:4532–4534.

[45] Kalinkin AV, Smirnov MY, Nizovskii AI, Bulthtiyarov VI. X-ray photoelectron spectra of platinum compounds excited with monochromatic AgL alpha irradiation. J Electron Spectrosc. 2010;177:15–18.

[46] Lin TH, Margossian T, Zheng LQ, Kumar S, Marozau I, Sereda O, Zemlyanov D, Shih CJ, Zenobi R, Baudouin D, De Micheli G, Gaillardon PE, Coperet C. Conformal deposition of conductive single-crystalline cobalt silicide layer on Si wafer via a molecular approach. Chem Mater. 2018;30:2168–2173.

[47] Lin TH, Margossian T, De Marchi M, Thammasack M, Zemlyanov D, Kumar S, Jagielski J, Zheng LQ, Shih CJ, Zenobi R, De Micheli G, Baudouin D, Gaillardon PE, Coperet C. Low-temperature wet conformal nickel silicide deposition for transistor technology through an organometallic approach. Acs Appl Mater Inter. 2017;9:4948–4955.

[48] Gilbert JB, Rubner MF., Cohen RE. Depth-profiling XPS analysis of diffusion. Proc Nat Acad Sci. 2013;110:6651–6656.

[49] Chen Y-Y, Yu B-Y, Wang W-B, Hsu M-F, Lin W-C, Lin Y-C, Jou J-H, Shyue J-J. X-ray photoelectron spectrometry depth profiling of organic thin films using C60 sputtering. Anal Chem. 2008;80:501–505.
[50] Shard AG, Havelund R, Spencer SJ, Gilmore IS, Alexander MR, Angerer TB, Aoyagi S, Barnes J-P, Benayad A, Bernasik A, Ceccone G, Counsell JDP, Deeks C, Fletcher JS, Graham DJ, Heuser C, Lee TG, Marie C, Marzec MM, Mishra G, Rading D, Renault O, Scurr DJ, Shon HK, Spampinato V, Tian H, Wang F, Winograd N, Wu K, Wucher A, Zhou Y, Zhu Z. Measuring compositions in organic depth profiling: Results from a VAMAS interlaboratory study. J Phys Chem B. 2015;119:10784–10797.
[51] Cumpson PJ, Portoles JF, Barlow AJ, Sano N, Birch M. Depth profiling organic/inorganic interfaces by argon gas cluster ion beams: Sputter yield data for biomaterials, in-vitro diagnostic and implant applications. Surf Interface Anal. 2013;45:1859–1868.
[52] Miyayama T, Sanada N, Suzuki M, Hammond JS, Si SQD, Takahara A. X-ray photoelectron spectroscopy study of polyimide thin films with Ar cluster ion depth profiling. J Vac Sci Technol A. 2010;28: L1–L4.
[53] Rafati A, Davies MC, Shard AG, Hutton S, Mishra G, Alexander MR. Quantitative XPS depth profiling of codeine loaded poly(l-lactic acid) films using a coronene ion sputter source. J Control Release. 2009;138:40–44.
[54] Fairley N. CasaXPS. http://www.casaxps.com. Casa Software Ltd.
[55] Luo W, Zemlyanov DY, Milligan CA, Du YC, Yang LM, Wu YQ, Ye PD. Surface chemistry of black phosphorus under a controlled oxidative environment. Nanotechnology. 2016;27:434002–434011.
[56] Zemlyanov DY, Jespersen M, Zakharov DN, Hu JJ, Paul R, Kumar A, Pacley S, Glavin N, Saenz D, Smith KC, Fisher TS, Voevodin AA. Versatile technique for assessing thickness of 2D layered materials by XPS. Nanotechnology. 2018;29:115705–115714.
[57] Paul R, Reifenberger RG, Fisher TS, Zemlyanov DY. Atomic layer deposition of FeO on Pt(111) by ferrocene adsorption and oxidation. Chem Mater. 2015;27:5915–5924.
[58] Gharachorlou A, Detwiler MD, Nartova AV, Lei Y, Lu JL, Elam JW, Delgass WN, Ribeiro FH, Zemlyanov DY. Palladium nanoparticle

formation on $TiO_2(110)$ by thermal decomposition of Palladium(II) hexafluoroacetylacetonate. ACS Appl Mater Inter. 2014;6: 14702–14711.

[59] Jablonski A, Zemek J. Overlayer thickness determination by XPS using the multiline approach. Surf Interface Anal. 2009;41:193–204.

[60] Cumpson PJ. The Thickogram: A method for easy film thickness measurement in XPS. Surf Interface Anal. 2000;29:403–406.

[61] Powell CJ. NIST Database for the Simulation of Electron Spectra for Surface Analysis (SESSA). In: Natl Std. Ref. Data Series (NIST NSRDS) Gaithersburg MD, 20899: National Institute of Standards and Technology; 2018.

[62] Shard AG. Detection limits in XPS for more than 6000 binary systems using Al and Mg Kα X-rays. Surf Interface Anal. 2014;46:175–185.

[63] Faradzhev NS, Hill SB, Powell CJ. Quantitative analysis of trace levels of surface contamination by X-ray photoelectron spectroscopy. Part II: Systematic uncertainties and absolute quantification. Surf Interface Anal. 2017;49:1214–1224.

[64] Hill SB, Faradzhev NS, Powell CJ. Quantitative analysis of trace levels of surface contamination by X-ray photoelectron spectroscopy. Part I: Statistical uncertainty near the detection limit. Surf Interface Anal. 2017;49:1187–1205.

[65] Manne R. Molecular orbitals and inner-electron-shell chemical shifts for sulfur and chlorine oxy-anions. J Chem Phys. 1967;46:4645.

[66] Moulder JF, Stickle WF, Sobol PE, Bomben K. *Handbook of X-ray Photoelectron Spectroscopy*. Eden Prairie, USA: Physical Electronics Inc., 1992.

[67] Zeng Z, Greeley J. Characterization of oxygenated species at water/ Pt(111) interfaces from DFT energetics and XPS simulations. Nano Energy. 2016;29:369–377.

[68] Zeng Z, Ma X, Ding W, Li W. First-principles calculation of core-level binding energy shift in surface chemical processes. Sci China Chem. 2010;53:402–410.

[69] Kittel C. *Introduction to Solid State Physics*. New York: Wiley; 2005.

[70] Shirley DA. *Advances in Chemical Physics*, (I Prigogine and SA Rice eds.) "ESCA", John Wiley & Sons, Inc., 1973, pp. 85–159.

[71] Gaarenstroom SW, Winograd N. Initial and final-state effects in esca spectra of cadmium and silver-oxides. J Chem Phys. 1977;67: 3500–3506.

[72] Ning X, Wang Z, Zhang Z. Fermi level shifting, charge transfer and induced magnetic coupling at $La_{0.7}Ca_{0.3}MnO_3/LaNiO_3$ interface. Sci Rep. 2015;5:8460.

[73] Szuber J, Bergignat E, Hollinger G, Polakowska A, Kościelniak P. XPS study of the surface Fermi level of $(NH_4)_2S_x$-passivated GaAs(100) surface. Vacuum. 2002;67:53–58.

[74] Debehets J, Homm P, Menghini M, Chambers SA, Marchiori C, Heyns M, Locquet JP, Seo JW. Detecting Fermi-level shifts by Auger electron spectroscopy in Si and GaAs. Appl Surf Sci. 2018;440:386–395.

[75] Nelin CJ, Bagus PS, Ilton ES, Chambers SA, Kuhlenbeck H, Freund HJ. Relationships between complex core level spectra and materials properties. Int J Quantum Chem. 2010;110:2752–2764.

[76] Wagner CD, Joshi A. The Auger parameter, its utility and advantages — A review. J Electron Spectrosc. 1988;47:283–313.

[77] Wagner CD. Auger lines in X-ray photoelectron spectrometry. Anal Chem. 1972;44:967–973.

[78] Nosova LV, Stenin MV, Nogin YN, Ryndin YA. Exafs and Xps studies of the influence of metal-particle size, nature of support and H_2 and Co adsorption on the structure and electronic-properties of palladium. Appl Surf Sci. 1992;55:43–48.

[79] Fritsch A, Legare P. Metal vapor-deposition on carbon — An Xps study of Pd and Au cluster growth. Surf Sci. 1985;162:742–746.

[80] Boyen HG, Ethirajan A, Kastle G, Weigl F, Ziemann P, Schmid G, Garnier MG, Buttner M, Oelhafen P. Alloy formation of supported gold nanoparticles at their transition from clusters to solids: Does size matter? Phys Rev Lett. 2005;94:16804–16807.

[81] Baer DR, Engelhard MH, Lea AS, Nachimuthu P, Droubay TC, Kim J, Lee B, Mathews C, Opila RL, Saraf LV, Stickle WF, Wallace RM, Wright BS. Comparison of the sputter rates of oxide films relative to the sputter rate of SiO_2. J Vac Sci Technol A. 2010;28:1060–1072.

[82] Wasa K. Handbook of sputtering technology (Second Edition), eds. Wasa, K., Kanno, I., and Kotera, H., "2 — Sputtering Phenomena," William Andrew Publishing, Published; 2012. pp. 41–75.

[83] Wagner CD, Davis LE, Zeller MV, Taylor JA, Raymond RH, Gale LH. Empirical atomic sensitivity factors for quantitative-analysis by electron-spectroscopy for chemical-analysis. Surf Interface Anal. 1981;3:211–225.

[84] Powell CJ. Elemental binding-energies for X-ray photoelectron-spectroscopy. Appl Surf Sci. 1995;89:141–149.

[85] Citrin PH, Hamann DR. Measurement and calculation of polarization and potential-energy effects on core-electron binding-energies in solids — X-ray photoemission of rare-gases implanted in noble-metals. Phys Rev B. 1974;10:4948–4963.

[86] Cumpson PJ, Seah MP, Spencer SJ. Simple procedure for precise peak maximum estimation for energy calibration in AES and XPS. Surf Interface Anal. 1996;24:687–694.

https://doi.org/10.1142/9781800613294_0003

Chapter 3

Introduction to Chemical State Analysis by XPS with Examples

George H. Major[*,§], Neal Fairley[†,¶], Vincent Fernandez[‡,||] and Matthew R. Linford[*,**]

[*]*Department of Chemistry and Biochemistry, Brigham Young University, Provo, UT 84602, USA*

[†]*Casa Software Ltd., Bay House, Teignmouth, UK*

[‡]*Université de Nantes, CNRS, Institute of Materials of Nantes Jean Rouxel, IMN, F-44000 Nantes, France*

[§]*georgemviolin@gmail.com*

[¶]*neal@casaxps.com*

[||]*vincent.fernandez@cnrs-imn.fr*

[**]*mrlinford@chem.byu.edu*

Abstract

X-ray photoelectron spectroscopy (XPS) data are quantified by applying bell-shaped curves to the spectra. As with most measurements, there are multiple levels of information that can be extracted from XPS analysis, including the most basic, e.g., elemental identification and quantitation, and more complex fitting that can yield in-depth chemical state composition. Here, we discuss considerations associated with peak fitting XPS spectra with synthetic peaks. In all levels of quantification, understanding the underlying XPS signal is imperative to proper analysis. This ideal/underlying XPS signal is a Lorentzian distribution, but peak broadening effects perturb this distribution. A convolution of the original Lorentzian distribution with another function (typically a Gaussian), which accounts for signal broadening, is often used for synthetic peak generation. Different core-level electrons have different sensitivities to photoionization from the X-ray source, so relative sensitivity factors are required for quantitative analysis. Here, we introduce these concepts with peak fits of real spectra from silver, sodium chloride, lithium fluoride, and lithium sulfate, which help reveal the underlying sample physics and chemistry.

3.1 Introduction

Chemical state analysis by X-ray Photoelectron Spectroscopy (XPS) is often seen as synonymous with fitting sets of bell-shaped (Gaussian) curves to spectra. These curves are referred to as components or synthetic peaks within a peak model, where each component is typically interpreted as corresponding to a specific chemical state. However, while the basic bell shape is convenient and simple to use, it is often not sufficient. The selection of shapes used as components in a peak model is important both in terms of the proper description of photoemission and for accurate quantification in XPS. There are more complex, varied, and often useful spectral shapes beyond simple Gaussians for XPS peak fitting. Therefore, a more general concept for chemical state analysis by XPS is components/synthetic peaks with appropriate properties tailored to the sample chemistry and spectral envelopes encountered in practice. Here, we consider the influence of curve selection (peak shape) on our ability to interpret the chemistry of a sample by XPS. We begin with a discussion of the importance of convolution in understanding XPS peak shapes and then show examples of fitting and quantifying XPS spectra of NaCl, Ag, LiF, and Li_2SO_4, where emphasis is placed on choosing appropriate peak shapes and backgrounds, as well as the role of sensitivity factors.

3.2 Context for Quantification by XPS

There are multiple levels of information that can be extracted from XPS measurements (see also Chapters 2 and 4). Elemental identification, which is often obtained from survey spectra [1], is arguably the most basic information acquired. In some cases, it may be the only information required for an analysis. Specific, quantitative, in-depth chemical state composition, which reflects the oxidation states of the elements in a material [2], is possibly the most complex output from XPS. Within these two extremes lie other layers of information and analysis, including the identification of binding energies and relative intensities of photoemission signals, which are both used to infer the chemical composition of a sample. XPS spectra represent the sum of photoemission events for a given X-ray source and sample. The desired information in the resultant spectra primarily consists of emissions of core-level electrons from a given element, although Auger and valence band signals are also observed. Within the sampling depth of a material, photoemission occurs for all elements in XPS, except for hydrogen and helium, although the presence of hydrogen manifests itself in valence band spectra and in the chemical shifts of the elements. Different chemically shifted peaks from the core-level electrons of an element add to the number/complexity of photoemission features. Since the widths of XPS signals and their chemical shifts are of comparable size, peak overlaps are common in XPS. Accordingly, peak fitting is often required in XPS data analysis to extract chemical information from spectra, where the necessary curves/functions/synthetic peaks/line shapes attempt to approximate shapes intrinsic to the photoemission process itself, along with other effects of the sample and the instrumentation. Mathematical optimization is used to fit an ensemble of such curves to spectral forms, where the inputs to such a model are the number of curves, their mathematical shapes, and their positions, and the outputs are their binding energies (positions), widths, and intensities (areas), each associated with a specific photoemission process/oxidation state.

As explained in detail in Chapter 2, XPS is performed by gathering and detecting photoelectrons ejected from a sample that has been excited by an X-ray source of known energy. This process has been reviewed and discussed multiple times, including in books and a recent set of tutorials/guides on this subject [3–6]. The motivation for this recent tutorial information is the current "reproducibility crisis" in science [7–9] that has manifested itself in XPS as a significant amount of incorrect peak fitting and data analysis entering the scientific literature [10,11]. Some of the recent guides noted above attempt to address these specific issues [12–15]. The photoionization of atoms at their core levels, most commonly in XPS from materials in the solid state, but increasingly from liquids, gases, and materials that lack high-vacuum compatibility [16], produces electron emission/photoelectrons with well-defined kinetic energies. Based on the physics of the photoelectric effect, the measured kinetic energies of the photoelectrons are converted into binding energies, which are representative of a material. The spectrometer work function is a relatively minor correction to this calculation. The binding or kinetic energies obtained in XPS are generally characteristic of atoms in a specific chemical state. However, XPS signals are not single lines (monochromatic signals) but rather peaks with widths determined by multiple factors, including quantum mechanics, the peak width of the exciting X-rays, the nature of the sample itself, and the instrument. The resulting peak shapes range from being essentially Lorentzian to having both Gaussian and Lorentzian characters. These signals can be modeled as Gaussian–Lorentzian sum (GLS) or product (GLP) functions, or the Voigt function. This latter function, which is the convolution of a Gaussian and a Lorentzian function, best represents the modification of the natural peak shape of photoemission (the Lorentzian) with other effects [17,18]. In the early days of XPS, the computers that were available struggled to perform the convolution needed to create the Voigt function, which led to the widespread use of the GLS and GLP functions. In many applications, both in the past and today, the GLS and GLP functions are adequate representations of the photoemission process in XPS. Nevertheless, the Voigt function appears to better represent the data being produced by more modern spectrometers [19]. The

GLS, GLP, and Voigt functions are relatively simple, symmetric peak shapes. More complex signals/shapes/envelopes that result from spin–orbit splitting, multiplet signals, Auger signals, sample charging, overlapping signals, shake-up signals, and asymmetry are often encountered in XPS. The backgrounds needed to separate the zero-loss signal from inelastically scattered electrons can also be complex [20,21]. It is not uncommon to encounter more than one of these issues/effects in a single XPS spectrum.

As noted, the instrument itself influences XPS peak shapes. That is, depending on instrument design, the process of collecting a photoemission signal may distort the intrinsic energy distribution to some degree. Typical XPS instruments gather signals using electron lens elements and apertures to select and guide photoemission into the energy analyzer. This analyzer filters/guides a subset of the photoelectrons onto a detection system, which produces a signal that is recorded by the computer software. The computer/software records electrons arriving at the detector and merges data streams to form an energy-resolved signal, which is presented to the user as a spectrum. Systems using a narrow entrance slit aperture, a hemispherical analyzer (HSA), and a low-pass energy offer the least deformation in photoemission energy distributions for core levels. As expected, photoemission with a broad intrinsic energy distribution exhibits less distortion than narrowly distributed photoemission. Materials with wider band gaps and broader energy distributions (wider peaks) provide examples of well-formed spectral shapes that are representative of the intrinsic photoemission signal. We will begin by analyzing such spectra.

3.3 Relative Sensitivity Factors, Backgrounds, and Line Shapes

3.3.1 *Convolution as an important underlying principle in XPS data collection and analysis*

Underlying any XPS signal is the intrinsic signal that would be produced and registered by a perfectly monochromatic source and an instrument that does not in any way broaden the signal. The effects that broaden the intrinsic peak shape can be modeled mathematically as convolution(s) of the intrinsic, Lorentzian, peak shape with other functions that increase its width. The question that might then arise is as follows: At what point does the intrinsic peak become wide enough, or the perturbing function(s) become narrow enough, such that there is no noticeable effect on the intrinsic peak shape? In other words, at what point does the spectrometer return the intrinsic peak shape? Figure 3.1 helps address this issue. It shows calculations of convolutions of Lorentzian and Gaussian functions, where the percent change in the full width at half maximum (FWHM) of the resulting Voigt function (compared to the original Lorentzian, which had an FWHM of unity) is plotted as a function of the FWHM of the Gaussian, which ranged from 10–50% of the width of the original Lorentzian. This plot suggests that as long as the width of the Gaussian is less than 20% of that of the Lorentzian, the FWHM of their convolution will be largely unchanged compared to the width of the original Lorentzian, i.e., there will be less

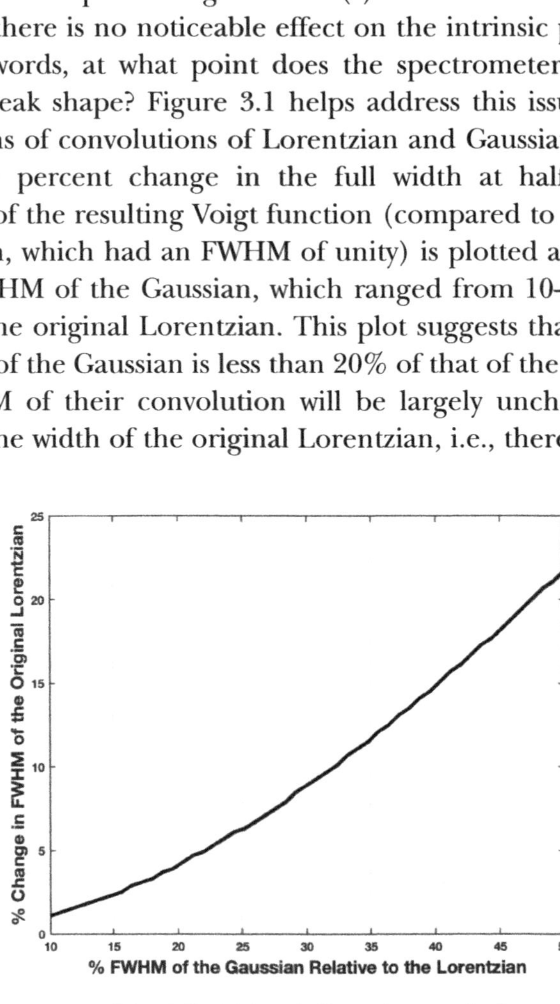

Fig. 3.1. The impact of the full width at half maximum of a Gaussian function on a Lorentzian function in convolution. The y-axis shows the percent change in the FWHM of the original Lorentzian function as a function of the percent FWHM of the Gaussian function relative to the original Lorentzian function (the x-axis).

than a 5% change in the FWHM of the original function. These results are confirmed by theory [22], which states that if two functions, f and g, are convolved to produce a new function, $f*g$, the variance (square of the standard deviation) of the new function is equal to the sum of the variances of the original functions, as follows:

$$\sigma_{f*g}^2 = \sigma_f^2 + \sigma_g^2. \tag{3.1}$$

For example, if the standard deviations of the functions f and g are unity and 0.2, respectively, the standard deviation of their convolution, σ_{f*g}, is 1.04, which is consistent with the results in Figure 3.1. Thus, we expect that sufficiently broad intrinsic signals in XPS will be (almost) faithfully returned by a high-quality instrument operated at high resolution (low-pass energy). Appendix 1 contains the MATLAB code used to make Figure 3.1.

3.3.2 *Photoemission and relative sensitivity factors: Cl 2s and 2p signals*

Sodium chloride is an example of a material where (i) only two elements contribute to the photoemission and (ii) a significant band gap exists compared to some of the photoemission peak widths. A wide band gap minimizes the influence of inelastic scattering of electrons within the sample, resulting in a relatively flat background beneath peaks. In other words, simpler backgrounds, which in many cases are linear, are more common for wider band gap materials. The Cl 2s and Cl 2p peaks (see Figures 3.2 and 3.3, respectively) appear with binding energies that are quite close together (ca. 268 and 198 eV, respectively), which means that instrumental conditions should have been similar for these peaks during data acquisition. These photoemission signals are notable for having a rather pronounced Lorentzian nature in the Cl 2s signal but a more compact, Gaussian shape in the Cl 2p signal. Charge compensation is necessary to acquire good XPS data from an insulator like sodium chloride (see Chapter 5), and the efficiency and uniformity of charge

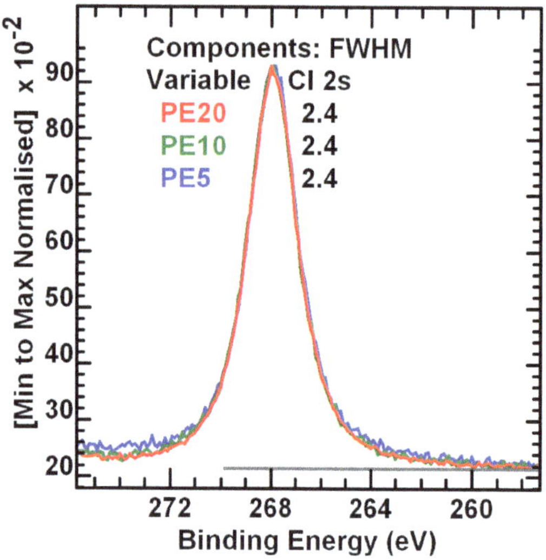

Fig. 3.2. Cl 2s spectra measured using pass energies of 20, 10, and 5 on a Kratos Axis Nova instrument at La Trobe University. These results were obtained by fitting a Voigt line shape relative to a flat background (see gray horizontal line) to the lower binding energy portion of the spectrum (up to almost 270 eV). This is the part of the spectrum that is not noticeably influenced by inelastic scattering.

compensation play a role in measured/acquired line shapes, adding another potential factor that can deform a signal. In the case of substantial distortion to a peak shape, e.g., by differential charging, it may be better not to peak fit it and to simply characterize it with a width function [23,24]. Charge compensation is used to maintain a constant potential at the sample by returning charge to the surface via means other than electrical conductivity. As expected, charge compensation was employed in the acquisition of the data in Figures 3.2 and 3.3.

The Cl 2s and Cl 2p spectra in Figures 3.2 and 3.3, respectively, are fitted to symmetric Voigt line shapes relative to constant backgrounds. In the case of Figure 3.2, this constant background is indicated by a flat gray line that only goes part of the way across the peak. Importantly, this spectrum is only fitted to this point to avoid the part of the signal where inelastic scattering (background signal) occurs, which causes a rise in the baseline.

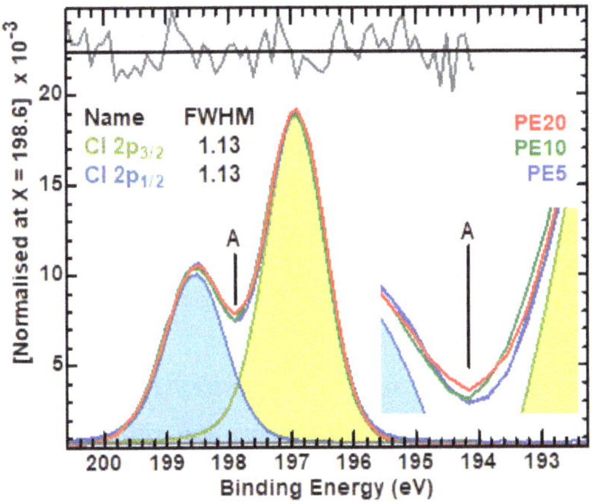

Fig. 3.3. Sodium chloride Cl 2p spectra measured using pass energies of 20, 10, and 5 on a Kratos Axis Nova instrument at La Trobe University. The valley between these Cl 2p doublet peaks indicates a small change in FWHM due to instrumental energy resolution performance resulting from pass energy reduction.

The table in Figure 3.2 records identical FWHM values independent of decreasing pass energy: 20, 10, and 5, which corresponds to progressively higher resolution. Lower-pass energy is the means by which energy dispersion is increased within an HSA energy filter. The observation that increased energy dispersion prior to signal detection does not influence the FWHM for Cl 2s suggests that these recorded spectra contain limited instrumental artifacts, such that they represent to a significant degree the intrinsic shape for photoemission from Cl 2s. Thus, the remaining influence on the peak width should be the effect of the finite energy width of the X-ray source. However, because this latter effect is small compared to the width of the peak (it is typically ca. 0.25 eV), the arguments on convolution above suggest that this Cl 2s narrow scan should be very close to the true intrinsic signal. Finally, note that these effects can be viewed as a sum of variances, where the width/FWHM of an XPS peak (σ_{FWHM}) is defined by its intrinsic width, broadening from the analyzer, and broadening from the X-ray source, as follows:

$$\sigma^2_{\text{FWHM}} = \sigma^2_{\text{intrinsic}} + \sigma^2_{\text{analyzer}} + \sigma^2_{\text{X-ray source}}. \tag{3.2}$$

In contrast, because it is noticeably narrower than the Cl 2s spectrum, the Cl 2p spectrum in Figure 3.3 is easily fitted with a linear baseline across its entire width. However, the spectra in Figure 3.3 show that the minimum between the spin–orbit components decreases as the pass energy decreases. That is, the narrower peak envelope/signal here is affected to a greater degree by instrument broadening.

3.3.3 *Photoemission and relative sensitivity factors: Ag 3d$_{5/2}$ spectrum and background selection for the Cl 2s and 2p peaks*

Similar results/effects to those shown in the previous section appear in the Ag 3d$_{5/2}$ spectrum in Figure 3.4. That is, the

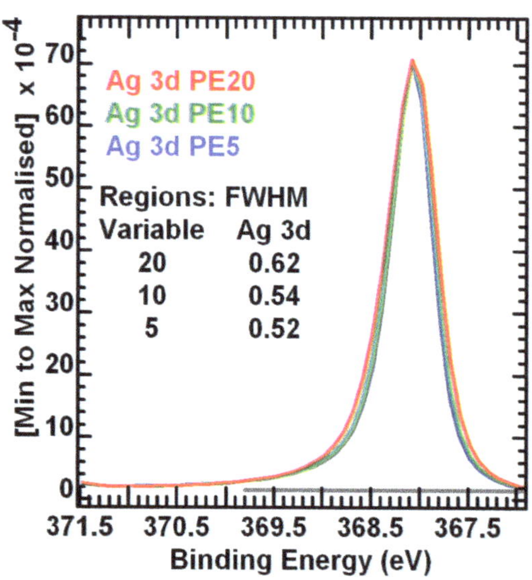

Fig. 3.4. Ag 3d$_{5/2}$ photoemission measured at pass energies of 20, 10, and 5 using the same instrument used to study sodium chloride in Figures 3.2 and 3.3. Note how reducing the pass energy from 20 to 10 yields a significant reduction in the measured FWHM. That is, the intrinsic FWHM for this signal may be narrower than the narrowest peak produced here because of the energy resolution limitations (effects of broadening) of the measurement.

changes to the pass energy here alter the measured FWHM; using pass energy 5 compared to 20 causes the FWHM value of this peak to decrease from 0.62 eV to 0.52 eV. These results suggest that the material's intrinsic line width may be narrower than the ability of the instrument operating at its lowest pass energy to measure.

Pass energy is only one means of altering the energy resolution of an XPS instrument. The data shown in Figures 3.2–3.4 were acquired using a slot-selected area aperture matching the width of the entrance aperture to the HSA. For a signal of a given energy, an HSA transfers an image of the entrance aperture to the exit plane for the HSA/detector. This energy is mixed with a spatially dispersed signal causing loss of energy resolution in the recorded spectra. Therefore, narrower entrance slits can enhance energy resolution. However, this narrowing reduces sensitivity to the sample as less energy/fewer photoelectrons pass through the slit. So, although the energy resolution for the Ag $3d_{5/2}$ spectra in Figure 3.4 can be enhanced by reducing the entrance aperture width, it would result in a loss of sensitivity (signal). However, for sodium chloride, factors other than the energy resolving power of the instrument are at play, which restricts the lower FWHM limit/value for this particular material. That is, the natural widths of XPS signals have implications for XPS quantification. In the case of sodium chloride, the intrinsically wide Cl 2s photoemission peak distributes signal over a larger energy interval than the Cl 2p peak, which leads to the (relatively speaking) significant Gaussian contribution to the Cl 2p peak shape.

A comparison of the photoemission signals from two different core levels of an atom requires the use of relative sensitivity factors. Theoretically determined photoionization cross-sections provide relative sensitivities for photoemission. The example of the Cl 2s and Cl 2p photoemission in Figure 3.5 illustrates how Scofield cross-sections can be applied to obtain a ca. 1:1 ratio for the Cl 2s:Cl 2p signals, but only if the underlying chemistry and physics of the sample are understood. The more compact Cl 2p signal is well modeled/fit/integrated with a linear baseline because the width of the Cl 2p signal is less than the band gap of the material. That is, the

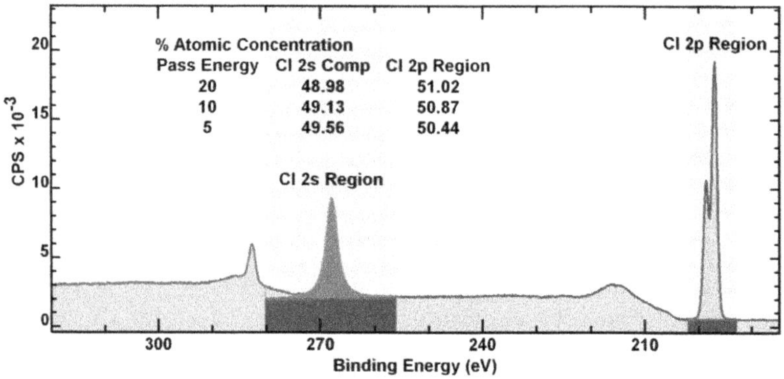

Fig. 3.5. Quantification of the Cl 2s and Cl 2p signals from NaCl. The Cl 2s signal is modeled with a Voigt line shape on a horizontal background defined by the lower binding energy portion of the signal. The Cl 2p doublet is integrated over a horizontal background that connects both of its sides. Scofield cross-sections and escape depth corrections based on effective attenuation lengths are applied. No transmission correction is used as the Kratos transmission function is not an absolute transmission function and therefore not appropriate for use with theoretical cross-section sensitivity factors.

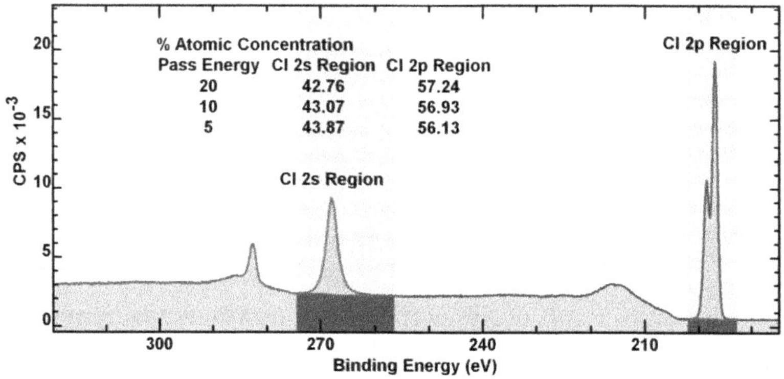

Fig. 3.6. Quantification of the same data shown in Figure 3.5 and with the same procedure used in Figure 3.5, except with Shirley backgrounds for both the Cl 2s and 2p regions. (The Shirley background for the Cl 2s region extends across the entire peak.) Note that Scofield cross-sections applied to these two peaks fail to return a corrected peak intensity in a 1:1 ratio. The source for these errors is the truncation of a Lorentzian photoemission distribution caused by the application of a Shirley algorithm to the Cl 2s data.

rise in the baseline due to inelastic scattering to the left (higher binding energy side) of the Cl 2p signal does not overlap to any significant extent with the main Cl 2p signal. In contrast, the baseline rise from inelastic scattering begins to occur on the high binding energy side of the much broader Cl 2s peak. As a result, the appropriate way to integrate the Cl 2s signal is with a horizontal (linear) baseline that is defined by the lower binding energy side of the signal. This approach (Figure 3.5) ignores the inelastically scattered electrons from the material and yields a good (nearly 1:1) peak area ratio for these two signals. In contrast, Figure 3.6 shows the results of using Shirley backgrounds on both the Cl 2s and 2p regions. While it is expected that either a linear or a Shirley background will appropriately model the essentially horizontal background of the Cl 2p signal, for the reasons just mentioned, the Shirley background does not reasonably represent the physics and chemistry of the Cl 2s peak, which leads to inappropriate quantitation. This analysis is discussed in greater detail in a recent article we published [25].

3.3.4 *Photoemission and relative sensitivity factors: Sensitivity factors applied to LiF and Li_2SO_4 spectra*

Quantification in XPS is not simply based on fitting the components of a signal using appropriate peak shapes. There are other factors involved, such as the escape depth correction, angular distribution correction, instrument-transmission response, and, as indicated in the last paragraph, background selection. All these factors influence one's ability to compare intensities from different photoemission lines. While the use of theoretical sensitivity factors requires these adjustments to be made, these same factors are also in force when determining empirical sensitivity factors. Empirical sensitivity factors are limited in scope by their very nature; these factors are specific to a given standard material and not applicable to other materials or spectra with similar photoemission lines. An example illustrating this danger is the sensitivity factor reported by Wagner for Li 1s (0.022) relative to F 1s based on lithium fluoride [26].

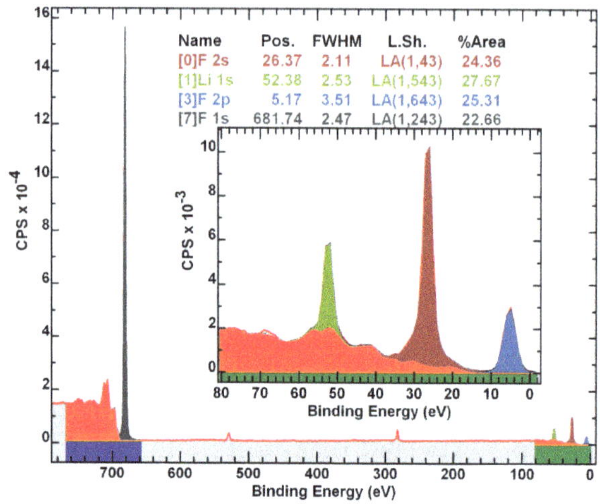

Fig. 3.7. XPS of lithium fluoride, where the expected stoichiometry of 1:1 (LiF) can only be achieved by making use of Scofield cross-sections when the loss structure associated with the F 2s signal is modeled making use of the loss structure that is clearly present for the F 1s signal to approximate the background beneath the Li 1s peak. These complex background shapes make LiF unsuitable as a standard material for computing empirical sensitivity factors.

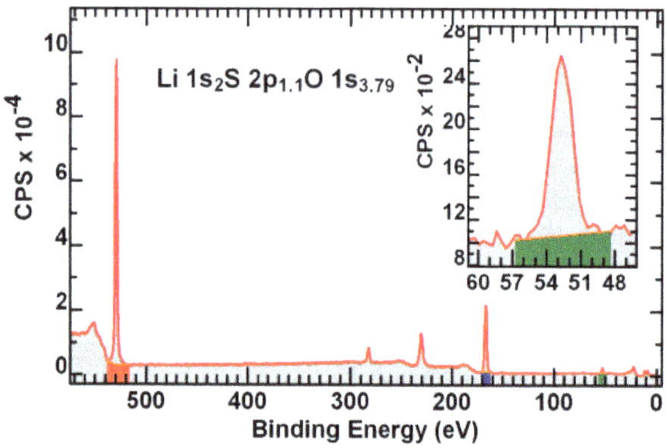

Fig. 3.8. XPS of lithium sulfate quantified using Scofield cross-sections corrected for effective attenuation lengths using an operating mode on a Kratos Axis Nova with near-flat transmission. The expected empirical formula, Li_2SO_4, is in reasonable agreement with XPS results based on the simple integration regions shown. Lithium sulfate differs from lithium fluoride shown in Figure 3.7 in that the Li 1s signal from Li_2SO_4 is relatively isolated with a simple background for photoemission compared to LiF.

The sensitivity factor for Li 1s reported by Wagner is about twice the Scofield cross-section (0.013 relative to F 1s) for Al anode X-rays. This discrepancy is due to significant background shapes and interferences between the F 2s background and the Li 1s photoemission signal, which results in a poorly defined Li 1s sensitivity factor based on data alone.

Figure 3.7 illustrates the influence of the F 2s signal on the background beneath the Li 1s peak for LiF. The background beneath the Li 1s signal is estimated using the F 1s loss structures to predict a possible background associated with the F 2s peak. While not necessarily accurate, the analysis in Figure 3.7 demonstrates the potential uncertainty in computing the Li 1s photoemission signal from LiF. By contrast, lithium sulfate (Figure 3.8) offers Li 1s photoemission with relatively benign background influences and supports the use of Scofield cross-sections as a traceable quantification procedure for quantification by XPS.

3.4 Concluding Remarks

The selection of appropriate peak shapes is important in XPS data analysis (peak fitting). For symmetric XPS peaks, Voigt functions generally provide the best representation of the natural peak shape as it is affected by other factors that broaden it. Convolution is an underlying mathematical principle that informs peak choice in XPS data analysis/interpretation. Sample physics and chemistry directly influence XPS backgrounds, where wider band gap materials generally show less complex backgrounds. We demonstrate the effects of lowering the pass energy on the XPS spectra of NaCl (the Cl 2s and 2p peaks) and Ag (the Ag $3d_{5/2}$ peak). The possible complexity associated with quantifying XPS results using sensitivity factors is illustrated using LiF and Li_2SO_4 spectra.

Acknowledgment

This work incorporates data from the Victorian node of the Australian National Fabrication Facility (ANFF), a company established under the National Collaborative Research Infrastructure Strategy to provide nano and microfabrication facilities for researchers in Australia, through the La Trobe University Centre for Materials and Surface Science. Data are reproduced under a Creative Commons license (CC BY-NC 4.0 International).

Appendix 1. MATLAB Code Used to Do and Plot the Calculations in Figure 3.1

```
%% Convolution Gaussian and Lorentzian
x = linspace(-5,5,10000); % defining a vector for data
gh = 1; gF = 1; gE = 0; % Variables labeled gX are for
Gaussian functions
lh = 1; lF = 1; lE = 0; % Variables labeled lX are for
Lorentzian functions
% These functions are used in CasaXPS
g = gh*exp((-4)*log(2)*((x-gE).^2)/(gF.^2)); % A basic
Gaussian distribution
l = lh./(1+4*(((x-lE).^2)/(lF.^2))); %A basic Lorentzian
distribution
voigt = conv(g,l,'same'); % This is the built-in con-
volution function from MATLAB
plot(voigt) %
shg

%% Loop Showing Narrowing Gaussians
gh = 1; gE = 0; % Variables for Gaussian
lh = 1; lF = 1; lE = 0; % Variables for Lorentzian

% Allocating arrays to fill from loop
voigt_width = zeros(0,10000);
gaus_width = zeros(0,10000);

%simple loop to change the width of the Gaussian
for gF = 0.1:0.008:0.5
    g = gh*exp((-4)*log(2)*((x-gE).^2)/(gF.^2));
    l = lh./(1+4*(((x-lE).^2)/(lF.^2)));
```

```
    voigt = conv(g,l,'same');
    std_voigt = voigt - (max(voigt)/2);
    voigt_width = [voigt_width; std_voigt]; %appending
voigt with the adjusted width (subtracted by half
height)
    gaus_width = [gaus_width; g]; %appending gaus_
width with new widths
    plot(std_voigt); % visual check for proper
execution
    hold on
end
hold off
shg
%% Finding FWHM of each set generated above
% Approach to find FWHM is, as maximum value is 1,
subtract
% by half and then set negative values to 0, positive
values
% to 1 and then added to find distance

%Gaussian FWHM
sign_gaus = gaus_width - 0.5;
sign_gaus(sign_gaus < 0) = 0;
sign_gaus(sign_gaus > 0) = 1;
fwhm_gaus = sum(sign_gaus.');

%Lorentzian FWHM
sign_lorentz = l - 0.5;
sign_lorentz(sign_lorentz < 0) = 0;
sign_lorentz(sign_lorentz > 0) = 1;
fwhm_lorentz = sum(sign_lorentz.');

%Voigt FWHM
sign_voigt = voigt_width;
sign_voigt(sign_voigt < 0) = 0;
sign_voigt(sign_voigt > 0) = 1;
fwhm_voigt = sum(sign_voigt.');
```

```
%% Creating percentage change in FWHM

%This normalizes the found fwhms to the Lorentzian
ave_voigt = (fwhm_voigt./(fwhm_lorentz)-1)*100;
gaus_x = fwhm_gaus./(fwhm_lorentz)*100;

%% Plotting the Functions

plot(gaus_x, ave_voigt,'k','LineWidth',3)
grid on; t = title(''); t.FontSize = 20; g = ylabel('%
Change in FWHM of the Original Lorentzian');
g.FontSize = 15; g.FontWeight = 'bold';
x = xlabel('% FWHM of the Gaussian Relative to the
Lorentzian');
x.FontSize = 15; x.FontWeight = 'bold'; ax = gca;
ax.FontWeight = 'bold';
ax.XAxis.LineWidth = 1; ax.YAxis.LineWidth = 1;
```

References

[1] Shah D, Patel DI, Roychowdhury T, Rayner GB, O'Toole N, Baer DR, Linford MR. Tutorial on interpreting X-ray photoelectron spectroscopy survey spectra: Questions and answers on spectra from the atomic layer deposition of Al_2O_3 on silicon. J Vac Sci Technol B. 2018;36(6):062902.

[2] Gupta V, Ganegoda H, Engelhard MH, Terry J, Linford MR. Assigning oxidation states to organic compounds via predictions from X-ray photoelectron spectroscopy: A discussion of approaches and recommended improvements. J Chem Educ. 2014;91(2):232–238.

[3] Van der Heide P. *X-ray Photoelectron Spectroscopy: An Introduction to Principles and Practices.* John Wiley & Sons: Hoboken, New Jersey, 2011.

[4] Baer DR, Artyushkova K, Richard Brundle C, Castle JE, Engelhard MH, Gaskell KJ, Grant JT, Haasch RT, Linford MR, Powell CJ, Shard AG, Sherwood PMA, Smentkowski VS. Practical guides for X-ray photoelectron spectroscopy: First steps in planning, conducting, and reporting XPS measurements. J Vac Sci Technol A. 2019;37(3):031401.

[5] Stevie FA, Donley CL. Introduction to X-ray photoelectron spectroscopy. J Vac Sci Technol A. 2020;38(6):063204.
[6] Baer DR, Artyushkova K, Brundle CR, Castle JE, Engelhard MH, Gaskell KJ, Grant JT, Haasch RT, Linford MR, Powell CJ. Practical guides for x-ray photoelectron spectroscopy: First steps in planning, conducting, and reporting XPS measurements. J Vac Sci Technol A. 2019;37:031401.
[7] Baker M. 1,500 scientists lift the lid on reproducibility. Nature. 2016, 533 (7604), 452-454.
[8] Sené M, Gilmore I, Janssen J-T. Metrology is key to reproducing results. Nature News. 2017;547(7664):397–399.
[9] Baer DR, Gilmore IS. Responding to the growing issue of research reproducibility. J Vac Sci A. 2018;36(6):068502.
[10] Linford MR, Smentkowski VS, Grant JT, Brundle CR, Sherwood PM, Biesinger MC, Terry J, Artyushkova K, Herrera-Gomez A, Tougaard S, Skinner W, Pireaux J-J, McConville CF, Easton CD, Gengenbach TR, Major GH, Dietrich P, Thissen A, Engelhard M, Powell CJ, Gaskell KJ, Baer DR. Proliferation of faulty materials data analysis in the literature. Microsc Microanal. 2020;26(1):1–2.
[11] Major GH, Avval TG, Moeini B, Pinto G, Shah D, Jain V, Carver V, Skinner W, Gengenbach TR, Easton CD, Herrera-Gomez A, Nunney TS, Baer DR, Linford MR. Assessment of the frequency and nature of erroneous X-ray photoelectron spectroscopy analyses in the scientific literature. J Vac Sci Technol A. 2020;38(6):061204.
[12] Major GH, Fairley N, Sherwood PM, Linford MR, Terry J, Fernandez V, Artyushkova K. Practical guide for curve fitting in X-ray photoelectron spectroscopy. J Vac Sci Technol A. 2020;38(6):061203.
[13] Gengenbach TR, Major GH, Linford MR, Easton CD. Practical guides for X-ray photoelectron spectroscopy (XPS): Interpreting the carbon 1s spectrum. J Vac Sci Technol A. 2020;38(1):061203.
[14] Easton CD, Kinnear C, McArthur SL, Gengenbach TR. Practical guides for X-ray photoelectron spectroscopy: Analysis of polymers. J Vac Sci Technol A. 2020;38(2):023207.
[15] Burrell MC. Method for correcting peak overlaps in quantitative Auger electron spectroscopy of Cr-containing oxides. J Vac Sci Technol A. 2020;38(1):013201.

[16] Patel DI, Roychowdhury T, Jain V, Shah D, Avval TG, Chatterjee S, Bahr S, Dietrich P, Meyer M, Thißen A, Linford MR. Introduction to near-ambient pressure X-ray photoelectron spectroscopy characterization of various materials. Surf. Sci. Surf Sci Spectra. 2019;26(1):016801.

[17] Jain V, Biesinger MC, Linford MR. The Gaussian-Lorentzian sum, product, and convolution (Voigt) functions in the context of peak fitting X-ray photoelectron spectroscopy (XPS) narrow scans. Appl Surf Sci. 2018;447:548–553.

[18] Sherwood APM. Rapid evaluation of the Voigt function and its use for interpreting X-ray photoelectron spectroscopic data. Surf Interface Anal. 2019;51(2):254–274.

[19] Major GH, Avval T, Patel DI, Shah D, Roychowdhury T, Barlow A, Pigram P, Greiner M, Fernandez V, Herrera-Gomez A, Linford MR. A discussion of approaches for fitting asymmetric signals in X-ray photoelectron spectroscopy (XPS), noting the importance of Voigt-like peak shapes. Surf Interface Anal. 2021, 53(8), 689–707.

[20] Engelhard MH, Baer DR, Herrera-Gomez A, Sherwood PMA. Introductory guide to backgrounds in XPS spectra and their impact on determining peak intensities. J Vac Sci Technol A. 2020;38(6):063203.

[21] Tougaard S. Practical guide to the use of backgrounds in quantitative XPS. J Vac Sci Technol. 2021;39:011201.

[22] Bracewell RN. *The Fourier Transform & Its Applications* (3rd edn.) McGraw-Hill Science: Columbus, OH., 1999.

[23] Singh B, Velázquez D, Terry J, Linford MR. Comparison of the equivalent width, the autocorrelation width, and the variance as figures of merit for XPS narrow scans. J Electron Spectros Relat Phenomena. 2014;197:112–117.

[24] Singh B, Velázquez D, Terry J, Linford MR. The equivalent width as a figure of merit for XPS narrow scans. J Electron Spectros Relat Phenomena. 2014;197:56–63.

[25] Major G, Fairley N, Fernandez V, Linford MR. Chemical state Analysis in XPS: A Case Study. Selecting the Right Baseline to Obtain the correct Cl 2s to Cl 2p area ratio from a sample of NaCl. Vac Technol Coating. 2022 February.

[26] Briggs D, Seah MP (eds.) *Practical Surface Analysis. Auger and X-Ray Photoelecton Spectroscory* (2nd edn., Vol. 1, pp. 151–152). Chichester: Wiley, 1990.

https://doi.org/10.1142/9781800613294_0004

Chapter 4

The Practical Dos and Don'ts of Using XPS to Qualify and Quantify Powder Catalysts

Saulius Kaciulis

ISMN-CNR, Rome, Italy

saulius.kaciulis@cnr.it

Abstract

The practical aspects of XPS analysis in the application for powder catalysts, including also nanopowders and nanoparticles, are revealed and discussed. Most of these features are applicable also to the composites with a mixture of catalytic powders in ceramic matrix. All the steps of typical XPS characterization are briefly discussed: various methods of sample preparation and mounting, spectra acquisition, processing, interpretation, and elemental quantification. This discussion is aimed at the beginners in XPS, trying to evidence the practical dos and don'ts in experimental procedures and data processing. The procedures of spectra processing and interpretation are illustrated by experimental examples. Following the recommendations presented in this chapter, it is possible to conduct a reliable XPS investigation of surface chemical composition before and after catalytic reactions or other sample treatments.

4.1 Introduction

XPS studies of metals and/or oxides with a specific planar form, the so-called model catalysts, are well known for many years [1,2] and will be specifically discussed in Chapter 7. However, from the material point of view, the majority of heterogeneous catalysts contain various metals and their oxides or salts. Most of them are used in the form of powders or even nanopowders and nanoparticles, as has been indicated in a recent review [3]. In addition, for industrial applications, the powder catalysts are often coated on a ceramic matrix. Therefore, catalyst characterization by XPS very often requires the analysis of materials in the form of powders or coating.

Undesirable effects occurring in powder samples, which might lead to erroneous XPS results, can be caused by the sample preparation, spectra acquisition (experimental factors), acquired data interpretation, and elemental quantification procedures. Again, most of these effects can occur in any powder sample, not only in the catalytic ones [4]. Therefore, in order to trace a correct roadmap of XPS analysis for powders, we shall overview the whole procedure, divided into four working steps: sample preparation and mounting, spectra acquisition, data processing and interpretation, and elemental quantification.

4.2 Sample Preparation and Mounting

Before the start, it is important to remember that some nanoparticles of metals and metal oxides (e.g., containing Ce, Ru, and Ti) could be toxic and this should be taken into account when manipulating and discarding those samples [5,6]. In XPS, like in any other experiment, the safety rules for working with hazardous materials must be rigorously respected. In Chapter 2 were shown some best practice examples of how to ship or transfer a flat sample from the preparation laboratory to the XPS location. Here will be demonstrated some procedures to mount the sample on an XPS instrument sample holder. There are a few different ways to prepare and fix the powders on the sample holder, as illustrated in Figure 4.1.

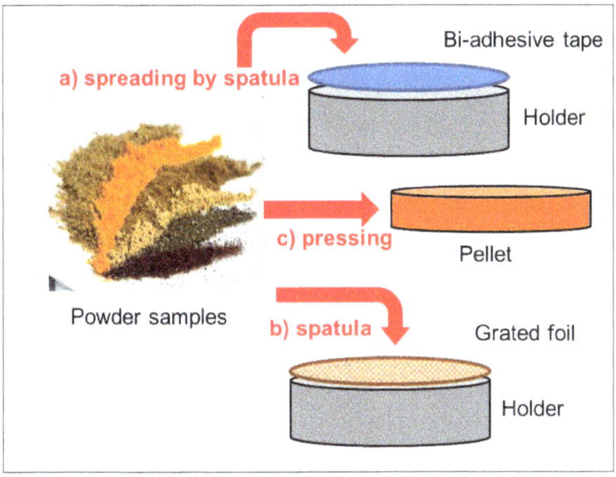

Fig. 4.1. Preparation of powder samples: (a) on bi-adhesive tape, (b) on soft metal foil, and (c) pressing into a pellet.

The first way (and the simplest one) is just to cover with powder the surface of a bi-adhesive tape by using a spatula. Care must be taken to create a uniform powder overlayer as thin as possible (see Figure 4.1(a)). The excess powder must be avoided because free powder particles migrating in the XPS vacuum system can be potentially very destructive to the instrument, particularly for the vacuum valves and the turbo or mechanical pumps. The bi-adhesive tape for powder support can be a simple Scotch-type, conductive carbon disk or metallic (Cu, Al, etc.) tape covered with a sticky substance. Although most of the commercial bi-adhesive tapes claim to be vacuum compatible, the main disadvantages of this method are a long degassing time in UHV and possible sample charging if an electrically insulating tape is used. The sample charging is much lower if the conductive carbon disk or metallic tape is used for support. However, even these samples could be charged in the spectrometers with a monochromatized X-ray source. The reason is the absence of higher energy *Bremsstrahlung* radiation, which promotes the compensation of surface charging by creating more inelastically scattered electrons onto the sample. Of course, this charging can be suppressed externally by using a flux of low-energy electrons (i.e., electron flood gun) or even better by combining the fluxes of electrons and ions, if these sources are available in the spectrometer. For more details about charge compensation, please refer to Chapter 5.

The second way to fix a powder sample (Figure 4.1(b)) is by pressing it using a spatula onto a soft metal foil, for example, In or Au. It works better if the foil was previously grated with a sharp steel stylus, creating a rough and soft metal surface. In this case, the possibility of sample charging is lower but remains the interference of photoemission signals from the foil and powder sample because it is impossible to completely cover the foil surface with powder. One of the best metals for the foil substrate is pure gold (at least 99.9% purity), which can be easily cleaned and grated, moreover, it generates only few Au photoemission peaks, with a minimum contribution of carbon and oxygen signals (see Figure 4.2). Of course, the choice of Au is not suitable if one of the powder constituent elements has a main photoemission peak very near to Au 4f (80–90 eV), Au 4d (330–360 eV), or Au $4p_{3/2}$ (540–560 eV) spectral regions. Another metal with even better mechanical properties to use as a substrate is indium foil, which is much softer, therefore it is

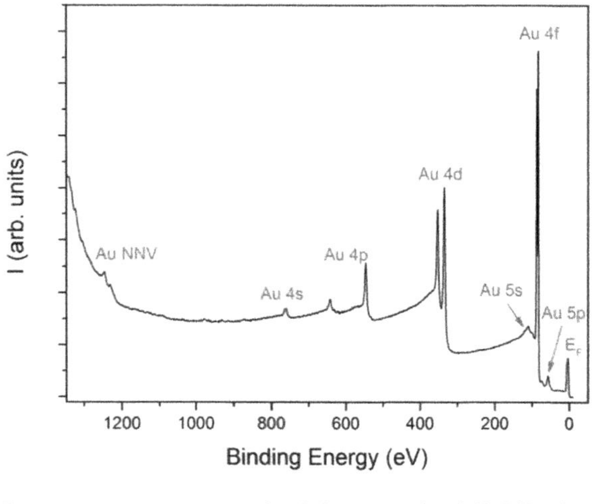

Fig. 4.2. Wide survey spectrum acquired for pure Au foil. NB: there is plenty of free space for the peaks of other elements (signals from overlayer of powder).

easier to embed the powder particles onto this substrate. The substrate of In yields the high-intensity photoemission peaks of 3d (440–460 eV), 3p (660–710 eV), and 4d (15–20 eV), but they are also accompanied by the peaks of O 1s (from In oxide) and C 1s (from adventitious carbon) that always are present on its surface. For this reason, the usage of In foil is not advisable, if the investigated powder contains oxygen or carbon related to the catalyst. The main problem of this method is the adhesion of the powder particles: some compounds don't adhere to any metal foil and are sucked out by vacuum pumps when introduced into the spectrometer, therefore, a different mounting method must be used for these samples. A pure Au foil can be used also for the nanoparticles diluted in a solvent: it is enough to put a drop of solution on the foil and dry it in air or even better under the flux of nitrogen or argon.

The third method is to prepare a pellet of investigated powder (Figure 4.1(c)) by using a mechanical press and a pellet press die. The smaller pellets are always better because it is easier to fix them on the sample holder with small clips or screws. The possibility of sample charging under X-rays depends on the conductivity of the pellet: if it is not conductive, the sample can charge significantly. In some cases, also the sample composition and morphology might be slightly modified by the pressing force, which is crushing the powder particles and possibly introducing more contamination of adventitious carbon and oxygen from the atmosphere.

When dealing with nanoparticles dispersed in some capping solvent, it is necessary to separate the particles from the solution, minimizing the residual contamination by solvent, but retaining any coating of the particles. As it was indicated above (Figure 4.1(b)), the easiest way is to drop the solution on a metal foil (or other solid support, e.g., a piece of Si wafer) using a pipette and let the sample dry. This approach is applicable only if the solvent can be easily evaporated (e.g., alcohol and acetone) and doesn't leave too much traces on the particles' surface. More sophisticated (although more time-consuming) separation methods, described in Ref. [7], are dialysis, centrifugation, diafiltration [8], or flash drying filtering, which is particularly recommended for the reactive particles [9].

It is necessary to remember that some amount of adventitious carbon inevitably remains on the particles' surface after any separation and cleaning processes.

Et voilà, finally, the samples are prepared, inserted in the UHV chamber of our spectrometer, and degassed. Now, we are ready for the acquisition of photoemission spectra.

4.3 Spectra Acquisition

As we are investigating powder or composite samples, it is better to acquire the spectra in large area mode to get an average photoemission signal from rough and potentially inhomogeneous sample surfaces. In modern spectrometers with a monochromatized X-ray source, the maximum analysis area is about 1 mm in diameter, which is enough for signals averaging over surface roughness and other micro-inhomogeneities. Always it is recommended to start with a wide survey spectrum in order to check for possible sample charging and to choose a suitable mode of its compensation. Only when we are sure that the charging is absent or minimized (to about a few eV) and stable, then it is possible to start the acquisition of elemental regions at high resolution, i.e., with a small energy step, which is typically of 0.05–0.1 eV.

It is not recommended to use ion sputtering for surface cleaning, especially for the samples mounted on an adhesive tape, where it can induce surface charging or contamination with organic compounds from an exposed tape in the gaps of the powder overlayer. Also, in the case of pellets or samples mounted on a metallic foil, the outcome of sputtering is quite unpredictable because the surface is rough and shadow effects will result in non-uniform removal of contaminants. If the samples are heavily contaminated with adventitious carbon, only a mild sputtering (at an energy of 1 keV or lower) for a short time is admissible to minimize the effects of non-uniformity and preferential sputtering. Other sample treatments *in situ*, such as annealing in UHV, are not advisable in the case of bi-adhesive support because the outcome of any treatment could be quite unpredictable: powder contamination with organic compounds, detachment of the particles from support, high increase in the vacuum pressure, etc. It is recommended to check the specifications of the bi-adhesive tape before attempting any annealing in vacuum. With only a few exceptions, the samples mounted on adhesive tapes should not be heated above 80°C.

Starting for the first time the analysis of some new compounds, it is recommended to acquire the scans in the so-called manner 1, 2, 3... 1, 2, 3..., i.e., to acquire every scan of elemental regions in a sequential way and to save separately the results of all the scans. This precaution is necessary for the detection of possible reduction of the oxidation state of some transition and rare-earth metals (e.g., Rh, Fe, Co, Cu, and Ce) under irradiation by X-rays [10–16]. When this effect is taking place, it is possible to restrict or eliminate it completely by cooling the sample, e.g., with liquid nitrogen or a Peltier cooler, if such an option is available in your instrument. Otherwise, only the first scan of the region with sensitive metal is valid for further data processing, and also its acquisition time must be minimized. It should be noted that the changes in oxidation state during the catalytic reactions can be observed directly by using near-ambient pressure XPS, as is discussed in the chapters dedicated to this technique.

4.4 Spectra Processing and Interpretation

The basic principles of spectra processing for powders and nanoparticles are exactly the same as for bulk materials, such as metals and their alloys recently reviewed in Ref. [2], where are presented numerous experimental cases of different metals and alloys. The main steps are the identification of photoemission peaks, background subtraction, and quantification by using the area of the elemental peaks.

Even if the binding energy (E_B) scale of the spectrometer is correctly calibrated by using the reference samples of clean metal foils and following an ISO standard [17], during the analysis of catalytic samples, their surface can positively charge due to the emission of electrons. For correct identification of the elemental peaks and their chemical state, it is very important to assure that the E_B scale is right, i.e., to re-calibrate it for sample charging, if it is present. For this operation, we need some reliable reference peak at a well-known value of E_B. Very often the aliphatic carbon peak of C 1s at E_B = 285.0 or 284.8 eV is used for this purpose, but sometimes this approach can give the erroneous results. For example, if nanoparticles in organic solution (or after purification with organic solvents) are deposited on a flat substrate, the main C 1s peak could not be aliphatic carbon but the bonds of C – O or C = O with E_B of around 286 eV or sometimes it could be even carboxylic carbon with E_B of about 288 eV. In this case, a more reliable reference is the main photoemission peak of some metal or metal oxide, which can be combined also with the O 1s peak characteristic for the same oxide. Another reliable reference for correction of the E_B scale is the Fermi level, which by definition is the E_B = 0 eV if the sample is conductive and makes good electrical contact with the sample holder connected to the spectrometer's earth (please refer to Chapter 2 for more details about energy level alignment). However, in the case of insulating or semiconducting samples, like most of the metal oxides, the photoemission signal starts from the valence band maximum, i.e., noticeably below the Fermi level, so the reference level is not "visible" by XPS and therefore cannot be used. An extensive description of sample charging under X-rays and detailed indications of how it can be eliminated are disclosed in recent reviews [18,19] and discussed in Chapter 5.

It should be also noted that the experimental attempts of charge compensation by using low-energy electron and ion sources are not always successful because it is quite difficult to select the right intensity of neutralizing fluxes. Some amount of residual charging can shift the E_B scale not only towards higher values but also to lower ones if the overall surface charging becomes negative. In this case, the apparent position of reference C 1s peak can be found at about 284–283 eV. In some samples, particularly of the nanoparticles supported in a porous and not conductive matrix, the so-called differential charging effect is observed. In particular, the particles in contact with the conductive substrate are not charging, while other particles without this contact (i.e., on the topmost surface layer) are charged. This effect is illustrated in Figure 4.3, which presents the Ag 3d spectrum acquired for Ag nanoparticles in the matrix of mesoporous silicate [20]. After correction of the apparent shift (about 4 eV) by using aliphatic C 1s peak as the reference and applying this shift for all the peaks except the metallic component of Ag 3d (E_B = 368.2 eV), the second spin–orbit doublet of Ag 3d

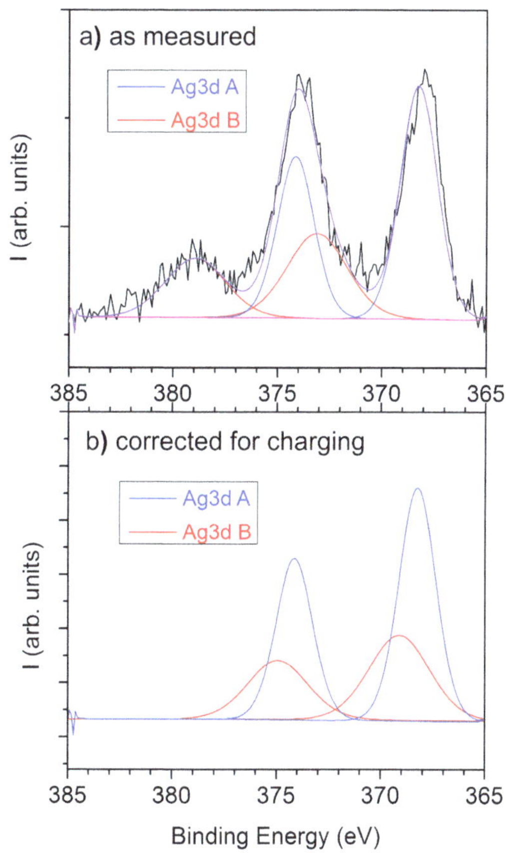

Fig. 4.3. Spectrum of Ag 3d region acquired for Ag nanoparticles in composite film: (a) peak fitting of acquired spectrum with two spin–orbit doublets A and B and (b) synthetic peaks of A and B doublets after differential charge correction (–4.0 eV) for the second doublet B. Adapted with permission from Ref. [20].

(E_B of about 369 eV) was attributed to the size shift in very small nanoparticles [21–23].

Obviously, the identification of the oxidation state is very important (perhaps, it is more important than elemental quantification), especially in the cases when the catalyst is characterized before and after catalytic reactions. However, in some metals, the chemical shift in different oxidation states is very small and invisible experimentally: for example, between Ag^0 and Ag^{1+} or Cu^0 and Cu^{1+}. In these cases, it is possible to determine the oxidation state from the modified Auger parameter (α'). As discussed in Chapter 2, α' was introduced by Wagner [24,25] in the 1980s and is calculated from the E_B of a photoemission peak and the kinetic energy (E_K) of an Auger peak of the same element.

For example, in the case of silver, the Ag $3d_{5/2}$ photoemission and the Ag $M_4N_{45}N_{45}$ Auger peaks are used, whereas for Cu, photoemission $2p_{3/2}$ and Auger $L_3M_{45}M_{45}$ peaks are used. The values of Auger parameter α' for many metals and their compounds can be found in textbooks [1,26] and databases dedicated to surface analysis, which are available online, e.g., Ref. [27–29].

In some metal alloys and compounds, the main photoemission peak of one of the constituent elements and the Auger signal from another one may overlap, for example, the signals of Fe 2p – Cu LMM, Cr 2p – Cu LMM, and Co 2p – Fe LMM [1,26–28]. This overlap is much worse than the interference of the photoemission peaks from different elements because it can't be easily resolved by peak fitting due to the complex shape of Auger peaks. However, if a dual X-ray anode is available, a simple change of the excitation source, for example, from Al Kα to Mg Kα, shifts the Auger peaks in the E_B scale by the energy difference between the two sources (it is 233 eV in the case of Al and Mg anodes) and allows the further processing of photoemission peaks.

Another difficult example of identification of oxidation state is Mn, which can have three different states of Mn^{2+}, Mn^{3+}, and Mn^{4+}, but the main photoemission peak of Mn 2p has a complex shape due to the effect of multiplet splitting [30,31]. The peak fitting of the 2p spectrum of Mn oxides is quite difficult and practically makes

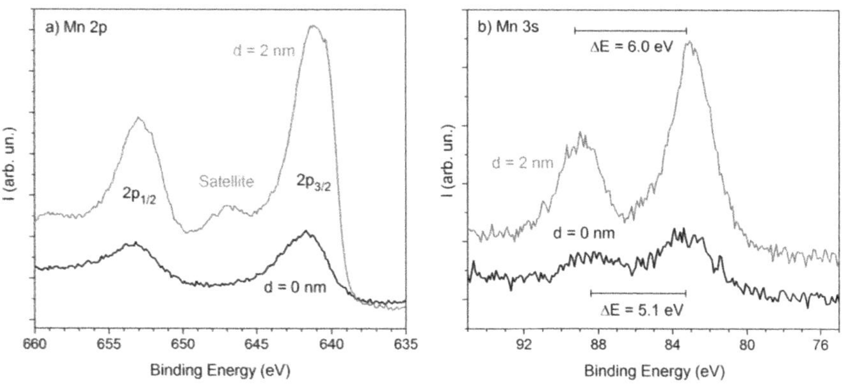

Fig. 4.4. Spectra of Mn 2p (a) and Mn 3s (b) regions at different sputtering depths (d = 0 and 2 nm) acquired for $La_{0.8}Sr_{0.2}(Mn)_zO_3$ film with Mn excess ($z > 1$). Adapted with permission from Ref. [34].

impossible the separation between the Mn^{3+} and Mn^{4+} states, whereas Mn^{2+} ions can be recognized from the presence of shake-up satellites (see Figure 4.4(a)). In this case, one can use the Mn 3s peak to identify the oxidation state. The Mn 3s peak has two, multiplet split components due to the coupling of non-ionized 3s electron with 3d valence-band electrons. From the distance between the splitted peaks, it is possible to estimate the Mn oxidation state [31], which in the case of Figure 4.4(b) corresponds to Mn^{3+}.

One more case of difficult oxidation state identification is La (and other rare-earth metals) in oxides and hydroxides, where the presence of very strong shake-up satellites (in some compounds, they are more intense than the main peak) complicates the accurate peak fitting. Moreover, the E_B values of La^{3+} in oxide and hydroxide are identical. However, as shown in Figure 4.5, it is possible to determine easily the distance between the main La $3d_{5/2}$ peak and its satellite and from its value (around 4 eV for $La(OH)_3$ and 4.6 eV for La_2O_3 [32,33]) to identify the La compound [34].

Another example of the oxidation state recognition from the shake-up satellites comes from the XPS characterization of a three-way automotive catalyzer [35]. As it has been revealed in Ref. [35], the oxidation state of La (identified from the La 3d peak) did not

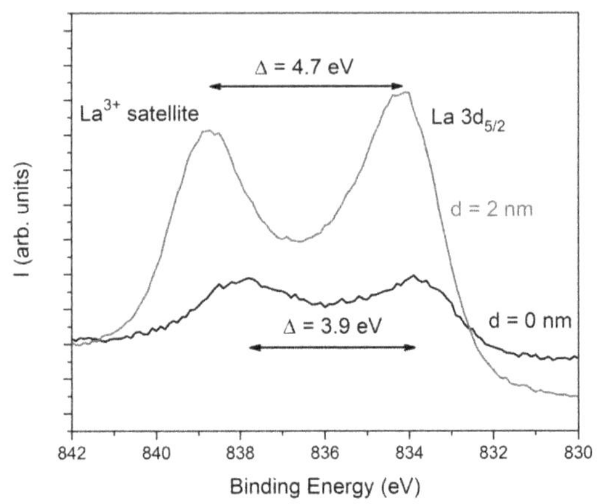

Fig. 4.5. Spectra of La $3d_{5/2}$ acquired for $La_{0.8}Sr_{0.2}(Mn)_zO_3$ film at different sputtering depths ($d = 0$ and 2 nm).

change after sample treatments and long operation time in working conditions. On the contrary, the Ce oxidation state was drastically modified after working for 500 h in real gas exhaust and afterwards was not restored after high-temperature treatment in humid atmosphere. As it is illustrated in Figure 4.6, the reduction of CeO_2 to a mixture of Ce^{3+} and Ce^{4+} states was identified from the intensity decrease of shake-up satellites v'', v''' and u'', u''' corresponding to the main peaks of v (Ce $3d_{5/2}$) and u (Ce $3d_{3/2}$), respectively [36]. This example of Ce oxidation state identifiable from the intensity of shake-up satellites is typical for many rare-earth compounds, where the main photoemission peaks of rare-earth metals are accompanied by high-intensity satellites due to the unfilled electronic structure of the uppermost 4f core level.

In some experimental cases, it is difficult to identify the metal oxidation state (as could happen in many catalysts modified during the catalytic reaction), for example, in Co, Fe, and Sn. Nevertheless, it is possible to use also the valence band spectra for this purpose. The characteristic shape of the valence band is different for Sn^{4+} and Sn^{2+} [37,38] or Fe^{3+} and Fe^{2+} species [39,40], and the same principle is valid for many other metals when the identification from the main

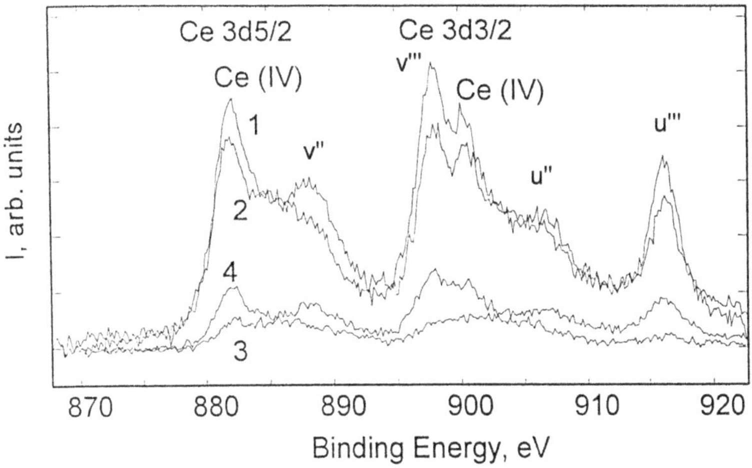

Fig. 4.6. Spectra of Ce 3d region acquired for the samples of three-way automotive catalyzer: 1 — new catalyzer, 2 — heated in UHV for 2 h at 600°C, 3 — after working for 500 h in gas exhaust, and 4 — heated in humid atmosphere for 8 h at 1000°C. Adapted with permission from Ref. [35].

photoemission peaks is difficult. Of course, it is necessary to remember that the information depth for the valence band is much higher due to high kinetic energy of photoelectrons, moreover, the elemental quantification from the valence band signals is impracticable without calibration using corresponding reference samples. One more problem could be encountered in the case of composites, where the signals from different elements can overlap in the valence band.

Once the main photoemission peaks are identified, it is possible to start the background subtraction, which is required for the calculation of the peak areas, i.e., for the quantification procedure. The origin of photoemission background is inelastic scattering of electrons that is slowly increasing with E_B (i.e., with decreasing E_K), as it is illustrated in Figure 4.2. Each photoemission peak is accompanied by a local increase of background due to a higher number of scattered electrons. The most commonly used types of backgrounds are Shirley [41,42] and linear. Linear background is very simple; it is just a straight line connecting two points before and after the peak; obviously, spikes in a noisy spectrum should be avoided. Of course, the selection of these points is important, because the whole background must be located under the peak, i.e., the line must not cross the spectrum at any point. More details about background subtraction can be found in Chapter 3.

Computation of the Shirley background is an iterative calculation based on the assumption that for every point of the spectrum the background intensity generated by a photoemission signal is proportional to the number of all photoelectrons with higher E_K (i.e., lower E_B). Of course, the shape of this background is more similar to the real contribution of inelastically scattered electrons, but it could generate a significant error in the peak area estimation, when applied to a large region with two or more photoemission peaks, e.g., spin–orbit split doublets.

Tougaard-type background [43,44] is more complex and its calculation is based on the so-called universal cross-section of electron energy losses. The main problem of this background is the assumption that the spectrum for which the background is determined is ideal, i.e., it is without any instrumental artifact. However, the experimental spectra cannot always be refined to this extent which is required for appropriate use of the Tougaard background, which is the reason why the linear and Shirley algorithms are mainly used in practical applications of XPS. The accuracy in Tougaard's approach can be better, but it is completely parameterized and, if one is not careful, the background can be bent anywhere, whereas in Shirley's model, the possibilities (and also the error sources) are much more limited. Therefore, from the simplicity and reliability of the Shirley background follows its wide acceptance. The choice between Shirley and linear backgrounds must be based on the evaluation of the spectral region width and peak intensity, i.e., signal-to-noise ratio. Generally, a linear background is recommended if the region is very large or the spectrum is noisy (low signal intensity). Consistency is very important: mixing of different backgrounds during the spectra processing must be absolutely avoided. More details to help choose between liner and Shirley backgrounds can be found in Chapter 3.

After background subtraction, follows the peak fitting if it is necessary to separate the different chemical species of the same element or the overlapping peaks of different elements. During the peak fitting, it is compulsory to use an asymmetric peak shape for the elements in metallic state, as it is indicated in Chapters 2 and 3, where are given the detailed instructions and examples for this procedure.

4.5 Elemental Quantification

Now, we are ready for the quantification of constituent elements and their different species. By the term "quantification" we mean the process of comparing two or more intensities in XPS spectra to determine the amount of material at the surface of a sample. Usually, the peak area ratios are used for relative quantification, since it is difficult to determine correctly the absolute values of the cross-section σ and inelastic mean free path of photoelectrons λ (see Eq. (4.1)). It should be noted that for correct quantification it is necessary to include in the determination of peak area all the losses (plasmons, shake-up, shake-down satellites, etc.) because the values of photoemission cross-section σ are calculated for the total excitation of core level, including also the photoelectrons that are losing energy for any excitation before escaping from the solid. Of course, here will be discussed only its simplified version because the quantification from the first principles (*ab initio*) [1] is quite difficult and is used only for very special cases. When using the same instrument under identical conditions, many component terms in the full quantification equation are canceled out and only the cross-section σ, inelastic mean free path λ, and transmission function G must be considered. The signal intensity, which is proportional to the atom density, can simply be written as follows:

$$I = N \cdot \sigma \cdot \lambda(E_K) \cdot G(E_K), \tag{4.1}$$

where I is the signal intensity (peak area), N is the atomic concentration, E_K is the photoelectron kinetic energy, σ is the cross-section of the relative core level, and λ is the inelastic mean free path of

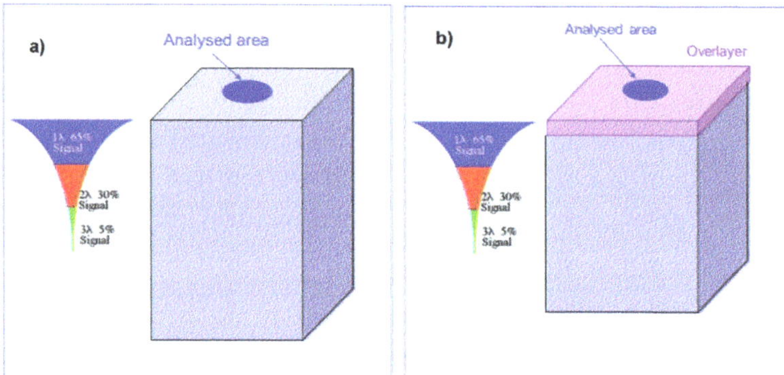

Fig. 4.7. Effective depth of XPS analysis: (a) homogeneous sample and (b) sample with a single overlayer.

photoelectrons. The transmission function $G(E_K)$ usually is already included in the software dedicated to data processing, therefore only σ and λ must be determined for correct quantification. Of course, it is necessary to remember that the quantification formula given above is valid only for a homogeneous surface, as illustrated in Figure 4.7. It is applicable also for a homogeneous contamination overlayer (Figure 4.7(b)), remembering that the values of σ and λ are different for contaminant and the investigated compound. From the ratio of photoemission signals registered for a single overlayer and the underneath located compound, it is possible to determine the thickness of the overlayer, d (contaminant or oxide), by using the Thickogram [43]. Another way is to use the iterative calculation based on the Beer–Lambert law [1], describing the energy losses due to inelastic scattering and dependence of photoemission intensity on the sampling depth with the main parameter of inelastic mean free path λ:

$$I(d) = I_0\, e^{-d/\lambda \cdot \cos(\theta)}, \tag{4.2}$$

where θ is the emission angle measured from surface normal. The sensitivity of XPS measurement increases if the analysis is carried out at a grazing angle because the information depth (or analyzed thickness, d) decreases with increasing emission angle, as $d(\theta) = d_0 \cos(\theta)$, where d_0 is the information depth at $\theta = 0°$, which is equal to about 3λ. If the two peaks from the overlayer and substrate have very similar kinetic energies (e.g., as in any metal and its oxide), then it is possible to calculate the overlayer thickness from a simple equation [1]:

$$t = \lambda \cos\theta \cdot \ln\left(1 + \frac{I_o \sigma_s}{I_s \sigma_o}\right), \tag{4.3}$$

where t is the overlayer thickness, I_0 is the peak area of the overlayer, and I_s is the peak area of the substrate, which becomes even simpler when the values of σ are identical or very similar (as in metal and its oxide).

Coming back to the elemental XPS quantification, we must clarify how the missing values of σ and λ can be correctly determined. The starting point to illustrate how the inelastic mean free path (IMFP) of electrons depends on the kinetic energy, i.e., the function $\lambda(E_K)$, is the so-called universal curve of IMFP [46], where are plotted the experimental values for most of the chemical elements. In the E_K range of soft X-rays (20–1500 eV), this logarithmic graph is linear, resulting in the dependence of $\lambda(E_K)$ commonly approximated to $E_K^{0.6}$. A more accurate approximation of the universal curve [46] resulted in dependencies $\lambda = 0.41(aE_K)^{0.5}$ for elemental materials (like Si and C) and $\lambda = 0.72(aE_K)^{0.5}$ for inorganic compounds, where a is the monolayer thickness in nm. Obviously, the calculation error of λ could be quite high if the wrong approximation is used, for example, it is 2.2 nm for elemental Si and 3.3 nm for SiO_2. Later have been developed more sophisticated methods for λ calculation [47–50] and experimental databases [27,51] for all the elements and numerous compounds that have been created.

The cross-section σ (or sensitivity factor) of an examined core level can be found by using two approaches: theoretical calculations or empirical data sets. For experiments using conventional X-ray sources (e.g., Al or Mg Kα), the most commonly used data sets are the calculated one by Scofield [52] and the empirical one created by Wagner [53]. The above-mentioned databases give the cross-section for fixed photon (X-rays) energies, therefore their use in experiments where tunable synchrotron X-ray radiation is used is rather limited. In this case, the photoionization cross-sections calculated by Yeh and Lindau [54] are used. Those data can be also found online in a handy database at the Elettra synchrotron website [55]. Of course, it is possible to create your own experimental data set by analyzing the reference samples, but this approach is very time-consuming and usually is applied only for the quantification of a narrow range of analyzed materials. The main advantage of Wagner (or any other empirical) data set is a higher precision of quantification because the transmission function $G(E_K)$ is already included, but this is valid only for the same or very similar instrument. Therefore, if you are not sure about the transmission function $G(E_K)$, it is recommended to use the Scofield data set, which usually is integrated into the data processing software of modern spectrometers.

It should be noted that all the considerations on the quantification procedure are valid only for homogeneous samples, especially if we need the absolute quantification. In any case, the relative error of absolute quantification, i.e., of determination of elemental atomic concentration, could be significant (till about 10%) even for homogeneous samples without the instrument calibration by using reference samples of the same compound. Relative quantification will give very reproducible information on similar samples or on the same sample before and after different treatments, including also catalytic reactions. However, most of the catalysts are in the form of powder or nanoparticles, therefore their samples are not perfectly homogeneous. Moreover, the pellets of powder unavoidably will have a rough surface, which must be considered in the quantification. In powder samples with big micrometric size particles, the average signals of photoemission from a large sample area will portray an average information on the chemical state of the catalyst surface and will result in a correct chemical composition, including also the contribution of inhomogeneous oxidized overlayer or surface contaminants.

In the case of powder catalyst samples, the XPS peaks of constituent elements usually are noticeably broadened in comparison with flat and uniform samples. The possible reasons for peak broadening, related to the sample, could be the following: (i) presence of multiple chemical states in different parts of the sample, (ii) slightly different charging of the particles, (iii) actual broadening of the electron orbitals caused by disorder and non-uniformities, and (iv) variations in the band bending at the rough surface resulting in the changes of the work function. If the size of the particles is big enough for the small-area XPS analysis (about 50 μm or more), then it is possible to investigate single particles and even to apply ion sputtering for their XPS depth profiling, as has been demonstrated for TiH_2 powders [56] and other similar samples.

In the case of nanoparticles with a size smaller than 10 nm, the XPS signals are derived from the whole volume of the particles, i.e., from the particle top to the center (or even bottom). Therefore, if the XPS information depth (which is about 3λ) exceeds the particle size as it is illustrated in Figure 4.8, it is possible to separate the signals from the outer shell (oxide or different coating) and core of the particles and from their intensity ratio I_{shell}/I_{core} to determine the shell thickness deploying modeling approaches (see Ref. [57]) and to accomplish the quantification by using the single sphere model for peak intensities [58]. If the nanoparticles with a known core diameter are investigated, a relatively simple method based on

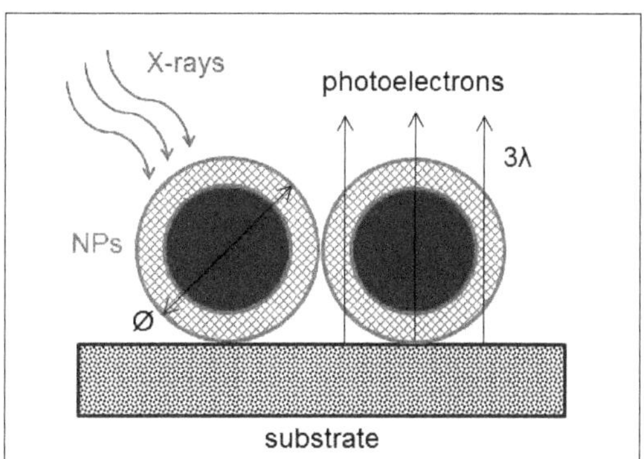

Fig. 4.8. Schematic drawing of the photoemission from core–shell nanoparticles with a diameter smaller than XPS information depth of 3λ.

effective attenuation lengths of photoelectrons [59] can be applied for the determination of the shell thickness from the measured intensity ratio I_{shell}/I_{core}. In conclusion, the XPS analysis, including also the elemental quantification based on appropriate modeling, can be successfully applied even for the investigation of complex nanoparticles.

4.6 Concluding Remarks

The main problems of preparation, mounting, and XPS measurements of catalyst samples (powders or nanoparticles) are related to their complex morphology. The surface of catalyst samples is almost always inhomogeneous and it is often covered with contaminants, i.e., it is quite different from the flat model catalysts. The same is valid also for photoemission spectra processing and interpretation, where the sample charging, photoemission signals broadening, and overlapping must be taken into account as well as the contribution of samples morphology to the correct quantification procedure. However, the recommendations presented in this chapter provide the main guidelines for a reliable XPS analysis of powder catalysts before and after the catalytic reaction or other types of treatments.

References

[1] *Practical Surface Analysis, Vol. 1, Auger and X-ray Photoelectron Spectroscopy.* D Briggs and MP Seah (Eds.). Chichester, UK: John Wiley & Sons, 1990.

[2] Bolli E, Kaciulis S, Mezzi A. ESCA as a tool for exploration of metals' surface. Coatings. 2020;10:1182.

[3] Venezia AM, La Parola V, Liotta LF. Structural and surface properties of heterogeneous catalysts: Nature of the oxide carrier and supported particle size effects. Catal Today. 2017;285:114–124.

[4] Barr TL. Applications of electron spectroscopy to heterogeneous catalysis. In D Briggs, MP Seah (Eds.), *Practical Surface Analysis, Auger and X-ray Photoelectron Spectroscopy* (Chapter 9, Vol.1). Chichester, UK: John Wiley & Sons, 1990.

[5] Maynard AD, Warheit DB, Philbert MA. The new toxicology of sophisticated materials: Nanotoxicology and beyond. Toxicol Sci. 2011; 120(1):S109–S129. Doi:10.1093/toxsci/kfq372.

[6] Handy RD, Owen R, Valsami-Jones E. The ecotoxicology of nanoparticles and nanomaterials: Current 558 status, knowledge gaps, challenges, and future needs. Ecotoxicology. 2008;17:315–325. Doi:10.1007/s10646-559 008-0206-0.

[7] Baer DR, Shutthanandan V. Nano-objects as biomaterials immense opportunities, significant challenges and the important use of surface analytical methods. In: Ducheyne P, Grainger D, Healy K, Hutmacher D, Kirkpatrick CJ, editors. Comprehensive Biomaterials II. Oxford: Elsevier; 2017, Chapter 4.6.

[8] Sweeney SF, Woehrle GH, Hutchison JE. Rapid purification and size separation of gold nanoparticles via diafiltration. J Am Chem Soc. 2006;128:3190–3197.

[9] Nurmi JT, Sarathy V, Tratnyek PT, Baer DR, Amonette JE, Karkamkar A. Recovery of iron/iron oxide nanoparticles from solution: Comparison of methods and their effects. J Nanoparticle Res. 2010;13:1937–1952.

[10] Knotek ML, Feibelman PJ. Stability of ionically bonded surfaces in ionizing environments. Surf Sci. 1979;90:78–90.

[11] Cazaux J. The influence of radiation damage (microscopic causes) on the sensitivity of Auger electron spectroscopy and X-ray photoelectron spectroscopy. Appl Surf Sci. 1985;20:457–471.

[12] Iijima Y, Niimura N, Hiraoka K. Prevention of the reduction of CuO during X-ray photoelectron spectroscopy analysis. Surf Interface Anal. 1996;24:193–197.

[13] Salmeron M. From surfaces to interfaces: Ambient pressure XPS and beyond. Topics Catal. 2018;61:2044–2051.

[14] Reyoy MN, Scott RWJ, Grosvenor AP. Ceria nanocubes: Dependence of the electronic structure on synthetic and experimental conditions. J Phys Chem C. 2013;117:10095–10105.

[15] Chenakin SP, Szukiewicz R, Barbosa R, Kruse N. Surface analysis of transition metal oxalates: Damage aspects. J Electron Spectrosc Related Phenom. 2016;209:66–77.

[16] Laiho T, Leiro JA, Heinonen MH, Mattila SS, Lukkari J. Photoelectron spectroscopy study of irradiation damage and metal-sulfur bonds of

thiol on silver and copper surfaces. J Electron Spectrosc Related Phenom. 2005;142:105–112.

[17] ISO 15472:2010. Surface chemical analysis — X-ray photoelectron spectrometers — Calibration of energy scales. Geneva: ISO; 2010.

[18] Greczynski G, Hultman L. X-ray photoelectron spectroscopy: Towards reliable binding energy referencing. Progress Mater Sci. 2020;107:100591.

[19] Baer DR, Artyushkova K, Cohen H, Easton Ch.D, Engelhard M, Gengenbach TR, Greczynski G, Mack P, Morgan DJ, Roberts A. XPS guide: Charge neutralization and binding energy referencing for insulating samples. J Vac Sci Technol A. 2020;38:031204.

[20] Ambrogi V, Donnadio A, Pietrella D, Latterini L, Alunni Proietti F, Marmottini F, Padeletti G, Kaciulis S, Giovagnoli S, Ricci M. Chitosan films containing mesoporous SBA-15 supported silver nanoparticles for wound dressing. J Mater Chem B. 2014;2:6054–6063.

[21] Lopez-Salido I, Lim D.Ch., Kim YD. Ag nanoparticles on highly ordered pyrolytic graphite (HOPG) surfaces studied using STM and XPS. Surf Sci. 2005;588:6–18.

[22] Luo K, St. Clair TP, Lai X, Goodman DW. Silver growth on TiO_2(110) (1 x 1) and (1 x 2). J Phys Chem B. 2000;104:3050–3057.

[23] Bolli E, Mezzi A, Burratti L, Prosposito P, Casciardi S, Kaciulis S. X-ray and UV photoelectron spectroscopy of Ag nanoclusters. Surf Interface Anal. 2020;52:1017–1022.

[24] Wagner CD. X-ray photoelectron spectroscopy with X-ray photons of higher energy. J Vac Sci Technol. 1978;15:518–523.

[25] Wagner CD. The Auger parameter, its utility and advantages: A review. J Electron Spectrosc Related Phenom. 1988;47:283–313.

[26] Moulder JF, Stickle WF, Sobol PE, Bomben KD. *Handbook of X-ray Photoelectron Spectroscopy*. Eden Prairie, MN, USA: Phys. Electronics Inc., 1992.

[27] http://www.lasurface.com/accueil/index.php.

[28] https://xpssimplified.com/periodictable.php.

[29] https://srdata.nist.gov/xps.

[30] Biesinger MC, Payne BP, Grosvenor AP, Leo Lau LWM, Gerson AR, Smart RSC. Resolving surface chemical states in XPS analysis of first

row transition metals, oxides and hydroxides: Cr, Mn, Fe, Co and Ni. Appl Surf Sci. 2011;257:2717–2730.

[31] Gao T, Norby P, Krumeich F, Okamoto H, Nesper R, Fjellv Håg. Synthesis and properties of layered-structured Mn_5O_8 nanorods. J Phys Chem C. 2010;114:922–928.

[32] Siegmann HC, Schlapbach L, Brundle CR. Self-restoring of the active surface in the hydrogen sponge $LaNi_5$. Phys Rev Lett. 1978;40: 972–975.

[33] Rabelo-Neto RC, Sales HBE, Inocêncio CVM, *et al.* CO_2 reforming of methane over supported $LaNiO_3$ perovskite-type oxides. Appl Catal B. 2018;221:349–361.

[34] Kaciulis S, Bolli E, Mezzi A, Vagner M, Plausinaitiene V, Kersulis S, Zurauskiene N, Lukose R. Surface and structural analysis of epitaxial $La_{1-x}Sr_x(Mn_{1-y}Co_y)zO_3$ films. Surf Interface Anal. 2020;52:900–906.

[35] Battistoni C, Cantelli V, Debenedetti M, Kaciulis S, Mattogno G, Napoli A. XPS study of ceramic three-way catalysts. Appl Surf Sci. 1999;144–145:390–394.

[36] Burroughs P, Hamnett A, Orchard AF, Thornton G. Satellite structure in the X-ray photoelectron spectra of some binary and mixed oxides of lanthanum and cerium. J Chem Soc Dalton Trans. 1976;17: 1686–1698.

[37] Gaggiotti G, Galdikas A, Kaciulis S, Mattogno G, Setkus A. Surface chemistry of tin oxide based gas sensors. J Appl Phys. 1994;76: 4467–4471.

[38] Themlin JM, Chtab M, Henrard L, Lambin P, Darville J, Gilles JM. Characterization of tin oxides by X-ray-photoemission spectroscopy. Phys Rev B. 1992;46:2460–2464.

[39] McIntyre NS, Zetaruk DG. X-ray photoelectron spectroscopic studies of iron oxides. Anal Chem. 1977;49:1521–1529.

[40] Xue M, Wang S, Wu K, Guo J, Guo Q. Surface structural evolution in iron oxide thin films. Langmuir. 2011;27:11–14.

[41] Shirley DA. High-resolution X-ray photoemission spectrum of the valence bands of gold. Phys Rev. 1972;5:4709–4714.

[42] Proctor A, Sherwood PMA. Data-analysis techniques in X-ray photoelectron spectroscopy. Anal Chem. 1982;54:13–19.

[43] Tougaard S. Inelastic background removal in X-ray excited photoelectron-spectra from homogeneous and inhomogeneous solids. J Vac Sci Technol A. 1987;5:1230–1234.
[44] Tougaard S, Sigmund P. Influence of elastic and inelastic scattering on energy spectra of electrons emitted from solids. Phys Rev B. 1982;25:4452–4466.
[45] Cumpson PJ. The Thickogram: A method for easy film thickness measurement in XPS. Surf Interface Anal. 2000;29:403–406.
[46] Seah MP, Dench WA. Quantitative electron spectroscopy of surfaces: A standard data base for electron inelastic mean free paths in solids. Surf Interface Anal. 1979;1:2–11.
[47] Tanuma S, Powell CJ, Penn DR. Calculations of electron inelastic mean free paths (IMFPs) IV. Evaluation of calculated IMFPs and of the predictive IMFP formula TPP-2 for electron energies between 50 and 2000 eV. Surf Interface Anal. 1993;20:77–89.
[48] Tanuma S, Powell CJ, Penn DR. Calculations of electron inelastic mean free paths (IMFPs). 6. Analysis of the Gries inelastic scattering model and predictive IMFP equation. Surf Interface Anal. 1997;25:25–35.
[49] Jablonski A, Tilinin IS, Powell CJ. Mean escape depth of signal photoelectrons from amorphous and polycrystalline solids. Phys Rev B. 1996;54:10927–10937.
[50] Jablonski A, Powell CJ. Evaluation of calculated and measured electron inelastic mean free paths near solid surfaces. J Phys Chem Ref Data. 1999;28:19–62.
[51] https://www.nist.gov/srd/nist-standard-reference-database-100.
[52] Scofield HF. Hartree-Slater subshell photoionization cross-sections at 1254 and 1487 eV. J Electron Spectrosc. 1976;8:129–137.
[53] Wagner CD. Sensitivity of detection of the elements by photoelectron spectrometry. Anal Chem. 1972;44:1050–1053.
[54] Yeh JJ, Lindau I. Atomic subshell photoionization cross sections and asymmetry parameters: 1 <Z< 103. At Data Nucl Data Tables. 1985;32(1):1–155. Doi:10.1016/0092-640X(85)90016-6.
[55] https://vuo.elettra.eu/services/elements/WebElements.html.
[56] Kaciulis S, Soltani P, Mezzi A, Montanari R, Lapi G, Richetta M, Varone A, Barbieri G. Oxidative treatment effect on TiH_2 powders. Surf Interface Anal. 2018;50:1195–1199.

[57] Baer DR, Wang YC, Castner DG. Use of XPS to quantify thickness of coatings on nanoparticles. Micros Today. 2016;24:40–45.
[58] Chudzicki M, Werner SMW, Shard AG, Wang Y-C, Castner DG, Powell CJ. Evaluating the Internal Structure of Core-Shell Nanoparticles Using X-ray Photoelectron Intensities and Simulated Spectra. J. Phys. Chem. C. 2015;119:17687–17696.
[59] Shard AG. A straightforward method for interpreting XPS data from core–shell nanoparticles. J Phys Chem C. 2012;116:16806–16813.

https://doi.org/10.1142/9781800613294_0005

Chapter 5

XPS Analysis of Electrically Insulating Catalytic Materials

David J. Morgan

Cardiff Catalysis Institute, Cardiff University, Cardiff CF10 3AT, UK

HarwellXPS: The EPSRC National Facility for X-ray Photoelectron Spectroscopy, Research Complex at Harwell (RCaH), Didcot, Oxon OX11 0FA, UK

morgandj3@cardiff.ac.uk

Abstract

Due to the highly insulating nature of most catalytic samples, some form of charge compensation is required to study such samples by X-ray photoelectron spectroscopy (XPS). This chapter provides an overview of charge build-up at surfaces and how such phenomena may be overcome, enabling the analyst to obtain meaningful spectral shapes and binding energies from the recorded photoelectron spectra. An overview of the variety of approaches used for charge correction in the literature is presented, with a critical view towards their use in catalytic materials.

5.1 Introduction

The collection of XPS spectra from electrically insulating catalytic materials differs from those collected from conductors, such as metals, since there is a build-up of charge following the photoelectric event at the surface, therefore differences in referencing to the vacuum level as opposed to the Fermi level used for conductors means accurate binding energy determination is more difficult. In this chapter, the background and methods relating to making measurements on such materials are discussed.

5.2 Catalysis and Catalytic Materials

Catalysis is inherently a surface phenomenon where generally a liquid or gas impinges on the surface of a solid material. Heterogeneous catalysts are typically solid materials comprised of nanoparticulate metals supported on a highly insulating material, such as an inorganic oxide, carbide, or nitride [1–4], such as that shown in Figure 5.1: a bright-field STEM image taken from a 5 wt% Pd/SnO_2 catalyst.

Given the mean free path of the analyzed photoelectrons [5], X-ray photoelectron spectroscopy (XPS) is inherently surface sensitivity and therefore ideally suited to the study of catalytic materials [6–9].

Fig. 5.1. Representative STEM bright-field image of a Pd nanoparticle on a SnO_2 support.

5.3 Charge Compensation

With few exceptions, catalytic materials are insulating, therefore their analysis by XPS raises two significant problems: first due to the unknown surface potential of the catalyst, measured electron kinetic energies (*c.f.* binding energies) will not match those values calculated or tabulated [10], while second, any surface potential may not be uniform across the sample, resulting in peak distortion or erroneous assignment of surface chemistry [11,12].

Charge compensation is required to overcome these problems, the theory and methodology of which are discussed in the following sections. For all XPS analyses on insulating material, the method and parameters of charge compensation should always be reported, details of which can be found in the ISO 19318:2021 and ASTM E1523-15 standards [13,14].

The reader should note that although this section is titled "charge compensation", the term "charge neutralization" is perhaps more commonly encountered. However, the chosen terminology is more accurate and consistent with the objective of achieving a uniform and stable charge surface potential required for successful XPS analysis.

5.3.1 *Charge compensation theory*

To allow the accurate measurement of the energy of emitted photoelectrons, there needs to be a way to reference the electrical potential of the emitting atom to the spectrometer measuring its energy. Photoelectron spectrometers are typically calibrated to measure the kinetic energy of electrons entering the energy analyzer by assumption that the electron will effectively have zero kinetic energy in the analysis chamber. Any excess charge on the electron when it enters the analyzer will mean it may accelerate or decelerate on its path to the energy analyzer and therefore will arrive at the analyzer and detector with a different energy from that which it escaped the surface with.

For conductors, the Fermi levels of the material and spectrometer align by an electrical connection (see diagram in Chapter 2), so the spectrometer measures electrons relative to its Fermi level, and hence the electron leaving the sample has the same energy relative to its Fermi level and discussed more in Section 5.4. For insulators, such a mechanism of equilibration does not exist and positive surface charges build up and hence the samples' potential, with respect to the spectrometer, changes with time and instrumental factors [12].

5.3.2 *Charge compensation methodology*

The conventional method of charge control is shown in Figure 5.2, where the photoemitted electrons are compensated by those from a low energy electron source, commonly termed a "flood gun", which is typically placed as close to normal to the sample surface as possible.

Most modern systems now employ monochromatic X-ray sources which introduce some complications with respect to charge compensation (see Section 5.3.3). Older, non-monochromatic sources could perhaps be considered ideal for charge compensation, since they are placed in close proximity to the sample surface and X-ray-induced excitation of the aluminum window and the sample holder typically provided an adequate source of secondary electrons to compensate for the positive charge at the sample surface, albeit with

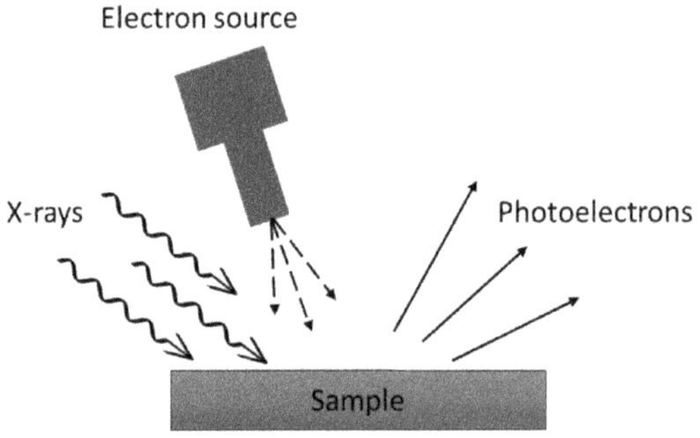

Fig. 5.2. Schematic of a common place method of controlling the surface charge of an insulating sample during photoemission, where a flood of low-energy electrons irradiates the surface.

a small upward shift (typically < 3 eV) in binding energy from that which may be expected based on tabulated values [10,15].

5.3.3 *Current charge compensation methodologies*

Currently, there are two general methods of charge compensation for systems utilizing monochromatic radiation, which facilitate neutralization of the analyzed area.

One method, used, for example, by Kratos Analytical in their AXIS series of spectrometers, employs a magnetic immersion lens (*aka* snorkel lens). In such systems (Figure 5.3), low-energy electrons are directed by magnetic field lines minimizing their deflection as they approach the sample. A negatively polarized charge balance plate situated below the analyzer input lens (therefore above the sample) allows the reflection of many low-energy electrons emitted from the sample back to the sample surface. As a lower number of electrons return to the surface than are emitted, additional electrons possessing a narrow energy distribution are injected through thermionic emission from a heated filament into the magnetic field lines.

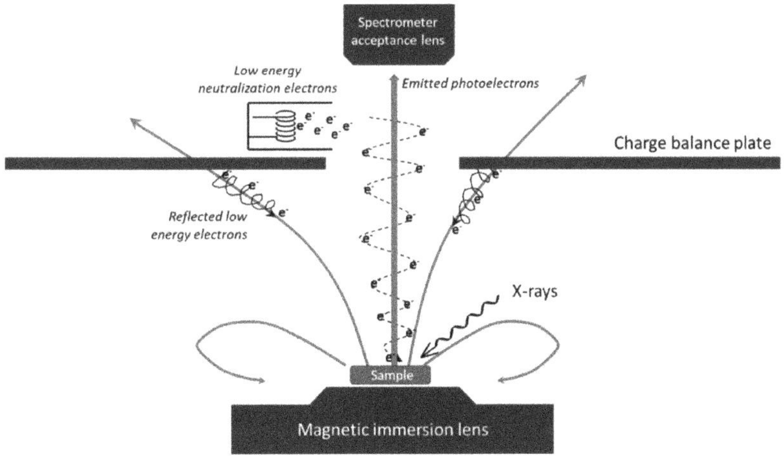

Fig. 5.3. Magnetic immersion lens charge compensation. A magnetic field generated by the immersion lens traps photoemitted electrons.

The optimum operation of such a neutralizer often shifts spectra to lower binding energy, which is typically within ca. 0.5 of the difference between the balance plate and filament bias voltages. Typically, the C 1s value of adventitious carbon is seen to shift to between ~282.5 and ~281.5 eV implying an overall negative surface charge which can be corrected to a suitable reference value.

A second method commonly used in systems with micro-focused X-ray sources is reliant on reducing charging in the peripheral area, which is unilluminated by the X-ray beam as shown in Figure 5.4. Versions of these sources are found on PHI and Thermo Scientific instruments.

In such sources, an electron flood source (ca. 1 eV) and a very low energy ion source (≤10 eV) provide excellent charge neutralization of all sample types. Such sources were borne out of the modeling of focused X-ray sources, where a range of surface potentials were identified during X-ray illumination. Using a low-current, low-energy ion beam, the potential of the unilluminated region, which can be charged negatively by several eV, is kept at a value close to that of the illuminated area so that more flooding electrons can reach the area undergoing photoemission. Providing the ion beam energy is kept low (<50 eV), then no argon ion-implantation is noted.

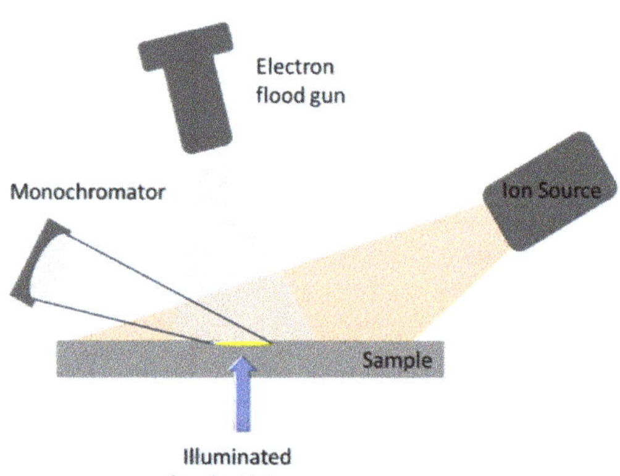

Fig. 5.4. Schematic of a dual ion–electron charge compensation as found in systems with micro-focused X-ray sources.

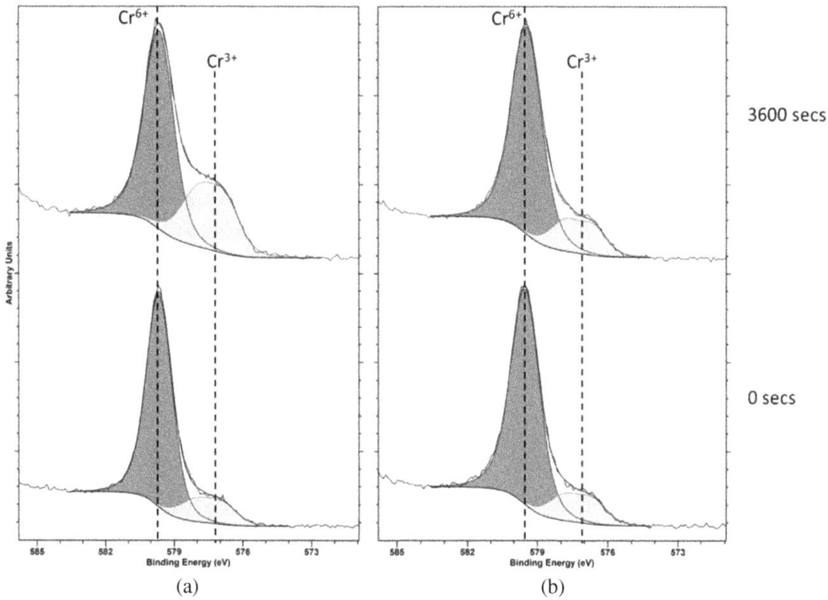

Fig. 5.5. Example of the reduction of CrO_3 using a dual electron–ion flood gun where the neutralizer operating conditions are (a) unoptimized and (b) optimized. The total time of neutralizer only irradiation between measurements was approximately 3600 sec (1 hr). In this case, the reduction in (a) is shown to be due to the preferential loss of adventitious carbon from the surface [16].

Despite excellent neutralization, it has been shown that without due consideration of the source operation, then sample damage can occur by mechanisms, such as stimulated carbon loss, an example of which is shown in Figure 5.5 and discussed further in Ref. [16].

5.3.4 *Recognizing charging problems*

Poorly mounted samples or improper setup of the charge compensation system can lead to differential charging which leads to broadening, lowering of peak intensity, deformation, or shifting of core-level peaks which can lead to erroneous assignment of chemical states. Figure 5.6 shows such an example of differential charging taken from the analysis of a PdV/TiO_2 catalyst, where the flood gun

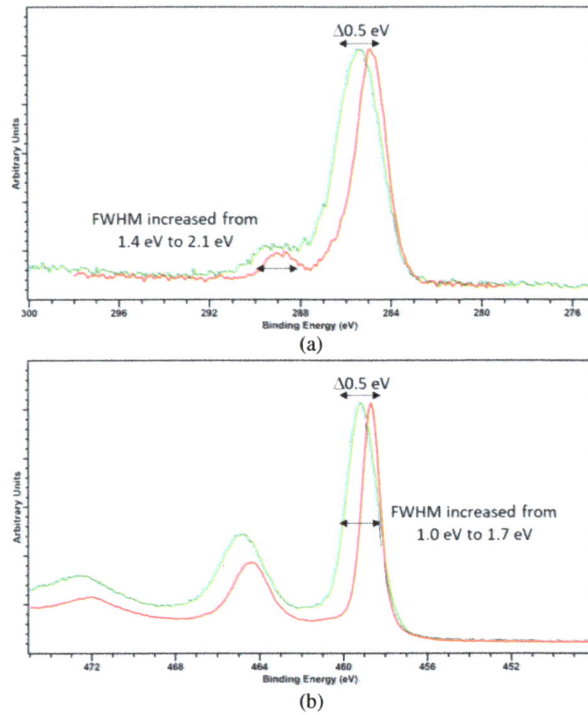

Fig. 5.6. (a) C 1s and (b) Ti 2p spectra take from a 1% Pd-V/TiO_2 catalyst recorded with (red) and without (green) optimized neutralizer settings. Note the similar peak broadening (0.7 eV) to both core levels. The spectral intensities have been normalized for clarity.

was not sufficiently optimized, leading to a broadening of all peaks by 0.7 eV from that expected under the standard operating conditions. This example highlights points that when acquiring or reviewing/analyzing data, the analysts should ask themselves questions such as follows:

- Are the peaks within ca. 5 eV of the expected relative binding energy?
- Are the peak shapes and FWHMs close to that expected for the element's chemical state and orbital?
- Could the material have changed during analysis?

- Was the flattest area chosen for analysis?
- Was the charge neutralizer optimization checked on a suitable material (e.g., polyethylene terephthalate, PET) before analysis?

5.3.5 *Practical aspects for charge compensation*

For conducting samples, contact between the sample and spectrometer is sufficient to obtain accurate binding energies (Section 5.4). For samples with an unknown conductivity, they only need to possess a conductivity $<10^{-10}\ \Omega^{-1}\,m^{-1}$ to experience charge build-up at the surface [17]. A more complex situation exists for insulating samples. During photoemission, the surface charge equilibrium potential will be affected by differences in conductivity through the sample [17,18] and also by those of the neutralizing species; therefore it is better to ensure such samples are electrically isolated from the spectrometer and control the surface potential alone using the flood source.

Despite the advances in charge compensation detailed in Section 5.3.3, some samples still prove difficult to charge compensate and several additional methodologies may be employed to facilitate improved analysis.

5.3.5.1 *Sample size and orientation*

Minimizing sample size will typically improve charge compensation since the number of flooding electrons repelled from the surface will be less, improving compensation significantly. For samples which cannot be made smaller without significant damage, such as those from a ceramic monolith, then the periphery of the area to be analyzed can be covered or wrapped in aluminum or copper foil which will generate secondary electrons to facilitate charge compensation. Mounting the samples so that the electrons from the flood source approach the sample normal to the analyzed surface significantly improves compensation also; this is especially useful for samples which are rough or concave in nature where shadowing effects limit the effectiveness of the neutralizer.

5.3.5.2 *Sample preparation*

A number of guides on sample preparation for catalysts, polymer, and nanoparticulate materials have been published [19–22], and all highlight that the topography of a sample is important in obtaining well neutralized, meaningful data. Generally, powders are best analyzed in their natural form rather than pelletized as such compaction can lead to limited diffusion of elements if *in situ* treatments are to be made [23]. Ideally, a uniform surface is required for good charge compensation and powders can generally be pushed into recesses or wells in a sample holder, and depending on the particle size, the powder may find a natural uniformity or may be flattened by a clean razor blade or lightly compacted using a simple press.

5.3.5.3 *Charging screens*

High-transparency mesh screens which are grounded to the spectrometer can be placed above the sample at a position outside the acceptance area of the analyzer (typically a few millimeters above the sample) and allow flooding electrons to approach the sample without deflection [24]. Typically, the mesh is made of nickel, but non-magnetic materials, such as gold, should be used for systems incorporating magnetic lenses.

5.3.5.4 *Chamber pressure*

Where the spectrometer design permits, introduction of argon in the region of 10^{-7} mbar (10^{-5} Pa) produces a small, but useful, flux of positively charged ions which decreases the charging in the periphery of the sample [25], yielding similar results to the dual sources discussed in Section 5.3.3. Such neutralization methods are also used in NAP-XPS systems, such as the EnviroESCA from SPECS, albeit at much higher gas pressures, e.g., 1 mbar (100 Pa), as discussed in Chapter 10 [26].

5.3.5.5 *In situ heating*

Used with success in the analysis of thermally stable zeolite materials [27,28], charge compensation is possible by heating the sample *in situ* to temperatures where cation, proton, or oxygen anion conductivity leads to a complete breakdown of the insulating properties of the sample. However, to successfully use this method, an inert sample holder, samples which are well outgassed prior to analysis, and a non-reactive vacuum environment are required, whilst a temperature of ca. 650 K is required for efficient compensation.

5.4 Elemental Binding Energies and Calibration

Spectral calibration is of great importance in understanding surface speciation and chemistry on catalytic samples. For conducting samples, such referencing is facile and allows the determination of an absolute binding energy scale independent of sample properties. This forms the basis of the ISO 15472 standard for spectrometer calibration [29] and should be performed for all spectrometers to ensure linearity across the energy range with defined periodic checks using Au, Ag, and Cu standards.

Figure 5.7 shows a diagram of the energy levels for both conducting and insulating samples within the spectrometer environment. Conducting samples are in direct electrical contact with the spectrometer, therefore their Fermi levels come into equilibrium due to a flow of charge. Therefore, the binding energy (E_{BE}) of any emitted photoelectron is measured with respect to the samples' Fermi level and ejected with a kinetic energy of $E_{KE} = h\nu - E_{BE} - \phi_s$, where $h\nu$ is the impinging photon energy and ϕ_s is the sample work function.

For the insulating sample, there is clearly no mechanism of conduction, so instead an equilibrium position is reached where the energy of the flooding electrons equals the sample vacuum potential [30]. Here, the kinetic energy $E_{KE} = h\nu - E_{BE} - \phi_s + E_e$, where E_e is the effective energy of the flood electrons relative to the Fermi level of

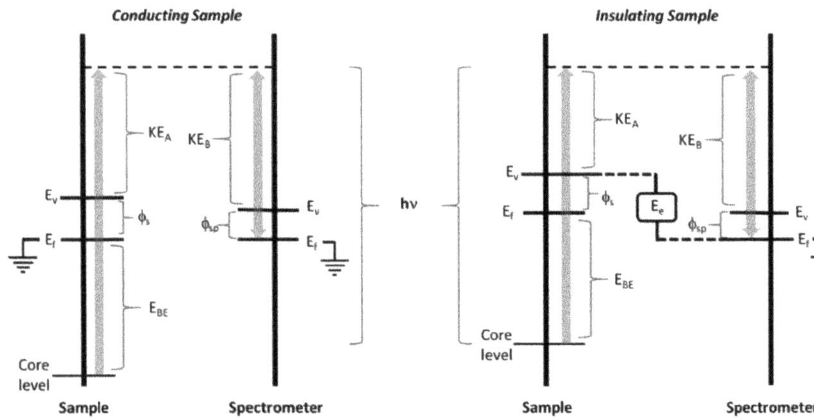

Fig. 5.7. Energy level diagrams for conducting (left) and insulating (right) samples. For the conducting sample, the equilibration of the sample and spectrometer Fermi levels can be seen. For the insulating samples, the vacuum level (E_v) becomes in equilibrium with the flooding electrons (E_e), and consequently, the measured kinetic energy is strongly dependent on the sample work function (ϕ_s).

the spectrometer. From this equation, even if E_e is known, the knowledge of the sample work function is necessary to determine the absolute value of the binding energy.

While the spectrometer calibration discussed earlier is highly important, it must be remembered that this is different from *charge referencing*, which is concerned with stabilizing and correcting spectral energies for insulating samples to a reference point allowing the determination of chemical states. In the following subsections, approaches to referencing these materials are discussed.

5.4.1 *Charge referencing: Conductive materials*

In the minority, some catalytic materials, such as IrO_2, RuO_2 [31], or graphitic carbon [32], are conducting. Therefore, providing the samples are suitably mounted allowing direct contact with the spectrometer, then as Figure 5.8 shows (and provided it is recorded), the Fermi level (E_F) may be used as a calibration point.

Conducting materials will typically have a high density of states (DOS) near the Fermi edge, but for samples with a low DOS, a band

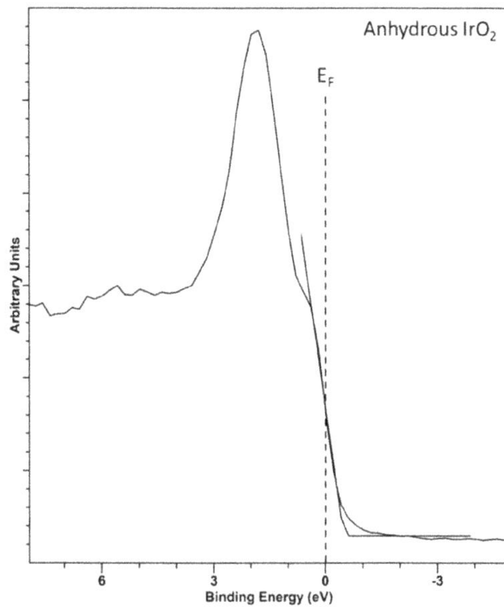

Fig. 5.8. Fermi level (E_F) referencing for the conducting material, anhydrous IrO_2.

gap, or those with significant charging, such a calibration may be ambiguous at best or worse, be entirely wrong.

5.4.2 *Charge referencing: Insulating materials*

Before introducing typical methods of charge referencing, the most important point must be remembered, specifically, regardless of the methods used for charge referencing, unless a satisfactory statement of how the energy referencing was performed, i.e., all reported binding energies should be treated with a degree of skepticism.

5.4.2.1 *Internal standard*

Such a method could be considered the most accurate as any bulk-like species with a well-known binding energy may serve as an absolute energy reference, especially for a catalytic support material; suitable examples are Si 2p in SiO_2 (103.5 eV), Ti $2p_{3/2}$ in TiO_2

(485.5 eV), or the u^{iii} peak for 4^+ species in the Ce 3d spectrum (916.7 eV). Such a methodology has been employed using, for example, lattice oxygen and La 4d core levels for oxides and lanthanum materials, respectively [33,34].

5.4.2.2 *Metal decoration*

To overcome a potential erroneous reference point, analysts may, through *in situ* evaporation, decorate a sample with either Au (or another noble metal) directly on the sample. In such cases, as shown by Hnatowich *et al.* [35], the noble metal comes into electrical equilibrium with the sample surface, and therefore compared with the same core-level peak of the noble metal sample which is in direct contact with the spectrometer, the charge state can be identified. This method is not without complications, for example, the need for an evaporation source, additional peaks appearing in spectra from the deposited metal, and more importantly, the variation in binding energy as a function of particle size [36–40]. Of course, such a method may not be universally applicable in the study of catalysts given the use of noble metals as the catalytically active species [41].

5.4.2.3 *Noble gas implantation*

Sputtering a sample with a noble gas such as argon or xenon will cause ion implantation [42,43]. Ensuring an incident ion energy <5 keV, the implantation depth will generally be of the order of the information depth inherent in XPS. Kohiki *et al.* [43,44] have shown that taking an Ar $2p_{3/2}$ binding energy of 242.3 eV (ca. 1.7 eV from gas phase argon [45]), an accuracy of ca. 0.4 eV in reported energies is obtained. However, argon implanted in conductive materials exhibits a 2.5 eV difference in binding energy to that of gas phase argon and 0.8 eV to that in insulators, hence a material with a partial conductive character may lie between the two values, illustrating that even the noble nature of the implanted atoms cannot provide surety of a stable energy reference. Although Kohiki *et al.* reported negligible ion-induced reduction of materials, the deformation of the substrate caused by the impact of the noble gas ions which in turn can form point defects through preferential sputtering, resulting in peak broadening and increasing the potential for observing reduced oxidation states, cannot be ignored.

5.4.2.4 *The Auger parameter*

From the early days of XPS analysis, Wagner championed the use of Auger transitions as an aid to chemical state identification, since Auger shifts can be much larger than those for core levels [46–48]. Such identification is one of the ultimate goals in XPS analysis, and the benefit of using such a method for identification is that there is no need for any form of formal charge referencing. The modified Auger parameter (α', now more commonly referred to as just the Auger parameter) is defined as the sum of the Auger kinetic energy and a core-level peak binding energy which is a quantity defined by the chemical state of the emitting atom. In this form, the Auger parameter is independent of excitation source energy (e.g., Al or Mg sources). More details about the calculation of α' are given in Chapter 2.

As any charge state of the surface would influence the core level and Auger electrons identically, the Auger parameter is unaffected by charging effects and provides a useful means of chemical state identification providing there is sufficient Auger intensity [47,49] and a suitable reference database [50] for comparison. However, such Auger parameter data are less readily available than the equivalent photoemission core levels, although a laboratory could establish an internal database through the measurements of well-characterized standards.

5.4.2.5 *Deliberate contamination*

A novel approach to referencing was proposed by Vesely *et al.* [51], where a new work function term (ϕ_{carbon}) is added to the Einstein equation so that $E_{BE} = h\nu - E_{KE} - (\phi_s + \phi_{carbon})$. By comparing the measured C 1s peak position of soot deposited from the flame of a candle to a value recorded for a thick layer of soot deposited in a similar way (283.8 eV). The difference is assigned as ($\phi_s + \phi_{carbon}$) and the C 1s spectrum is corrected by this amount to serve as an absolute reference to align the core levels. Although a significant assumption that any charging in the carbon overlayer and the sample is identical, peak positions within the accepted levels of uncertainly (± 0.2 eV) were found. Similarly, perfluorocyclohexane has been used as a deliberate contaminant applied via condensation, where the C-F peak is taken as the reference (C 1s = 292.9 eV) [52].

5.4.2.6 *Adventitious carbon*

Purposely left until last, referencing using the C 1s line of adventitious carbon (AdC) is the most common method of charge correction since its introduction by Siegbahn and co-workers [53]. Despite its simplicity and all samples handled in an ambient atmosphere having such carbon contamination, the use of this internal reference is fraught with difficulties, primarily due to the unknown surface potentials, or gradients thereof and the nature of the carbon.

The initial issue of AdC referencing is choosing a suitable reference energy. The ISO and ASTM standards [29,54], for example, assume the C-C/C-H peak of the measured C 1s spectrum to have a binding energy in the range 284.6–285.0 eV from which any correction can be calculated and hence applied to all core levels, however clearly, these standards already highlight a 0.4 eV uncertainty. As far back as 1982, Swift [55] highlighted the uncertainties with AdC referencing while, more recently, Jacquemin *et al.* [56] elegantly demonstrated how the C 1s binding energy varies according to the substrate composition, but it is logical to assume also that not only will the AdC vary with substrate type but also the environment and time of exposure [57].

Returning to conducting substrates, AdC on such surfaces has been shown to vary by 2 eV or more [58], and taking an arbitrary C 1s value, such as 284.8 eV, can result in a non-zero density of states above the Fermi level [57]. Clearly, despite the belief that the AdC does not change, the differences observed for conducting substrates clearly indicate that the binding energy of the carbon is not a property solely inherent to the AdC layer alone. In observing such variance, Greczynski and Hultman have given a detailed

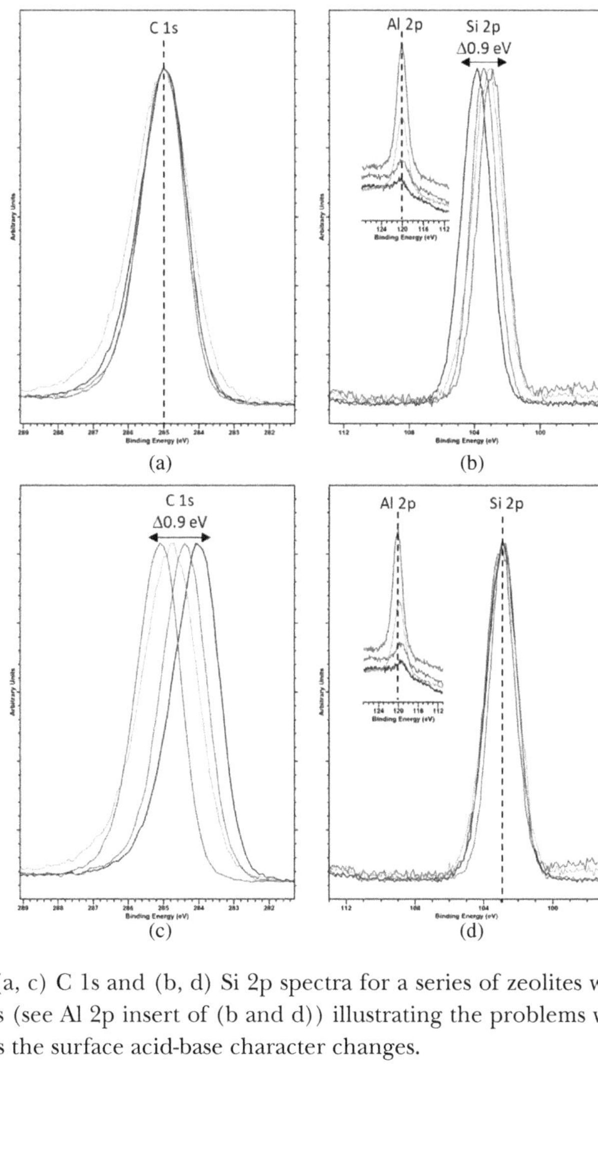

Fig. 5.9. (a, c) C 1s and (b, d) Si 2p spectra for a series of zeolites with different Al contents (see Al 2p insert of (b and d)) illustrating the problems with AdC referencing as the surface acid-base character changes.

historical perspective on binding energy referencing [57,59,60] and proposed a new method for referencing taking into account the sample work function, however this method has only been demonstrated to work well for conducing and thin-film materials to date.

An example of the issues with AdC referencing can be seen in Figure 5.9, where zeolitic materials have been prepared with varying Si/Al ratios. Calibrating all spectra to the C 1s peak, here taken to be 285.0 eV (Figure 5.9(a)), the Si 2p peaks can be seen to vary by 0.9 eV, whereas the Al 2p (insert of Figures 5.9(b) and 5.9(d)) remains constant at ca. 120 eV. Conversely, assuming a typical Si 2p value of 103 eV as recorded for SiO_2 (Figure 5.9(d)), then both C 1s and Al 2p core levels move by 0.9 eV.

As the Si/Al ratio increases within the zeolite framework, the level of surface acidity changes, and the interaction of carbon with the surface would be expected to be different and therefore cause a shift in the C 1s binding energy. An argument could possibly be made for the C 1s remaining constant, and the Si 2p is moving, however, even for cation exchange of Al into a Si tetrahedral site, the majority of the analyzed volume of the silica lattice would be SiO_2-like and shifts as large as almost 1 eV would not be expected [56].

5.5 Concluding Remarks

There have been great improvements in both the understanding and control of charging in the analysis of insulating materials by XPS, and instrument manufacturers have been quick to embrace technologies to facilitate a stable surface potential to enable the collection of repeatable, high-resolution data on a wide range of insulating materials. However, the need for a well-identified and reliable reference point still exists, and it is open to discussion whether this panacea can be found for all sample types. However, the use of the C 1s signal for AdC, while wildly used, should be used with care and with consideration of potential chemical or physical changes of the material, and therefore used in conjunction with a secondary internal reference (e.g., substrate peak).

References

[1] Ertl G, Knözinger H, Weitkamp J. *Handbook of Heterogeneous Catalysis.* Weinheim: Wiley-VCH Verlag GmbH & Co; Wiley-VCH., 2008. Doi:10.1524/zpch.1999.208.part_1_2.274.

[2] Védrine JC. *Metal Oxides in Heterogeneous Catalysis.* Elsevier, 2018. Doi: 0.1016/B978-0-12-811631-9.00014-4.

[3] Twigg MV. *Catalyst Handbook* (2nd edn.). Frome: Wolfe Publishing Ltd., 1989. Doi:10.1201/9781315138862.

[4] Hargreaves JSJ. Heterogeneous catalysis with metal nitrides. Coord Chem Rev. 2013;257:2015–2031. Doi:10.1016/j.ccr.2012.10.005.

[5] Powell CJ, Jablonski A. Surface sensitivity of X-ray photoelectron spectroscopy. Nucl Instruments Methods Phys Res Sect A Accel Spectrometers Detect Assoc Equip. 2009;601:54–65. Doi:10.1016/j.nima.2008.12.103.

[6] Lefebvre J, Galli F, Bianchi CL, Patience GS, Boffito DC. Experimental methods in chemical engineering: X-ray photoelectron spectroscopy-XPS. Can J Chem Eng. 2019;97:2588–2593. Doi:10.1002/cjce.23530.

[7] Palomino RM, Hamlyn R, Liu Z, Grinter DC, Waluyo I, Rodriguez JA, Senanayake SD. Interfaces in heterogeneous catalytic reactions: Ambient pressure XPS as a tool to unravel surface chemistry. J Electron Spectros Relat Phenomena. 2017;221:28–43. Doi:10.1016/j.elspec.2017.04.006.

[8] Feng DM, Li YW, Zhou JL, Zhang BJ. XPS study of iron catalysts for Fischer-Tropsch synthesis. Fuel Sci Technol Int. 1995;13:559–567. Doi:10.1080/08843759508947694.

[9] Beccat P, Da Silva P, Huiban Y, Kasztelan S, Quantitative surface analysis by XPS: Application to hydrotreating catalysts. 1999.

[10] National Institute of Standards and Technology. X-ray photoelectron spectroscopy database. www.srdata.nist.gov/xps/Default.aspx; Doi:10.18434/T4T88K.

[11] Metson JB. Charge compensation and binding energy referencing in XPS analysis. Surf Interface Anal. 1999;27:1069–1072. Doi:10.1002/(sici)1096-9918(199912)27:12<1069::aid-sia677>3.0.co;2-a.

[12] Cazaux J. Mechanisms of charging in electron spectroscopy. J Electron Spectros Relat Phenomena. 1999;105:155–185. Doi:10.1016/s0368-2048(99)00068-7.

[13] ISO 19318:2021. *Surface Chemical Analysis — Reporting of Methods Used for Charge Control and Charge Correction.* Geneva: International Organization for Standardization, 2004.

[14] ASTM E1523-15. *Standard Guide to Charge Control and Charge Referencing Techniques in X-ray Photoelectron Spectroscopy.* West Conshohocken, PA: ASTM International, 2015.

[15] Surface Science Spectra. http://avs.scitation.org/journal/sss.

[16] Edwards L, Mack P, Morgan DJ. Recent advances in dual mode charge compensation for XPS analysis. Surf Interface Anal. 2019;51:925–933. Doi:10.1002/sia.6680.

[17] Baer DR, Artyushkova K, Cohen H, Easton CD, Engelhard M, Gengenbach TR, Greczynski G, Mack P, Morgan DJ, Roberts A. XPS guide: Charge neutralization and binding energy referencing for insulating samples. J Vac Sci Technol A. 2020;38:031204. Doi:10.1116/6.0000057.

[18] Gijzeman OLJ, Mens AJM, Van Lenthe, JH. Mortier WJ, Weckhuysen BM. The effect of chemical composition and structure on XPS binding energies in zeolites. J Phys Chem B. 2003;107:678–684. Doi:10.1021/jp021948d.

[19] Davies PR, Morgan DJ. Practical guide for X-ray photoelectron spectroscopy: Applications to the study of catalysts. J Vac Sci Technol A. 2020;38:033204. Doi:10.1116/1.5140747.

[20] Baer DR. Guide to making XPS measurements on nanoparticles. J Vac Sci Technol A. 2020;38:031201. Doi:10.1116/1.5141419.

[21] Easton CD, Kinnear C, McArthur SL, Gengenbach TR. Practical guides for X-ray photoelectron spectroscopy: Analysis of polymers. J Vac Sci Technol A. 2020;38:023207. Doi: 10.1116/1.5140587.

[22] Stevie FA, Garcia R, Shallenberger J, Newman JG, Donley CL. Sample handling, preparation and mounting for XPS and other surface analytical techniques. J Vac Sci Technol A. 2020;38:063202. Doi:10.1116/6.0000421.

[23] Hellgardt K, Chadwick D. Design of sample holders for surface analysis of powders. Rev Sci Instrum. 1996;67:4025–4026. Doi:10.1063/1.1147270.

[24] Bryson CE. Surface potential control in XPS. Surf Sci. 1987; 189–190:50–58. Doi:10.1016/S0039-6028(87)80414-4.

[25] Yu XR, Hantsche H. Pressure dependence of the charging effect in monochromatized small spot X-ray photoelectron spectroscopy. J Electron Spectros Relat Phenomena. 1990;50:19–29. Doi:10.1016/0368-2048(90)80002-R.

[26] Shah D, Bahr S, Dietrich P, Meyer M, Thißen A, Linford MR. Coffee bean, by near-ambient pressure XPS. Surf Sci Spectra. 2019;26:24006. Doi:10.1116/1.5110902.

[27] Grünert W, Schlogl R, Photoelectron spectroscopy of zeolites. In HG Karge, J Weitkamp (Eds.), *Molecular Sieves* (pp. 467–515). Berlin, Heidelberg: Springer, 2004. Doi:10.1007/b94241.

[28] Gruenert W, Schloegl R, Karge HG. Investigations of zeolites by photoelectron and ion scattering spectroscopy. 1. New applications of surface spectroscopic methods to zeolites by a high-temperature measurement technique. J Phys Chem. 1993;97:8638–8645. Doi:10.1021/j100135a017.

[29] ISO Standards. (n.d.). https://www.iso.org/committee/54618.html.

[30] Lewis RT, Kelly MA. Binding-energy reference in X-ray photoelectron spectroscopy of insulators. J Electron Spectros Relat Phenomena. 1980;20:105–115. Doi:10.1016/0368-2048(80)85010-9.

[31] Morgan DJ. Resolving ruthenium: XPS studies of common ruthenium materials. Surf Interface Anal. 2015;47:1072–1079. Doi:10.1002/sia.5852.

[32] Morgan DJ. Cluster cleaned HOPG by XPS. Surf Sci Spectra. 2017;24:024003. Doi:10.1116/1.4993771.

[33] Li JPH, Zhou X, Pang Y, Zhu L, Vovk EI, Cong L, van Bavel AP, Li S, Yang Y. Understanding of binding energy calibration in XPS of lanthanum oxide by in situ treatment. Phys Chem Chem Phys. 2019;21:22351–22358. Doi:10.1039/C9CP04187G.

[34] Ardizzone S, Bianchi CL, Fadoni M, Vercelli B. Magnesium salts and oxide: An XPS overview. Appl Surf Sci. 1997;119:253–259. Doi:10.1016/S0169-4332(97)00180-3.

[35] Hnatowich DJ, Hudis J, Perlman ML, Ragaini RC. Determination of charging effect in photoelectron spectroscopy of nonconducting solids. J Appl Phys. 1971;42:4883–4886. Doi:10.1063/1.1659869.

[36] Unger WES, Gross T, Böse O, Lippitz A, Fritz T, Gelius U. VAMAS TWA2 project A2: Evaluation of static charge stabilization and

determination methods in XPS on non-conducting samples. Report on an inter-laboratory comparison. Surf Interface Anal. 2000;29: 535–543. Doi:10.1002/1096-9918(200008)29:8<535::AID-SIA900> 3.0.CO;2-5.

[37] Radnik J, Mohr C, Claus P. On the origin of binding energy shifts of core levels of supported gold nanoparticles and dependence of pretreatment and material synthesis. Phys Chem Chem Phys. 2003; 5:172–177. Doi:10.1039/b207290d.

[38] Aruna I, Mehta BR, Malhotra LK, Shivaprasad SM. Size dependence of core and valence binding energies in Pd nanoparticles: Interplay of quantum confinement and coordination reduction. J Appl Phys. 2008;104:064308-1–064308-5. Doi:10.1063/1.2973682.

[39] DiCenzo SB, Berry SD, Hartford EH. Photoelectron spectroscopy of single-size Au clusters collected on a substrate. Phys Rev B. 1988; 38:8465–8468. Doi:10.1103/PhysRevB.38.8465.

[40] Wertheim GK, Dicenzo SB, Youngquist SE. Unit charge on supported gold clusters in photoemission final state. Phys Rev Lett. 1983;51: 2310–2313. Doi:10.1103/PhysRevLett.51.2310.

[41] Hutchings GJ. Heterogeneous gold catalysis. ACS Cent Sci. 2018;4:1095–1101. Doi:10.1021/acscentsci.8b00306.

[42] Barbieri PF, Landers R, Marques FC. Electronic and structural properties of implanted xenon in amorphous silicon. Appl Phys Lett. 2007;90:164104-1–164104-3. https://doi.org/10.1063/1.2723072.

[43] Kohiki S, Ohmura T, Kusao K. A new charge-correction method in X-ray photoelectron spectroscopy. J Electron Spectros Relat Phenomena. 1983;28:229–237. Doi:10.1016/0368-2048(83)80001-2.

[44] Kohiki S, Ohmura T, Kusao K. Appraisal of a new charge correction method in X-ray photoelectron spectroscopy. J Electron Spectros Relat Phenomena. 1983;31:85–90. Doi:10.1016/0368-2048(83) 85017-8.

[45] Patel DI, Bahr S, Dietrich P, Meyer M, Thißen A, Linford MR. Argon gas, by near-ambient pressure XPS. Surf Sci Spectra. 2019;26:014024. Doi:10.1116/1.5110408.

[46] Wagner CD. Chemical shifts of Auger lines, and the Auger parameter. Faraday Discuss Chem Soc. 1975;60:291–300. Doi:10.1039/ DC9756000291.

[47] Wagner CD, Joshi A. The auger parameter, its utility and advantages: A review. J Electron Spectros Relat Phenomena. 1988;47:283–313. Doi:10.1016/0368-2048(88)85018-7.

[48] Wagner CD. Auger parameter in electron spectroscopy for the identification of chemical species. Anal Chem. 1975;47:1201–1203. Doi:10.1021/ac60357a021.

[49] Moretti G. Auger parameter and Wagner plot in the characterization of chemical states by X-ray photoelectron spectroscopy: A review. J Electron Spectros Relat Phenomena. 1998;95:95–144. Doi:10.1016/s0368-2048(98)00249-7.

[50] Powell CJ. Recommended Auger parameters for 42 elemental solids. J Electron Spectros Relat Phenomena. 2012;185:1–3. Doi:10.1016/j.elspec.2011.12.001.

[51] Vesely CJ, Langer DW. Electronic core levels of the IIB-VIA compounds. Phys Rev B. 1971;4:451–462. Doi:10.1103/PhysRevB.4.451.

[52] Connor JA. Handbook of X-ray and ultraviolet photoelectron spectroscopy. In D Briggs (Ed.), *Handbook of X-ray Ultraviolet Photoelectron Spectroscopy* (p. 183). London: Heyden, 1977.

[53] Fahlman A, Nordling C, Siegbahn K. ESCA : Atomic, molecular and solid state structure studied by means of electron spectroscopy. Uppsala : Almqvist and Wiksell; 1967. http://lib.ugent.be/catalog/rug01:001343982.

[54] ASTM Standards. https://www.astm.org/COMMIT/SUBCOMMIT/E4203.htm.

[55] Swift P. Adventitious carbon — The panacea for energy referencing? Surf Interface Anal. 1982;4:47–51. Doi:10.1002/sia.740040204.

[56] Jacquemin M, Genet MJ, Gaigneaux EM, Debecker DP. Calibration of the X-ray photoelectron spectroscopy binding energy scale for the characterization of heterogeneous catalysts: Is everything really under control? ChemPhysChem. 2013;14:3618–3626. Doi:10.1002/cphc.201300411.

[57] Greczynski G, Hultman L. X-ray photoelectron spectroscopy: Towards reliable binding energy referencing. Prog Mater Sci. 2020;107:100591-1–100591-46. Doi:10.1016/j.pmatsci.2019.100591.

[58] Tielsch BJ, Fulghum JE. Differential charging in XPS. Part III. A comparison of charging in thin polymer overlayers on conducting and

non-conducting substrates. Surf Interface Anal. 1997;25:904–912. Doi:10.1002/(sici)1096-9918(199710)25:11<904::aid-sia316>3.0.co;2-d.

[59] Greczynski G, Hultman L. C 1s peak of adventitious carbon aligns to the vacuum level: Dire consequences for material's bonding assignment by photoelectron spectroscopy. ChemPhysChem. 2017; 18:1507–1512. Doi:10.1002/cphc.201700126.

[60] Greczynski G, Hultman L. Reliable determination of chemical state in X-ray photoelectron spectroscopy based on sample-work-function referencing to adventitious carbon: Resolving the myth of apparent constant binding energy of the C 1s peak. Appl Surf Sci. 2018;451: 99–103. Doi:10.1016/j.apsusc.2018.04.226.

https://doi.org/10.1142/9781800613294_0006

Chapter 6

Assigning XPS Peaks to Chemical Environments Using First-Principles Calculations

Gabriel L. S. Rodrigues[*,‡], Mikael Valter[*,§], Peter Amann[*,†,¶] and Lars G. M. Pettersson[*,||]

[*]*Department of Physics, AlbaNova University Center, Stockholm University, Stockholm S-106 91, Sweden*

[†]*Scienta Omicron AB, Danmarksgatan 22, Uppsala 75323, Sweden*

[‡]*glibanio@kth.se*

[§]*mikael.valter@fysik.su.se*

[¶]*peter.amann@scientaomicron.com*

[||]*lgm@fysik.su.se*

Abstract

We discuss a simple intuitive chemical interpretation of XPS shifts in terms of the Z+1 approximation and the chemical bonding in the final core-ionized state. We show how this applies to vibrational excitations and discuss the effects of metallic screening as well as when the Z+1 approximation can be expected to break down. Approaches to compute XPS shifts are discussed and exemplified with a focus on the case of CO/Ni(100). We discuss new experimental advances to study heterogeneous catalytic reactions — CO oxidation and CO hydrogenation — using XPS at high pressure (~1 bar) where reactants, intermediates, and products all contribute to the XPS signal. To disentangle the spectra, we perform microkinetic modeling of the reactions combined with a genetic algorithm to determine which initial data need to be re-examined due to DFT inaccuracy, adsorbate–adsorbate interaction, side reactions that are not included, or uncontrolled experimental features. Combined with computed XPS positions, the resulting coverages of various species can then be used to build theoretical XPS spectra to be directly compared with the experiment. Here, we describe the first steps towards this goal.

6.1 Introduction

This chapter, as indicated by the title, will focus on determining the chemical environment of the probed atom in a molecule based on computed XPS binding energies in conjunction with experimental measurements. For the theoretician to be successful in this endeavor, we find that a good, intuitive — perhaps even pictorial — understanding of the process is essential. We will thus, before going into the details of actually computing XPS spectra, go through some essential aspects for understanding the origin of the shifts in the measured binding energies that result from probing the same species in different chemical environments. Our focus will be to relate as closely as possible to experiments that are becoming ever more complex and approaching the conditions of actual industrial heterogeneous catalytic processes. Thus, neither surfaces nor surface coverages are necessarily well defined and characterized and may furthermore be dynamically evolving in time as a catalyst is activated.

The experiments use high pressure and temperature and reactions take place, forming transient species that can be spectators, part of a side reaction, or lead to the final products. The high temperature means that surface adsorbates are not necessarily stationary but can move across the catalytic surface and be found in a distribution of sites that is determined by their energy differences. This entails developing libraries of calculated XPS peak positions of all possible participants in, e.g., the reaction between H_2 and CO or CO_2 over a specific catalytic transition-metal surface. It also requires an understanding of the experimental conditions to which we will devote one subsection of the chapter.

Under reaction conditions, even the surface can become flexible [1], restructure, or change composition [2]. The coverage of different species in the reaction may be such that computed XPS for the isolated species may not be relevant for the actual situation due to the changed chemical environment, and XPS binding energies must thus be computed for various assumed cases in order to cover all possible situations.

Measured XPS peaks under these conditions are, particularly for more complex reactions with a broader distribution of species, due to these circumstances rather broad and contain contributions from a range of components. In many cases, it can be sufficient to predict the energy at which various species should be found, but for a more detailed understanding and interpretation of the data, more information from theory is needed.

Here, we will describe an approach using microkinetic modeling of the assumed reaction sequence to determine the expected coverages in comparison with the integrated coverage of each element (C, O, N, etc.) from the temperature-dependent experiment. We apply machine learning through a genetic algorithm to the ingredients (chemisorption energies and barriers) in the model to find values that need to be re-evaluated to come to an agreement with the experiment, or surface-active sites and additional reactions that were not expected. The results from the microkinetic model, after passing the genetic algorithm, can then be used to construct theoretical XPS spectra to be directly compared with the experiment in an iterative process if necessary.

In the following, we will go through the computational approach and some aspects of the high-pressure experiments that should help understand the specific challenges to theory and then discuss the microkinetic modeling and genetic algorithm. First, however, we will build an intuitive picture of the origin of XPS binding energy shifts.

6.2 Chemical Interpretation — Z+1 Approximation

The core electrons form a closed electronic shell and as such it forms a spherically symmetric charge distribution. It is well known from electrostatics (and gravity) that this charge (mass) distribution can be replaced by a point charge (point mass) at the origin when considering the interaction with charges (masses) outside the sphere. Since the outer valence electron distribution is mainly located beyond the core region, this amounts to the core electrons reducing the effective nuclear charge that is seen by the valence electrons. From the view point of the valence electrons, the XPS process of removing a screening core electron is thus equivalent to increasing, by essentially one unit, the effective nuclear charge that the valence electrons see. This corresponds to changing the valence structure to that of the next element to the right in the periodic table and is known as the Z+1 or equivalent core approximation [3–5]. This provides a simple chemical understanding of XPS shifts in terms of differences in the chemical bonding between different final states. One example is the chemical shift of 1.3 eV [6] between the two nitrogen atoms of N_2 chemisorbed on-top at Ni(100), which directly corresponds to the difference in chemisorption energy between NO bound through nitrogen or oxygen, still in the geometry of N_2/Ni(100) [6,7].

6.2.1 *Electronic effects in the final core-hole state and the ESCA molecule*

The ESCA molecule (Figure 6.1) is important in electron spectroscopy because, as we can see from Figure 6.1 and Table 6.1, the XPS binding energies of the four inequivalent carbons span an impressive range of more than 7 eV; in fact, it was originally selected to show the power of Electron Spectroscopy for Chemical Analysis (ESCA) [8]. This large range is due to the strongly distinct chemical

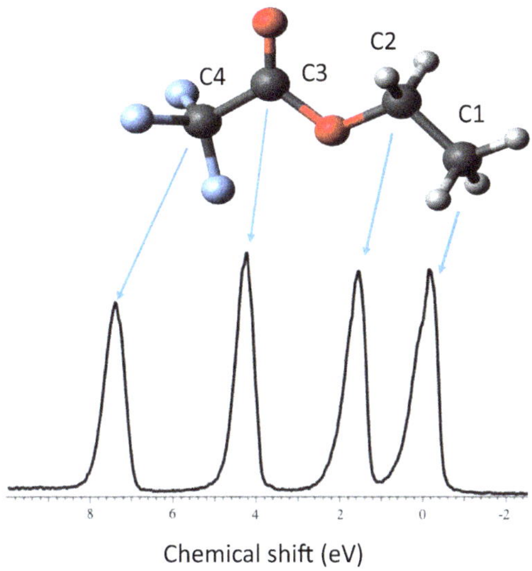

Fig. 6.1. Ethyltrifluoroacetate or the "ESCA molecule". Legend: hydrogen — white; carbon — gray; oxygen — red; fluorine — blue. Adapted with permission from Ref. [5].

Table 6.1. Experimental [9] and calculated [5] C 1s chemical shifts (eV) for the different carbons in ethyltrifluoroacetate using explicit core hole (CH) or the Z+1 approximation. Values using the CAM-B3LYP exchange-correlation functional.

	C1	C2	C3	C4
Exp.	0.00	1.72	4.33	7.46
CH	0.00	1.67	4.49	7.46
Z+1	0.00	1.68	4.37	7.50

environments, with large differences in electronegativity of the elements bonded to each carbon.

We note that the XPS binding energy computationally is defined as the difference in total energy between the core-hole state and the ground state, i.e., $BE = E(CH) - E(GS)$. The shift, $\Delta BE(i, j)$, between atoms i and j is then simply $\Delta BE(i, j) = BE(i) - BE(j)$ where we take atom j as the reference (carbon C1 in the table). Since the ground state energy is the same in this molecule (and in coadsorbates in surface problems), it cancels out and the shifts will be entirely determined by the energy difference between the final states through $\Delta BE(i, j) = E(CH(i)) - E(CH(j))$. Due to the subtraction, much of the inherent inaccuracies in the computational procedure cancel out, and good estimates of shifts can typically be obtained using regular DFT functionals. However, for the ESCA molecule, it has been challenging to obtain the shifts quantitatively and was shown by van den Bossche *et al.* [10] to require employing hybrid functionals, i.e., with a measure of exact exchange included. In the table, we show results from Ref. [5] using the CAM-B3LYP functional for both the full core-hole (CH) calculations and for the Z+1 counterpart. The latter was computed simply as the valence-ionized molecule with the targeted carbon replaced by a nitrogen atom; in the gas phase, there is no channel available to provide the extra valence electron to the nitrogen. It is clear that the Z+1 picture works excellently in this case.

Starting with C1, this carbon has the lowest BE since it is directly bonded only to H and C, which are atoms with the smallest electronegativity in the molecule. Indeed, core ionization of the methyl carbon corresponds in the Z+1 picture to NH_3^+ connected to the C2 carbon which is not so unfavorable. Going to C2, despite being bonded to a methyl, which is considered an electron-donating group (especially when hyperconjugation is possible) which should provide efficient screening of the core hole, it is also bonded to a very electronegative O, with electron-withdrawing effects that should surpass that of $-CH_3$ donation. This is confirmed by the increase of 1.72 eV in the BE. The trend here continues as C3 has higher BE than C2, since C3 is bonded to two O atoms and a C from a $-CF_3$ group. Therefore, the binding energy increases again to 4.33 eV. Finally, C4 is bonded to three F atoms, with fluorine being the most electronegative element in the periodic table. Then, the difference in the chemical environment is so high compared to C1 that the chemical shift reaches 7.46 eV.

In the example above, the Z+1 calculation (replacing carbon with nitrogen and taking total energy differences between the resulting new molecules) gave quantitative agreement with the full core-hole calculation. This is also the case for surface core-level shifts in metals [4,5,11,12] and their adsorbate-coverage dependence [5]. There is a caveat, however, and there are situations where this simple picture cannot be expected to be quantitative. Although the fundamental physics behind the equivalent core picture is solid, it is clear that, since the equivalent core is closed-shell, it cannot quantitatively describe resulting open-shell systems where final-state multiplet splittings are important and involve coupling to the now unpaired electron in the final-state core level [7,13]. However, the fundamental concept of reduced screening and the step to the right still applies. The Z+1 state still determines the qualitative shift, but a more stringent calculation including the core hole and its spin coupling with the valence unpaired electrons has to be performed to be quantitative [13]. We will return to this aspect when discussing the case of CO/Ni(100).

6.2.2 *Vibrational excitations and metallic screening*

The XPS process results in vibrational excitations if the potential energy surfaces (PES) of the initial and final (core-hole) states are sufficiently different. For a quantitative calculation of the resulting Franck–Condon (FC) profile, the PES of the initial and core-hole states need to be computed along the relevant degrees of freedom and the vibrational wave functions determined. The FC profile is then given by the projection of the ground state vibrational wave function onto each of the vibrational wave functions of the core-ionized state. Before undertaking such an endeavor, a qualitative estimate of the importance of vibrational broadening can be obtained through the Z+1 picture.

Taking gas-phase and chemisorbed CO as examples, we note that after the creation of a core hole in the gas phase, the proper comparison would be to NO^+ and CF^+ for C 1s and O 1s, respectively, while when chemisorbed on a metal surface, the screening electron from the metal neutralizes the charge and the comparison would instead be to NO and CF. In Table 6.2, we give the intermolecular

Table 6.2. Equilibrium bond distance (Å) for CO, core-ionized or core-excited, and the corresponding Z+1 analogs.

	r_e	Z+1	r_e
CO	1.128		
C*O (C $1s^{-1}$)	1.063	NO^+	1.063
CO* (O $1s^{-1}$)	1.153	CF+	1.154
C*O (C $1s^{-1}2\pi^1$)	1.153	NO	1.151
CO* (O $1s^{-1}2\pi^1$)	1.291	CF	1.272

distances in each case, either from the Huber–Herzberg tables [14] for the ground state and valence-ionized cases or from geometry optimization for the core-excited states. When these are very different from that of the initial CO, it will be important to take the vibrational profile into account. From the table, it is clear that no large effects should be expected in the gas phase, as the geometries of NO^+ and CF^+ are not dramatically different from that of CO as directly demonstrated by the low number of vibrational final states being involved (Figure 6.2(a)). However, for O 1s ionization when chemisorbed on a metal surface, the transfer of screening charge into the π^* results in an electronic structure corresponding to CF with a bond elongation of 0.14 Å, and a very broad FC profile can be expected. This is verified in Figure 6.2(c) with O 1s XPS of CO/Ni(100). This picture of the electronic structure effects is very elegantly verified experimentally [7] by the comparison with the Franck–Condon profile of O 1s excited into the π^* of gas-phase CO (Figure 6.2(b)). When generating theoretical XPS spectra also these considerations of the width should go into the modeling described in Section 6.5.

In conclusion, we note that the vibrational profile can be very well understood through the final-state picture, where the change in bond distance of the resulting Z+1 state compared to that of the original molecule determines the extent of the vibrational profile. Here, it is also clearly important to consider the screening of the core hole that provides an extra electron to the LUMO of the Z+1 state, as is clear from the CO* case above when comparing the gas

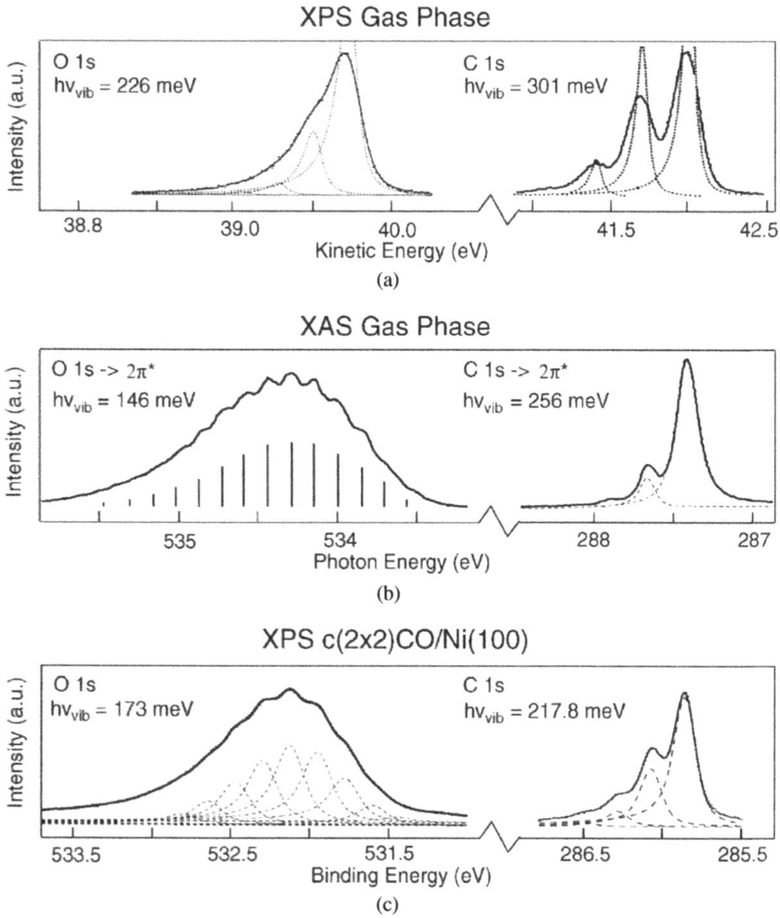

Fig. 6.2. (a) Measured XPS profiles for gas-phase CO. (b) Experimental 1s→π^* XAS spectra of gas-phase CO. (c) Experimental XPS spectra of CO/Ni(100). Adapted with permission from Ref. [7].

phase and chemisorbed vibrational progressions with that of the gas-phase XAS.

6.3 Computational Aspects

Although core-electron binding energies (CEBEs) can be computed in terms of absolute values within DFT, they strongly depend on the functional [15,16]. In Figure 6.3, we show the average (AD) and

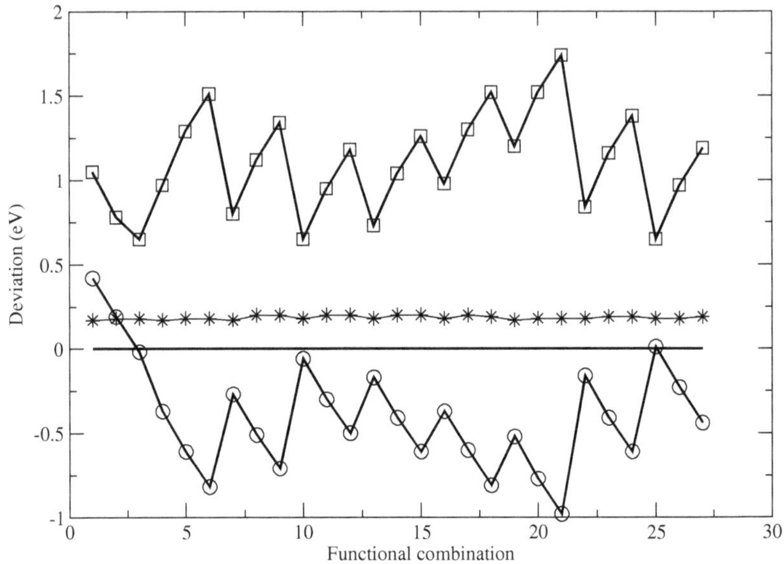

Fig. 6.3. Average (AD, circles) and maximum (MAD, squares) deviation of computed absolute CEBEs for the 27 combinations of nine exchange and three correlation functionals compared to experiment for 18 gas-phase molecules. The deviations (ΔE) from experimental term values are given by the starred curve. Data from Ref. [15].

maximum (MAD) deviation of computed absolute CEBEs for the 27 combinations of nine exchange and three correlation functionals compared to the experiment for 18 gas-phase molecules [15]. It is clear that the spread can be large and it could be tempting to search for a functional that gives good results for the system of interest, e.g., combinations 3, 10, or 25 in the figure that give the lowest deviations for these specific molecules.

However, we also include in the figure the deviation (ΔE, stars) from the experiment of the three first-term values (energy difference between successive excited states) in the computed XAS spectrum. Here, we use the lowest core-excited state as a reference which largely eliminates the dependence on the functional and shows that most of the difficulty lies in the description of the high-density core region. This implies that XPS *shifts* can be computed much more accurately than the absolute binding energies since deficiencies in the description of the core-hole state cancel out. For very accurate results, however, also the shifts can be challenging to compute as exemplified by the ESCA molecule, where hybrid functionals were found to be necessary [10].

To place the computed spectra on an absolute scale and relate to the experiment, a common reference is thus needed, which typically is a known experimental peak in a mixed adsorbate system. An alternative is to calibrate against the molecule in gas phase to assess the accuracy of the calculations and correct for the functional [17].

The example we are going to use here for the calculation of XPS chemical shifts of adsorbates at surfaces is CO adsorbed on Ni(100), for which a wide range of experimental data are available [18–21]. When CO is the only adsorbate in a CO/Ni(100)c(2×2) phase, CO molecules only occupy the top site on the surface and the C 1s binding energy is 285.9 eV. Coadsorption of hydrogen allows CO to form high-coverage phases and occupy other sites. It has been shown experimentally [19] that the presence of hydrogen in hollow sites decreases the CO-Ni bond strength sufficiently to not only permit CO to move to the bridge and hollow sites but also increase the C 1s binding energy by 0.4 eV to 286.3 eV for CO on-top. In the following sections, we will discuss the origin of and how to compute these shifts.

6.3.1 *Electronic effects in the final core-hole state*

In terms of theory and computation, the XPS binding energy can be understood as the difference in electronic energy of the two states involved in the process: the initial, ground state, described by an n-electron wave function, and the final, core-ionized state with its distribution of vibrational excitations, described by an $(n-1)$-electron wave function. In the case of molecules adsorbed on conductor surfaces (normally grounded during the experiments), the number of electrons can be considered constant, as the removal of a core electron is compensated by a conduction band electron (screening charge) added at the Fermi level, preserving system charge neutrality. The screened final state thus equals the lowest XAS state and the process corresponds to excitation from the core to the Fermi level.

In an all-electron approach, the core hole can be included directly by changing the occupation and converging to the core-ionized (gas phase or adsorbate on an insulator) or first core-excited state (adsorbate on a conducting substrate). This forms the so-called ΔSCF (or ΔKS for Kohn–Sham DFT) approach, where we calculate the total electronic energy of the system with the desired core hole and subtract it by the total energy of the ground state.

This approach has the advantage of including the important relaxation effects in the electronic structure. In the case of periodic boundary calculations using pseudopotentials or projector augmented waves (PAW) [22], we have full relaxation effects in the all-electron valence, while the core orbitals in the pseudopotential or PAW are only relaxed in the free-atom calculation used to build the description. Thus, the major effects are included in this approach, but typically the core hole is spin-averaged in the PAW or pseudopotential. Thus, if there is an exchange interaction with an open-shell valence and the remaining core electron, then this interaction may be underestimated.

When we think about a normal XPS experimental setup with a given adsorbate occupying different sites of the sample, there is only one initial state, which is the state of the system before being probed. Therefore, the binding energy will change according to the different final states that are created in the system, as illustrated in Figure 6.4.

The chemical shifts will be produced by the differences in chemical bonding experienced by the core-ionized element due to the different chemical environments where this element is present. Therefore, thinking in terms of comparing to the reference state, if the chemical environment provides additional stabilization of the Z+1 state, there is a lowering of the measured binding energy. Contrarily, if the final state is destabilized, the binding energy is increased. Since the creation of the core hole produces a positive net charge in the ionized atom, a chemical environment rich in

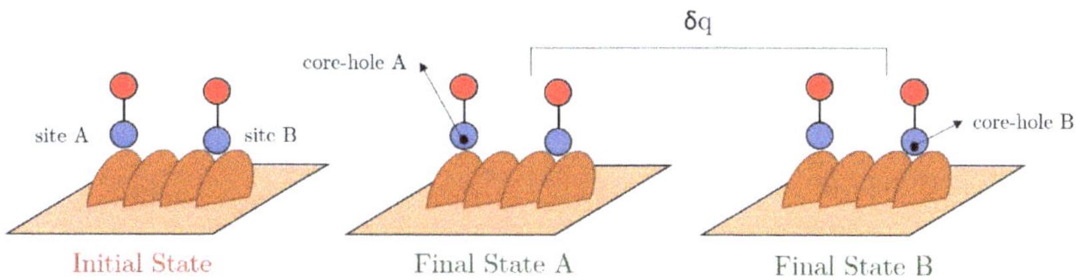

Fig. 6.4. Representation of a surface system with two identical molecules (composed of one red and one blue atom) adsorbed at different sites. The δq quantity represents the chemical shift produced by the different chemical environments to sites A and B experienced by the respective final state.

electron density will transfer charge and stabilize the final state more easily. Inversely, when the chemical environment withdraws the electronic density of an atom, the final state should be destabilized. So, if we consider only inductive effects, the BE should increase when an atom binds to other electronegative atoms (electron-density withdrawing environment) and decrease when it binds to atoms with lower or equal electronegativity (electron-density donating/releasing environment). This can be compared to the stabilization of carbocations in organic chemistry, which become more stable if electron-density donating groups lie closer to the positive carbon.

Going back to our CO on Ni example, we can now interpret the increase in C 1s binding energy upon hydrogen coadsorption. As the CO-Ni bond is weakened in CO/H/Ni(100), we should expect that the electron sharing (or orbital overlap) between Ni and CO is reduced and the final state in the C 1s XPS is destabilized compared to CO/Ni(100), increasing the XPS binding energy. Besides the change in the binding energy of on-top/CO, the hydrogen coadsorption also allows the formation of a $p(2\sqrt{2} \times \sqrt{2})$ phase that has CO adsorbed on both top and hollow sites. C 1s experimental binding energies of CO on Ni(100) are shown in Table 6.3. We can see that moving from on-top/CO to hollow/CO results in a significant decrease in the C 1s BE. After core ionization of the carbon in CO, the resulting equivalent core atom is NO, which indeed prefers a higher coordination when chemisorbed on nickel [23]. The final

Table 6.3. Experimental C 1s binding energies taken from Ref. [20].

System	Phase	Site	C 1s BE (eV)
CO/Ni(100)	c(2 × 2)	on-top	285.9
CO/H/Ni(100)	c(2 × 2)	on-top	286.3
	p(2√2 × √2)	on-top	286.2
		hollow	285.1

state is thus more stabilized in the hollow position than on-top such that the XPS peak from carbon appears at lower binding energy.

6.3.2 *CO on Ni(100) and computing chemical shifts*

Continuing with the CO/Ni(100) example, we will describe the computational procedure for XPS binding energies in periodic systems following a ΔKS approach.

One important consideration when computing XPS binding energies in extended systems is the possible interactions of the created core hole with its own images in the system with periodic boundary conditions (PBC). In a typical experiment, the probability of core-hole creation is not high enough to generate a surface with a large density of core-ionized molecules, so it is safe to assume that no significant interactions between the core-ionized molecules should appear in our model. This can be achieved by the extension of the supercell beyond the computationally cost-effective size of typical geometry optimizations. In our experience, when computing XPS binding energies for adsorbates on metals, it is usually safe to have a spatial distance of 8 Å or more between adjacent core holes in all dimensions. Figure 6.5 illustrates an example of supercells that could be used for the optimization and further XPS calculations of a diatomic molecule on top of an FCC(100) metallic surface using a 0.25 monolayer (ML) coverage. In this periodic slab model, we consider that a sufficient vacuum layer is added in the vertical direction (commonly *z*-axis) to avoid the interaction of the surface with its periodic images.

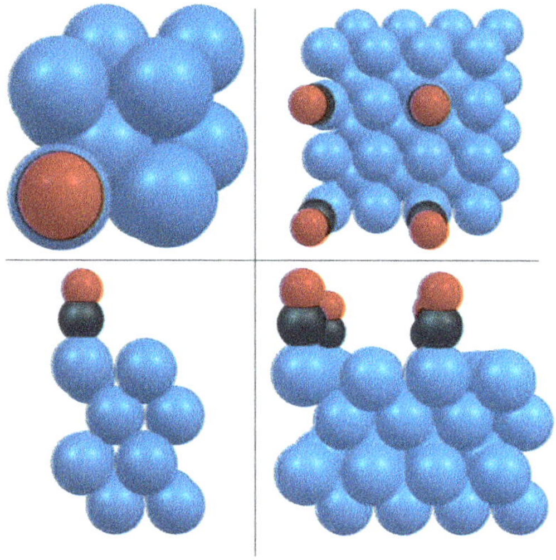

Fig. 6.5. Top and side views of slab models exemplifying PBC calculations of a diatomic molecule adsorbed on the top site of a (100) metallic surface using a 0.25 ML coverage. The smaller 2×2 structure on the left is more efficient for geometry optimizations, while the bigger 4×4 structure on the right is safer to avoid interactions between core holes in core-ionized calculations.

Another important aspect of the core-excited calculation is the preservation of system neutrality upon the removal of a core electron by the addition of an extra electron to the Fermi level. This is important because it accounts for the charge screening, a physical effect promoted by electrons from the conductive surface in the presence of the positive charge from the core ionization. Furthermore, at conductive surfaces, the calculation of the final state for XPS chemical shifts should correspond to the first core-excited state, also making ΔKS binding energies useful to accurately energy-calibrate XAS spectra computed using approximate methods, such as the half-core-hole transition potential approach [24]. Here, we note that, while the XAS process for lighter atoms does not allow spin-flips, there are no such restrictions applicable to the screening of the created core hole. The fully screened XPS state should thus be computed as the lowest accessible state, irrespective of spin or whether it is dipole-allowed.

Summarizing, to do proper ΔSCF or ΔKS calculations of surface adsorbates on metals using PBC systems:

- Calculate the single-point energy of the ground state of your system making sure you have a large enough supercell to avoid interaction between core holes.
- Calculate the single-point energy of the system with a core hole and in the same geometry as the ground state. Make sure the final state corresponds to the first core-excited state and that charge neutrality is preserved. Or, for Z+1 calculations, replace the desired core-ionized atom with the next element in the periodic table.
- Subtract the total electronic energy of the core-excited and ground states to get XPS binding energies.
- Subtract the CEBEs from the reference to obtain the desired chemical shifts. One must observe that if the initial state is the same, that is, we have the same supercell for different core-excited states, the ground state energy of each core-hole system will cancel each other making the ground state calculations unnecessary.

Generally, calculated Z+1 chemical shifts are less affected by the size of the supercell. Nevertheless, one should still avoid possible strong interactions between the replaced Z+1 atoms. An important aspect of the effective core approximation and its recommendation here is that it provides qualitatively reliable estimates for trends in the binding energy shifts. Therefore, Z+1 calculations can be a useful way to identify "ill-behaved" systems where the traditional state-averaged core-hole calculations seem to fail, e.g., when multiplet effects come into play.

Although in some quantum chemistry packages it is possible to get absolute values for DFT core-level binding energies, they are strongly functional-dependent [15] and consequently not reliably comparable to experimental values. On the other hand, the chemical shifts rely on error cancellation upon the subtraction of total energies that usually make them functional-independent. Based on this, an experimental reference is important to calibrate the system and shift the theoretical data accordingly. Regarding the example being used here, we will use the CO/Ni(100)c(2 × 2) phase as our reference system and subtract all experimental and calculated XPS binding energies of CO at other sites on Ni(100) from our reference, generating a series of chemical shifts that can be an extremely useful tool for either confirmation or prediction of new experimental results. A numerical example of the capabilities of calculated DFT chemical shifts for CO on Ni(100) can be seen in Table 6.4, taken from Ref. [25], where the RPBE exchange-correlation functional was used. Accordingly, the experimental comparison is taken from Ref. [20], the same as before.

From Table 6.4, we see that the computed shifts are consistent with experimental data, with deviations smaller than 0.2 eV, even for the larger O 1s chemical shifts, which can be as high as 1.9 eV. We can easily understand the trends in the shifts by considering the final-state equivalent-core molecules, i.e., NO for C 1*s* ionization and CF for O 1*s*. Both these molecules prefer a higher coordination when chemisorbed on Ni, with a driving force that is larger for CF than for NO as also indicated by the significantly larger shift. The larger driving force for CF has been discussed by Nilsson [7] in terms of the large difference in electronegativity resulting in a more unsaturated, atomic-like carbon; atomic carbon strongly prefers adsorption in hollow sites.

Table 6.4. Calculated and experimental C 1s and O 1s chemical shifts for different phases of CO on Ni(100) taken from Ref. [25]. Z+1 values in parenthesis.

	CO/Ni(100)		CO/H/Ni(100)	
	c(2 × 2) on-top	p(3√2 × √2) bridge	p(2√2 × √2) on-top	p(2√2 × √2) hollow
C 1s (exp.)	0	−0.4	0.3	−0.8
C 1s (calc.)	0	−0.47(−0.36)	0.19(0.13)	−0.60(−0.28)
O 1s (exp.)	0	−0.9	0.6	−1.9
O 1s (calc.)	0	−1.10(−1.09)	0.43(0.41)	−1.86(−1.83)

As we see in the on-top cases, the inclusion of hydrogen brings a considerable change in XPS energies by weakening the chemisorption bond in the initial state. The Z+1 approximation shows excellent performance compared to the CH approach for O 1s BE shifts with only marginal differences. However, considering the C 1s BE shifts, the Z+1 describes poorly the hollow and bridge chemical shifts while it is close to the CH description for the on-top site.

Due to the difference in electronegativity, the 1π orbital is polarized towards the oxygen in CO, and, as a consequence, the π^* is instead polarized towards carbon. Since the π^* is a valence orbital, this results in a significant splitting, 1.455 eV [26], between the singlet and triplet C 1*s* to π^* excitations. This splitting is much smaller when excited from oxygen.

We can estimate the effect of this exchange interaction for CO chemisorbed at the different sites by actually computing the core-excited state as either singlet or triplet. However, this requires that the 1*s* core electron is present, which precludes the use of a standard core-hole pseudopotential or PAW setup. It also requires some consideration since a standard ΔKS calculation of the first core-excited state as an open-shell singlet gives a 50:50 mixture of the singlet and triplet excitation due to the fact that the Kohn–Sham state corresponds to a single-determinant wave function. However, the triplet state, as any highest-spin state, can be described as a single determinant, which allows obtaining the correct singlet energy as $E(\text{singlet}) = 2E(\text{singlet}') - E(\text{triplet})$, where singlet′ indicates the singlet energy computed as a mixed spin state.

To simply investigate the effect of chemisorption at different sites on the exchange coupling in the screened state, we take a Pd_5 pyramidal cluster that can represent both on-top and hollow adsorption of CO. Palladium was chosen rather than Nickel due to difficulties with all-electron calculations on 3d transition metals with multiple open shells. The calculations used the RPBE functional [27] using the StoBe–deMon code [28] which is specially designed for cluster and X-ray spectroscopy calculations. For gas-phase CO, the singlet/triplet splitting was obtained as 1.11 eV in good agreement with the experiment, while for on-top and hollow CO, this was reduced to 0.65 and 0.35 eV, respectively.

The failure to quantitatively describe the site-dependence of the carbon XPS in the case of the Z+1 picture is thus most likely related to that the Z+1 approximation, and its closed-shell core, does not include the exchange interactions between the core hole and the new valence screening electron (in the case of CO added to its π^* orbital). For on-top in both $c(2 \times 2)$ and $p(2\sqrt{2} \times \sqrt{2})$ phases, where coordination is similar, the exchange interaction should also be similar canceling out any of its strong effects, and the Z+1 approximation behaves well. In the other two sites, where the coordination and orbital overlap are very different (compared to the reference on-top site), any difference in the exchange interaction due to the changes in the π^* orbital distribution is not included in a closed-core shell picture, making Z+1 a poorer approximation.

6.4 High-Pressure XPS — Relating Theory to Experiment

Before we move to the final topic, which is to connect directly to experiments on heterogeneous catalysis under high-pressure conditions, we will briefly discuss the specifics of the measurements. Near-ambient pressure XPS (NAP-XPS) is a technique that can probe the surface in an atom-sensitive way, but *operando* measurements present challenges. The involved challenges have been termed the *pressure gap*, the *materials gap*, and the *complexity gap*. These also represent challenges to the theoretical modeling. Using hard X-rays and the Scienta Omicron R4000 HiPP-2 electron spectrometer has allowed atmospheric pressure to be reached [29,30]. This allows closing the so-called *pressure gap* that previously described the inability of the technique to reach industrial process pressures. The basic principles and the applications of the NAP-XPS method using *soft* and *hard X-rays* are extensively discussed in Chapter 10, however for the sake of consistency, some relevant information is provided in the following.

Catalyst materials have been the subject of extensive experimental investigation and have emerged as highly sophisticated but complex systems with mixtures of multiple components and nanoparticles being finely dispersed on metal oxide supports. Their properties have been optimized for high turnover rates, high selectivity as well as enhanced lifetime, being in balance with industrial profitability. The use of NAP-XPS in combination with such materials presents a real challenge termed the *materials gap* and is often related to the charging effect that occurs in the XPS process due to the non-conducting nature of the metal oxide support.

The *complexity gap* relates to the transient behavior of catalyst materials, and time-resolved studies are necessary to investigate the dynamics of the reaction. Pump-probe experiments performed at free-electron lasers (FEL) allow us to investigate the fs to ps time dynamics and give unique insights into the bond-breaking and bond-formation during the catalytic reaction. Out-of-equilibrium dynamics, e.g., the important initial step to activate a raw catalyst, reach timescales from subseconds to hours. To be able to experimentally access this and determine what changes occur, short integration times with efficient electron detection are necessary.

One of the biggest challenges from a technology point of view is related to the short inelastic mean free path (IMFP) of photoelectrons inside the gas phase. This is strongly dependent on pressure and reduces from 1.25 mm at 0.85 mbar to 0.21 mm at 5 mbar for N_2 gas and electron kinetic energy of 1400 eV [31]. At a pressure of several hundred mbars, this is expected to be in the 10 s of the micrometer regime. This limits the gas layer to a thickness of the order of the IMFP. Researchers at Stockholm University have developed the virtual cell approach to reach the bar pressure regime [30]. In this approach, a gas channel system is integrated into the spectrometer front cone that allows directing the gas onto the sample surface and creating a confined high-pressure zone. With this, a pressure of several bars can be achieved on the sample, while the chamber is still in the mbar regime. Using grazing incidence hard X-rays and a small sample-to-aperture distance of 20–30 μm, photoelectrons can be detected with high efficiency allowing for routine measurements in the hundreds of mbar pressure. By taking longer integration times into account, exploring into the bar pressure regime is also possible.

Figure 6.6(a) shows a schematic representation of the virtual cell being located at the apex of the front cone. A ca. 70 μm thick

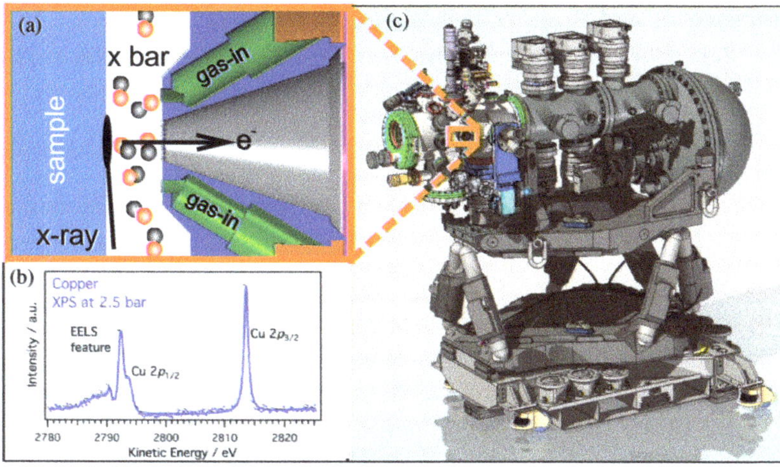

Fig. 6.6. APXPS system dedicated for high pressures utilizing hard X-ray, grazing incidence photoemission spectroscopy. (a) Schematic of the virtual cell approach being located inside the main chamber of (b). (c) Cu 2p XP spectra obtained at a pressure of 2.5 bar of He.

perforated Ti foil separates the high-pressure zone from the differential pumping of the spectrometer. The instrument as depicted in Figure 6.6(c) builds around the Scienta Omicron Hipp 2 analyzer being situated on a high-precision customized Newport Hexapod that allows for precise alignment of the instrument with respect to the incoming X-ray beam. The high-pressure capabilities of the instrument were initially tested on a Cu single crystal under 2.5 bar of He pressure. The results of these experiments are shown in Figure 6.6(b) and clearly show the characteristic electron energy loss (EELS) features that arise from electron scattering inside the gas phase in addition to the Cu $2p_{1/2}$ and $2p_{3/2}$ peaks.

To keep the angular acceptance of the electrons high, the spectrometer is equipped with a lens inside the differential pumping section. The lens refocuses the electrons, and nodal planes are created that allow placing further apertures for the differential pumping. Ray tracing calculations allow for optimized electron trajectories with high transmission and high energy resolution [32] and have resulted in a very efficient electron detection. The use of APXPS in combination with hard X-rays turns out to be valuable to study the gas–solid interface under high pressures. The field is entering into pressure regimes that mimic industrial processes more closely, and further progress is expected in the future to overcome the pressure gap.

Here, we will present two applications that illustrate the capability of this experimental development in a well-defined reaction that is the oxidation of CO with O_2 on Pd(100) and in a much more complex reaction that is the hydrogenation of CO with H_2 with rhodium as catalyst comparing two different facets.

Blomberg and coworkers [33] have studied the CO oxidation reaction on Pd(100) under reaction conditions. Figure 6.7 shows experimental data of the C 1*s* region gathered using a mixture of CO and O_2 gas at a ratio of 1:1 with 6% CO and O_2 in He. The total flow was adjusted to reach the indicated pressure for the reaction,

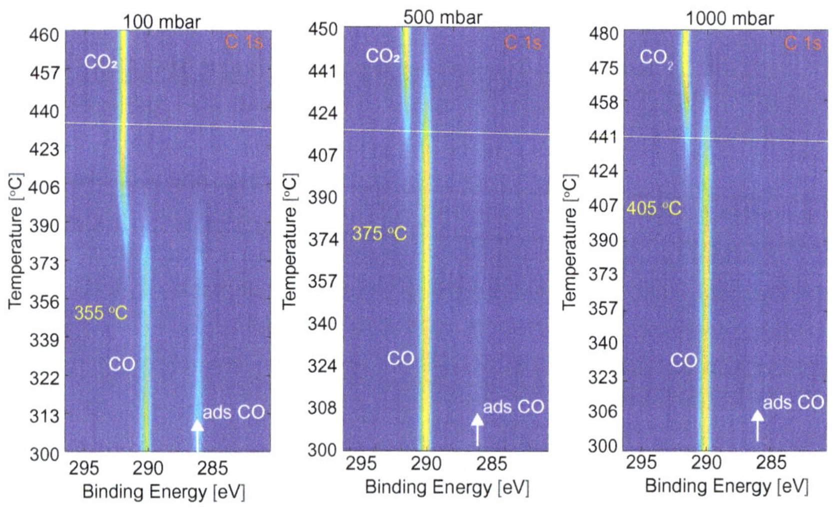

Fig. 6.7. XP spectra of the C 1s region studied at different total pressures of 100, 500, and 1000 mbar. A 1:1 gas mixture of CO:O_2 in He was supplied to the surface. The temperature was ramped at 6°C/min and the regime around the light-off temperature is captured. High spectral intensity corresponds to yellow and low intensity to blue colors in the image. Figure courtesy of S. Blomberg (Lund University).

being 100, 500, and 1000 mbar, respectively. The low relative contribution of adsorbates to the spectral intensity makes such experiments challenging.

Hard X-rays of 4600 eV were used to increase the IMFP. Grazing incidence photons at 0.6° allow for increased surface sensitivity. The temperature was increased continuously at a rate of 6°C/min and three core levels, Pd $3d_{5/2}$, O 1*s*, and C 1*s* spectra, were collected sequentially. This corresponds to a total acquisition time of ca. 30 minutes for three core levels and one temperature sweep, highlighting the performance of this instrument. In C 1*s*, adsorbed CO appears at a binding energy of 286 eV and the gas-phase peaks of CO and CO_2 are visible between 290–293 eV. At low temperature, the Pd surface is CO poisoned and thus inactive. However, with increasing temperature, the formation of CO_2 can be observed spectroscopically.

With higher temperature, the CO_2 signal gradually increases, and a large overlapping regime is recognized where CO gas and CO_2 gas are observed simultaneously. The observation of an overlapping regime is unusual for APXPS and was related to the specific geometry of the virtual cell. CO and O_2 are guided into the cell volume at a high gas flow and close distance to the surface. Therefore, the reaction can be followed in the kinetic regime and is not mass-transfer limited, which is believed to be important for future experimental studies. With increasing pressure, the light-off temperature increases from 355°C at 100 mbar to 405°C at 1000 mbar as indicated in the figure, showing the usage of APXPS for exploring higher pressure regimes.

In Figure 6.8, we move to the real challenge of a theoretical assignment of contributions to the measured XPS signal. This is the C 1s region measured under ongoing hydrogenation of CO with H_2 at 150 mbar pressure and 225°C, comparing Rh(111) and Rh(211) as catalysts. A 1:2 ratio between CO and H_2 was used in the gas stream and the measurements used 4600 eV photon energy at 0.4° grazing incidence. At the (111) surface, two maxima are clearly visible where the lower-intensity peak at 285.8 eV is assigned to chemisorbed CO and the other at 284 eV to various carbon species at the

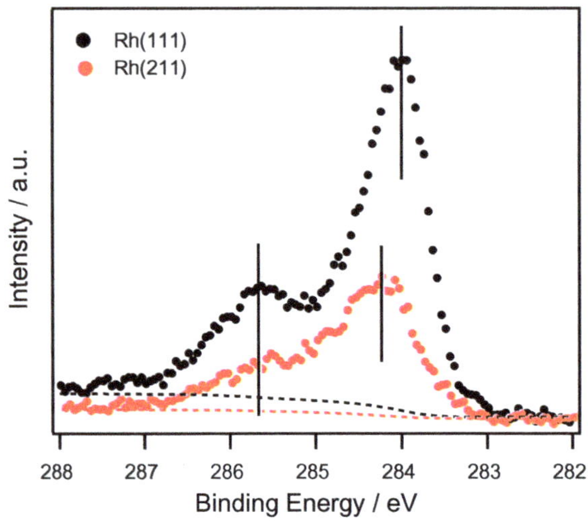

Fig. 6.8. Comparison of XP spectra accumulated on Rh (111) and Rh (211) under *operando* conditions at 150 mbar and 225°C. Figure courtesy of David Degerman (Stockholm University).

surface, mainly atomic carbon and CH. On the more reactive, stepped (211) surface, the coverage (total signal) is lower and the second peak shifts to slightly higher energy (284.3 eV), which indicates a larger contribution from more hydrogenated species. This is consistent with the expectation that both CO dissociation and hydrogenation of carbon will be more efficient on (211) than on (111).

The spectrum indicates some of the difficulties associated with high-pressure XPS. The first is that the spectra become broader and associated with more noise than in conventional XPS measurements. The high-pressure gas phase furthermore contributes to the background through scattering. With the high-energy X-rays that are used to generate electrons of sufficient energy to reach the analyzer, there is also the recoil effect to take into account when analyzing the XPS data. In addition, hydrogenation does not lead to large shifts in peak positions, and various intermediates, such as COH, HCO, and CO at different sites, are all computed to contribute in the range 284.5–285.5 eV. To further compound the situation when attempting to disentangle the contributions, the XPS binding energy of a given species may depend on stabilizing or destabilizing interactions with other adsorbates. It will thus be necessary to develop an idea of what the coverage of various species should be during the reaction and determine computationally how that affects the XPS binding energies. This will be the topic of the next section.

6.5 Relating to Experiment — Microkinetic Modeling and Genetic Algorithms

At normal XPS conditions with low pressures and well-characterized surfaces, and simple reactions with a limited number of intermediates, core-level shifts can alone be used to understand reaction mechanisms successfully. However, when the number of intermediates is so large that it is no longer possible to distinguish single adsorbates, and when the surface is less well-characterized, as in the CO hydrogenation study in Figure 6.8, more atomistic data are needed.

Here, an approach combining microkinetic modeling (MKM) with machine learning can give more input. The first step in this process is DFT calculations of the full reaction landscape with elementary reaction energies and energy barriers. Using transition-state theory, the MKM can be constructed based on the rates over the barriers (forward and backward), which can be expressed as coupled differential equations, from adsorption of the reactant(s) to desorption of the product(s). By solving these equations at steady-state conditions, the MKM provides coverages of each intermediate and reaction rates to the products, which can be compared with the XPS data. The more element-specific experimental XPS data that are available and the more temperatures and pressures that have been included, the more constraining the comparison will be, but also total elemental coverages can be integrated and compared if deconvolution is not possible.

In order to know what is happening in the experiment, we would like to reproduce the experimental coverages and/or rates.

However, chemisorption energies and barriers may be affected by coadsorbates at high coverage. The surface may reconstruct and new sites appear, and furthermore, unexpected side reactions may come into play. For a complex reaction, it is thus not obvious how the input to the MKM and its underlying assumptions should be changed to achieve that; we can include more pathways, consider surface kinks, step edges, and other defects, and reinvestigate values with more accurate calculation methods, but it is hard to manually pinpoint the issues.

Instead, machine learning in the form of a genetic algorithm (GA) can be used as a search method. The GA is a way to optimize a many-dimensional problem by natural selection. How the GA can be applied to the MKM is shown in Figure 6.9. The program starts by initializing a number of individuals (typically hundreds for enough diversity). In our case, an individual is a full set of elementary reaction rates and barriers, randomized in an interval around the DFT-calculated energies. Each individual is evaluated by

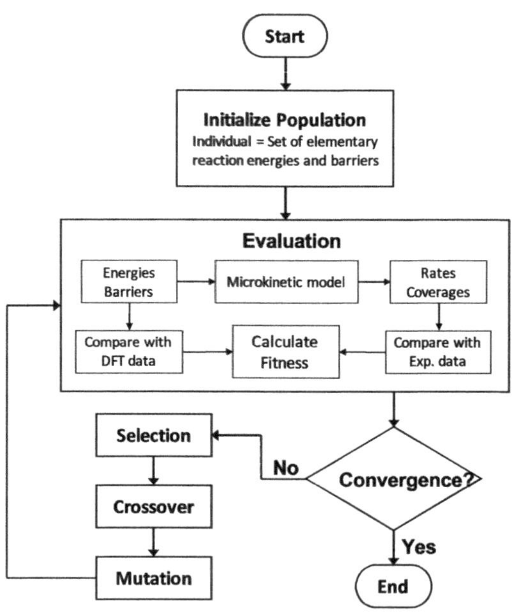

Fig. 6.9. Genetic algorithm applied to microkinetic modeling.

calculating a fitness in the form of a single number. The fittest individuals are selected to have children, inheriting parts of their parents' properties (cross-over). The children are subject to random mutations (energies are changed), and the new population is evaluated again. This continues until a given stopping condition is met, such as a preset good-enough fitness, no improvement in a preset number of iterations or simply a maximum number of iterations.

The choice of fitness function is paramount for this scheme to work. Obviously, deviation from experimental data must give penalty. As an example with experimental total coverages θ_{exp} of carbon and oxygen, we can calculate

$$b = \sum_{T,p;C,O} \left| \log\theta_{exp} - \log\theta_{calc} \right|. \tag{6.1}$$

This assumes that the measurements have been performed at several temperatures (T) and pressures (p) and gives a measure of the overall deviation of predicted coverages from the experimental coverages. However, if experimental comparison is the only fitness criterion, there is very little control of individual energies; a vast number of mathematically optimal solutions can be generated that are completely unphysical. In particular, from a chemical perspective, a reaction path typically has a small number of rate-determining steps, while the other energies may be altered without changing the outcome. Thus, it is necessary to penalize deviation from the DFT energies as well to find the most significant parameters. Such a penalty may be calculated as

$$p = \sum_{i}^{N} \left| E_{i,\mathrm{GA}} - E_{i,\mathrm{DFT}} \right|, \tag{6.2}$$

where N is the number of independent energies. The fitness number f can be calculated as a weighted sum of b and p.

As a part of analyzing the result of the GA, energies that the GA finds to be critical and need large changes may be investigated with a higher-level technique, such as hybrid functionals. If, e.g., a barrier is indicated to be too high, but the original value is verified through higher-level calculations, then a different mechanism, such as a path with an edge-mediated transition state, may be required. If the corresponding energies are recalculated, the updated values are inserted in the DFT-energy reference and the GA can be run again.

6.5.1 *Example: CO_2 hydrogenation on Rh(111)*

The MKM-GA method was developed with CO_2 hydrogenation on Rh(111) as a test reaction. We used the CatMAP package [34] for the MKM and based the GA on implementation in Python by Ryan Mohammad Solgi [35]. The experimental coverages used in this section were preliminary and later turned out to have some inaccuracies, so conclusions on specific results should not be drawn. The point is to show the workflow given experimental data.

In Figure 6.10, we see the deviation between GA and DFT energies. The energies are split into three categories: adsorbates are blue, transition states are green, and gas-phase molecules are red. H_2(g), H_2O(g), and CO(g) are absent as they are chosen as references with zero energy by definition. We can see a positive deviation for the atomic oxygen chemisorption energy, suggesting that the GA would prefer a lower oxygen chemisorption energy for an optimal solution. Furthermore, the GA gives a lower energy barrier for CO_2 dissociation. On the computational side, the conclusion is to investigate these two states.

How well did the GA reproduce the experimental coverages? In Figure 6.11, we show the logarithmic difference between GA and experimental integrated carbon and oxygen coverages as a function of temperature. The calculated difference in the order of magnitude (b in Eq. (6.1)) is 0.24, which is relatively low for experimental-computational comparison, but the deviation is very uneven as can be seen in the figure. This causes some concern and raises the question if the fitness function should be modified to penalize this behavior. The exact form of the fitness function will probably need to be tailored to the specific application going

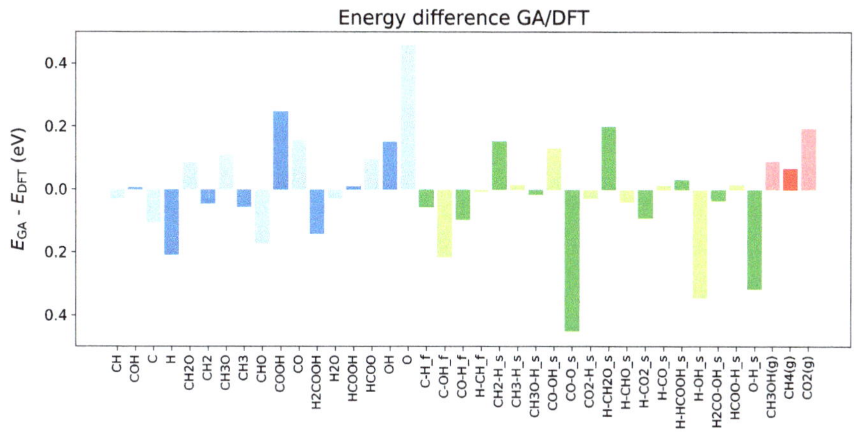

Fig. 6.10. Deviation between DFT energies and GA-generated energies. Adsorbates are shown in blue, transition states in green, and gas-phase molecules in red.

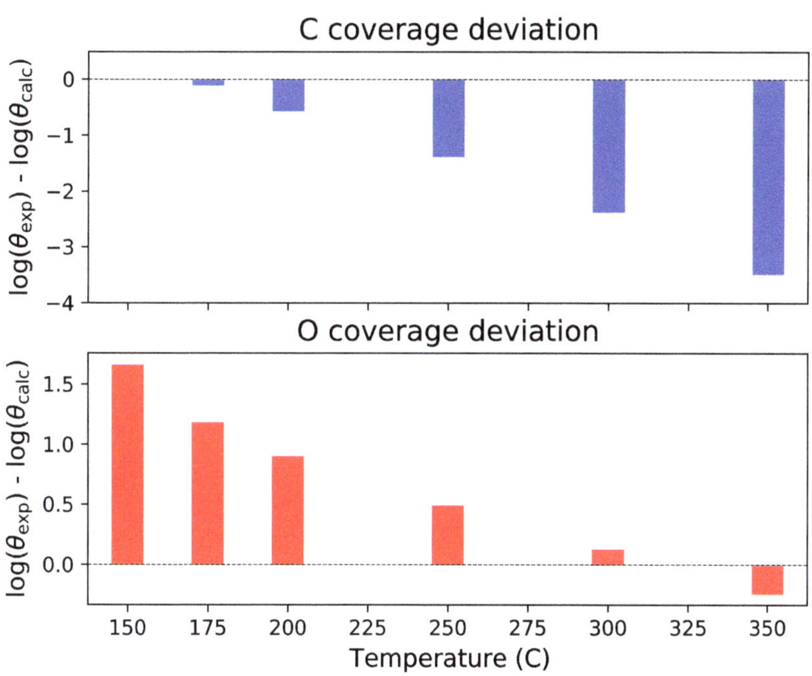

Fig. 6.11. Deviation between experimental coverages and coverages from microkinetic model with GA-generated energies. Total C coverage is shown in blue and total O coverage in red.

forward. We also note that there are uncertainties in the extracted experimental coverages due to the difficulties mentioned in Section 6.4.

As a further check on the derived coverages, an XPS spectrum can be generated based on the computed binding energies weighted by the proposed coverages. This can then be directly compared to the actual measured data. Clearly, this will be an iterative process, introducing more and more complexity in terms of side reactions, unexpected active sites, as well as adsorbate–adsorbate interactions that modify chemisorption energies and barriers, as well as the XPS binding energies. Going down this long and winding path will be a fascinating and potentially very rewarding journey.

6.6 Concluding Remarks

In this chapter, we have focused on a chemically intuitive interpretation of XPS in terms of the Z+1 or equivalent core picture [3–5] where the shifts in XPS binding energies correspond to differences in chemical adsorption energy in the final state. Replacing the carbon with nitrogen at each of the four positions in the ESCA molecule (ethyltrifluoroacetate) and taking the energy differences give a quantitative representation of the XPS chemical shifts, of similar quality as by introducing the explicit core hole at the respective positions. However, there are cases, as exemplified by C 1*s* XPS of CO at different sites on Ni(100), where the exchange interaction or multiplet effects make it necessary to consider the explicit core hole for quantitative accuracy. However, the Z+1 picture is an intuitively useful tool to understand trends and effects, giving a general direction of shifts with respect to a given reference and, in specific cases, identify when to expect vibrational excitations to affect the peak profile and width or when a state-averaged picture of the core hole is not a good enough representation of the different final states. Summarizing, although XPS calculations of multiple species, as in the case of building a BE database, may seem rather "black box", it is always important to understand the chemical nature of the system to identify possible caveats. This is especially true when dealing with binding energy calculations of heavier atoms, such as metals, where multiplet effects can play a role. But even in such cases, the Z+1 approximation can be a useful way to start.

Recent developments in XPS instrumentation now enable *operando* studies at pressures reaching ~1 bar, which allows investigating reactions in heterogeneous catalysis under more realistic conditions in terms of both pressure and temperature. This opens for different sites to be populated, as well as reactions to occur leading to various intermediates and products. At elevated temperatures, the adsorbates also become increasingly mobile and probe different non-equilibrium sites which further complicates the interpretation of XPS at higher pressures. Here, libraries of computed XPS binding energies considering probable species and sites at the surface provide invaluable assistance in the interpretation of the experimental data.

In combination with microkinetic modeling of the full reaction sequence that is investigated, further bounds on the interpretation of the data can be obtained. Since the microkinetic modeling relies on assumptions on the catalytic sites, the accuracy of computed energetics, as well as the reaction pathway, we explore combining the modeling with a genetic algorithm (GA) targeted with reproducing the experimental data. With this, we determine which computed energetics (adsorption energies and barriers) need to be investigated further, e.g., that the GA prefers to either increase or decrease significantly. They can be computed at a higher level and verified or corrected. If verified, the discrepancy may be due to adsorbate–adsorbate interactions that can also be computed and included in the microkinetic modeling. If this is insufficient to reach an agreement with the experimental data, it is an indication that the surface under reaction conditions may be different from what is assumed or that unexpected side reactions become important.

Having the predicted coverage of each species at the surface under the conditions of the experiment and their computed XPS positions, it will be possible to generate computed XPS spectra for the reacting mixture to compare directly with the experiment. However, this will require significant computational effort but will be extremely rewarding as the combination of true high pressure *operando* XPS and reliable modeling can provide significant new insights into industrially critical processes.

Acknowledgments

We are grateful for support from the Swedish Foundation for Strategic Research (SSF) through grant number ITM 17-0034 and from the Knut & Alice Wallenberg (KAW, grant nr. 2016.0042) foundation. We thank David Degerman, Stockholm University, for preparing Figure 6.8 and Sara Blomberg, Lund University, for providing Figure 6.7. The Swedish National Infrastructure for Computing (SNIC) provided computational resources at the PDC and NSC centers, which are partially funded by the Swedish Research Council through grant agreement no. 2016-07213. The spectra in Figures 6.7 and 6.8 were obtained at the P22 beamline of Petra III at DESY (Hamburg, Germany).

References

[1] Somorjai GA. The flexible surface. Correlation between reactivity and restructuring ability. Langmuir. 1991;7:3176–3182.

[2] Öberg H. *et al.* Stability of Pt-modified Cu(111) in the presence of oxygen and its implication on the overall electronic structure. J Phys Chem C. 2013;117:16371–16380.

[3] Mårtensson N, Nilsson A. On the origin of core level binding energy shifts. J El Spec Rel Phenom. 1995;75:209.

[4] Johansson B, Mårtensson N. Core-level binding-energy shifts for the metallic elements. Phys Rev B. 1980;21:4427–4457.

[5] Delesma FA *et al.* A chemical view on X-ray photoelectron spectroscopy: The ESCA molecule and surface-to-bulk XPS shifts. ChemPhysChem. 2018;19:169–174.

[6] Egelhoff WF. N_2 on Ni(100): Angular dependence of the N_{1s} XPS peaks. Surf Sci. 1984;141:L324.

[7] Nilsson A. Application of core level spectroscopy to adsorbates. J El Spec Rel Phenom. 2002;126:3.

[8] Siegbahn K *et al. ESCA, Atomic, Molecular and Solid State Structure Studied by Means of Electron Spectroscopy*. Uppsala: Almqvist and Wiksell, 1967, p. 283.

[9] Travnikova O *et al.* The ESCA molecule — Historical remarks and new results. J El Spec Rel Phenom. 2012;185:191–197.

[10] Van den Bossche M *et al.* Effects of non-local exchange on core level shifts for gas-phase and adsorbed molecules. J Chem Phys. 2014;141:034706.

[11] Aldén M, Skriver HL, Johansson B. Ab initio surface core-level shifts and surface segregation energies. Phys Rev Lett. 1993;71: 2449–2452.

[12] Rosengren A, Johansson B. Calculated transition-metal surface core-level binding-energy shifts. Phys Rev B. 1980;22:3706–3709.

[13] Bagus PS, Sousa C, Illas F. Limitations of the equivalent core model for understanding core-level spectroscopies. Phys Chem Chem Phys. 2020;22:22617–22626.

[14] Huber KP, Herzberg G. *Constants of Diatomic Molecules. Molecular Spectra and Molecular Structure* (Vol. IV). New York: Van Nostrand Reinhold, 1979.

[15] Takahashi O, Pettersson LGM. Functional dependence of core-excitation energies. J Chem Phys. 2004;121(21):10339–10345.

[16] Triguero L *et al.* Separate state versus transition state Kohn-Sham calculations of X-ray photoelectron binding energies and chemical shifts. J El Spec Rel Phen. 1999;104:195.

[17] Leetmaa M *et al.* Theoretical approximations to X-ray absorption spectroscopy of liquid water and ice. J El Spec Rel Phen. 2010; 177:135–157.

[18] Koel BE, Peebles DE, White JM. Low temperature coadsorption of hydrogen and carbon monoxide on Ni(100). II. ELS and XPS studies. Surf Sci. 1983;125:739–761.

[19] Westerlund L, Jönsson L, Andersson S. Hydrogen recombination and Σ-desorption from the Ni(100)/H/CO coadsorption system. Surf Sci. 1988;199:109–120.

[20] Antonsson H *et al.* Vibrational motion and geometrical structure in adsorbed CO studied by core level photoelectron spectroscopy. J El Spec Rel Phenom. 1990;54/55:601–613.

[21] Tillborg H *et al.* Adsorption-site-dependent X-ray-absorption spectroscopy: CO/H,H_2/Ni(100). Phys Rev B. 1993;47:1699–1702.

[22] Blöchl PE. Projector augmented-wave method. Phys Rev B. 1994;50:17953.

[23] Brown WA, King DA. NO chemisorption and reactions on metal surfaces: A new perspective. J Phys Chem B. 2000;104:2578–2595.

[24] Triguero L, Pettersson LGM, Ågren H. Calculations of near-edge X-ray absorption spectra of gas phase and chemisorbed molecules by means of density functional and transition potential theory. Phys Rev B. 1998;58:8097.

[25] Diesen E *et al.* Accuracy of XAS theory for unraveling structural changes of adsorbates: CO on Ni(100). AIP Adv. 2020;10:115014.

[26] Shaw DA *et al.* Electron impact excitation of inner-shell excited states of CO. J Phys B. 1984;17:2091–2100.

[27] Hammer B, Hansen LB, Nørskov JK. Improved adsorption energetics within density-functional theory using revised Perdew-Burke-Ernzerhof functionals. Phys Rev B. 1999;59:7413–7421.

[28] Hermann K *et al. StoBe-deMon.* Stockholm-Berlin: deMon Software, 2014.

[29] Takagi Y *et al.* X-ray photoelectron spectroscopy under real ambient pressure conditions. Appl Phys Express. 2017;10:8–11.

[30] Amann P *et al.* A high-pressure X-ray photoelectron spectroscopy instrument for studies of industrially relevant catalytic reactions at pressures of several bars. Rev Sci Instrum. 2019;90:103102.

[31] Tougaard S, Greiner M. Method to correct ambient pressure XPS for the distortion caused by the gas. Appl Surf Sci. 2020;530:147243.

[32] Wannberg B. Electron optics development for photo-electron spectrometers. Nucl Instrum Methods Phys Res A. 2009;601:182–194.

[33] Blomberg S *et al.* Bridging the pressure gap in CO oxidation. ACS Catal. 2021;11:9128–9135.

[34] Medford AJ *et al.* CatMAP: A software package for descriptor-based microkinetic mapping of catalytic trends. Catal Lett. 2015;145: 794–807.

[35] Solgi RM. Geneticalgorithm 1.0.2. [cited 2020 Mar 19]. Available from: https://pypi.org/project/geneticalgorithm/.

https://doi.org/10.1142/9781800613294_0007

Chapter 7

Application of XPS in Studies of Model Catalysts: From Single Crystals to Supported Nanoparticles

Günther Rupprechter

Institute of Materials Chemistry, Technische Universität Wien, Vienna 1060, Austria

guenther.rupprechter@tuwien.ac.at

Abstract

This chapter describes X-ray photoelectron spectroscopy (XPS) studies of model catalysts, covering both ultrahigh vacuum (UHV) and near-ambient pressure (NAP) conditions, as well as *operando* characterization. A variety of UHV-grown model catalysts are presented, including single-crystal surfaces, surface alloys, thin oxide films, and supported nanoparticles. In addition to area-averaging measurements, a brief account of spatially resolved XPS by scanning photoelectron microscopy (SPEM) is given. The XPS model studies are typically paralleled by complementary experimental and computational methods, aiding in the full understanding of the XPS characterization.

7.1 Introduction: XPS in Model Catalysis

X-ray photoelectron spectroscopy (XPS) is one of the most powerful surface characterization methods in materials science, which was recognized by the Nobel Prize awarded to Kai Siegbahn in 1981 [1]. XPS also has a prominent role in heterogeneous catalysis research, as it is able to reveal the elemental composition and oxidation state of solid catalysts [2–6]. Traditionally, XPS was used to characterize catalysts before and after reaction to identify not only compositional changes but also reasons for possible deactivation, such as carbon coking or sulphur poisoning. Gas/reactant adsorption at cryogenic temperature and temperature-programmed reactions were typically studied under ultrahigh vacuum (UHV). Since its early beginnings, XPS was scarcely carried out under non-UHV conditions by employing strong differential pumping [2,7–9]. The number of such high-pressure or, more recently termed, (near-) ambient (or atmospheric) pressure (N)AP-XPS studies has particularly increased in the last two decades, as combining focusing electrostatic lenses and synchrotron radiation enabled much better signal-to-noise at reasonable acquisition times [10,11]. As (more) realistic gas environments become accessible, NAP-XPS allows us to characterize the active catalyst state and, in favorable cases, adsorbed reactant/product molecules or even intermediates. Clearly, improvements in sensitivity, energy resolution, and time resolution are beneficial for elucidating mechanistic steps of catalytic surface reactions [12].

In this chapter, the focus is on XPS studies of *model catalysts*, both under UHV and NAP conditions [13–22] including *operando* characterization [23]. A variety of UHV-grown model catalysts will be discussed, including not only single-crystal surfaces, surface alloys, thin oxide films, and supported nanoparticles but also "high surface area model catalysts", i.e., well-defined nanoparticles supported on commercial oxides [22,23] (for technological catalysts, refer to Chapter 10). The XPS model studies are typically paralleled by complementary experimental and computational methods (see examples in Chapter 6), aiding in the full understanding of the XPS results.

7.2 XPS Basics and Operation Modes

XPS analysis is based on the photoelectric effect. The catalyst surface is illuminated by an X-ray beam $h\nu$ that creates photoelectrons, and by measuring their kinetic energy E_{kin}, their binding energy BE can be calculated (Figure 7.1). This yields information not only on the catalyst elemental composition (core levels) and valence band structure but also on adsorbed species and, if present, the surrounding gas phase. For laboratory XPS, dual-anode X-ray guns are most frequently used (Mg Kα = 1253.6 eV; Al Kα = 1486.6 eV; linewidth ~1 eV), and for the typical catalyst materials, photoelectrons with kinetic energy in the range of 250–1400 eV are obtained. The corresponding inelastic mean free path (IMFP) of the photoelectrons is a few nm, which makes XPS inherently surface-sensitive, which can be further increased by angle-dependent measurements. In Figure 7.1, the take-off angle is zero degrees from the surface normal, but when the sample is tilted for a take-off angle of, e.g., 60°, the surface sensitivity is even higher. Using synchrotron radiation of high *brilliance* (flux, beam divergence, and linewidth) not only increases the signal/noise ratio but also enables the reduction of the X-ray photon energy and thus the photoelectron kinetic energy and the resulting IMFP, which maximizes the surface sensitivity. Vice versa, varying the photon energy from low to high enables depth-profiling of the near-surface region (important for subsurface species [24], layered [25], or core–shell structures). Furthermore, the tunability of synchrotron radiation also enables resonantly enhanced photoemission, which can strongly increase sensitivity. Standard hemispherical electron energy analyzers have a range of up to 3500 eV with a resolution of ~0.25 eV, but using monochromated X-ray guns or synchrotron excitation calls for using analyzers with a resolution below 0.1 eV.

For catalytic studies, the near-ambient (atmospheric) pressure mode (NAP-XPS) is most useful, as it can be applied during ongoing processes or reactions in the few mbar pressure ranges, enabling us to monitor changes in surface composition (e.g., oxidation, reduction, segregation, alloying, coking, adsorption, and desorption).

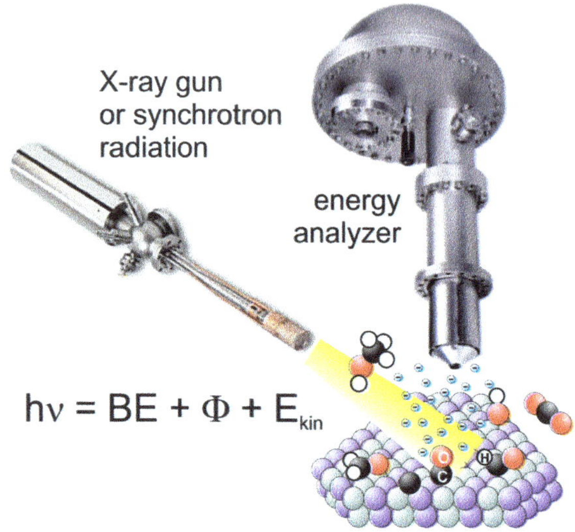

Fig. 7.1. Illustration of X-ray photoelectron spectroscopy (XPS) characterization of a catalyst surface and reactive species.

Early NAP-XPS setups were lab-based [2,7–9], but around/after the year 2000, the development of differentially pumped electrostatic lenses and the combination with synchrotron radiation made NAP-XPS "take-off". The number of NAP-XPS studies strongly increased since then [4–6,10,11,26–31]. Currently, operating pressures are mostly limited to tens of mbar, but there are ongoing efforts to extend the pressure range. Further operation modes include micro-spectroscopy and XPS microscopy [32–34] (Chapter 11) and time-resolved (Chapter 9) or even ultrafast spectroscopy [12].

7.3 Model Catalysts and Experimental Setups

In the following, the most frequently examined model catalysts are discussed (Figure 7.2). Nobel metal single crystals have been used since the early days of "surface science model catalysis", as they provide nearly perfect test structures [35–38]. The clean and well-ordered surfaces can be thoroughly characterized by low energy electron diffraction (LEED), Auger electron spectroscopy (AES),

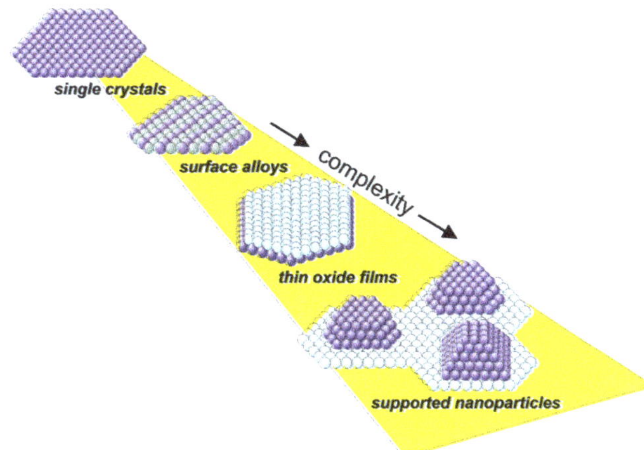

Fig. 7.2. Illustration of model catalysts most frequently used for fundamental studies: smooth and stepped noble metal single crystals, bimetallic single crystals and surface alloys, surface oxides and thin oxide films, and supported nanoparticles. Typically, these systems are prepared under ultrahigh vacuum (UHV) to ensure control of catalyst morphology and composition (cleanliness).

and XPS, and the same methods can also be used to investigate molecular interactions. Single crystal studies are often criticized as being "too simple" (material gap), but they still provide very detailed molecular-level insight into surface processes and elementary reaction steps. Bimetallic single crystals or surface alloys provide the next level of complexity. Recently, they have received increased interest in the form of single-atom alloys [39–41]. In this case, scanning tunneling microscopy (STM) is indispensable to confirm the "site isolation". Oxide surfaces, apparently of huge importance in catalysis, are often studied in the form of thin films grown on suitable substrates [42], as the limited conductivity of oxide single crystals may create difficulties for electron-based techniques. Thin oxide films also provide suitable supports for metal deposits [13,14,43,44], together creating nanoparticle model catalysts, ideally characterized by XPS and STM.

The UHV preparation of model catalysts mostly requires the typical surface science methodology of sputtering, annealing, oxidation/reduction, and metal evaporation. Figure 7.3(a) shows

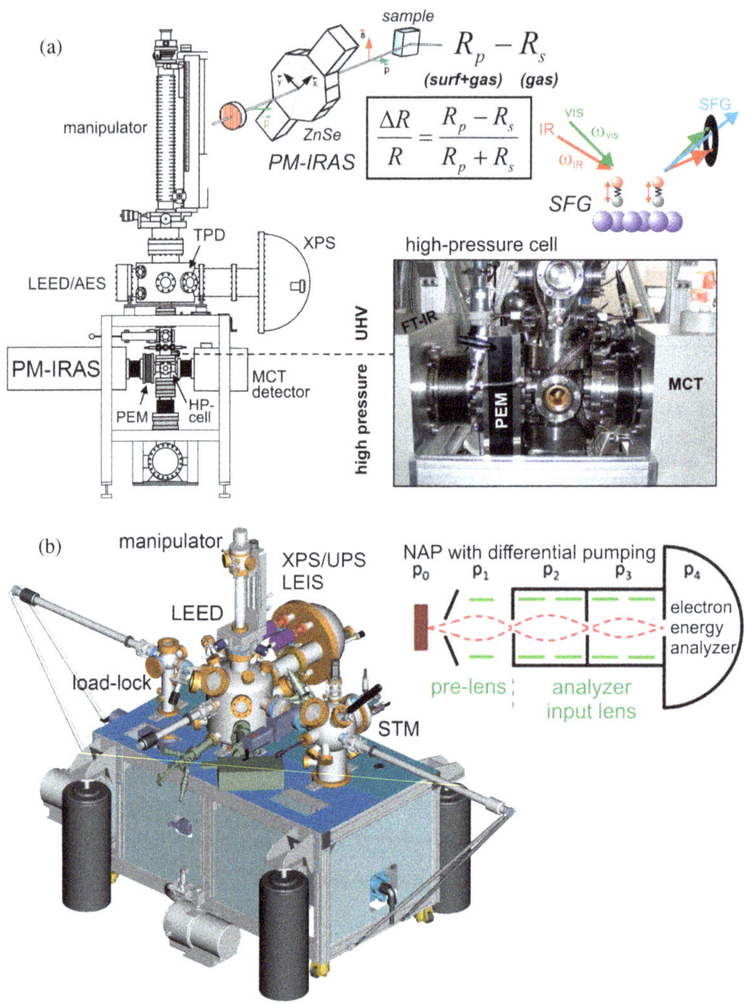

Fig. 7.3. Schematics of UHV surface preparation and analysis systems for catalytic model studies. The system in (a) combines XPS, LEED/AES, and TPD in UHV with PM-IRAS (or SFG) and kinetic tests in UHV-compatible high-pressure cell shown as inset. Principles of PM-IRAS and SFG are included. (a) Adapted with permission from Ref. [19]. Copyright (2007) Academic Press. The system in (b) enables us to perform model catalyst characterization by XPS, UPS, LEIS, LEED, and STM. The inset shows the configuration of NAP-XPS. Adapted with permission from Ref. [6]. Copyright (2013) The Royal Society of Chemistry.

such a setup, combining sample preparation with characterization by XPS, LEED/AES, thermal desorption spectroscopy (TDS), and vibrational spectroscopy [45]. The latter, either as polarization-modulation infrared reflection absorption spectroscopy (PM-IRAS) or sum frequency generation (SFG) [21,46,47], is performed using a UHV-compatible high-pressure cell, whereas the other methods are carried out under UHV conditions. A similar apparatus is displayed in Figure 7.3(b), combining XPS with UV-photoelectron spectroscopy (UPS), low energy ion scattering (LEIS), and STM. The right inset of Figure 7.3(b) displays the principle of a NAP-XPS cell, with strong differential pumping and focusing electrostatic lenses [6].

7.4 Case Studies

In the following, case studies of surface processes on model catalysts of increasing complexity are discussed, intended to bridge the materials gap between single crystals and complex nanostructured catalysts. Apart from XPS, different complementary methods are usually invoked for more complete insight. In several cases, UHV-XPS is extended to NAP-XPS to examine reaction-induced compositional changes. For NAP-XPS in applied catalysis [22,23], please refer to Chapter 10.

7.4.1 *Single crystals*

The interaction of CO with noble metal single crystals is the "drosophila" system of model catalysis, as molecular and possible dissociative adsorption are relevant for CO oxidation, hydrogenation, water gas shift, methanol synthesis, Fischer–Tropsch, etc. [19,48]. CO adsorbs in different configurations, i.e., single (linearly, on-top), double (bridging), and triple (hollow) bonded, which can be differentiated by XPS, especially when using synchrotron excitation [7,49]. Nevertheless, complementary characterization by, e.g., vibrational spectroscopy is advisable.

7.4.1.1 *CO adsorption and dissociation*

Figure 7.4 illustrates a study of CO adsorption on Pd(111) at room temperature, complementing XPS with SFG [7,50]. In the early days of "high pressure" surface science, it was speculated that high-pressure adsorption configurations may exist. For this reason, SFG

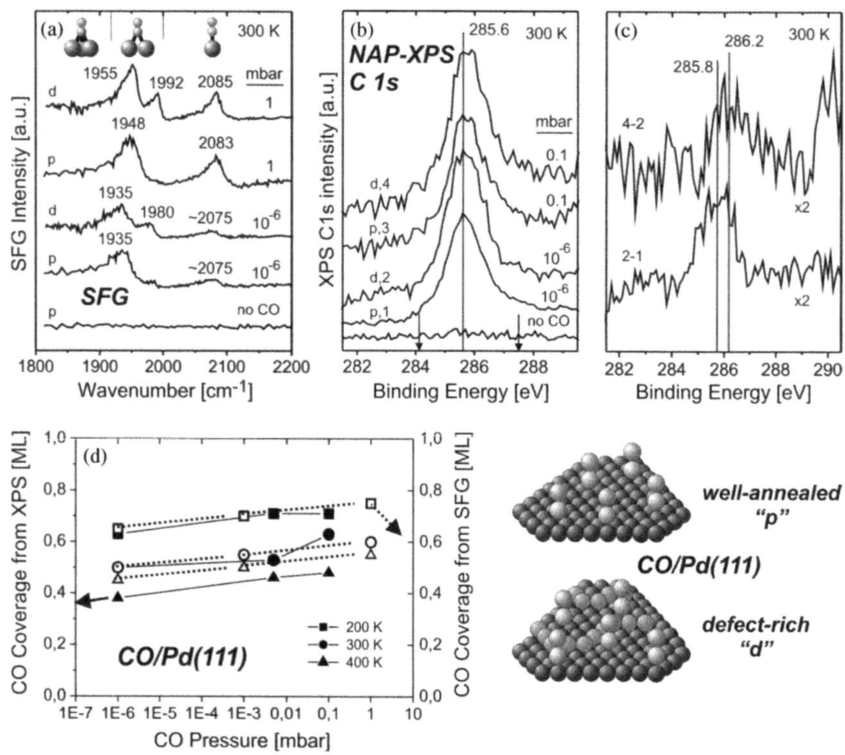

Fig. 7.4. SFG (a) and NAP-XPS (b) studies of CO adsorption on Pd single crystal surfaces at 300 K. The labels "p" and "d" refer to well-annealed (perfect) and defect-rich (ion-bombarded) Pd(111). (c) The difference spectra reveal CO adsorption on steps/defects (2-1) and on on-top sites (4-2). XPS spectra were normalized to the Pd3d$_{5/2}$ integral intensity at 334.9 eV. (d) The pressure-dependent CO coverages determined by XPS (full symbols, full lines) agree well with those deduced from the SFG vibrational frequencies (open symbols, dashed lines). (a–c) Adapted with permission from Ref. [52]. Copyright (2004) Elsevier. (d) Adapted with permission from Ref. [7]. Copyright (2003) American Chemical Society.

(Figure 7.4(a)) and NAP-XPS (Figure 7.4(b)) spectra were acquired up to 1 mbar, but only the linear/bridge/hollow CO species well known from UHV studies were observed. The features at 1935–1955 cm^{-1} and 2075–2085 cm^{-1} originate from bridge and on-top bonded CO, respectively.

This also holds true when rougher Pd surfaces were created by Ar ion sputtering, apart from an additional defect-related CO species detected both by XPS (Figure 7.4(b)) and SFG (Figure 7.4(a); 1980–1992 cm^{-1} attributed to CO bridge bonded on steps). Figure 7.4(d) displays the CO coverage depending on CO pressure and temperature, directly determined by XPS and indirectly deduced from the SFG vibrational frequencies. Overall, no new CO species were observed, but especially for adsorption at cryogenic temperature, less-ordered non-equilibrium structures may be obtained, as CO mobility can be limited [51]. The XPS spectra in Figure 7.4(b) also show that no amorphous or graphitic carbon (~284 eV; marked by an arrow in Figure 7.4(b)) is present, ruling out significant CO dissociation both on smooth and rough Pd surfaces, at least under these conditions. Pd carbonyl species (287–288 eV) and carbides (<283.5 eV) were also not detected.

Indeed, CO dissociation on rough noble metal surfaces represents a long-standing debate in surface science [52]. Somorjai and co-workers reported CO dissociation on low and high Miller-index Pt single crystals at mbar CO pressure and around 550 K [53]. However, DFT calculations reported quite high barriers, rendering dissociation unlikely [54]. These seemingly contradicting results could recently be explained by combining NAP-XPS, SFG, and DFT [46]. As will be discussed in Section 7.4.4, "CO dissociation" in fact proceeds via Boudouard CO disproportionation $2\,CO \rightarrow C + CO_2$ but only at high CO coverage on low-coordinated sites.

7.4.1.2 *Carbonaceous species*

Carbon-containing deposits/layers (CH_x) may originate from hydrocarbon adsorption, decomposition, and reaction, and their surface concentration can be quantified by XPS. For example, during selective hydrogenation of 1,3-butadiene, CH_x deposits may be formed, leading to catalyst deactivation [55]. During selective hydrogenation, Pd-carbide-like phases (Pd_xC_y), with carbon located in the subsurface region and Pd acting as a carbon "sponge", were formed during hydrogenation reactions, and they favorably affected selectivity [24,56–58].

Figure 7.5 illustrates a study of reactive methanol adsorption on Pd(111), studied by (NAP-)XPS and PM-IRAS [59–61]. As evident

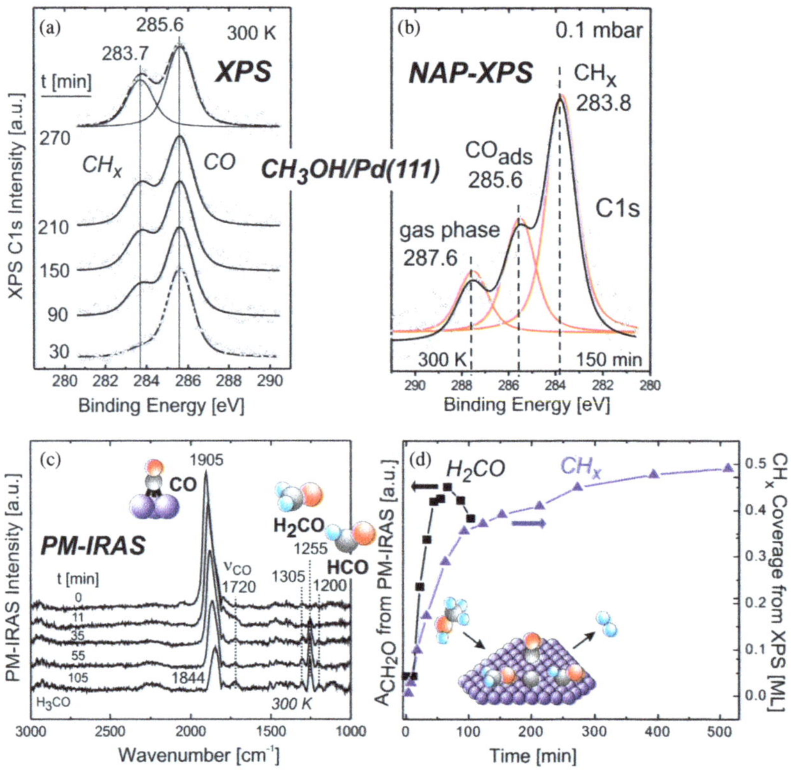

Fig. 7.5. XPS (a), NAP-XPS (b), and PM-IRAS (c) spectra of room temperature CH_3OH decomposition on Pd(111), at 5×10^{-7} (a), 0.1 (b), and 10^{-5} mbar (c). (d) Comparison of the time-dependent evolution of CH_2O (integrated intensity from (c)) and of CH_x deduced from XPS. (a,b) Adapted with permission from Ref. [59]. Copyright (2004) American Chemical Society. (c,d) Adapted with permission from Ref. [60]. Copyright (2005) American Chemical Society.

from XPS and vibrational spectroscopy, dissociative adsorption forms methoxy (CH_3O), formaldehyde (CH_2O), formyl (CHO), CO, and CH_x. As the previous section has shown that CO dissociation does not occur on Pd(111) under these conditions, the CH_x species presumably originates from C–O bond cleavage within flat-lying CH_2O. Comparing pre- and post-reaction, SFG spectra of CO adsorption indicated that mainly Pd hollow sites were blocked, suggesting the presence of C or CH species [59]. The analysis of the time-dependent evolution of CH_x (from XPS) and of CH_2O (from PM-IRAS) revealed another interesting effect (Figure 7.5(d)): a partially CH_x-poisoned surface was beneficial for CH_2O formation, as its decomposition requires empty hollow sites for both CHO and CO, whereas formaldehyde can still be formed via hydroxymethyl (CH_2OH) on top sites [19,60].

7.4.2 *Surface alloys*

The catalytic properties of bimetallic surfaces are typically not just the sum of the properties of the individual constituents, due to electronic interactions that modify adsorption and reaction characteristics [22,23,62]. Along these lines, "site isolation" of an active metal in an inert or less active matrix, recently called "single atom alloys" [39,41], can be used to control selectivity in hydrocarbon hydrogenation (PdAg [63] and PdAu [64,65] are prominent examples) or dehydrogenation (e.g., PtCu [66,67], PdGa [40,68], and RhGa [69]). Clearly, well-defined bimetallic surfaces are beneficial for fundamental studies, suggesting the use of bimetallic single crystals. However, during cleaning/annealing, surface segregation of one constituent may occur, so preparing surface alloys may be a better way.

Figure 7.6 shows how a 1:1 PdZn surface alloy is prepared, that is of interest for hydrogen generation, as discussed in the following [70–72]. XPS and LEED are used to monitor the different preparation steps (Figures 7.6(b) and 7.6(c)). Starting with a clean Pd(111) surface (335.0 eV), three monolayers (ML) of Zn are deposited at 90 K. The amorphous Zn overlayer weakens the LEED and dampens

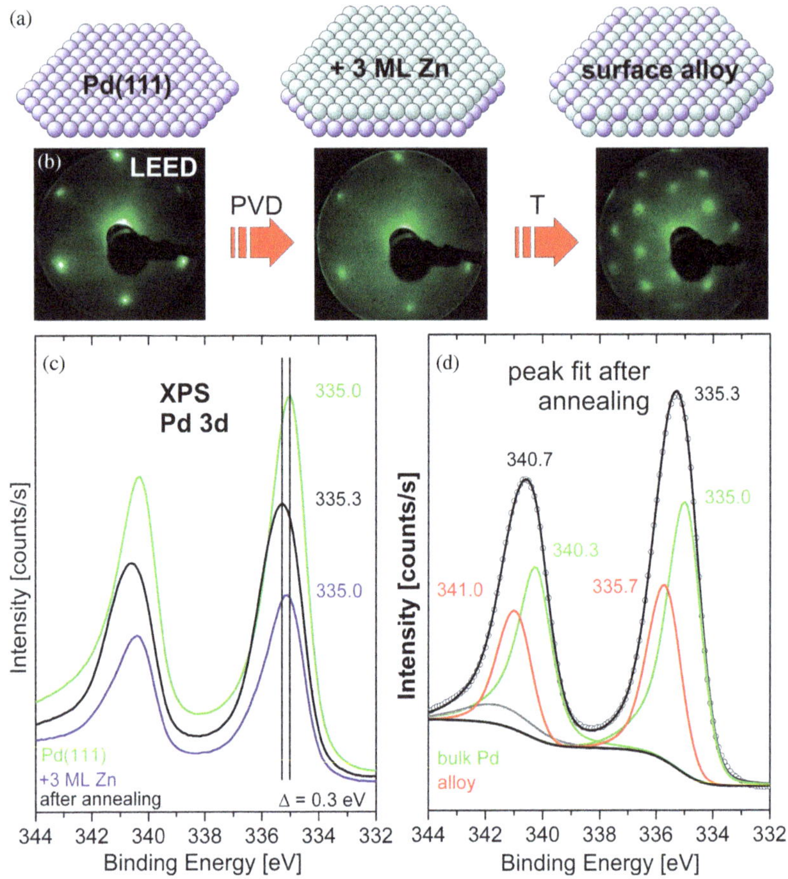

Fig. 7.6. (a) UHV preparation of a well-defined PdZn 1:1 surface alloy by Zn deposition (3 ML) on Pd(111) and subsequent annealing to 550 K. (b) LEED patterns (at 100 eV) and XPS (c) enable us to monitor each step and show the ordering and composition of the surface alloy, respectively. (d) Peak fit of an XPS spectrum acquired after annealing. Adapted with permission from Ref. [72]. Copyright (2014) Springer.

the XPS intensity, but only UHV annealing to 550 K produces a well-ordered PdZn surface alloy, indicated by LEED and a +0.3 eV shift in Pd 3d (corroborated by Zn 2p [73]; not shown). The observed peak shift originates from charge transfer from Zn to Pd, as predicted by DFT [74,75], creating $Zn^{0.4+}Pd^{0.4-}$. The 1:1 PdZn/Pd(111) alloy has a p(2 × 1) structure with alternating rows of Pd and Zn (Figure 7.6(a)). The superposition of the diffraction patterns of three (2 × 1) surface domains with different rotational orientations (rotated by 60°) then creates the "2 × 2" LEED pattern (Figure 7.6(b) [72]). Figure 7.6(d) shows a peak fit, differentiating the PdZn surface (335.7 eV) from the underlying Pd bulk that strongly contributes to the spectrum, as a laboratory X-ray gun was used.

According to combined AES and LEIS measurements, the so obtained PdZn surface alloy is a "multilayer" (up to layers) with a corrugated surface with Zn slightly "sticking out" [70,73]. Further annealing to 600 K or higher leads to loss of Zn from the second and deeper layers into the bulk, producing a PdZn "monolayer" with a smooth (non-corrugated) surface. DFT modeling showed that the surface 1:1 PdZn layer is more stable than the deeper "subsurface" layers [74,75]. Note that the multilayer and monolayer exhibit the same surface composition and structure, but the coordination of the surface atoms is markedly different.

Clearly, such a bimetallic model catalyst is perfect for catalytic studies (with PdZn being in fact an intermetallics), which is why PdZn surface alloys were employed for methanol steam reforming (MSR) to H_2 and CO_2 (Figure 7.7) [70]. If selective catalysts are available, easy-to-handle liquid methanol could serve as a source of hydrogen, which is less easy to store (compressed, chemical, and cryogenic). Atmospheric pressure kinetic tests of MSR on PdZn indicated that only the multilayer was selective to H_2/CO_2, whereas the monolayer produced undesired H_2/CO (Figure 7.7(d); CO would poison, e.g., a fuel cell catalyst).

Synchrotron NAP-XPS has significantly contributed to gain insights into the selectivity and a possible reaction mechanism. The *operando* Pd 3d and Zn 3d spectra (650 eV photon energy), acquired under reaction conditions (Figures 7.7(a)–7.7(c)) with simultaneous MS gas phase analysis, once more confirm the presence of the monolayer and multilayer PdZn surface alloys but clearly with higher surface sensitivity than in Figure 7.6 (lab-source Mg Kα = 1253.6 eV). Interestingly, the valence band (VB) spectrum of the PdZn monolayer is quite similar to that of Pd(111), despite the

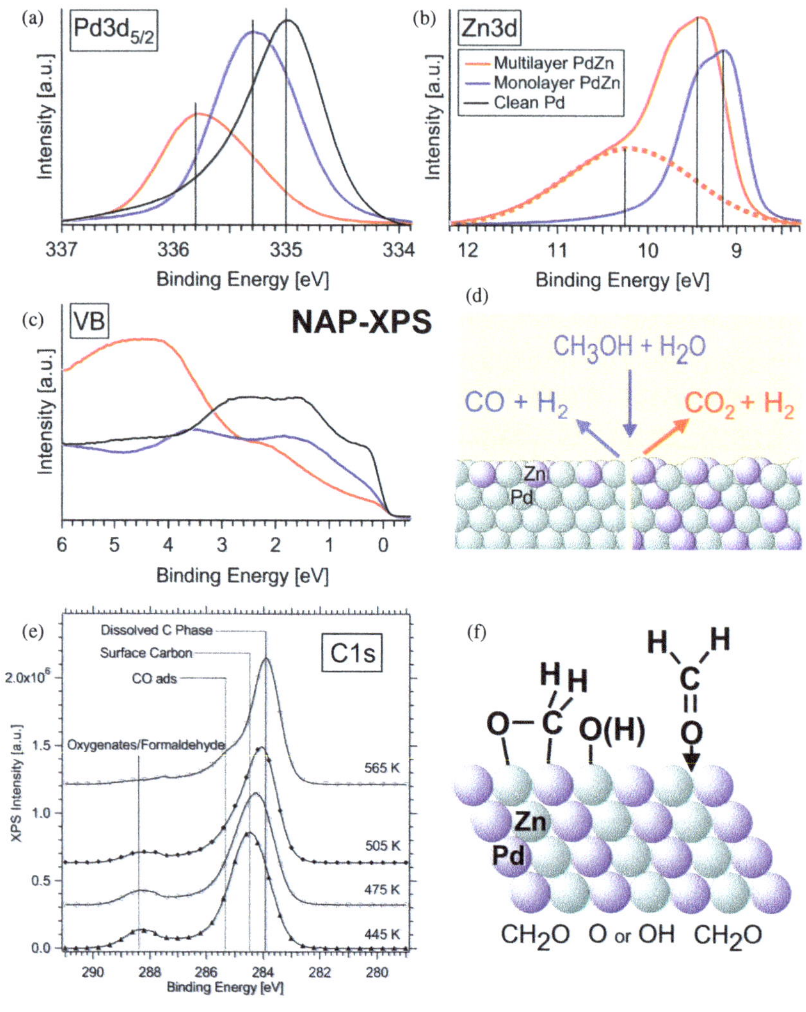

Fig. 7.7. *Operando* NAP-XPS spectra (Pd 3d (a), Zn3d (b), valence-band (VB) (c) and C1s (e) regions; BESSY II) acquired during MSR on the PdZn 1:1 multilayer (red traces) and monolayer alloy (blue traces). For comparison, the corresponding spectra for "pure" Pd are included (black traces). The oxidized ZnOH component is highlighted by the dotted red line (b). To obtain similar information depth, the Pd3d and C1s spectra were recorded with 650 eV photon energy and the Zn3d and valence-band regions with 120 eV. Reaction conditions: 0.12 mbar methanol, 0.24 mbar water, and 553 K. In (d and f), a model illustrates the CO_2 selectivity of the PdZn multilayer based on a bifunctional mechanism. Adapted with permission from Ref. [70]. Copyright (2010) John Wiley & Sons.

presence of 50% Zn on the surface (Figure 7.7(c)). In contrast, the VB of the multilayer is significantly different and resembles that of Cu (which is also a good MSR catalyst). The modified VB may be responsible for water activation, which is evident from the broad Zn 3d feature around 10.3 eV (Figure 7.7(b)). The multilayer thus activates methanol, likely as flat-lying formaldehyde, as well as water as OH, and both can react to CO_2 and H_2 (Figure 7.7(d)). The PdZn monolayer does not activate water and can only decompose methanol to CO and H_2. The proposed mechanism was further supported by *operando* C1s spectra, which showed oxygenates/formaldehyde only for reaction on the multilayer (Figure 7.7(e)) [70].

7.4.3 *Surface oxides and thin oxide films*

Depending on reaction conditions, a catalyst surface may be modified by the reactants. Apart from the already discussed CH_x deposition, hydrides or oxides may be formed during hydrogenation or oxidation, respectively. Clearly, the stability of such phases depends on the applied reaction conditions, calling for *in situ* or *operando* [23] NAP-XPS characterization.

7.4.3.1 *Surface oxides*

Accordingly, the formation and reactivity of surface oxides were intensively studied, and surface-sensitive XPS is the apparent method of choice [76–83]. Figure 7.8 displays high-resolution XPS spectra of a two-dimensional Pd_5O_4 surface oxide grown on Pd(111) and a corresponding STM image, with the information of both combined with DFT calculations yielding the structure model in Figure 7.8(c). Molecular beam studies of CO oxidation at ~200–400 K indicated that Pd_5O_4 was about five times less active than an O-covered Pd surface, but under oxygen-rich conditions, Pd_5O_4 may constitute the active phase [78]. For methanol oxidation (5 mbar CH_3OH and 5 mbar O_2) up to 400 K pre- and post-reaction XPS, as well as *operando* PM-IRAS, did not indicate surface oxidation of Pd(111) under these conditions [21].

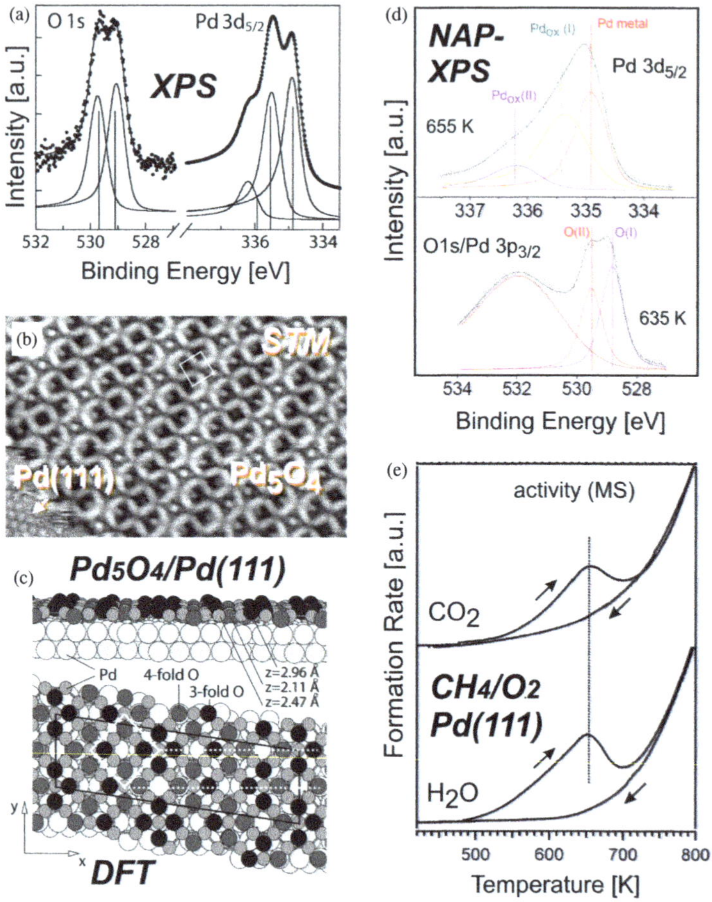

Fig. 7.8. (a) High-resolution O 1s and Pd3d$_{5/2}$ XPS of Pd_5O_4 surface oxide. The different types of O and Pd atoms can be resolved using synchrotron radiation. (b) Corresponding STM image (10 nm × 6 nm). (c) DFT-derived structure model of the Pd_5O_4 surface oxide. (d,e) *Operando* NAP-XPS (d) and MS (e) studies of CH_4 combustion on Pd(111). (a–c) Adapted with permission from Ref. [80]. Copyright (2002) American Physical Society. (d,e) Adapted with permission from Ref. [86]. Copyright (2011) American Chemical Society.

The valence state of Pd catalysts was vividly discussed for oxidation reactions at higher temperatures, such as methane combustion. For "low temperature" (<800°C) methane combustion, pre- and post-reaction characterization suggested PdO (or PdO_x) as active phase in Pd/Al_2O_3 [79,84,85]. However, the Pd particles sinter and re-facet, activity increases with time-on-stream, and there is strong hysteresis upon heating/cooling.

NAP-XPS is thus crucial to identify the active phase present under specific reaction conditions [86]. Figure 7.8(d) displays *operando* spectra (total pressure of 0.3 mbar; CH_4:O_2 = 1:5), acquired simultaneous to MS activity analysis (Figure 7.8(e)), to directly identify the active phase of CH_4 combustion on Pd(111). The occurrence of the oxide phase sensitively depends on pressure and temperature, once more with a pronounced redox hysteresis upon temperature variation. At the activity maximum at 650 K, *operando* NAP-XPS detected a new surface phase of PdO seeds on a Pd_5O_4 surface oxide, created by a delicate balance of seed formation and reduction by CH_4. Below 650 K, this active phase was not yet fully developed, and above 650 K, PdO is replaced by metallic Pd and subsurface oxygen. Clearly, this complex behavior cannot be understood by *ex situ* measurements. Even more so, as upon cool-down, neither the activity maximum nor the PdO seeds/Pd_5O_4 phase was re-established [86].

7.4.3.2 *Thin oxide films*

The oxidation of metal surfaces does not always lead to well-ordered surface oxides, and if the oxide layer becomes too thick and insulating, the application of surface science methods is difficult. Oxide single crystals are an alternative but with limited availability and doping may be required to increase conductivity. For these reasons, oxides are often studied as 2D crystalline thin films, grown on suitable conducting substrates [43,87]. Their thickness of a few atomic layers to several nanometers facilitates analysis by spectroscopic/microscopic techniques. A variety of model oxides is available, including Al_2O_3/NiAl(110) or /Ni_3Al(111), SiO_2/Mo(112), CeO_2/Ru(0001), MgO/Ag(100), V_xO_y/Au(111), Fe_3O_4/Pt(111), Nb_2O_5/Au(100), ZrO_2/Pt_3Zr(0001), and Co_3O_4/Ir(100), to name just a few [43,87]. The oxide films may serve as catalysts or active supports for metal nanoparticles, especially for oxidation reactions.

In the following, the preparation of a ZrO_2 thin film model support is described, with synchrotron XPS being key for characterizing the composition and structure via depth profiling [25,88]. Subsequently, the model ZrO_2 film is employed to study CO_2 activation, which is important for catalytic reactions, such as methanol synthesis, methane dry reforming, reverse water gas shift, or CO_2 hydrogenation to hydrocarbons [30,31]. A single crystal of $Pt_3Zr(0001)$ is used as a substrate, and normal emission XPS spectra in the Zr 3d and O 1s ranges were recorded with photon energies (PE) of 320 and 670 eV, respectively. Accordingly, the photoelectrons have a kinetic energy (KE) of ~140 eV (IMFP of 0.54 nm), so mainly the two topmost layers are probed. The Zr 3d spectrum of the clean alloy (Figure 7.9(a), bottom) exhibits a pronounced

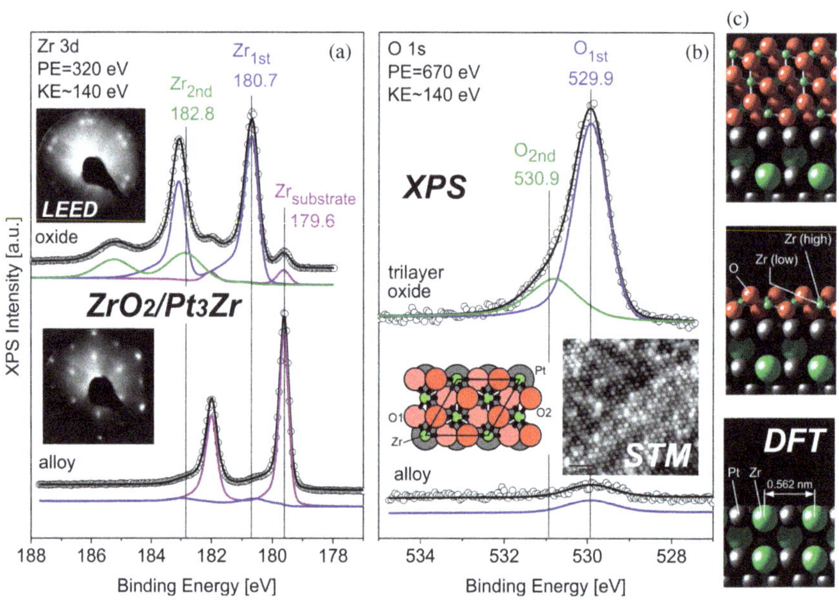

Fig. 7.9. Zr 3d (a) and O 1s (b) XP spectra of the clean alloy (bottom) and of the oxide annealed at 1023 K (top; spectra taken at 300 K; kinetic energy ≈ 140 eV). The insets show LEED patterns, and an STM image and top-view structure model of the trilayer oxide. (c) Side views of the Pt_3Zr substrate (bottom), the trilayer oxide film (middle), and the three-trilayer film (top) used for DFT calculations. Adapted with permission from Ref. [25]. Copyright (2015) American Chemical Society.

doublet at 179.6 eV and 182.0 eV, and the characteristic of alloyed (metallic) Zr in Pt_3Zr (Zr(0001) has a BE of 178.6 eV) [25]. The tiny Zr 3d features were assigned to Zr bound to residual oxygen that showed up in O 1s (Figure 7.9(b), bottom).

The ZrO_2 thin film is then grown by oxidation (1×10^{-7} mbar O_2 at 673 K), followed by UHV annealing to 1073–1173 K, as evident from LEED (Figure 7.9(a), upper inset), STM (Figure 7.9(b), inset), and DFT (Figure 7.9(c), middle [25]). A small XPS peak of the substrate ($Zr_{substrate}$) was still present, but two doublets of oxidic species appeared (Zr_{1st} and Zr_{2nd}). The main Zr_{1st} peak at 180.7 eV is shifted by +2.1 eV relative to pure Zr and can be assigned to an O-Zr-O trilayer oxide (see structure models in top and side views). The weaker Zr_{2nd} species at 182.8 eV is shifted by +4.2 eV with respect to pure Zr and is characteristic of bulk ZrO_2 (around 183 eV). Two species were also observed in the O 1s region. For both types of oxides, quantitative analysis indicated a Zr to O ratio close to 0.50, so they should be assigned to ZrO_2 rather than to a suboxide.

XPS depth profiling, i.e., varying the incident PE energy, was employed to determine the thickness of the two oxidic species (yielding IMFPs of 0.54, 0.72, and 0.90 nm). This indicated that Zr_{1st} results from a thin O-Zr-O tri-layer oxide film, whereas Zr_{2nd} is due to thicker "bulk" oxide islands (Figure 7.9(c), top; 5–20 nm in size), with the latter confirmed by STM and DFT. This is also in line with the O 1s spectra (Figure 7.9(b)).

The ultrathin ZrO_2 film was then used for a catalytic study. Utilizing CO_2 as a sustainable carbon source requires its efficient activation by catalytically active oxides, which bind CO_2 via carbonaceous species. We have thus investigated the interaction of CO_2 with the ZrO_2 trilayer surface. For adsorption in UHV, TPD indicated a weak interaction (T_{des} of 117 K). Increasing the CO_2 pressure up to ~0.1 mbar did not lead to measurable adsorption, but the presence of water strongly promoted the formation of surface-bound species. The interaction of co-adsorbed CO_2 + H_2O is evident from (lab-based) *in situ* NAP-XPS shown in Figure 7.10(a) (Zr3d) and Figure 7.10(b) (C 1s) [31].

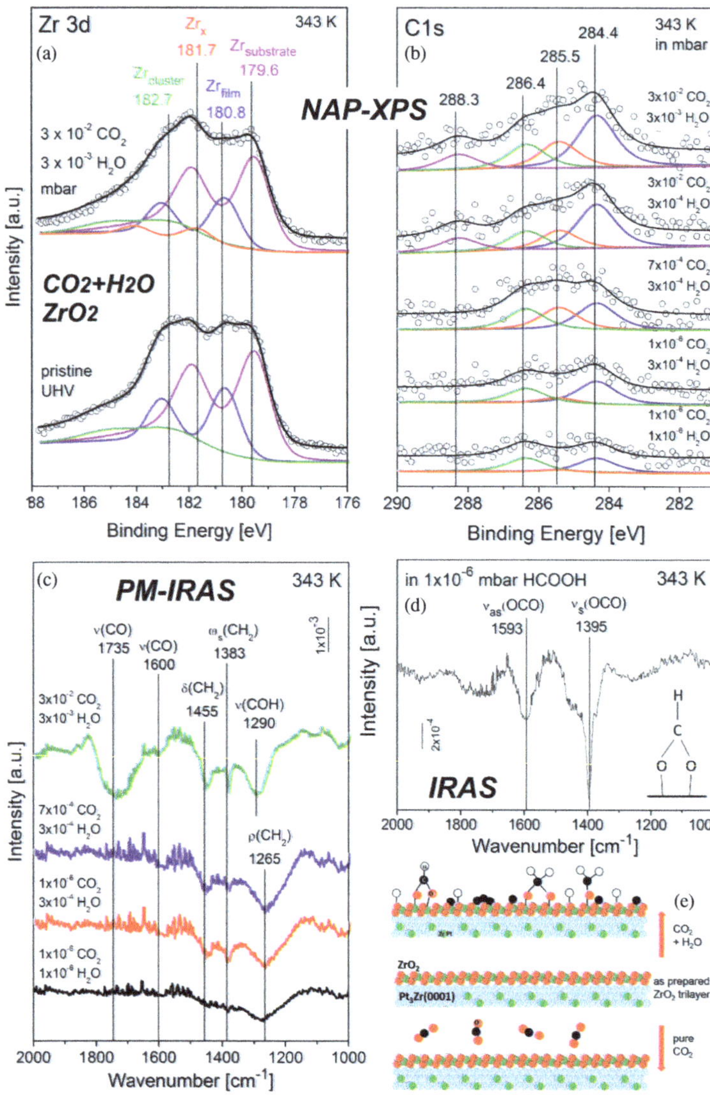

Fig. 7.10. Lab-based NAP-XPS of the ZrO_2 trilayer film at various CO_2 and H_2O partial pressures: (a) Zr 3d spectra; (b) C 1s spectra of ZrO_2 trilayer. Conditions are indicated and all spectra are taken with Al Kα. (c) Corresponding *in situ* IRAS of different CO_2 and H_2O mixtures on ultrathin ZrO_2. (d) Reference IRAS spectrum of HCOOH adsorption on ZrO_2 thin film. (e) Illustration of CO_2 and CO_2 + H_2O exposure of the ZrO_2 trilayer surface. Adapted with permission from Ref. [31]. Copyright (2019) Elsevier.

Although several carbonaceous species were observed, their identification required additional input by complementary PM-IRAS spectra, acquired under the same conditions (Figure 7.10(c)). Taking into account reference spectra of formic acid (Figure 7.10(d)) and formaldehyde (not shown), a mechanistic model could be deduced (Figure 7.10(e)). In the presence of water, or of surface hydroxyl groups, formates, formaldehyde, carbon, and possibly dioxomethylene were observed. These species, that were absent for "dry CO_2", highlight the ability of ZrO_2 as active support and may be intermediates of upconverting CO_2 to methanol or other products. This was paralleled by partial transformation of Zr^{4+} into a new Zr_x species that awaits further identification (Figure 7.10(a)).

7.4.4 *Supported nanoparticles*

Transition metal oxides are not only important catalysts but also serve as supports for metal nanoparticles. They may strongly anchor them counteracting sintering, but for active supports, some specific catalytic steps or reactions may indeed occur on the oxide surface or at the metal/oxide interface [20,23,89–93].

Figure 7.11 illustrates how combining NAP-XPS with SFG can be utilized to reveal the reaction mechanism of CO oxidation on nano-structured $Pt/ZrO_2/Si(100)$ [46,47]. The model catalysts were prepared by atomic layer deposition (ALD), and depending on the number of ALD cycles, either larger smooth Pt islands (~40 nm in size, ~10 nm in thickness) or smaller (~6–8 nm) rougher Pt nano-particles were obtained, as shown by the cross-section TEM insets in Figures 7.11(a) and 7.11(b). CO oxidation was carried out at 30 mbar ($CO:O_2 = 1:2$), using a high-pressure reaction cell coupled to MS analysis [94,95]. The *operando* SFG spectra in Figures 7.11(a) and 7.11(b) indicated ignition temperatures of 673 and 723 K for smooth and rough Pt, respectively. The disappearance of adsorbed CO shows when the Pt surface changes from the CO-poisoned inactive to the O-covered active state [96,97]. Nevertheless, it is surprising that the rougher surfaces seem less active, contrary to the general expectation.

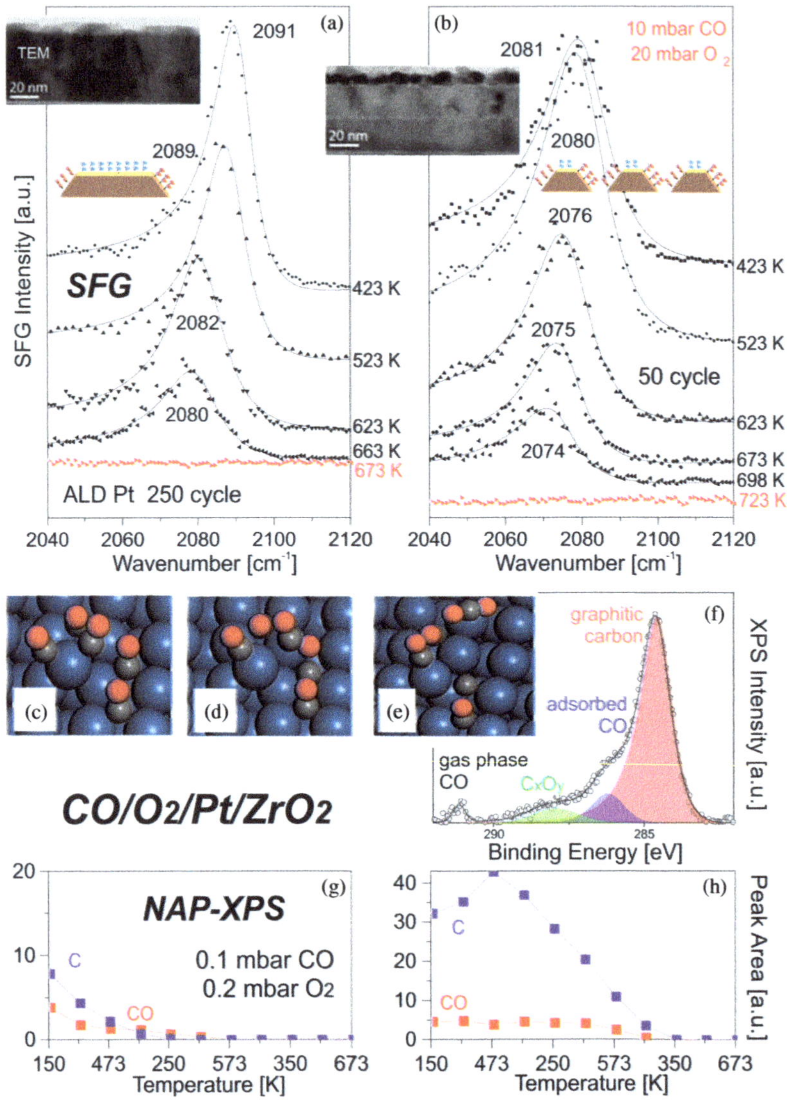

Fig. 7.11. (a,b) *Operando* SFG spectra (ppp-polarization) acquired during CO oxidation of 250 and 50 cycle ALD Pt samples, with cross-section TEM micrographs as insets. (g,h) Shows the corresponding temperature-dependent evolution of adsorbed CO and carbon deduced from *operando* C 1s NAP-XPS, such as the one shown in (f) for the 50 cycle ALD Pt film at 423 K, (c–e) are initial, transition, and final states for dissociation at an adatom on Pt(100), respectively. Adapted with permission from Ref. [46]. Copyright (2021) American Chemical Society.

As already mentioned in Section 7.4.1, CO dissociation on Pt surfaces had been controversially discussed. A *direct* proof of the presence of carbon during CO oxidation, with its amount depending on the Pt morphology and the reaction conditions, was provided by NAP-XPS (1 mbar CO and 2 mbar O_2; Figure 7.11(f)). Apparently, on smaller rougher Pt nanoparticles, more carbon is present (Figure 7.11(h)) than on smooth Pt islands (Figure 7.11(g)), shifting the reaction onset to higher temperatures. Thus, under the current conditions, initial CO oxidation is found to be strongly influenced by the removal of carbon deposits rather than being determined by the CO and oxygen inherent activity. Accordingly, at variance with the general expectation, rough Pt nanoparticles are seemingly less active than smoother Pt films.

DFT calculations were key to explain the origin of the carbon deposits [46]. Rather than direct CO dissociation, which is energetically unfavorable, CO disproportionation may occur on low-coordinated Pt sites via a disproportionation mechanism (Figures 7.11(c)–7.11(e)). However, this is only feasible at high CO coverage (pressure) and when the remaining C-atom is stabilized by a favorable coordination (Figures 7.11(e)). The CO reaction onset is thus determined by a delicate balance between CO disproportionation (Boudouard reaction) and oxidation. Throughout the reaction, the Pt surfaces remained metallic, as evident from *operando* NAP-XPS Pt 4f spectra [46].

$Pd/Al_2O_3/NiAl(110)$ is likely the most studied nanoparticle model catalyst [17,18,21]. Returning to methanol oxidation, Figure 7.12 collects results of XPS/SFG studies of Pd nanoparticles (about 6 nm in size) and contrasts them to Pd(111). Under reaction conditions, SFG and PM-IRAS detected adsorbed CO species in both cases, but comparing pre- and post-reaction, Pd 3d spectra revealed that while Pd(111) remained metallic, the Pd nanoparticles were oxidized [18]. The ~0.4 eV binding energy shift for Pd particles suggests an oxidation state of "PdO_x" (between Pd and PdO, for which the BE shift would be 1.5 eV [85]). Still, the vibrational frequencies of adsorbed CO were characteristic of Pd^0, pointing to *partial* surface oxidation. Indeed, using CO as a probe molecule for

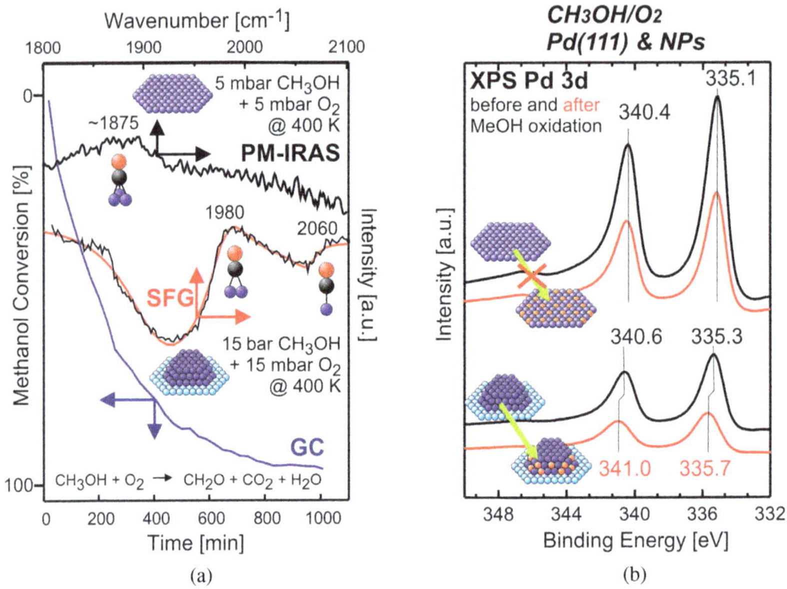

Fig. 7.12. Methanol oxidation on Pd(111) and Pd/Al_2O_3 (mean particle size of 6 nm). (a) *In situ* steady-state PM-IRAS and SFG spectra, displayed together with the CH_3OH conversion as monitored by gas chromatography. (b) Comparison of Pd 3d XP spectra obtained before (upper) and after (lower) reaction. Adapted with permission from Ref. [18]. Copyright (2007) PCCP Owner Societies.

post-reaction characterization revealed that about half of the particle surface was oxidized. Partially, oxidized Pd/PdO_x nanoparticles were also reported by Schalow *et al.* [98].

7.4.5 *Scanning photoelectron microscopy (SPEM) and recent developments*

So far, all XPS spectra reported herein were area-averaging, which means integrating over a sample surface area of a few mm^2. This is appropriate for structurally homogenous surfaces such as single crystals, well-ordered thin films, and well-dispersed nanoparticles of narrow size distribution. However, for structurally heterogeneous catalyst surfaces, the heterogeneity is "smeared out" by averaging, making such integral spectra less meaningful. This limitation can be overcome by microspectroscopy, acquiring XPS spectra with an X-ray spot size down to ~130 nm, or by scanning photoelectron microscopy (SPEM; "ESCA microscopy"), described in detail in Chapter 11 [32–34].

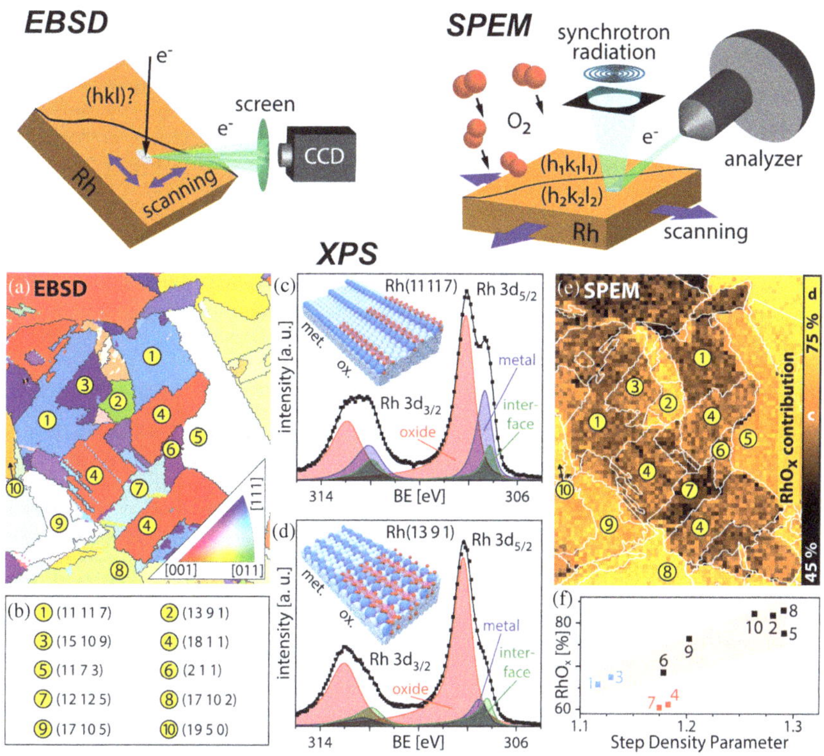

Fig. 7.13. Illustration of electron backscatter diffraction (EBSD; top left) and scanning photoelectron microscopy (SPEM; top right). (a) EBSD image of a polycrystalline Rh foil (512 × 600 μm); (b) crystallographic orientations (Miller indices) of the Rh(hkl) domains; (c) and (d) show examples of local XPS spectra (PE of 650.9 eV, take-off angle of 60°) and ball models of the metallic and oxidized Rh(11 11 7) and Rh(13 9 1) domains, respectively. Squares: measured values; solid lines: sum of the deconvoluted components; (e) *oxidation map* of the region in (a) after oxidation ($T = 623$ K, $P_{O2} = 2.5 \times 10^{-4}$ mbar, $t = 90$ min), with the right edge of the map displaying the color scale for the RhO_x contribution; (f) shows the RhO_x contribution versus the step density parameter (SDP; squares: measured values; shaded area serves as a guide for the eye). Adapted with permission from Ref. [34]. Copyright (2021) Springer Nature.

As illustrated in Figure 7.13 (top), to acquire a SPEM image, it is apparently easier to move the sample and collect the local XPS signal point-by-point, instead of scanning the focused X-ray beam over the sample.

Figure 7.13(a) shows such a synchrotron SPEM study of the oxidation of a heterogeneous surface, i.e., a polycrystalline Rh foil [34]. It exposes many grains of different crystallographic orientations, which were characterized by electron backscatter diffraction (EBSD) (Figure 7.13(a)). Indeed, the polycrystalline Rh foil represents a "surface structure library" [99,100], exposing a variety of surface orientations (Figure 7.13(b)) that can be exposed to identical oxidation conditions.

Using SPEM, the surface oxide on the various grains was characterized, detecting the formation of a trilayer O-Rh-O surface oxide (Figures 7.13(c) and 7.13(d)), in agreement with previous studies of low-index single crystals [77,80,101–104]. However, using SPEM, the anisotropy of Rh surface oxidation could be directly imaged (Figure 7.13(e)), with the extent of surface oxidation correlating with the density of steps on a respective surface grain (Figure 7.13(f)). Rh surface oxidation is mediated by steps (insets in Figures 7.13(c) and 7.13(d)) and kinetically controlled, explaining the correlation with the step density. The type of step is also important, explaining the outliers in Figure 7.13(f). The locally resolved XPS characterization was further combined with photoemission electron microscopy (PEEM) [99,105,106], to determine the role of Rh surface oxides in catalytic hydrogen oxidation. Based on the analysis of propagating reaction fronts as indicators of surface reactivity, a high transient activity of Rh surface oxide was observed [34]. SPEM was also used for *in situ* imaging of hydrogen oxidation on polycrystalline Rh, revealing a coexistence of four different states on adjacent differently oriented domains of the same Rh sample at the exactly same conditions [107]. Apart from catalytically active and inactive steady states, multifrequential oscillating states were imaged via "XPS movies".

Another aspect, not touched so far, is the time-resolution of XPS. When using synchrotron radiation and advanced detection schemes, acquisition times can be significantly reduced, enabling us to monitor surface processes in real time [12,108]. NAP-XPS has also been applied to studies of polarized perovskite electrodes, important for solid-state electrochemistry of, e.g., water splitting or CO_2 electrolysis [29,109]. The effect of polarization on $LaSrFeO_x$ perovskite surface composition (segregation) and on the adsorbed species (carbonates, C) was directly revealed.

7.5 Concluding Remarks

XPS and its high (mbar) pressure variant, NAP-XPS, are indispensable tools for model catalysis research, especially when synchrotron radiation is employed. This enables us to monitor the preparation of model surfaces, characterizing composition, oxidation state, and cleanliness. The interaction with reactants can be examined both under UHV and *operando* conditions, yielding correlations between catalyst state, activity, and selectivity. A major advantage of XPS is that it provides quantitative information on surface species, but their chemical identity is not always directly revealed. Thus, combining XPS with complementary methods, such as LEED, TPD, SFG, (PM-)IRAS, and STM, and paralleled by kinetic studies is the most successful. XPS will remain to be a major workhorse of catalysis research, and its application at increasingly higher pressures and temperatures, under polarization, in resonantly enhanced and locally resolved modes, holds promise for exciting breakthroughs in the future.

References

[1] Siegbahn K. Electron spectroscopy for atoms, molecules, and condensed matter. Science. 1982;217:111–121.

[2] Joyner RW, Roberts MW, Yates K. A "high-pressure" electron spectrometer for surface studies. Surf Sci. 1979;87:501–509.

[3] Bonzel HP, Kleint C. On the history of photoemission. Prog Surf Sci. 1995;49:107–153.

[4] Salmeron M, Schlögl R. Ambient pressure photoelectron spectroscopy: A new tool for surface science and nanotechnology. Surf Sci Rep. 2008;63:169–199.

[5] Knop-Gericke A, Kleimenov E, Hävecker M, Blume R, Teschner D, Zafeiratos S, Schlögl R, Bukhtiyarov VI, Kaichev VV, Prosvirin IP, Nizovskii AI, Bluhm H, Barinov A, Dudin P, Kiskinova M. *Chapter 5 X-ray Photoelectron Spectroscopy for Investigation of Heterogeneous Catalytic Processes. Advances in Catalysis.* Academic Press, 2009, pp. 213–272.

[6] Starr DE, Liu Z, Hävecker M, Knop-Gericke A, Bluhm H. Investigation of solid/vapor interfaces using ambient pressure X-ray photoelectron spectroscopy. Chem Soc Rev. 2013;42:5833–5857.

[7] Kaichev VV, Prosvirin IP, Bukhtiyarov VI, Unterhalt H, Rupprechter G, Freund H-J. High-pressure studies of CO adsorption on Pd(111) by X-ray photoelectron spectroscopy and sum-frequency generation. J Phys Chem B. 2003;107:3522–3527.

[8] Pantförder J, Pöllmann S, Zhu JF, Borgmann D, Denecke R, Steinrück H-P. New setup for *in situ* X-ray photoelectron spectroscopy from ultrahigh vacuum to 1mbar. Rev Sci Instrum. 2005;76: 014102.

[9] Ruppender HJ, Grunze M, Kong CW, Wilmers M. In situ X-ray photoelectron spectroscopy of surfaces at pressures up to 1 mbar. Surf Interface Anal. 1990;15:245–253.

[10] Ogletree DF, Bluhm H, Lebedev G, Fadley CS, Hussain Z, Salmeron M. A differentially pumped electrostatic lens system for photoemission studies in the millibar range. Rev Sci Instrum. 2002;73: 3872–3877.

[11] Bluhm H. Photoelectron spectroscopy of surfaces under humid conditions. J Electron Spectrosc Relat Phenom. 2010;177: 71–84.

[12] Shavorskiy A, Neppl S, Slaughter DS, Cryan JP, Siefermann KR, Weise F, Lin M-F, Bacellar C, Ziemkiewicz MP, Zegkinoglou I, Fraund MW, Khurmi C, Hertlein MP, Wright TW, Huse N, Schoenlein RW, Tyliszczak T, Coslovich G, Robinson J, Kaindl RA, Rude BS, Ölsner A, Mähl S, Bluhm H, Gessner O. Sub-nanosecond time-resolved ambient-pressure X-ray photoelectron spectroscopy setup for pulsed and constant wave X-ray light sources. Rev Sci Instrum. 2014;85:093102.

[13] Freund H-J. Adsorption of gases on complex solid surfaces. Angew Chemie Int Ed English. 1997;36:452–475.

[14] Bäumer M, Freund H-J. Metal deposits on well-ordered oxide films. Prog Surf Sci. 1999;61:127–198.

[15] Freund H-J. Clusters and islands on oxides: From catalysis via electronics and magnetism to optics. Surf Sci. 2002;500:271–299.

[16] Goodman DW. Model catalysts: From imagining to imaging a working surface. J Catal. 2003;216:213–222.

[17] Freund HJ, Bäumer M, Libuda J, Risse T, Rupprechter G, Shaikhutdinov S. Preparation and characterization of model catalysts: From ultrahigh vacuum to in situ conditions at the atomic dimension. J Catal. 2003;216:223–235.

[18] Bäumer M, Libuda J, Neyman KM, Rösch N, Rupprechter G, Freund H-J. Adsorption and reaction of methanol on supported palladium catalysts: Microscopic-level studies from ultrahigh vacuum to ambient pressure conditions. Phys Chem Chem Phys. 2007;9:3541–3558.

[19] Rupprechter G. Sum frequency generation and polarization–modulation infrared reflection absorption spectroscopy of functioning model catalysts from Ultrahigh vacuum to ambient pressure. In BC Gates, H Knözinger (Eds.). *Advances in catalysis.* Academic Press, 2007, pp. 133–263.

[20] Freund H-J, Pacchioni G. Oxide ultra-thin films on metals: New materials for the design of supported metal catalysts. Chem Soc Rev. 2008;37:2224–2242.

[21] Rupprechter G, Weilach C. Spectroscopic studies of surface–gas interactions and catalyst restructuring at ambient pressure: Mind the gap! J Phys Condens Matter. 2008;20:184019.

[22] Föttinger K, Rupprechter G. In Situ spectroscopy of complex surface reactions on supported Pd–Zn, Pd–Ga, and Pd(Pt)–Cu nanoparticles. Acc Chem Res. 2014;47:3071–3079.

[23] Rupprechter G. Operando surface spectroscopy and microscopy during catalytic reactions: From clusters via nanoparticles to mesoscale aggregates. Small. 2021;2004289.

[24] Armbrüster M, Behrens M, Cinquini F, Föttinger K, Grin Y, Haghofer A, Klötzer B, Knop-Gericke A, Lorenz H, Ota A, Penner S, Prinz J, Rameshan C, Révay Z, Rosenthal D, Rupprechter G, Sautet P, Schlögl R, Shao L, Szentmiklósi L, Teschner D, Torres D,

Wagner R, Widmer R, Wowsnick G. How to control the selectivity of palladium-based catalysts in hydrogenation reactions: The role of subsurface chemistry. ChemCatChem. 2012;4:1048–1063.

[25] Li H, Choi J-IJ, Mayr-Schmölzer W, Weilach C, Rameshan C, Mittendorfer F, Redinger J, Schmid M, Rupprechter G. Growth of an ultrathin zirconia film on Pt_3Zr examined by high-resolution X-ray photoelectron spectroscopy, temperature-programmed desorption, scanning tunneling microscopy, and density functional theory. J Phys Chem C. 2015;119:2462–2470.

[26] Salmeron M. From surfaces to interfaces: Ambient pressure XPS and beyond. Top Catal. 2018;61:2044–2051.

[27] Tao F, Grass ME, Zhang Y, Butcher DR, Renzas JR, Liu Z, Chung JY, Mun BS, Salmeron M, Somorjai GA. Reaction-driven restructuring of Rh-Pd and Pt-Pd core-shell nanoparticles. Science. 2008;322:932–934.

[28] Nguyen L, Tao FF, Tang Y, Dou J, Bao X-J. Understanding catalyst surfaces during catalysis through near ambient pressure X-ray photoelectron spectroscopy. Chem Rev. 2019;119:6822–6905.

[29] Opitz AK, Nenning A, Rameshan C, Rameshan R, Blume R, Hävecker M, Knop-Gericke A, Rupprechter G, Fleig J, Klötzer B. Enhancing electrochemical water-splitting kinetics by polarization-driven formation of near-surface iron(0): An in situ XPS study on perovskite-type electrodes. Angew Chemie Int Ed. 2015;54:2628–2632.

[30] Rameshan C, Li H, Anic K, Roiaz M, Pramhaas V, Rameshan R, Blume R, Hävecker M, Knudsen J, Knop-Gericke A, Rupprechter G. In situ NAP-XPS spectroscopy during methane dry reforming on ZrO_2/Pt(1 1 1) inverse model catalyst. J Phys Condens Matter. 2018;30:264007.

[31] Li H, Rameshan C, Bukhtiyarov AV, Prosvirin IP, Bukhtiyarov VI, Rupprechter G. CO_2 activation on ultrathin ZrO_2 film by H_2O co-adsorption: In situ NAP-XPS and IRAS studies. Surf Sci. 2019;679:139–146.

[32] S. Günther, Kaulich B, Gregoratti L, Kiskinova M. Photoelectron microscopy and applications in surface and materials science. Prog Surf Sci. 2002;70:187–260.

[33] Amati M, Barinov A, Feyer V, Gregoratti L, Al-Hada M, Locatelli A, Mentes TO, Sezen H, Schneider CM, Kiskinova M. Photoelectron

microscopy at Elettra: Recent advances and perspectives. J Electron Spectrosc Relat Phenom. 2018;224:59–67.

[34] Winkler P, Zeininger J, Suchorski Y, Stöger-Pollach M, Zeller P, Amati M, Gregoratti L, Rupprechter G. How the anisotropy of surface oxide formation influences the transient activity of a surface reaction. Nature Commun. 2021;12:69.

[35] Somorjai GA. The surface science of heterogeneous catalysis. Surf Sci. 1994;299-300:849–866.

[36] Somorjai GA, Rupprechter G. The flexible surface: Molecular studies explain the extraordinary diversity of surface chemical properties. J Chem Educ. 1998;75:161.

[37] Ertl G. Reactions at well-defined surfaces. Surf Sci. 1994;299–300: 742–754.

[38] King DA. Chemisorption on metals: A personal review. Surf Sci. 1994;299–300:678–689.

[39] Lucci FR, Liu J, Marcinkowski MD, Yang M, Allard LF, Flytzani-Stephanopoulos M, Sykes ECH. Selective hydrogenation of 1,3-butadiene on platinum–copper alloys at the single-atom limit. Nature Commun. 2015;6:8550.

[40] Taccardi N, Grabau M, Debuschewitz J, Distaso M, Brandl M, Hock R, Maier F, Papp C, Erhard J, Neiss C, Peukert W, Görling A, Steinrück HP, Wasserscheid P. Gallium-rich Pd–Ga phases as supported liquid metal catalysts. Nature Chem. 2017;9:862–867.

[41] Rupprechter G. Popping up to the surface. Nature Chem. 2017;9:833–834.

[42] Freund H-J, Bäumer M, Kuhlenbeck H. *Catalysis and Surface Science: What Do We Learn from Studies of Oxide-Supported Cluster Model Systems? Advances in Catalysis.* Academic Press, 2000, pp. 333–384.

[43] Campbell CT. Ultrathin metal films and particles on oxide surfaces: Structural, electronic and chemisorptive properties. Surf Sci Rep. 1997;27:1–111.

[44] Hansen KH, Worren T, Stempel S, Lægsgaard E, Bäumer M, Freund H-J, Besenbacher F, Stensgaard I. Palladium nanocrystals on Al_2O_3: Structure and adhesion energy. Phys Rev Lett. 1999;83: 4120–4123.

[45] Rupprechter G. Surface vibrational spectroscopy from ultrahigh vacuum to atmospheric pressure: Adsorption and reactions on single crystals and nanoparticle model catalysts monitored by sum frequency generation spectroscopy. Phys Chem Chem Phys. 2001;3:4621–4632.

[46] Pramhaas V, Roiaz M, Bosio N, Corva M, Rameshan C, Vesselli E, Grönbeck H, Rupprechter G. Interplay between CO disproportionation and oxidation: On the origin of the CO reaction onset on atomic layer deposition-grown Pt/ZrO_2 model catalysts. ACS Catal. 2021;11:208–214.

[47] Li X, Rupprechter G. Sum frequency generation spectroscopy in heterogeneous model catalysis: A minireview of CO-related processes. Catal Sci Technol. 2021;11:12–26.

[48] Rupprechter G. Surface science approach to heterogeneous catalysis. In IK Wandelt (Ed.), *Surface and Interface Science: Solid-Gas Interfaces I* (pp. 459–527). Weinheim: Wiley-VCH, 2016.

[49] Surnev S, Sock M, Ramsey MG, Netzer FP, Wiklund M, Borg M, Andersen JN. CO adsorption on Pd(111): A high-resolution core level photoemission and electron energy loss spectroscopy study. Surf Sci. 2000;470:171–185.

[50] Rupprechter G, Kaichev VV, Unterhalt H, Morkel M, Bukhtiyarov VI. CO dissociation and CO hydrogenation on smooth and ion-bombarded Pd(111): SFG and XPS spectroscopy at mbar pressures. Appl Surf Sci. 2004;235:26–31.

[51] Morkel M, Unterhalt H, Salmeron M, Rupprechter G, Freund H-J. SFG spectroscopy from 10^{-8} to 1000 mbar: Less-ordered CO structures and coadsorption on Pd(111). Surf Sci. 2003;532-535: 103–107.

[52] Kaichev VV, Morkel M, Unterhalt H, Prosvirin IP, Bukhtiyarov VI, Rupprechter G, Freund H-J. C–O bond scission on "defect-rich and perfect" Pd(111)? Surf Sci. 2004;566–568:1024–1029.

[53] McCrea K, Parker JS, Chen P, Somorjai G. Surface structure sensitivity of high-pressure CO dissociation on Pt(557), Pt(100) and Pt(111) using sum frequency generation surface vibrational spectroscopy. Surf Sci. 2001;494:238–250.

[54] Vandichel M, Grönbeck H. A dimer path for CO dissociation on PtSn. Catal Sci Technol. 2019;9:695–701.

[55] Silvestre-Albero J, Borasio M, Rupprechter G, Freund H-J. Combined UHV and ambient pressure studies of 1,3-butadiene adsorption and reaction on Pd(111) by GC, IRAS and XPS. Catal Commun. 2007;8:292–298.

[56] Teschner D, Vass E, Hävecker M, Zafeiratos S, Schnörch P, Sauer H, Knop-Gericke A, Schlögl R, Chamam M, Wootsch A, Canning AS, Gamman JJ, Jackson SD, McGregor J, Gladden LF. Alkyne hydrogenation over Pd catalysts: A new paradigm. J Catal. 2006;242:26–37.

[57] Bowker M, Morgan C, Perkins N, Holroyd R, Fourre E, Grillo F, MacDowall A. Ethene adsorption, dehydrogenation and reaction with Pd(110): Pd as a carbon 'sponge'. J Phys Chem B. 200;109(5): 2377–2386.

[58] Gabasch H, Hayek K, Klötzer B, Knop-Gericke A, Schlögl R. Carbon incorporation in Pd(111) by adsorption and dehydrogenation of ethene. J Phys Chem B. 2006;110:4947–4952.

[59] Morkel M, Kaichev VV, Rupprechter G, Freund H-J, Prosvirin IP, Bukhtiyarov VI. Methanol dehydrogenation and formation of carbonaceous overlayers on Pd(111) studied by high-pressure SFG and XPS spectroscopy. J Phys Chem B. 2004;108:12955–12961.

[60] Borasio M, Rodríguez de la Fuente O, Rupprechter G, Freund H-J. In situ studies of methanol decomposition and oxidation on Pd(111) by PM-IRAS and XPS spectroscopy. J Phys Chem B. 2005;109:17791–17794.

[61] Rodríguez de la Fuente O, Borasio M, Galletto P, Rupprechter G, Freund H-J. The influence of surface defects on methanol decomposition on Pd(111) studied by XPS and PM-IRAS. Surf Sci. 2004;566–568:740–745.

[62] Zafeiratos S, Piccinin S, Teschner D. Alloys in catalysis: Phase separation and surface segregation phenomena in response to the reactive environment. Catal Sci Technol. 2012;2:1787–1801.

[63] Khan NA, Uhl A, Shaikhutdinov S, Freund H-J. Alumina supported model Pd–Ag catalysts: A combined STM, XPS, TPD and IRAS study. Surf Sci. 2006;600:1849–1853.

[64] Chen M, Kumar D, Yi C-W, Goodman DW. The promotional effect of gold in catalysis by palladium-gold. Science. 2005;310:291–293.

[65] Garcia C, Truttmann V, Lopez I, Haunold T, Marini C, Rameshan C, Pittenauer E, Kregsamer P, Dobrezberger K, Stöger-Pollach M,

Barrabés N, Rupprechter G. Dynamics of Pd dopant atoms inside Au nanoclusters during catalytic CO oxidation. J Phys Chem C. 2020;124:23626–23636.

[66] Barrabés N, Cornado D, Föttinger K, Dafinov A, Llorca J, Medina F, Rupprechter G. Hydrodechlorination of trichloroethylene on noble metal promoted Cu-hydrotalcite-derived catalysts. J Catal. 2009;263:239–246.

[67] Barrabés N, Frare A, Föttinger K, Urakawa A, Llorca J, Rupprechter G, Tichit D. Pt-Cu bimetallic catalysts obtained from layered double hydroxides by an anion-exchange route. Appl Clay Sci. 2012; 69:1–10.

[68] Haghofer A, Ferri D, Föttinger K, Rupprechter G. Who is doing the job? Unraveling the role of Ga_2O_3 in methanol steam reforming on Pd_2Ga/Ga_2O_3. ACS Catal. 2012;2:2305–2315.

[69] Raman N, Maisel S, Grabau M, Taccardi N, Debuschewitz J, Wolf M, Wittkämper H, Bauer T, Wu M, Haumann M, Papp C, Görling A, Spiecker E, Libuda J, Steinrück H-P, Wasserscheid P. Highly effective propane dehydrogenation using Ga–Rh supported catalytically active liquid metal solutions. ACS Catal. 2019;9:9499–9507.

[70] Rameshan C, Stadlmayr W, Weilach C, Penner S, Lorenz H, Hävecker M, Blume R, Rocha T, Teschner D, Knop-Gericke A, Schlögl R, Memmel N, Zemlyanov D, Rupprechter G, Klötzer B. Subsurface-controlled CO_2 selectivity of PdZn near-surface alloys in H_2 generation by methanol steam reforming. Angew Chemie Int Ed. 2010;49:3224–3227.

[71] Weilach C, Kozlov SM, Holzapfel HH, Föttinger K, Neyman KM, Rupprechter G. Geometric arrangement of components in bimetallic PdZn/Pd(111) surfaces modified by CO adsorption: A combined study by density functional calculations, polarization-modulated infrared reflection absorption spectroscopy, and temperature-programmed desorption. J Phys Chem C. 2012;116: 18768–18778.

[72] Holzapfel HH, Wolfbeisser A, Rameshan C, Weilach C, Rupprechter G. PdZn surface alloys as models of methanol steam reforming catalysts: Molecular studies by LEED, XPS, TPD and PM-IRAS. Top Catal. 2014;57:1218–1228.

[73] Stadlmayr W, Rameshan C, Weilach C, Lorenz H, Hävecker M, Blume R, Rocha T, Teschner D, Knop-Gericke A, Zemlyanov D, Penner S, Schlögl R, Rupprechter G, Klötzer B, Memmel N. Temperature-induced modifications of PdZn layers on Pd(111). J Phys Chem C. 2010;114:10850–10856.

[74] Neyman KM, Lim KH, Chen Z-X, Moskaleva LV, Bayer A, Reindl A, Borgmann D, Denecke R, Steinrück H-P, Rösch N. Microscopic models of PdZn alloy catalysts: Structure and reactivity in methanol decomposition. Phys Chem Chem Phys. 2007;9:3470–3482.

[75] Bayer A, Flechtner K, Denecke R, Steinrück H, Neyman KM, Rösch N. Electronic properties of thin Zn layers on Pd(111) during growth and alloying. Surf Sci. 2006;600:78–94.

[76] Zorn K, Giorgio S, Halwax E, Henry CR, Grönbeck H, Rupprechter G. CO oxidation on technological Pd–Al_2O_3 catalysts: Oxidation state and activity. J Phys Chem C. 2011;115:1103–1111.

[77] Lundgren E, Mikkelsen A, Andersen JN, Kresse G, Schmid M, Varga P. Surface oxides on close-packed surfaces of late transition metals. J Phys Condens Matter. 2006;18:R481–R499.

[78] Gabasch H, Knop-Gericke A, Schlögl R, Borasio M, Weilach C, Rupprechter G, Penner S, Jenewein B, Hayek K, Klötzer B. Comparison of the reactivity of different Pd–O species in CO oxidation. Phys Chem Chem Phys. 2007;9:533–540.

[79] Salomonsson P, Johansson S, Kasemo B. Methane oxidation over PdO_x: On the mechanism for the hysteresis in activity and oxygen content. Catal Lett. 1995;33:1–13.

[80] Lundgren E, Kresse G, Klein C, Borg M, Andersen JN, De Santis M, Gauthier Y, Konvicka C, Schmid M, Varga P. Two-dimensional oxide on Pd(111). Phys Rev Lett. 2002;88:246103.

[81] Gabasch H, Unterberger W, Hayek K, Klötzer B, Kresse G, Klein C, Schmid M, Varga P. Growth and decay of the Pd(111)–Pd_5O_4 surface oxide: Pressure-dependent kinetics and structural aspects. Surf Sci. 2006;600:205–218.

[82] Zemlyanov D, Aszalos-Kiss B, Kleimenov E, Teschner D, Zafeiratos S, Hävecker M, Knop-Gericke A, Schlögl R, Gabasch H, Unterberger W, Hayek K, Klötzer B. In situ XPS study of Pd(111) oxidation. Part 1: 2D oxide formation in 10^{-3} mbar O_2. Surf Sci. 2006;600:983–994.

[83] Ketteler G, Ogletree DF, Bluhm H, Liu H, Hebenstreit ELD, Salmeron M. In situ spectroscopic study of the oxidation and reduction of Pd(111). J Am Chem Soc. 2005;127:18269–18273.

[84] Datye AK, Bravo J, Nelson TR, Atanasova P, Lyubovsky M, Pfefferle L. Catalyst microstructure and methane oxidation reactivity during the Pd↔PdO transformation on alumina supports. Appl Catal A. 2000;198:179–196.

[85] Demoulin O, Rupprechter G, Seunier I, Le Clef B, Navez M, Ruiz P. Investigation of parameters influencing the activation of a Pd/ γ-alumina catalyst during methane combustion. J Phys Chem B. 2005;109:20454–20462.

[86] Gabasch H, Hayek K, Klötzer B, Unterberger W, Kleimenov E, Teschner D, Zafeiratos S, Hävecker M, Knop-Gericke A, Schlögl R, Aszalos-Kiss B, Zemlyanov D. Methane oxidation on Pd(111): In situ XPS identification of active phase. J Phys Chem C. 2007;111: 7957–7962.

[87] Kuhlenbeck H, Shaikhutdinov S, Freund H-J. Well-ordered transition metal oxide layers in model catalysis — A series of case studies. Chem Rev. 2013;113:3986–4034.

[88] Antlanger M, Mayr-Schmölzer W, Pavelec J, Mittendorfer F, Redinger J, Varga P, Diebold U, Schmid M. $Pt_3Zr(0001)$: A substrate for growing well-ordered ultrathin zirconia films by oxidation. Phys Rev B. 2012;86:035451.

[89] Tauster SJ, Fung SC, Garten RL. Strong metal-support interactions. Group 8 noble metals supported on titanium dioxide. J Am Chem Soc. 1978;100:170–175.

[90] Hayek K, Kramer R, Paál Z. Metal-support boundary sites in catalysis. Appl Catal A. 1997;162:1–15.

[91] Hayek K, Fuchs M, Klötzer B, Reichl W, Rupprechter G. Studies of metal–support interactions with "real" and "inverted" model systems: Reactions of CO and small hydrocarbons with hydrogen on noble metals in contact with oxides. Top Catal. 2000;13: 55–66.

[92] Campbell CT. Electronic perturbations. Nature Chem. 2012;4: 597–598.

[93] Suchorski Y, Kozlov SM, Bespalov I, Datler M, Vogel D, Budinska Z, Neyman KM, Rupprechter G. The role of metal/oxide interfaces for long-range metal particle activation during CO oxidation. Nature Mater. 2018;17:519–522.

[94] Roiaz M, Pramhaas V, Li X, Rameshan C, Rupprechter G. Atmospheric pressure reaction cell for operando sum frequency generation spectroscopy of ultrahigh vacuum grown model catalysts. Rev Sci Instrum. 2018;89:045104.

[95] Rupprechter G, Dellwig T, Unterhalt H, Freund H-J. CO adsorption on Ni(100) and Pt(111) studied by infrared–visible sum frequency generation spectroscopy: Design and application of an SFG-compatible UHV–high-pressure reaction cell. Top Catal. 2001;15:19–26.

[96] Vogel D, Spiel C, Suchorski Y, Trinchero A, Schlögl R, Grönbeck H, Rupprechter G. Local catalytic ignition during CO oxidation on low-index Pt and Pd surfaces: A combined PEEM, MS, and DFT study. Angew Chemie Int Ed. 2012;51:10041–10044.

[97] Suchorski Y, Rupprechter G. Catalysis by imaging: From meso- to nano-scale. Top Catal. 2020;63:1532–1544.

[98] Schalow T, Brandt B, Laurin M, Schauermann S, Libuda J, Freund H-J. CO oxidation on partially oxidized Pd nanoparticles. J Catal. 2006;242:58–70.

[99] Suchorski Y, Rupprechter G. Heterogeneous surfaces as structure and particle size libraries of model catalysts. Catal Lett. 2018;148: 2947–2956.

[100] Suchorski Y, Datler M, Bespalov I, Zeininger J, Stöger-Pollach M, Bernardi J, Grönbeck H, Rupprechter G. Visualizing catalyst heterogeneity by a multifrequential oscillating reaction. Nature Commun. 2018;9:600.

[101] Todorova M, Li WX, Ganduglia-Pirovano MV, Stampfl C, Reuter K, Scheffler M. Role of subsurface oxygen in oxide formation at transition metal surfaces. Phys Rev Lett. 2002;89:096103.

[102] Gustafson J, Mikkelsen A, Borg M, Lundgren E, Köhler L, Kresse G, Schmid M, Varga P, Yuhara J, Torrelles X, Quirós C, Andersen JN. Self-limited growth of a thin oxide layer on Rh(111). Phys Rev Lett. 2004;92:126102.

[103] Flege JI, Hrbek J, Sutter P. Structural imaging of surface oxidation and oxidation catalysis on Ru(0001). Phys Rev B. 2008;78:165407.

[104] Diebold U, Li S-C, Schmid M. Oxide surface science. Ann Rev Phys Chem. 2010;61:129–148.

[105] Suchorski Y, Spiel C, Vogel D, Drachsel W, Schlögl R, Rupprechter G. Local reaction kinetics by imaging: CO oxidation on polycrystalline platinum. Chemphyschem. 2010;11:3231–3235.

[106] Suchorski Y, Rupprechter G. Local reaction kinetics by imaging. Surf Sci. 2016;643:52–58.

[107] Winkler P, Zeininger J, Raab M, Suchorski Y, Steiger-Thirsfeld A, Stöger-Pollach M, Amati M, Gregoratti L, Grönbeck H, Rupprechter G. Coexisting multi-states in catalytic hydrogen oxidation on rhodium. Nat Commun. 2021;12(2021): 6517.

[108] Höfert O, Gleichweit C, Steinrück H-P, Papp C. Ultrafast X-ray photoelectron spectroscopy in the microsecond time domain. Rev Sci Instrum. 2013;84:093103.

[109] Opitz AK, Nenning A, Rameshan C, Kubicek M, Götsch T, Blume R, Hävecker M, Knop-Gericke A, Rupprechter G, Klötzer B, Fleig J. Surface chemistry of perovskite-type electrodes during high temperature CO_2 electrolysis investigated by operando photoelectron spectroscopy. ACS Appl Mater Interfaces. 2017;9:35847–35860.

https://doi.org/10.1142/9781800613294_0008

Chapter 8

Application of Photoelectron Spectroscopy to Align the Energy Levels of Photocatalysts

Andreas Klein

Technische Universität Darmstadt, Institute of Materials Science, Electronic Structure of Materials,
Otto-Berndt-Str. 3, Darmstadt 64287, Germany
andreas.klein@tu-darmstadt.de

Abstract

The absolute position of the energy band edges is an important quantity affecting charge transfer processes at interfaces and thereby catalytic properties. This chapter describes how photoelectron spectroscopy can be applied to determine the energy level alignment of photocatalysts. The description focuses on semiconducting oxides and emphasizes the related opportunities and challenges. Comprehensive surface potential measurements are provided for a series of binary oxides illuminating the dependence of the Fermi energy, the electron affinity, and the work function on sample doping and preparation conditions. The experimental procedure for the direct determination of Schottky barrier heights and semiconductor energy band alignment using photoelectron spectroscopy as well as different effects resulting in a modification of band alignment will be described. The power of photoelectron spectroscopy for identifying the modifications and their guidelines for deriving an intrinsic alignment of compounds will be provided.

8.1 Introduction

8.1.1 *Importance of energy level alignment for photocatalysis*

Energy band alignment is a measure for the absolute positions of the energy bands of a material. In metallic compounds, the Fermi energy, E_F, can be used as a reference for the energy levels of all electronic states as it is fixed by the concentration of electrons and difficult to change due to a high density of states at the Fermi energy. In non-metallic compounds, which exhibit an energy gap between the highest occupied and lowest unoccupied electronic states, the Fermi energy is at a position where the density of states is low, ideally zero, making it very easy to change. Instead of the Fermi energy, the highest occupied valence band state and the lowest unoccupied conduction band state can be taken as a reference level for the energies of all electronic states in a non-metallic material. The latter are the valence band maximum, E_{VB}, and the conduction band minimum, E_{CB}, respectively.

Aligning the energy bands of two materials corresponds to an alignment of all electronic states of one material relative to those of the other. In electrical contact, electrons can be exchanged between materials, resulting in an alignment of the electrochemical potential, which is taken to be the same as the Fermi energy here. For a contact between two metals, the alignment of the energy bands is hence fixed by the alignment of the Fermi energy. For a contact between a metallic and a non-metallic material, the alignment corresponds to the position of the Fermi energy at the interface relative to E_{VB} and E_{CB}. These two quantities are identical to the Schottky barrier heights for holes, $\Phi_{B,p}$, and electrons, $\Phi_{B,n}$, respectively. At a contact between two non-metallic materials, the alignment results in a discontinuity (also called offset) in the valence band, ΔE_{VB}, and in the conduction band, ΔE_{CB}. Together with the Fermi energies far away from the contact between two materials, which are determined by doping, the Schottky barrier heights and band offsets determine the course of the electrostatic potential across the interface. This is crucial for the charge transport properties and determines the function of semiconductor devices [1]. Knowing the relative positions of the energy bands of materials is therefore of fundamental importance.

In photocatalytic applications, solid/liquid and solid/gas interfaces are involved [2–4]. For such interfaces, charge transfer processes from the substrate to respective energy levels in the liquids or gas molecules are of key importance for the catalytic properties. Therefore, the position of the energy levels of a solid relative to those of the redox potentials of the active species in a liquid or of (adsorbed) gas molecules is a key issue for understanding the catalytic behavior. The energetic situations for single and heterojunction photocatalysts are illustrated in Figure 8.1.

The top panel in Figure 8.1 shows the different steps in a photocatalytic process, which involve photogeneration of electron–hole pairs, their transport to the surface, and finally the transfer to the reactant. A reduction process would require the electron in the catalyst to have an energy higher than the redox level of the reactant, while an oxidation of the reactant requires that the hole has an energy lower than the respective redox level. In most cases, the energies of the electrons and holes are associated with the minimum of the conduction band, E_{CB}, and the maximum of the valence band, E_{VB}, respectively, as shown on the left side of Figure 8.1(a). This is not always the case, however. As illustrated on the right side of Figure 8.1(a), electrons and holes may be trapped and form polaronic states. This is not only a problem for the transport of the charges to the surface, which occurs by a low mobility hopping process, it also affects the energy levels of the electrons and holes in the catalyst. Such a situation can explain the low ability of Fe_2O_3 to reduced water, as photogenerated electrons are trapped and reduce Fe to Fe^{2+}. The energy of the electrons is then ~0.5 eV below the conduction band which is below the hydrogen evolution potential [5].

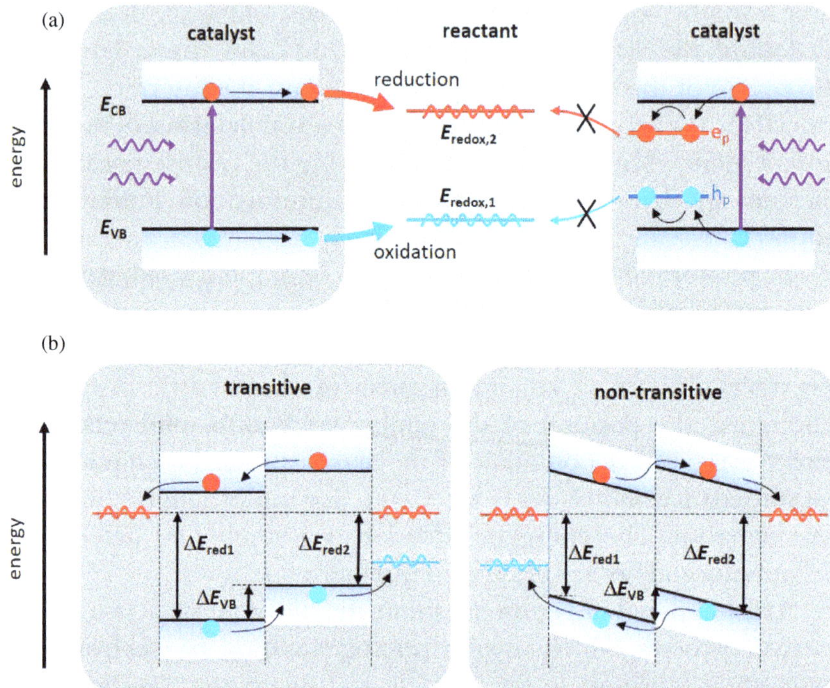

Fig. 8.1. Panel (a) shows the energy bands of a semiconductor relative to two redox potentials of either gaseous or liquid reactants. The left part shows the situation where the photoexcited carriers reside in the energy bands, where they are transported to the surface/interface. They can reduce/oxidize the reactant if the conduction/valence band is above/below the corresponding redox level. The right part shows the situation where both electrons and holes are trapped and form polarons. Reduction or oxidation of the reactant is then no more determined by the position of the band edges but by the polaronic energy levels, e_p and h_p. Such a situation has been observed to limit water reduction of Fe_2O_3 [5]. Panel (b) shows energy bands of a photocatalyst consisting of a pair of materials forming a heterojunction with a valence band discontinuity, ΔE_{VB}. Three interfaces are involved in the photocatalytic process. The left part shows the situation with a transitive band alignment and the right with non-transitive alignment. In the latter case, $\Delta E_{red1} \neq \Delta E_{VB} + \Delta E_{red2}$, which results in a built-in potential. A non-transitive alignment is encountered for mixed rutile/anatase TiO_2 particles [6]. Charge separation can be opposite for transitive and non-transitive band alignments.

Figure 8.1(a) does not include a driving source for charge separation of the photogenerated electrons and holes. It is required, however, to have the oxidation and reduction occurring at different locations. One possibility to separate electrons and holes is the use of a heterojunction catalyst, as shown in Figure 8.1(b). Alternative solutions are to generate a built-in electric field, for example, by partially coating the catalyst [7–9] or by utilizing facetted crystallites having different surface potentials [10].

In order to provide general guidelines for selecting photocatalytic materials, it is desirable to align all materials on a common energy scale. If this is possible, it will not be necessary to determine the energy band alignment between all possible combinations of materials but only the position of the energy levels of each material on this common energy scale. The alignment of the energy bands between any two materials, A and B, can then be obtained from their position relative to any reference level on the common energy scale employing transitivity:

$$E_A - E_B = (E_A - E_{ref}) - (E_B - E_{ref}). \tag{8.1}$$

However, transitivity of energy level alignment is not always observed. A deviation may indicate that the alignment is modified by some extrinsic contributions. Possible reasons for this will be discussed in Section 8.3.

Several energy levels are used as reference levels for aligning energy bands of solids. Among them are the vacuum energy, charge neutrality levels, or the standard hydrogen electrode potential. The first and the latter are related for aqueous solutions by a difference of 4.44 eV [11]. Charge neutrality levels of solids are used to align energy bands of semiconductors in cases of a high density of induced gap states, which is mostly the case for covalently bonded materials [12]. Charge neutrality levels can be calculated in different ways revealing partially different results [13–15]. The use of the vacuum energy as a reference level is also considered to be critical, as the respective potentials depend on surface condition (see Section 8.2). A similar argument holds for the standard hydrogen electrode potential, as dipole layers involving ionic species will form on the solid surface in contact to an electrolyte, which can shift the energy levels of the solid substantially with respect to those of the liquid [16].

An insightful test case for energy band alignment is given by anatase and rutile TiO_2. Nanoparticles containing mixtures of these two polytypes have been identified as being more efficient photocatalysts as single phase materials [17]. The synergistic effect of the mixed systems has been attributed to a built-in driving force for separation of photogenerated charge carriers. Such a driving force may result from either a built-in electric field or from energy barriers blocking charge transfer at the interface between anatase and rutile. The latter situation is illustrated in Figure 8.1(b). In order to determine the potential barriers, Kavan *et al.* have performed electrochemical measurements that located the conduction band edge of anatase 0.2 eV above that of rutile [18]. In contrast, more recent theoretical and experimental approaches agree that the band alignment at anatase/rutile solid-state junctions is the opposite, with the conduction band of rutile being 0.5 eV above that of anatase [6,19, 20]. Given that the results of the electrochemical measurements are valid for solid/electrolyte interfaces, the energy band alignment at the electrolyte/anatase/rutile/electrolyte interfaces becomes non-transitive. This non-transitivity could be beneficial for the photocatalytic properties, as it induces an electrical field for charge separation. The anticipated situation is illustrated on the right side of Figure 8.1(b).

Although the alignment of energy bands on a common energy scale with a single reference level is appealing, it is currently not well settled whether such an alignment is possible and reasonable. Systematic data are therefore required to obtain an accurate description of energy level alignment.

8.1.2 *Techniques to determine energy level alignment*

A series of techniques are available to determine the energy level alignment at interfaces [1,21,22]. Schottky barrier heights are most accurately determined from electrical measurements, which involve either current measurements in dependence on voltage and/or temperature or capacitance-voltage measurements. The former relies on the diode equation with an exponential dependence of the forward current on applied voltage and the latter determines the capacitance of the space-charge region. Using the Mott–Schottky analysis of such measurements, where $1/C^2$ is plotted in dependence on voltage, one can obtain both the band bending and the doping concentration. The barrier heights are then obtained by combining the bulk Fermi energy, which can be calculated from the doping concentration and the band bending. The analysis of band discontinuities of semiconductor heterojunctions by electrical measurements is more difficult. Detailed descriptions can be found in the book of Capasso and Margaritondo [21].

A disadvantage of electrical measurements is that bulk transport properties must be well controlled and that ohmic contacts are required. This may be difficult to realize for new compounds. For insulating samples, a different approach has been used for example with thin film ferroelectric materials, which exhibit barrier heights around 1 eV [23]. The leakage current through such metal–insulator–metal junctions has been proposed to be limited by charge injection. With a high electric field applied, the injection barrier is lowered by the image charge potential, resulting in a dependence of the barrier height on the square root of the applied voltage [22].

Schottky barrier heights and band offsets can be derived from internal photoemission measurements, where the photo-induced current is measured in dependence on photon energy [1,21,22]. A related technique is ballistic electron emission microscopy, where the photo-induced current is detected locally using a scanning tunneling microscope [24].

Frequently used in literature is the alignment of energy bands using electrochemical measurements, which mostly use impedance spectroscopy of solid/electrolyte interfaces for the determination of the flat band potentials using a Mott–Schottky analysis [25]. The energy band positions of the studied material are then obtained by assuming the Fermi energy in the bulk, which is not directly observed in such measurements. If charge transport does not occur in the conduction band but by polaron hopping, the measurements may reveal the carrier trapping level rather than the band edge energy as in the case of hematite Fe_2O_3 [5]. In the case of hematite, different electrochemical measurements have also revealed different flat band positions [26]. Once the band edge energies of a material are derived, the energy bands of different materials can be compared with each other by referencing the measured band edge positions to a standard electrode, if transitivity is valid.

In this chapter, the determination of surface potentials and energy band alignment by means of electron spectroscopy is described. In principle, this technique offers two approaches based (i) on the quantification of surface potentials and (ii) on the direct determination of barrier heights by interface analysis. The two approaches and their application to oxides are described in detail in the following two sections.

8.2 Surface Potentials

One possibility to align energy bands of different materials is their alignment with respect to the vacuum energy. This involves the work function, ϕ, the electron affinity, χ, and the ionization potential, I_P, of the materials. The ionization potential is the lowest photon energy required to remove an electron from a material by illumination. It corresponds to the transfer of an electron from the topmost occupied state at the valence band maximum to the vacuum energy: $I_P = E_{vac} - E_{VB}$. The electron affinity is defined by the energy gained when transferring an electron from vacuum to the lowest unoccupied state at the conduction band minimum: $\chi = E_{vac} - E_{CB} = I_P - E_g$. Finally, the work function is defined by the thermal removal of carriers from a material and is measured by the distance between the Fermi energy and the vacuum level: $\phi = E_{vac} - E_F$. For metallic samples, the three potentials coincide: $\chi = \phi = I_P$. The different surface potentials and their determination using photoelectron spectra are illustrated for non-metallic materials in Figure 8.2.

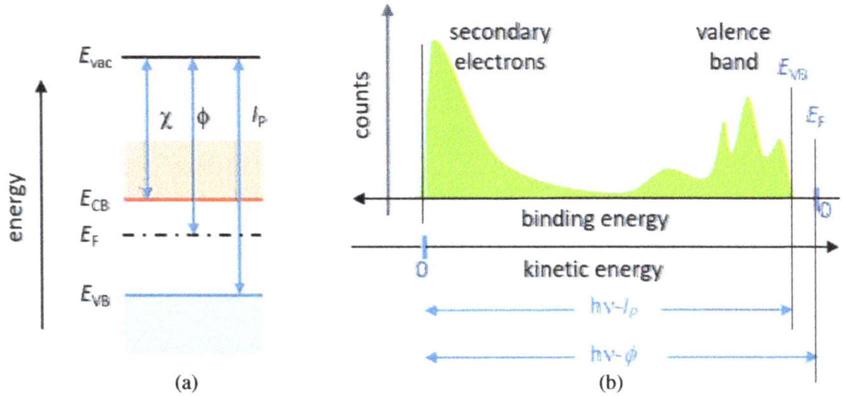

Fig. 8.2. Surface potentials of non-metallic materials including the electron affinity χ, the work function ϕ, and the ionization potential I_P (a). A schematic ultraviolet photoelectron spectrum showing the valence band states and the secondary emissions is shown in (b). The binding energy scale can be calibrated using a metallic sample to set the zero of binding energy to the Fermi energy E_F. The work function and ionization potential can then be obtained from the secondary electron cutoff and the onset of the valence band emissions as indicated in (b).

An alignment of energy bands of solids using the vacuum energy is very limited, as the surface potentials can vary substantially with surface condition. Changes of the Fermi energy can occur upon doping of the material or due to surface band bending. In addition, the electron affinity and ionization potential can change when the surface dipole is changing. The work function of metals, for example, depends on the orientation of the surface [27]. Substantial variations of more than 0.5 eV occur for certain elemental materials. For semiconductors, Ranke has determined a variation of the electron affinity of ~0.3 eV for Si using photoelectron spectroscopy of a cylindrical single crystal [28]. For the more polar semiconductor GaAs, a variation of the electron affinity of ~0.5 eV has been obtained using the same approach [29]. The variation might be even larger when different surface terminations are taken into account [30]. For the (100) surface of GaAs, four different surface reconstructions are known depending on the relative activity of Ga and As. The reason for the variation of surface reconstructions is the electrostatic instability of polar surfaces [31,32]. A well-known example is the (111) surface of GaAs, where the polarity is stabilized by a 2×2 reconstructed surface, where every fourth Ga atom is missing [33].

More relevant for catalytic applications than the conventional tetrahedrally bonded semiconductors are oxides [34,35]. Due to their polar bonding, a strong dependence of the surface potentials on surface orientation has to be expected. Reported surface potentials of single crystal ZnO exhibit a similar dependence on surface orientation as GaAs. The ionization potential is lowest for the cation-terminated (0001) surface with $I_P = 7.0$ eV and 0.8 eV higher for other surface orientations such as the $(000\bar{1})$ and the $(10\bar{1}0)$ surfaces [36,37]. The variation of the ionization potential of polycrystalline ZnO thin films is within the same limits and can thus be explained by the dependence of surface orientation on deposition conditions [38,39]. A clear dependence of ionization potential on surface orientation is also obtained for In_2O_3 [40,41]. In this case, the nonpolar (110) and the polar but dipole-free (111) surfaces exhibit no variation of surface termination within the chemical stability window, whereas the polar (100) surface can be stabilized by a variety of terminations with calculated ionization potentials varying by almost 2 eV.

The variation of surface potentials for SnO_2, TiO_2, and CeO_2 is shown in Figure 8.3. All data have been recorded directly after sample or surface preparation in an UHV cluster tool, which combines several thin film deposition chambers and a multitechnique surface analysis system by an ultrahigh vacuum sample transfer system [42,43]. By plotting the work function in dependence on the Fermi energy, samples with the same ionization potential will lie on a line with a slope of -1, indicating that the variation of the work function is only caused by a variation of the Fermi level position within the band gap. Such a variation can be obtained either by preparation of materials with different doping or with different preparation conditions or post-deposition treatments. For oxides, a variation of the Fermi energy can be obtained by a variation of the oxygen activity.

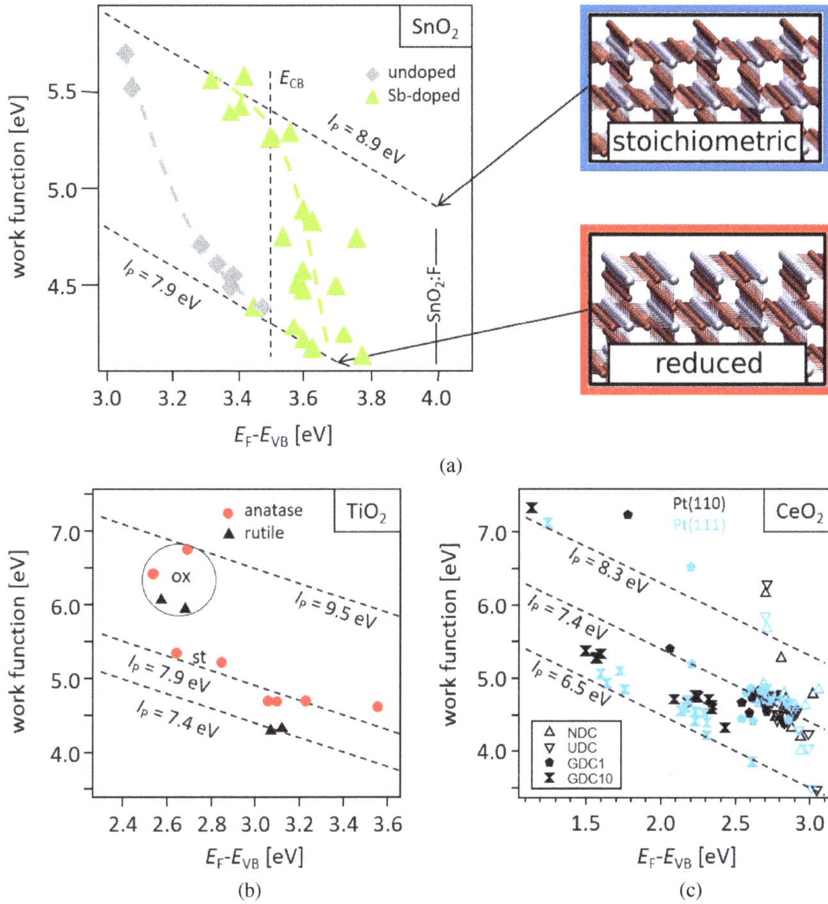

Fig. 8.3. Surface potentials of SnO_2 (a), TiO_2 (b), and CeO_2 [44] (c), as obtained using photoelectron spectroscopy with *in situ* sample preparation. Undoped and donor-doped (Sb) thin films' samples have been studied for SnO_2 [45] and single crystals for TiO_2 [46]. In the case of CeO_2, nominally undoped films (UDC), donor-doped films containing 1% Nb (NDC), and acceptor-doped films containing 1% (GDC1) or 10% Gd (GDC10) have been studied [44,47].

The data shown in Figure 8.3 have been obtained for different doped thin films prepared by magnetron sputtering (SnO_2 and CeO_2) or for single crystals (TiO_2). The approach is even suitable for nominally insulating materials. For CeO_2, which has a rather low conductivity at room temperature, the data have been obtained from thin films with a thickness of 40–50 nm, which have been deposited on Pt-coated single crystalline substrates [44,47]. With such a configuration, charging of the samples can be avoided. This is a prerequisite for a quantification of surface potentials.

For SnO_2, which is a classical gas sensor material, the ionization potential varies between 7.9 and 8.9 eV. These limits can be assigned to stoichiometric and reduced surface terminations, which are characterized by Sn^{4+} and Sn^{2+} oxidation states in the topmost surface layers [48], as indicated at the right part of Figure 8.3(a). A comparable variation of the ionization potential has been observed on different low-index SnO_2 surfaces [49]. It is furthermore evident that the ionization potential increases with more oxidizing conditions (lower Fermi energies). This variation is directly connected to the surface oxygen exchange behavior [50]. For undoped SnO_2, the work function of oxidized SnO_2 samples is too high to facilitate electron transfer from the substrate to oxygen molecules at the surface, thereby preventing the formation of adsorbed O_2^- or O_2^{2-} species and limiting oxygen incorporation. This is beneficial for gas sensor operation, as changes induced by varying the gas compositions are limited to the very surface and do not result in a (slower) change of the bulk defect concentration. This is consistent with a sluggish surface oxygen exchange reaction [51].

The work functions of TiO_2 after oxygen plasma treatment, indicated by "ox" in Figure 8.3(b), are rather high. These high values are explained by a surface dipole induced by the formation of adsorbed negatively charged oxygen species [46]. For other preparation conditions, anatase TiO_2 exhibits a rather constant ionization potential of ~7.9 eV. Evidently, TiO_2 does not exhibit the same variation of the cation oxidation state as Sn. The rutile modification of TiO_2 has not been studied as intensively as the anatase one. The obtained ionization potential is about 0.5 eV lower than that of anatase, which corresponds very well with the higher lying energy bands of rutile (see also discussion of energy band alignment in the following) [6].

The largest variation of the Fermi energy of almost 2 eV is obtained for CeO_2. This huge variation is related to the fact that both donor- and acceptor-doped materials have been studied. The resulting Fermi energy positions are close to what is expected from defect chemistry calculations, indicating the measured Fermi energies are representative of the bulk values and not affected by Fermi level pinning at surface states [52]. Such a behavior has also been observed for transparent conducting oxides [42] and for transition metal oxides with partially filled d-shells, such as Fe_2O_3 [5].

Except for the highest values, which are again obtained after oxygen plasma treatments, the ionization potential of the CeO_2 films depends hardly on the Fermi energy and shows no systematic variation with surface orientation [44,47]. However, the ionization potentials of the films doped with 10% Gd are ~0.7 eV lower than those of undoped and of Nb-doped films. This difference can be assigned to a surface space-charge region [44]. For high defect concentrations, as in the case of 10% doping, the extension of the surface space-charge region becomes less than 1 nm [53]. The effect of such a space-charge region on the work function is comparable to that of a surface dipole, while electron spectroscopy still probes the Fermi energy in the bulk. This results in an apparent reduction of the ionization potential, which is beneficial for oxygen incorporation as it lowers the redox level of adsorbed oxygen and thereby facilitates electron transfer, opposite to the effect observed for SnO_2 [52].

Unfortunately, systematic studies of the surface potentials as those presented above are only available for very few materials. The discussed results indicate, however, that single values of work functions are not representative of surface behavior. The dependencies of the surface potentials on doping and defect concentrations as well as that of surface orientation and termination have to be taken into account. According to these dependencies, an alignment of energy bands with respect to the vacuum energy is not reasonable. This does not exclude that a vacuum level alignment provides the same energy band positions as that obtained by measurements at interfaces. Such a behavior has been observed for anatase and rutile TiO_2 [46] and for $SrTiO_3$ and $PbTiO_3$ [54]. In these cases, the agreement of the different alignments might be related to the similar crystallographic structures of these combinations, but the available data are not sufficient to draw such a conclusion.

8.3 Interface Studies

8.3.1 *Determination of energy band alignment using XPS*

The standard method to obtain the energy band alignment between two materials by means of photoelectron spectroscopy has been proposed more than 40 years ago by Kraut and coworkers [55]. The method consists of at least three steps, involving the measurement of (i) the bare substrate surface, (ii) the interface with a signal from the substrate and a thin overlayer, and (iii) the thick overlayer. The XPS signal in Steps (i) and (iii) contains exclusively signals from the substrate and the overlayer, respectively. The signal of Step (ii) contains a mixture of the substrate and overlayer emissions. This implies that the overlayer thickness is of similar magnitude as the inelastic mean free path of the photoelectrons, which depends on their kinetic energy and is typically 1 – 3 nm for conventional laboratory-based instruments with Mg or Al X-ray anodes. At synchrotrons, also shorter and longer inelastic mean free paths are accessible due to the wider range of photon energies. A schematic series of spectra recorded during such an interface experiment is shown in Figure 8.4.

With such experiments, the valence band offset, ΔE_{VB}, between two semiconductors is derived according to (see Figure 8.4)

$$\Delta E_{VB} = E_{CL}^{VB}(\text{sub}) - E_{CL}^{VB}(\text{film}) - \Delta E_{CL}, \tag{8.2}$$

with $E_{CL}^{VB}(\text{sub})$ being the binding energy difference between a core level and the valence band maximum of the substrate, $E_{CL}^{VB}(\text{film})$ the same for a core level of the overlayer, and ΔE_{CL} the binding energy difference between the substrate and the overlayer core levels. The values are obtained from the substrate, the overlayer, and the interface sample, respectively.

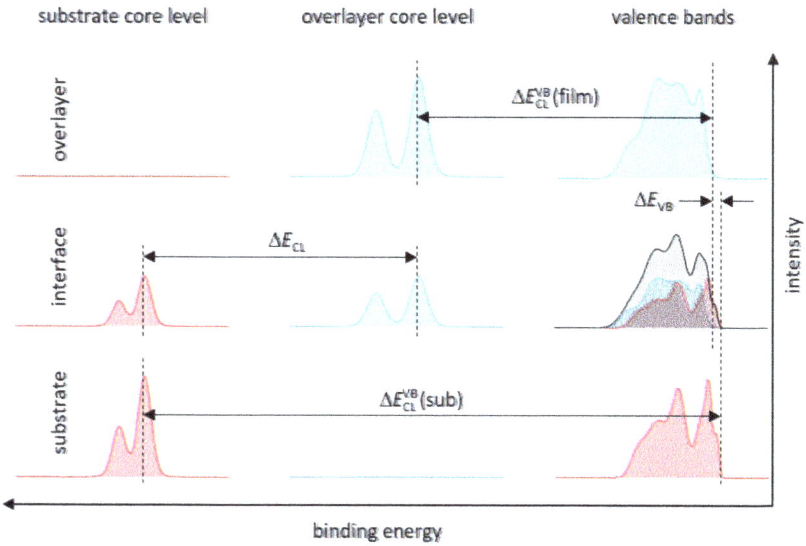

Fig. 8.4. Schematic illustration of the determination of energy band alignment between two semiconductors using photoelectron spectroscopy. Binding energy shifts due to the evolution of space-charge regions in the course of interface formation are neglected.

This procedure is quite general and allows the determination of the band alignment for many material combinations, as long as not all core levels are common in the substrate and the overlayer. However, the alignment determined from a single interface sample can be quite misleading. The importance of interfacial and bulk defects will be described in the following sections. In addition to the influence of defects, one has to assure that binding energy shifts of the substrate and the overlayer are in parallel in order to avoid an influence of space-charge regions in the overlayer. Different core levels of a material may also exhibit non-parallel binding energy shifts, which may be caused by a thickness dependent (electronic) structure or screening of the core levels [56–58]. In order to exclude such uncertainties, it is recommended to record all accessible energy levels and to follow their evolution in dependence on overlayer thickness.

8.3.2 *Modification of band alignment by interface defects*

This section will describe the influence of extrinsic (crystallographic) point defects at the interface on the alignment at semiconductor/metal interfaces, that is on the Schottky barrier heights. This effect should not be mixed up with the Fermi level pinning by intrinsic interface states [59,60]. The latter, which are also referred to as virtual gap states, are intrinsic to all interfaces and mostly pronounced for more covalently bonded semiconductors, such as Si or GaAs, which show a rather small variation of Schottky barrier height with metal electronegativity or work function [12,61].

Brillson has pointed out already in 1978 that Schottky barrier heights are pinned if the deposited metal reacts with the substrate [62]. Oxides are an ideal test case to study the dependence of barrier height on reactivity. While large variations of the interfacial Fermi energy are observed when metallic oxides with different work functions are used as barrier material [43,63,64], the Fermi level is pinned when interfacial reactions occur [58,65,66].

A chemical reduction of the substrate surface may occur even for the deposition of noble metals, such as Pt or Au [67,68], which has been assigned to the release of the condensation energy of the deposited atoms. Such a reduction can consequently be avoided if the metals are either deposited in the presence of oxygen [58,66, 69–71] or post-annealed in oxidizing atmosphere [67,72,73].

The reversible oxidation/reduction of an oxide/metal contact can be demonstrated using near-ambient pressure XPS, which allows measurements in pressures of up to a few millibar [74]. Variations of the Schottky barrier height of ~1 eV have been observed for $SrTiO_3$/Pt contacts by changing the atmosphere from vacuum to 1 mbar O_2 at T ≥ 100°C [43].

Figure 8.5 shows the Sn$3d_{5/2}$ emission of a SnO_2 thin film sample with a ~3 nm thick Pt contact. When the sample is stored in vacuum at 150°C, the Sn^{4+} peak of the SnO_2 substrate is observed together with a metallic Sn^0 component. With the introduction of oxygen, the metallic Sn component disappears. Concurrently, the binding energy of the Sn^{4+} emission is shifting to lower binding

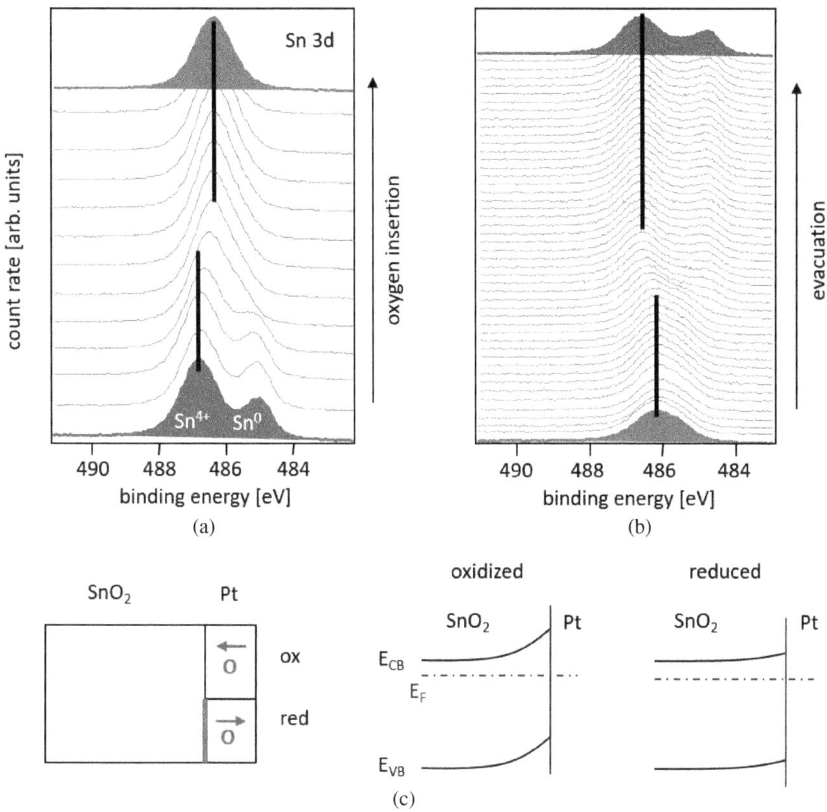

Fig. 8.5. (a) Near-ambient pressure $Sn3d_{5/2}$ photoemission spectra of a SnO_2 thin film sample coated with ~3 nm Pt at a temperature of 150°C. After deposition of Pt, the SnO_2 surface is partially reduced and exhibits a metallic Sn component [72]. With increasing oxygen pressure in the chamber, the metallic Sn component disappears and the binding energy of the Sn^{4+} component, which represents the Fermi energy at the SnO_2 surface, is shifting to lower binding energy. In (b), spectra recorded consecutively with a time step of ~20s during evacuation of the chamber are displayed. With decreasing oxygen pressure, the metallic Sn component reappears and the Fermi energy of SnO_2 raises concurrently. The oxidation and reduction process and the changes in the energy band diagram are shown in (c).

energy, which corresponds to a lowering of the Fermi energy (an increase of the Schottky barrier height for electrons). The spectra shown in Figure 8.5(b) are recorded sequentially with a time interval of ~20s during removal of oxygen by evacuation of the chamber.

Once the oxygen is mostly removed, the metallic Sn peaks reappear and the Schottky barrier is lowered again.

The reversible oxidation and reduction of the interface and the accompanying change of barrier height, which is illustrated in Figure 8.5(c), have been observed at temperatures as low as 100°C. With increasing temperature, the overall Sn signal increases and the Pt signal decreases, indicating de-wetting of the Pt film and the formation of islands. The de-wetting is also observed in a similar experiment at the $SrTiO_3$/Pt interface [43]. Eventually, the metallic Sn component is no more observed when the oxygen pressure chamber is changed at higher temperature. The metallic Sn peak remains absent after cooling the sample with the Pt islands. This observation indicates that the oxidation and reduction of Sn occur exclusively at the SnO_2/Pt interfaces. The thickness of the formed Pt islands is evidently too high for the photoelectrons to penetrate.

Due to the modification of Schottky barrier heights by interfacial defects, the transitivity rule cannot be applied to determine the energy band alignment between semiconductors if such defects cannot be excluded. As the formation of defects is quite frequently induced by the deposition of elementary metals, studies involving such interfaces have to be taken with care. In the case of oxides, the energy band alignment is best obtained using conductive oxides as contact materials. RuO_2 and ITO, which are easy to prepare even at room temperature, have been proven to be suitable for this approach [75].

8.3.3 *Modification of band alignment by bulk defects*

A modification of Schottky barrier heights and energy band alignment can also be induced by a high concentration of bulk defects. This is intentionally used in semiconductor technology in band alignment engineering. For the formation of ohmic contacts to semiconductors or for the formation of tunnel junctions in multijunction solar cells, the regions close to the interface are highly doped for narrowing the space-charge region in order to facilitate tunneling [1]. The extension of the space-charge region is then reduced to a few nanometers, which is of the same magnitude as the inelastic mean free path of the photoelectrons generated by X-ray absorption [76]. As a consequence, the binding energies of the photoelectron emission lines do no more reflect the Fermi level position directly at the interface. In contrast, the peak shape is convoluted with the potential profile in the space-charge region [53,77–80]. The situation is similar to the peak shift and broadening caused by lateral potential variations for inhomogeneous interfaces, which may occur at surfaces covered with metallic islands [81,82].

The situation for a p-type semiconductor with a high concentration of deep electron traps is illustrated in Figure 8.6. If the energy level of the trap state crosses the Fermi energy inside the space-charge region, the space-charge density increases abruptly, resulting in a very sharp potential drop. For defect concentrations ~1%, which have to be expected in many ionic compounds, an effective modification of the barrier height, Φ_B, occurs. In such a situation, the spectral shape, $I(E)$, is a convolution of the shape of the original spectra, $I(E_0)$, which is present for a small variation of the potential

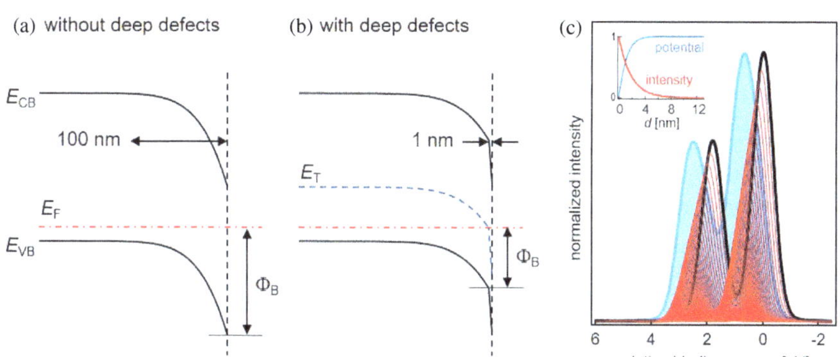

Fig. 8.6. Potential profiles at a semiconductor/metal interface without (a) and with a deep trap state (b). If the concentration of the trap state is $>10^{20}cm^{-3}$, a sharp potential drop near the interface occurs, which results in an effective modification of the barrier height Φ_B. The blue filled spectrum in (c) results from the potential profile and intensity attenuation with distance *d* from the interface, which is shown in the inset. An inelastic mean free path of 1.9 nm has been assumed in the simulation. The black spectrum corresponds to a constant potential of $\varphi = 0$.

within the probing depth, with the potential, $\varphi(x)$, and attenuation length, λ, according to

$$I(E) = \int I\left(E_0 - \varphi(x)\right) \cdot \exp(-x/\lambda) \mathrm{d}x. \tag{8.3}$$

An example for such a simulation is shown in Figure 8.6(c). A classical example for a deep defect is the oxygen vacancy in ZnO, which exhibits a 2+/0 transition ~ at 0.7–0.8 eV below the conduction band minimum [83,84]. These defects can lead to a substantial variation of the Schottky barrier height at ZnO/metal contacts [58,83]. They are also responsible for the variation of the energy band alignment at Cu_2O/ZnO heterocontacts, which exhibits a variation of valence band offset by more than 1 eV depending on interface preparation [85]. A variation of band alignment with interface preparation has also been observed for several interfaces involving ionically bonded thin film solar cell materials [56,86].

Narrow space-charge regions also occur at boundaries of highly doped semiconductors, such as the degenerate *n*-type transparent electrode materials [87]. An approach of evaluating the Fermi level at surfaces of degenerately doped SnO_2 thin films in the presence of surface space-charge regions based on the spectra simulations introduced above has been developed by Weidner [88]. The approach makes use of the dependence of binding energy shifts and peak broadening on the escape depth of the photoelectrons, which depends on their kinetic energy.

Several oxides including $SrTiO_3$, ZnO, and In_2O_3 have been reported to exhibit electron accumulation layers at the surface [89–92]. In these cases, the determination of the potential profile requires us to take the quantum confinement of the electron states into account, as the conduction band is bent below the Fermi energy at the surface, thereby forming a quantum well.

8.3.4 *Fermi level confinement by valence changes*

The determination of the energy band positions can also be affected by valence changes of the atoms of a compound. This is particularly the case for compounds containing transition metals. During the study of interface properties, the Fermi energy may be raised or lowered in the band gap at the interface. Raising/lowering the Fermi energy corresponds to the addition/removal of electrons, which formally corresponds to a reduction/oxidation. In classical semiconductors, such as Si or GaAs, raising or lowering the Fermi energy will eventually result in the generation of electrons in the conduction band or of holes in the valence band. The reduction or oxidation of species is a competing process. For example, raising/lowering the Fermi energy may result in a reduction ($Fe^{3+} \rightarrow Fe^{2+}$) or oxidation ($Fe^{3+} \rightarrow Fe^{2+}$) of the iron in hematite Fe_2O_3. This may occur for Fermi energies well inside the energy gap [93,94].

Figure 8.7(a) shows Fe2$p_{3/2}$ and Bi4f spectra of a $BiFeO_3$ thin film grown on top of a conducting (La,Sr)MnO_3 layer on a single crystal $SrTiO_3$ substrate after surface cleaning, oxygen plasma

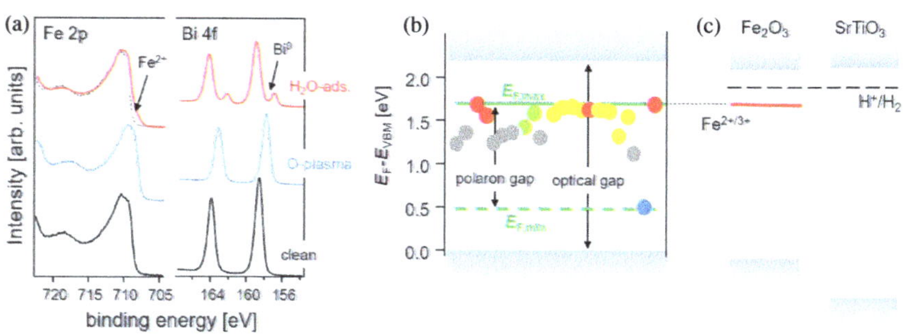

Fig. 8.7. (a) Fe2$p_{3/2}$ and Bi4f spectra of a $BiFeO_3$ thin film after surface cleaning, oxygen plasma treatment, and water exposure at room temperature. The different treatments result in a variation of binding energies, which corresponds to a variation of the Fermi energy; (b) variation of the Fermi energy of differently treated $BiFeO_3$ samples. Red data points correspond to water exposed samples and yellow data points to samples coated with a thin layer of Sn-doped In_2O_3. The upper limit of the Fermi energy is reached either by water exposure or by ITO deposition. In all cases, a reduction of the sample is evident from the appearance of reduced Fe and Bi emissions [95]; (c) comparison with energy levels and band alignment of hematite Fe_2O_3 and $SrTiO_3$ and the hydrogen evolution potential.

treatment, and water exposure at room temperature, respectively [95]. $BiFeO_3$ is a multiferroelectric material [96], which has been considered as a (photo-)catalytic material due to its band gap of 2.2 eV and its ferroelectric nature [97–99]. The variation of the Fermi energy of a series of differently treated samples is shown in Figure 8.7(b). The overall variation amounts to ~1.2 eV. After water exposure, the highest Fermi energies of $E_F - E_{VB} = 1.7$ eV are reached. This level constitutes an upper limit, which is evident from the observation of Fe^{2+} and Bi^0 species. In electrochemistry, this level is referred to as reduction potential [94]. The same Fermi energy and reduced species are observed when a thin layer of Sn-doped In_2O_3 is deposited on top of a clean $BiFeO_3$ surface. The upper limit of the Fermi energy is well below the conduction band minimum. $BiFeO_3$ will be unstable and decompose if the Fermi energy is driven above the limit, for example, by external boundary conditions. The lowest Fermi energy is observed after oxygen plasma treatment. In this case, the spectra do not reveal an oxidation of the sample, but the identification of oxidized species, such as Fe^{4+}, Bi^{5+}, or O^-, may be difficult using XPS. However, there is evidence that the oxidation of Fe occurs [100].

Due to the limitation of the Fermi energy by oxidation or reduction of constituents, a determination of energy band alignment is only reasonable if the Fermi energy remains within these limits. For the example of $BiFeO_3$, ITO cannot be used for a proper alignment, as it pushes the Fermi level to the upper limit. Limitations of the Fermi level by valence changes are also important when the Fermi level is determined by electrochemical methods, as carrier transport will not take place inside the energy bands. Electronic carriers will rather be trapped and transported by hopping between neighboring sites. In $BiFeO_3$, for example, electrons can hop between Fe sites. In such cases, the alignment obtained from electrochemical measurements refers to the reduction potential and not to the conduction band minimum.

The XPS analysis of hematite Fe_2O_3 has revealed such a situation and demonstrated that the efficiency of hematite for the reduction of water is not caused by a conduction band minimum below the hydrogen evolution potential [5]. In Fe_2O_3, the upper limit of the Fermi energy is also caused by the reduction of Fe^{3+} ions, which occurs for Fermi energies being ~0.5 eV below the conduction band minimum. A careful comparison of band alignment revealed that the hydrogen evolution potential is actually in between the reduction potential and the conduction band minimum of hematite. The comparison with $BiFeO_3$, which is given in Figure 8.7(c), has further shown that the reduction of Fe^{3+} occurs at the same energy as in Fe_2O_3, when the energy bands are properly aligned [95]. This compares well with the alignment of defect energy levels of transition metal impurities in III-V semiconductors, which has also been suggested to provide a measure of the band alignment between these compounds [101,102].

8.3.5 *The intrinsic band alignment*

As described in the previous sections, the determination of energy band alignment using photoelectron spectroscopy is affected by different phenomena, which can result in a substantial modification of the band alignment. In particular, interfacial and bulk defects and the oxidation/reduction of lattice species have to be considered. These phenomena are not specific to the determination of band alignment using photoelectron spectroscopy but fundamental to interface formation in general and affect also other determination techniques. A major advantage of photoelectron spectroscopy for the determination of the energy band alignment is that it simultaneously provides information on the chemical nature of the interface. Although not always possible, XPS can identify the oxidation or reduction of the substrate in the course of interface formation. An *in situ* preparation of the interface is highly beneficial in such experiments, as it allows eliminating extrinsic contributions to interface chemistry. For example, the formation of surface space-charge regions by adsorbates might be avoided [42].

The main challenge in the positioning of the energy bands is to obtain a band alignment, which is intrinsic to the material combination and not affected by the formation of any kind of defects.

This requires (i) avoiding interface reactions, (ii) keeping bulk defect concentrations as low as possible, and (iii) keeping the Fermi energy within the limits determined by the oxidation and reduction potentials. In the case of oxides, interface reactions can be easily avoided if oxides are used as contact partners. At higher temperatures, oxygen, and sometimes also cations, may diffuse across the interface. Therefore, the deposition temperature should be kept as low as possible. The influence of bulk defects can be studied by varying the preparation parameters of the substrate and the overlayer and evaluating any variation of the obtained band alignment. The presence of narrow space-charge regions might be detected if different core levels from the substrate, or from the overlayer, exhibit non-parallel binding energy shifts or varying peak shapes and widths. Regarding the third issue, the band alignment is considered to be reliable if lower/higher Fermi energies are observed with other contact materials or sample treatments. For this purpose, oxygen plasma treatments are effective in lowering the Fermi energy. With our setup, the highest Fermi energies are often obtained by room temperature water exposure or low-pressure atomic layer deposited Al_2O_3 films [95,103,104].

Following these recommendations, it is obvious that a reliable band alignment cannot be obtained from a single interface. It is mandatory to perform a series of experiments with different preparation parameters and different contact materials. For oxide interfaces, many materials with a large variation in work function are available for this purpose. In the experiment performed in the author's group, mainly Sn-doped In_2O_3 (ITO) and RuO_2, which exhibit work functions of 4.5 and 5.6 eV,[1] have been used [6,42,43, 63,64,75]. Other oxides, which reveal the lowest interfacial Fermi energies when used as contact material, are NiO and Co_3O_4 [64,105]. These are not very conducting and can only be used with conducting substrates or with thin films deposited onto conducting

[1]The work functions measured using ultraviolet photoelectron spectroscopy (UPS) of *in situ* prepared RuO_2 thin films exhibit values of up to 6.5 eV, but the Schottky barrier heights correspond to an effective work function of ~5.6 eV.

substrates. In contrast, ITO and RuO_2 can also be used with insulating substrates, as charging can be avoided when the sample is electrically contacted via the thin top electrode [106].

In agreement with the guidelines described in this section, the intrinsic energy band alignment of a number of oxides has been obtained using a large set of experiments [75]. A set of data for different oxides is shown in Figure 8.8. A similar plot for chalcogenides related to thin film solar cell materials has also been given in literature [86]. The energy bands of the different materials included in Figure 8.8 are all referenced to the valence band maximum of SnO_2. This is an arbitrary selection and chosen because the valence bands in SnO_2 are almost exclusively formed by the $2p$ orbitals of oxygen.

The left part of Figure 8.8 shows the energy bands of oxides, which do not have substantial contributions of cation orbitals in the valence band states. Most of them are at similar energy. Thus, corresponds to the common anion rule for semiconductor band alignment [107]. The common anion rule states that the valence band maxima of compounds sharing the same anion are at similar energy. This rule has originally been suggested for III-V semiconductors. It has later been found that the Fermi level pinning by induced gap

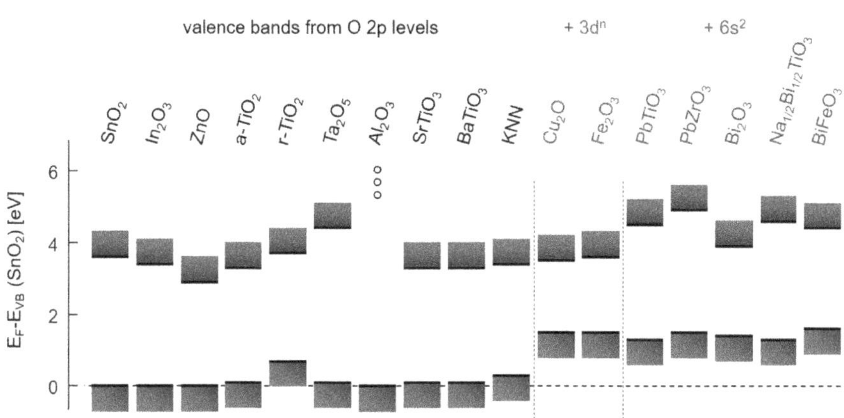

Fig. 8.8. Intrinsic energy band alignment of different oxide materials as obtained using interface experiments with photoelectron spectroscopy [75]. All energies are referenced to the valence band maximum of SnO_2.

states can lead to substantial deviations from the common anion rule [108]. However, for more ionic compounds, such as oxides and chalcogenides, the Fermi level pinning by induced gap states is not strong and the common anion rule is approximately valid as long as the valence band states are formed mostly by the anion *p*-states. Nevertheless, there are also noticeable exceptions from the common anion rule. A prominent example is the alignment between the rutile and anatase modifications of TiO_2. The valence band maximum of rutile is ~0.7 eV higher than that of anatase, which is caused by the different overlap of the non-bonding $O2p_z$-orbitals [6].

Considerably higher valence band maxima of oxides are observed when cation states hybridize with the $O2p$ states. This is the case for transition metals with partially occupied d-shells, such as in Fe_2O_3, where Fe has an electronic d^5 configuration. Moreover, oxides with occupied but shallow *d*-levels, such as oxides containing Cu or Ag, also exhibit higher valence band maxima. A high valence band maximum of a compound is important for achieving p-type conductivity. This is related to the self-compensation of acceptor dopants by intrinsic donor defects [109], which is caused by dependence of the defect formation energies on the Fermi energy [110]. In Gd-doped CeO_2, for example, the addition of Gd_{Ce} acceptor defects results in a charge compensation by oxygen vacancies [111] and not in a compensation by holes in the valence band. In 10 % Gd-doped CeO_2, Fermi energies are always more than ~1.2 eV above the valence band maximum as shown in Figure 8.3(c). In contrast, Cu_2O is a well-known p-type conductor [112] and shows a Fermi level close to the valence band maximum in XPS [85]. The hybridization between Cu *d* and chalcogen *p* states is also the reason for the higher valence band maxima of Cu-chalcopyrites as compared to that of their binary II-VI analogs. The effect is referred to as p–d repulsion [113]. It is important for the p-type conductivity of the Cu-chalcopyrite semiconductors, which are key materials for high-efficiency thin film solar cells [114]. Cu-chalcopyrite semiconductors also exhibit Fermi energies close to the valence band maximum [86].

Higher valence band maxima than oxides with $O2p$ dominated valence bands are also observed for oxides with cations having an s^2

configuration, such as Pb^{2+}, Bi^{3+}, and Sn^{2+} [54,115]. In these cases, there is also hybridization between the cation states and the O2*p* states [98], which results in an upward shift of the valence band maximum. Due to their potential p-type conductivity and lower energy gaps, the Sn^{2+} compounds SnO and SnS are candidates for p-type transistors and thin films solar cells, respectively [116,117].

8.4 Concluding Remarks

Photoelectron spectroscopy with X-ray and UV irradiation is a versatile technique to quantify the positions of energy bands and compare them between different materials. Among the various energy levels, which can be used for aligning energy bands, either the vacuum energy or the band edges of any reference material are a suitable reference for electron spectroscopy. The vacuum energy is directly obtained from the secondary electron cutoff at zero kinetic energy. With a proper binding energy calibration using the Fermi edge of a metallic sample, which provides the energy distance between the valence band maximum and the Fermi energy, the work function, ionization potential, and electron affinity can be obtained. It has been demonstrated that a single measurement of these quantities is not representative of the surface behavior, as they all depend on substrate doping and surface orientation and termination. Surface space-charge regions may also have a pronounced effect. Together, their dependence on sample and surface preparation can result in a variation of Fermi energy and work function by more than 2 eV. A systematic assessment of surface properties in dependence on the activity of the components is thus mandatory to reveal this dependence. This is straightforward in the case of oxides, for which the oxygen activity can be easily varied in a vacuum environment. By using *in situ* surface preparation, surface contaminations can be largely avoided. This is required for revealing the unmodified surfaces' properties.

The surface properties are important for interactions at solid/gas and partially also at solid/liquid interfaces. However, a positioning of the energy bands along the vacuum energy is not reasonable for obtaining the alignment at solid/solid interfaces, as the dependence of the energy band alignment at solid/solid interfaces on surface condition is not very pronounced, in contrast to the surface properties. To identify a dependence of band alignment on surface condition, for example, on surface orientation, other effects, which can strongly affect the band alignment, have to be avoided. Among the latter effects are deposition induced interface defects, high defect concentrations in the subsurface region of materials and valence changes of the lattice species. Photoelectron spectroscopy can sometimes directly detect the presence of such effects, as it is also sensitive to chemical changes at the interface, as long as the concentration of modified species is higher than the detection limit of the technique. Similar to the case of the surface properties, an intrinsic alignment of the energy bands of two materials can only be obtained from an interface experiment if an influence of the mentioned effects on the band alignment can be avoided. Performing interface experiments with a sequence of overlayer thickness and performing different experiments with a variation of substrate and film preparation and/or a variation of overlayer material are essential for identifying such dependencies. To reveal these dependencies and their origins is a unique strength of photoelectron spectroscopy, which makes it particularly useful for such studies.

References

[1] Sze SM, Ng KK. *Physics of Semiconductor Devices.* Hoboken: John Wiley & Sons, 2007.

[2] Fujishima A, Zhang X, Tryk DA. TiO_2 photocatalysis and related surface phenomena. Surf Sci Rep. 2008;63:515–582.

[3] Linsebigler AL, Lu G, Yates JT. Photocatalysis on TiO_2 surfaces: Principles, mechanisms, and selected results. Chem Rev. 1995;95 (3):735–758.

[4] Schneider J, Bahnemann D, Ye J, Puma GL, Dionysiou DD, (Eds.). *Photocatalysis: Fundamentals and Perspectives.* Cambridge: The Royal Society of Chemistry, 2016.

[5] Lohaus C, Klein A, Jaegermann W. Limitation of Fermi level shifts by polaron defect states in hematite photoelectrodes. Nature Commun. 2018;9:4309.

[6] Pfeifer V, Erhart P, Li S, Rachut K, Morasch J, Brötz J, Reckers P, Mayer T, Rühle S, Zaban A, Seró IM, Bisquert J, Jaegermann W, Klein A. Energy band alignment between anatase and rutile TiO_2. J Phys Chem Lett. 2013;4:0 4182–4187.

[7] Uddin MT, Nicolas Y, Olivier C, Toupance T, Mller M, Kleebe H-J, Rachut K, Ziegler J, Klein A, Jaegermann W. Preparation of RuO_2/ TiO_2 mesoporous heterostructures and rationalization of their enhanced photocatalytic properties by band alignment investigations. J Phys Chem C. 2013;117:22098–22110.

[8] Kashiwaya S, Aymonier C, Majimel J, Olivier C, Klein A, Jaegermann W, Toupance T. Supercritical CO_2-assisted deposition of NiO on (101)-anatase-TiO_2 for efficient facet engineered photocatalysts. New J Chem. 2018;42:18649–18658.

[9] Hermans Y, Olivier C, Junge H, Klein A, Jaegermann W, Toupance T. Sun light selective photodeposition of $CoO_x(OH)_y$ and $NiO_x(OH)_y$ on truncated bipyramidal $BiVO_4$ for highly efficient photocatalysis. ACS Appl Mater Interfaces. 2020;12:53910–53920.

[10] Kashiwaya S, Toupance T, Klein A, Jaegermann W. Fermi level positions and induced band bending at single crystalline anatase (101) and (001) surfaces: Origin of the enhanced photocatalytic activity of facet engineered crystals. Adv Energy Mater. 2018;8:1802195.

[11] Trasatti S. The absolute electrode potential: An explanatory note (Recommendations 1986). J Electroanal Chem. 1986;209:417–428.

[12] Kurtin S, McGill TC, Mead CA. Fundamental transition in the electronic nature of solids. Phys Rev Lett. 1969;22:1433–1436.

[13] Tersoff J. Schottky barrier heights and the continuum of gap states. Phys Rev Lett. 1984;520(6):465–468.

[14] Cardona M, Christensen NE. Acoustic deformation potentials and heterostructure band offsets in semiconductors. Phys Rev B. 1987;350(12):6182–6194.

[15] Mönch W. Empirical tight-binding calculation of the branch-point energy of the continuum of interface-induced gap states. J Appl Phys. 1996;80(9):5076–5082.

[16] Jaegermann W. The semiconductor/electrolyte interface: A surface science approach. In RE White (Ed.), *Modern Aspects of Electrochemistry* (Vol. 30, pp. 1–185). New York: Plenum Press, 1996.

[17] Ohno T, Sarukawa K, Tokieda K, Matsumura M. Morphology of a TiO_2 photocatalyst (Degussa, P-25) consisting of anatase and rutile crystalline phases. J Catal. 2001;203:82–86.

[18] Kavan L, Grätzel M, Gilbert SE, Klemenz C, Scheel HJ. Electrochemical and photoelectrochemical investigation of single-crystal anatase. J Am Chem So. 1996;118:6716–6723.

[19] Deák P, Aradi Bl, Frauenheim T. Band lineup and charge carrier separation in mixed rutile-anatase systems. J Phys Chem C. 2011; 115:3443–3446.

[20] Scanlon DO, Dunnill CW, Buckeridge J, Shevlin SA, Logsdail AJ, Woodley SM, Catlow CRA, Powell MJ, Palgrave RG, Parkin IP, Watson GW, Keal TW, Sherwood P, Walsh A, Sokol AA. Band alignment of rutile and anatase TiO_2. Nature Mater. 2013;12:798.

[21] Capasso F, Margaritondo G, (Eds.). *Heterojunction Band Discontinuities.* Amsterdam: North-Holland, 1987.

[22] Schroder DK. *Semiconductor Material and Device Characterization.* Hoboken: John Wiley & Sons, 2006.

[23] Scott JF. *Ferroelectric Memories.* Berlin: Springer Verlag, 2000.

[24] Bell LD, Kaiser WJ. Observation of interface band structure by ballistic-electron-emission microscopy. Phys Rev Lett. 1988;61:2368.

[25] Gelderman K, Lee L, Donne SW. Flat-band potential of a semiconductor: Using the Mott–Schottky equation. J Chem Educ. 2007;84:685.

[26] Hankin A, Bedoya-Lora FE, Alexander JC, Regoutz A, Kelsall GH. Flat band potential determination: Avoiding the pitfalls. J Mater Chem A. 2019;7:26162–26176.

[27] Lide DR, editor. *CRC Handbook of Chemistry and Physics* (85 edn.). Boca Raton: CRC Press, 2005.

[28] Ranke W, Xing YR. Surface dipole and Fermi-level position on clean, oxygen-, and water-covered cylindrical Si crystals: A photoelectron spectroscopy study. Phys Rev B. 1985;31:2246–2253.

[29] Ranke W. Ultraviolet photoelectron spectroscopy investigation of electron affinity and polarity on a cylindrical GaAs single crystal. Phys Rev B. 1983;27:7807–7810.

[30] Moll M, Kley A, Pehlke E, Scheffler M. GaAs equilibrium crystal shape from first principles. Phys Rev B. 1996;54:8844–8855.

[31] Tasker PW. The stability of ionic crystal surfaces. J Phys C. 1979;12:4977.

[32] Noguera C. Polar oxide surfaces. J Phys Cond Matter. 2000;12:R367–R410.

[33] Chadi DJ. Vacancy-induced 2 × 2 reconstruction of the Ga(111) surface of GaAs. Phys Rev Lett. 1984;52:1911.

[34] Diebold U. The surface science of titanium dioxide. Surf Sci Rep. 2003;48:53–229.

[35] Weiss W, Ranke W. Surface chemistry and catalysis on well-defined epitaxial iron-oxide layers. Prog Surf Sci. 2002;l70:1–151.

[36] Swank RK. Surface properties of II-VI compounds. Phys Rev. 1967;153:844–849.

[37] Jacobi K, Zwicker G, Gutmann A. Work function, electron affinity and band bending of zinc oxide surfaces. Surf Sci. 1984;141:109–125.

[38] Klein A, Körber C, Wachau A, Säuberlich, Gassenbauer Y, Schafranek R, Harvey SP, Mason TO. Surface potentials of magnetron sputtered transparent conducting oxides. Thin Solid Films. 2009; 518:1197–1203.

[39] Klein A, Körber C, Wachau A, Säuberlich, Gassenbauer Y, Harvey SP, Proffit DE, Mason TO. Transparent conducting oxides for photovoltaics: Manipulation of Fermi level, work function, and energy band alignment. Materials. 2010;3:4892–4914.

[40] Walsh A, Catlow CRA. Structure, stability and work functions of the low index surfaces of pure indium oxide and Sn-doped indium oxide (ITO) from density functional theory. J Mater Chem. 2010;20:10438–10444.

[41] Hohmann MV, Ágoston P, Wachau A, Bayer TJM, Brötz J, Albe K, Klein A. Orientation dependent ionization potential of In_2O_3: A natural source for inhomogeneous Barrier formation at electrode interfaces in organic electronics. J Phys Cond Matter. 2011;23:334203.

[42] Klein A. Transparent conducting oxides: Electronic structure property relationship from photoelectron spectroscopy with *in situ* sample preparation. J Am Ceram Soc. 2013;96:331–345.

[43] Klein A. Interface properties of dielectric oxides. J Am Ceram Soc. 2016;99:369–387.

[44] Wardenga H, Schuldt K, Waldow S, De Souza R, Klein A. Surface potentials and oxygen exchange of differently doped CeO_2. Phys Chem Chem Phys. 2022;24:1072–1080.

[45] Körber C, Ágoston P, Klein A. Surface and bulk properties of sputter deposited intrinsic and doped SnO_2 thin films. Sens Actuators B. 2009;139:665–672.

[46] Kashiwaya S, Morasch J, Streibel V, Toupance T, Jaegermann W, Klein A. The work function of TiO_2. Surfaces. 2018;1:73–89.

[47] Wardenga H, Klein A. Surface potentials of (111), (110) and (100) oriented CeO_{2-x} thin films. Appl Surf Sci. 2016;377:1–8.

[48] Batzill M, Diebold U. The surface and materials science of tin oxide. Prog Surf Sci. 2005;79:47–154.

[49] Rachut K, Körber C, Brötz J, Klein A. Growth and surface properties of epitaxial SnO_2. Phys Stat Sol (a). 2014;211:1997–2004.

[50] Körber C, Wachau A, Ágoston P, Albe K, Klein A. Self-limited oxygen exchange kinetics at SnO_2 surfaces. Phys Chem Chem Phys. 2011;13:3223–3226.

[51] Jamnik J, Kamp B, Merkle R, Maier J. Space charge influenced oxygen incorporation in oxides: In how far does it contribute to the drift of Taguchi sensors? Solid State Ionics. 2002;150:157–166.

[52] Wardenga H. Surface potentials of ceria and their influence on the surface exchange of oxygen. Thesis, Technische Universität Darmstadt; 2019. https://tuprints.ulb.tu-darmstadt.de/id/eprint/8969.

[53] Zurhelle AF, Tong X, Klein A, Mebane DS, De Souza RA. A space-charge treatment of the increased concentration of reactive Ce^{3+} at the surface of a ceria solid solution. Angew Chem Int Ed. 2017;56:14516–14520.

[54] Schafranek R, Li S, Chen F, Wu W, Klein A. $PbTiO_3/SrTiO_3$ interface: Energy band alignment and its relation to the limits of Fermi level variation. Phys Rev B. 2011;84:045317.

[55] Kraut EA, Grant RW, Waldrop JR, Kowalczyk SP. Precise determination of the valence-band edge in X-ray photoemission spectra: Application to measurement of semiconductor interface potentials. Phys Rev Lett. 1980;44:1620–1623.

[56] Klein A, Säuberlich F, Späth B, Schulmeyer T, Kraft D. Non-stoichiometry and electronic properties of interfaces. J Mater Sci. 2007;42:1890–1900.

[57] Deuermeier J, Fortunato E, Martins R, Klein A. Energy band alignment at the nanoscale. Appl Phys Lett. 2017;110:051603.

[58] Wendel P, Periyannan S, Jaegermann W, Klein A. Polarization dependence of ZnO Schottky barriers revealed by photoelectron spectroscopy. Phys Rev Mater. 2020;4:084604.

[59] Heine V. Theory of surface states. Phys Rev. 1965;138:A1689–A1696.

[60] Louie SG, Chelikowsky JR, Cohen ML. Ionicity and the theory of Schottky barriers. Phys Rev B. 1977;15:2154–2162.

[61] Mönch W. Mechanisms of barrier formation in Schottky contacts: Metal induced surface and interface states. Appl Surf Sci. 1989;41/42:128–138.

[62] Brillson LJ. Transition in Schottky barrier formation with chemical reactivity. Phys Rev Lett. 1978;40:260–263.

[63] Chen F, Schafranek R, Li S, Wu W, Klein A. Energy band alignment between $Pb(Zr,Ti)O_3$ and high and low work function conducting oxides — From hole to electron injection. J Phys D Appl Phys. 2010;43:295301.

[64] Hermans Y, Murcia-López S, Klein A, van de Krol R, Andreu T, Morante JR, Toupance T, Jaegermann W. Analysis of the interfacial characteristics of $BiVO_4$/metal oxide heterostructures and its implication on their junction properties. Phys Chem Chem Phys. 2019;21:5086–5096.

[65] Chen F, Schafranek R, Wu W, Klein A. Reduction induced Fermi level pinning at the interfaces between $Pb(Zr,Ti)O_3$ and Pt, Cu and Ag metal electrodes. J Phys D Appl Phys. 2011;44:255301.

[66] Schultz T, Vogt S, Schlupp P, von Wenckstern H, Koch N, Grundmann M. Influence of oxygen deficiency on the rectifying behavior of transparent-semiconducting-oxide–metal interfaces. Phys Rev Appl. 2018;9:064001.

[67] Schafranek R, Payan S, Maglione M, Klein A. Barrier heights at $(Ba,Sr)TiO_3$/Pt interfaces studied by photoemission. Phys Rev B. 2008;77:195310.

[68] Schafranek R, Klein A. *In situ* photoemission study of the contact formation of (Ba,Sr)TiO_3 with Cu and Au. Solid State Ionics. 2006;177:1659–1664.

[69] Michel J, Splith D, Rombach J, Papadogianni A, Berthold T, Krischok S, Grundmann M, Bierwagen O, von Wenckstern H, Himmerlich M. Processing strategies for high-performance Schottky contacts on n-type oxide semiconductors: Insights from In_2O_3. ACS Appl Mater Interfaces. 2019;11:27073–27087.

[70] Hyland AM, Makin RA, Durbin SM, Allen MW. Giant improvement in the rectifying performance of oxidized Schottky contacts to ZnO. J Appl Phys. 2017;121:10.

[71] Lajn A, Wenckstern HV, Zhang Z, Czekalla C, Biehne G, Lenzner J, Hochmuth H, Lorenz M, Grundmann M, Wickert S, Vogt C, Denecke R. Properties of reactively sputtered Ag, Au, Pd, and Pt Schottky contacts on n-type ZnO. J Vac Sci Technol B. 2009;27:1769–1773.

[72] Körber C, Harvey SP, Mason TO, Klein A. Barrier heights at the SnO_3/Pt interface: *In situ* photoemission and electrical properties. Surf Sci. 2008;602:3246–3252.

[73] Hanys P, Janecek P, Matolin V, Korotcenkov G, Nehasil V. XPS and TPD study of Rh/SnO_2 system — Reversible process of substrate oxidation and reduction. Surf Sci. 2006;600:4233–4238.

[74] Bluhm H, Hävecker M, Knop-Gericke A, Kiskinova M, Schlögl R, Salmeron M. *In situ* X-ray photoelectron spectroscopy studies of gas-solid interfaces at near-ambient conditions. MRS Bull. 2007; 34:1022–1030.

[75] Li S, Chen F, Schafranek R, Bayer TJM, Rachut K, Fuchs A, Siol S, Weidner M, Hohmann M, Pfeifer V, Morasch J, Ghinea C, Arveux E, Günzler R, Gassmann J, Körber C, Gassenbauer Y, Säuberlich F, Rao GV, Payan S, Maglione M, Chirila C, Pintilie L, Jia L, Ellmer K, Naderer M, Reichmann K, Böttger U, Schmelzer S, Frunza RC, Urši H, Malič B, Wu W-B, Erhart P, Klein A. Intrinsic energy band alignment of functional oxides. Phys Stat Sol (rrl). 2014;8:571–576.

[76] Tanuma S, Powell CJ, Penn DR. Proposed formula for electron inelastic mean free paths based on calculations for 31 materials. Surf Sci. 1987;192:L849.

[77] Ohashi N, Yoshikawa H, Yamashita Y, Ueda S, Li J, Okushi H, Kobayashi K, Haneda H. Determination of Schottky barrier profile at Pt/$SrTiO_3$:Nb junction by X-ray photoemission. Appl Phys Lett. 2012;101:251911.

[78] Hirose S, Yoshikawa H, Yanagi H, Ando A, Ueda S, Ohashi N. Impact of electrode oxidation on the current transport properties at platinum/(niobium-doped strontium-titanate) Schottky junctions. ECS J Solid State Sci Technol. 2014;3:243–248.

[79] Hirose S, Okushi H, Ueda S, Yoshikawa H, Adachi Y, Ando A, Ohsawa T, Haneda H, Ohashi N. Electric field and temperature dependence of dielectric permittivity in strontium titanate investigated by a photoemission study on Pt/$SrTiO_3$:Nb junctions. Appl Phys Lett. 2015;106:191602.

[80] Giesecke R, Hertwig R, Bayer T, Randall CA, Klein A. Modification of the cathodic Schottky Barrier height at the RuO_2 cathode during resistance degradation of Fe-doped $SrTiO_3$. J Am Ceram Soc. 2017;100:4590–4601.

[81] Miyano KE, King DM, Spindt CJ, Kendelewicz T, Cao R, Yu Z, Lindau I, Spicer WE. Potential-barrier measurements at clustered metal-semiconductor interfaces. Phys Rev B. 1991;43:11806–11814.

[82] Schlaf R, Klein A, Pettenkofer C, Jaegermann W. Laterally inhomogeneous surface potential distribution and photovoltage at clustered In/WSe_2 (0001) interfaces. Phys Rev B. 1993;48:14242–14252.

[83] Allen MW, Durbin SM. Influence of oxygen vacancies on Schottky contacts to ZnO. Appl Phys Lett. 2008;92:122110.

[84] Ágoston P, Albe K, Nieminen RM, Puska MJ. Intrinsic n-type behavior in transparent conducting oxides: A comparative hybrid-functional study of In_2O_3, SnO_2, and ZnO. Phys Rev Lett. 2009;103:245501.

[85] Siol S, Hellmann JC, Tilley SD, Graetzel M, Morasch J, Deuermeier J, Jaegermann W, Klein A. Band alignment engineering at Cu_2O/ZnO heterointerfaces. ACS Appl Mater Interfaces. 2016; 8:21824–21831.

[86] Klein A. Energy band alignment in Chalcogenide thin film solar cells from photoelectron spectroscopy. J Phys Cond Matter. 2015; 27:134201.

[87] Ginley DS, Hosono H, Paine DC, (Eds.). *Handbook of Transparent Conductors*. New York: Springer; 2010.

[88] Weidner M. Fermi level determination in tin oxide by photoelectron spectroscopy: Relation to optoelectronic properties; Band bending at surfaces and interfaces; Modulation doping. Thesis, Technische Universität Darmstadt; 2016. https://tuprints.ulb.tu-darmstadt.de/id/eprint/5224.

[89] Rödel TC, Fortuna F, Sengupta S, Frantzeskakis E, Fèvre PL, Bertran F, Mercey B, Matzen S, Agnus G, Maroutian T, Lecoeur P, Santander-Syro AF. Universal fabrication of 2D electron systems in functional oxides. Adv Mater. 2016;28:1976–1980.

[90] King PDC, Veal TD, Payne DJ, Bourlange A, Egdell RG, McConville CF. Surface electron accumulation and the charge neutrality level in In2O3. Phys Rev Lett. 2008;101:116808.

[91] Santander-Syro AF, Copie O, Kondo T, Fortuna F, Pailhs S, Weht R, Qiu XG, Bertran F, Nicolaou A, Taleb-Ibrahimi A, Fèvre PL, Herranz G, Bibes M, Reyren N, Apertet Y, Lecoeur P, Barthélémy A, Rozenberg MJ. Two-dimensional electron gas with universal subbands at the surface of $SrTiO_3$. Nature. 2011;469:189.

[92] Heinhold R, Cooil SP, Evans DA, Allen MW. Stability of the surface electron accumulation layers on the nonpolar $(10\bar{1}0)$ and $(11\bar{2}0)$ faces of ZnO. J Phys Chem C. 2014;118:24575–24582.

[93] Schultze JW, Lohrengel MM. Stability, reactivity and breakdown of passive films. Problems of recent and future research. Electrochimica Acta. 2000;45:2499–2513.

[94] Chen S, Wang L-W. Thermodynamic oxidation and reduction potentials of photocatalytic semiconductors in aqueous solution. Chem Mater. 2012;24:3659–3666.

[95] Bein NS, Machado P, Coll M, Chen F, Makarovic M, Rojac T, Klein A. Electrochemical reduction of undoped and cobalt-doped $BiFeO_3$ induced by water exposure: Quantitative determination of reduction potentials and defect energy levels using photoelectron spectroscopy. J Phys Chem Lett. 2019;10:7071–7076.

[96] Catalan G, Scott JF. Physics and applications of Bismuth ferrite. Adv Mater. 2009;21:2463–2485.

[97] Gao F, Chen XY, Yin KB, Dong S, Ren ZF, Yuan F, Yu T, Zou Z, Liu JM. Visible-light photocatalytic properties of weak magnetic $BiFeO_3$ nanoparticles. Adv Mater. 2007;19:2889.

[98] Walsh A, Payne DJ, Egdell RG, Watson GW. Stereochemistry of post-transition metal oxides: Revision of the classical lone pair model. Chem Soc Rev. 2011;40:4455–4463.

[99] Soltani T, Tayyebi A, Lee BK. $BiFeO_3/BiVO_4$ p-n heterojunction for efficient and stable photocatalytic and photoelectrochemical water splitting under visible-light irradiation. Catal. Today. 2020;340:188–196.

[100] Rojac T, Bencan A, Drazic G, Sakamoto N, Ursic H, Jancar B, Tavcar G, Makarovic M, Walker J, Malic B, Damjanovic D. Domain-wall conduction in ferroelectric $BiFeO_3$ controlled by accumulation of charged defects. Nature Mater. 2017;16:322.

[101] Langer JM, Delerue C, Lannoo M, Heinrich H. Transition-metal impurities in semiconductors and heterojunction band lineups. Phys Rev B. 1988;38:7723–7739.

[102] Zunger A. Theory of 3d transition atom impurities in semiconductors. Ann Rev Mater Sci. 1985;15:411–453.

[103] Hermans Y, Klein A, Murcia-López S, Jaegermann W. The interface properties of the water — Bismuth vanadate interaction. ACS Energy Lett. 2019;4:2522–2528.

[104] Weidner M, Fuchs A, Bayer TJM, Rachut K, Schnell P, Deyu GK, Klein A. Defect modulation doping. Adv Funct Mater. 2019;29: 1807906.

[105] Long DM, Klein A, Dickey EC. Barrier formation at $BaTiO_3$ interfaces with Ni and NiO. Appl Surf Sci. 2019;466:472–476.

[106] Chen F, Schafranek R, Wachau A, Zhukov S, Glaum J, Granzow T, von Seggern H, Klein A. Barrier heights, polarization switching and electrical fatigue in $Pb(Zr,Ti)O_3$ ceramics with different electrodes. J Appl Phys. 2010;108:104106.

[107] McCaldin JO, McGill TC and Mead CA. Correlation for III-V and II-VI semiconductors of the Au Schottky barrier energy with anion electronegativity. Phys Rev Lett. 1976;36:56–58.

[108] Tersoff J. Band lineups at II-VI heterojunctions: Failure of the common-anion rule. Phys Rev Lett. 1986;56:2755–2758.

[109] Walukiewicz W. Intrinsic limitations to the doping of wide-gap semiconductors. Physica B. 2001;302–303:123–134.

[110] Zhang SB, Northrup JE. Chemical potential dependence of defect formation energies in GaAs: Application to Ga self-diffusion. Phys Rev Lett. 1991;67:2339–2342.

[111] Nakayama M, Martin M. First-principles study on defect chemistry and migration of oxide ions in ceria doped with rare-earth cations. Phys Chem Chem Phys. 2009;11:3241–3249.

[112] Meyer BK, Polity A, Reppin D, Becker M, Hering P, Klar PJ, Sander T, Reindl C, Benz J, Eickhoff M, Heiliger C, Heinemann M, Bläsing J, Krost A, Shokovets S, Müller C, Ronning C. Binary copper oxide semiconductors: From materials towards devices. Phys Stat Sol (b). 2012;249:1487–1509.

[113] Jaffe JE, Zunger A. Theory of the band-gap anomaly in ABC_2 chalcopyrite semiconductors. Phys Rev B. 1984;29:1882.

[114] Nakamura M, Yamaguchi K, Kimoto Y, Yasaki Y, Kato T, Sugimoto H. Cd-free Cu(In,Ga)$(Se,S)_2$ thin-film solar cell with record efficiency of 23.35%. IEEE J Photovolt. 2019;9:1863–1867.

[115] Li S, Morasch J, Klein A, Chirila C, Pintilie L, Jia L, Ellmer K, Naderer M, Reichmann K, Gröting M, Albe K. Influence of orbital contributions to valence band alignment of Bi_2O_3, Fe_2O_3, $BiFeO_3$, and $Bi_{0.5}Na_{0.5}TiO_3$. Phys Rev B. 2013;88:045428.

[116] Ogo Y, Hiramatsu H, Nomura K, Yanagi H, Kamiya T, Hirano M, Hosono H. p-channel thin-film transistor using p-type oxide semiconductor, SnO. Appl Phys Lett. 2008;93:032113.

[117] Sinsermsuksakul P, Sun LZ, Lee SW, Park HH, Kim SB, Yang CX, Gordon RG. Overcoming efficiency limitations of SnS-based solar cells. Adv Energy Mater. 2014;4:7.

https://doi.org/10.1142/9781800613294_0009

Chapter 9

Time-Resolved X-ray Photoelectron Spectroscopy for Understanding of the Photocatalytic Phenomenon

Kenichi Ozawa

Institute of Materials Structure Science,
High Energy Accelerator Research Organization (KEK),
Tsukuba, Ibaraki 305-0801, Japan

kenichi.ozawa@kek.jp

Abstract

Time-resolved X-ray photoelectron spectroscopy is a powerful tool to study the lifetime of photoexcited carriers at solid surfaces via transient energy shifts of valence and core levels of the constituent elements. This chapter devotes a method to determine the lifetime of photoexcited carriers on rutile and anatase TiO_2 surfaces by a pump-probe technique utilizing an ultraviolet laser as a pump light to generate the carriers and a soft X-ray synchrotron radiation to probe the transient energy shift of core-level peaks. X-ray photoelectron spectroscopy is also utilized to assess a rate of photocatalytic decomposition and desorption of carboxylic acid molecules. The decomposition rate, that reflects photocatalytic activity, is found to be linearly correlated with the carrier lifetime. The activity-lifetime correlation gives important information about the role of the photocarriers in the photocatalytic reactions.

9.1 Introduction

Outstanding photocatalytic activity of titanium dioxide (TiO_2) has been attracting considerable attention since Fujishima and Honda discovered electrochemical photolysis of water at a TiO_2 electrode in the late 1960s [1,2]. Their work has stimulated literally thousands of works which devote to clarify the functionality of TiO_2 under light illumination and has eventually established one of scientific fields: *photocatalysis.* TiO_2 photocatalysis is not limited to water photolysis [1–3], but elimination of pollutants in the air and in solutions [4,5], inactivation of bacteria like *Escherichia coli* [6], superhydrophilicity [7], etc. have also been discovered as other functionalities that TiO_2 exhibits.

In TiO_2 photocatalysis, a key factor is electrons and holes generated upon absorption of the light whose energy exceeds the bandgap energy of TiO_2 (3.1–3.2 and 2.9–3.0 eV for anatase and rutile TiO_2, respectively [8]). Fractions of the generated electrons and holes are transported to the surface and initiate reduction and oxidation, respectively, while others are lost by electron–hole (e–h) recombination during the transport. Photocatalytic reactions in ambient conditions are considered to involve the so-called reactive oxygen species (ROSs), i.e., $O_2^{\bullet-}$, HO_2, •OH, and $O^{\bullet}$ [9,10], which result from the interaction of the photocarriers (photogenerated electrons and holes) with O_2 and H_2O molecules. The ROSs, then, interact with foreign molecules to initiate a series of chain reactions for a total decomposition.

Although details of each elemental step in the ROS-initiated chain reactions have not been fully understood yet, it is essential to focus on the dynamics of the photocarriers because they are essential for photocatalysis. Development of laser-assisted time-resolved measurement techniques has facilitated advances in our understanding of the carrier dynamics in a time domain down to femtoseconds. As summarized in Figure 9.1 [10,11], photoabsorption,

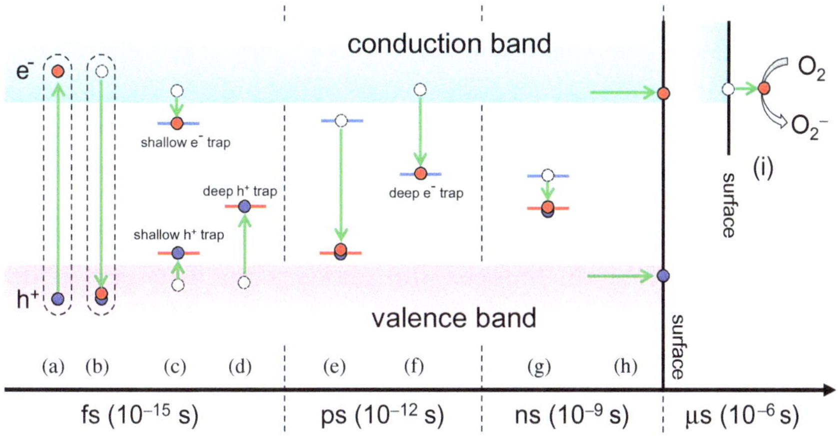

Fig. 9.1. Timescale of photoexcited carriers [10,11]: (a) exciton formation (fs), (b) exciton recombination (fs), (c) traps of a hole (50–170 fs) and an electron (100–250 fs) by shallow-trap states, (d) a hole trap by a deep-trap state (200 fs), (e) e–h recombination between shallow traps (1–10 ps), (f) an electron trap by a deep-trap state (50 ps), (g) e–h recombination between deep traps (>20 ns), (h) carrier transport to the surface (ns–μs), and (i) an electron transfer to adsorbed O_2 (10–100 μs).

generation of excitons (loosely bound e–h pairs), and quenching of the excitons by an e–h recombination are completed in femtoseconds. The excitons that escape the e–h recombination dissociate into free electrons and holes. These photocarriers have chances to be trapped by "shallow" and "deep" traps, which are, respectively, related to O defects and unsaturated Ti atoms in the case of the TiO_2 photocatalyst [12], during migration in the bulk of the crystal. This process takes femtoseconds to picoseconds [10,11]. The carrier traps greatly affect the carrier dynamics because a fraction of the trapped carriers is lost via a trap-mediated recombination and because the diffusion of the photocarriers is suppressed by trapping–detrapping processes by shallow- and deep-trap states in the band gap [13,14]. Surviving photocarriers reach the crystal surface in nanoseconds to microseconds and transfer to adsorbed species on the surface in tens to hundreds of microseconds. The photocarriers which participate in photocatalytic reactions are estimated to be only 10% of all photoexcited carriers [10,11,15], and the rest is quenched by various recombination processes.

Since the photocarrier transport from the bulk to the surface of the photocatalysts takes nanoseconds to microseconds (*cf.* Figure 9.1), photoresponsivity of the photocatalysts in this time domain should be related to the bulk-to-surface carrier transport and, hence, to photocatalytic phenomena. This chapter is devoted to clarify the relation between the photocarrier dynamics in this time domain and the photocatalytic activity of anatase and rutile TiO_2 by employing a surface science approach. Time-resolved X-ray photoelectron spectroscopy (TR-XPS) utilizing an ultraviolet (UV) laser pump and synchrotron radiation probe technique was used to evaluate the lifetime of photoexcited carriers, while the photocatalytic decomposition and desorption of a model molecule, acetic acid, on well-defined TiO_2 single crystal surfaces was determined by conventional X-ray photoelectron spectroscopy (XPS). A linear and positive correlation has been emerged between the photocatalytic reaction rate and the carrier lifetime, providing direct evidence for the involvement of the photocarriers in the photocatalytic reaction.

9.2 TR-XPS for Photocarrier Lifetime Evaluation

9.2.1 *Measurement technique*

Photoelectron spectroscopy including XPS is a well-established experimental technique for characterizing material properties in a non-destructive manner. Characteristic X-rays of Al Kα (1486.6 eV) and Mg Kα (1253.6 eV) lines are used as laboratory light sources, while synchrotron radiation (SR) in soft and hard X-ray regions is also frequently used. One of the advantages of the SR light over the laboratory X-rays is that the photon energy is tunable so that desired photon energies can be chosen.

SR is an inherently pulsed radiation light since it originates from individual electron bunches when they are radially accelerated by a magnet or by periodically arranged short-dipole magnets, like a wiggler or an undulator, in a storage ring [16]. Such pulsed lights can offer a good opportunity to conduct time-resolved measurements using a pump-probe technique. In the pump-probe technique, the pump pulse generates an excited state of the system, and the probe pulse monitors a relaxation process of the exited state. A unique point of this technique is that it can track the time evolution of the system in real time by adjusting a delay time, i.e., a time difference between the arrival times of the pump and probe pulses at a target material.

Two-photon photoelectron (2PPE) spectroscopy is a popular time-resolved photoelectron spectroscopy technique, in which femtosecond or picosecond laser pulses in the visible and UV regions are utilized for both pump and probe lights [17]. 2PPE spectroscopy allows us to directly probe the electrons excited in the conduction bands with a femtosecond time resolution. By combining the 2PPE with an angle-resolved measurement technique, an ultrafast relaxation process of the excited electrons in the conduction bands can be visualized in the energy-wavenumber space, as demonstrated in, for example, Refs. [18,19].

In the TR-XPS experiments, on the other hand, the SR pulses in the X-ray region are used as the probe pulses to excite core electrons of the elements in the system, which is in an excited state by the pump light. Unlike the 2PPE experiment, TR-XPS enables us to capture a local view of the excited system since core electrons are more localized at an atom than the electrons in the conduction band. TR-XPS is, thus, suitable to investigate the systems that are composed of multiple elements or components [20–22]. A TR-XPS measurement for a copper phthalocyanine/C_{60}/TiO_2(110) system [22], for example, allows us to individually follow the time evolution of pump-light-induced energy shifts of the C 1s components of C_{60} and phthalocyanine, as shown in Figure 9.2.

Figure 9.3 shows a schematic of a TR-XPS measurement system [23,24] installed on beamline BL07LSU of SPring-8 [25], an SR facility in Hyogo, Japan. The system consisted of a laser system for the pump light, an undulator beamline for the probe light, an electron spectrometer, and a timing control unit. Although details are different, the basic setup is common among the TR-XPS systems in operation in other SR facilities around the world [26–28].

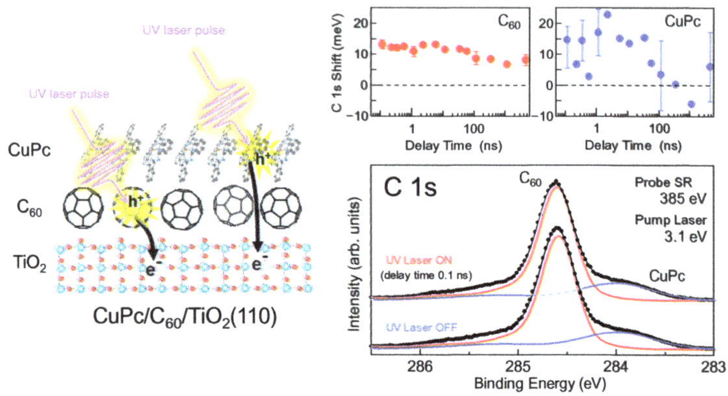

Fig. 9.2. C 1s spectra of a model solar cell system composed of monolayer-thick copper phthalocyanine (CuPc) and C_{60} layers deposited on $TiO_2(110)$, as depicted on the left-hand side. The spectra were acquired with and without the pump UV laser. Solid lines are the results of least-square fitting by Gaussian functions, which represent the C 1s peaks of CuPc and C_{60}. The upper panels on the right-hand side show UV-induced energy shifts (to the higher binding energies as a result of an electron transfer upon UV laser absorption) of the C 1s peaks of C_{60} and CuPc as a function of the delay time. Differences in the magnitude of the shift and the delay-time dependence are observed for both molecules. TR-XPS enables us to investigate the photoresponses separately for each composite.

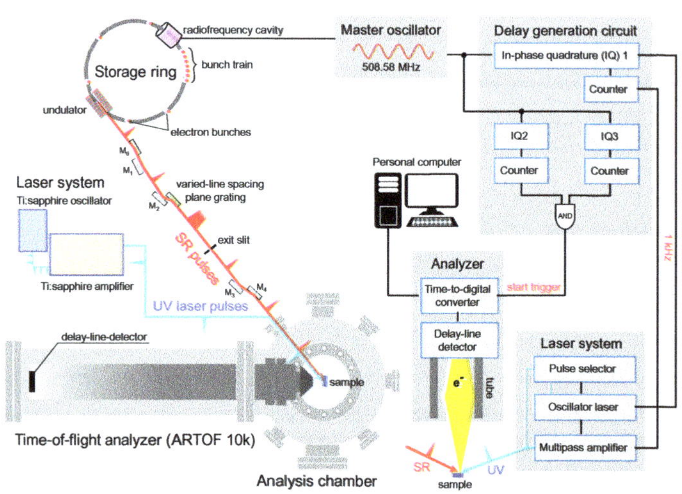

Fig. 9.3. Schematic illustration of the TR-XPS system at SPring-8 BL07LSU (on the left-hand side) and a block diagram of the timing control unit (on the right-hand side).

BL07LSU of SPring-8 was an entrance-slitless high-brilliant soft X-ray beamline with a 27-m-long undulator [25]. The storage ring was operated with one of several bunch modes from "A-mode" to "H-mode". Among these, hybrid operation modes, whose filling pattern was characterized by several isolated electron bunches and a bunch train, were suitable for time-resolved experiments. The SR pulses generated by the isolated electron bunches were used for the TR-XPS measurement. The pulse duration was 50 ps, which was larger than the laser pulse width (35 fs) and, thus, determined the time resolution of the measurement. The SR beam at BL07LSU was monochromatized by a variable-included-angle Monk–Gillieson mounting monochromator with a varied-line spacing plane grating. The photon energy from 250 to 2000 eV was available with a high energy resolving power (>10000). The beam was focused on the sample with a spot size of 10 μm (vertical) × 50 μm (horizontal) using two post-focusing mirrors (M_3 and M_4 in Figure 9.3).

For the TR-XPS experiments, a second harmonic of a Ti:sapphire laser (Odin-II HE, Oquntronix) was used with a repetition rate of 1 kHz and a pulse duration of 35 fs with a typical spot size of 700 μm (vertical) and 500 μm (horizontal). The available power density was several 10 mJ cm^{-2}, which was required to induce measurable surface photovoltages (SPVs) on the TiO_2 surfaces. The electron spectrometer installed in an analysis chamber was a commercial VG-Scienta ARTOF 10 k, which consisted of a time-of-flight tube, a microchannel plate, and a two-dimensional delay-line-detector. Photoelectrons generated by the SR pulses and emitted from the sample surface were recorded as their arrival times and positions on the detector, which were converted to kinetic energies and emission angles, respectively. The maximum acceptance angle of the spectrometer was ±15°.

A timing control and synchronization between the pump laser pulse and the probe SR pulse are essential for the pump-probe TR-XPS measurements. A sine wave of 508.58 MHz from a master oscillator was used to regulate a radiofrequency cavity of the storage ring of SPring-8. At BL07LSU, the same sine wave was used as a reference signal and was delivered into a delay-generation circuit (the right-hand side of Figure 9.3). The signals were, then, sent to an oscillator laser and a multipass amplifier of the laser system as well as to the analyzer system for a start trigger. The timing control between the laser and SR pulses was achieved by a phase shift of the reference signal using in-phase quadrature modulators in the delay-generation circuit. More details for the timing control unit are found elsewhere [23,24].

9.2.2 *Surface photovoltage (SPV)*

One of fundamental optical responses of semiconductors is a generation of an SPV. Necessary conditions for the SPV generation are that the energy of the irradiated light should be larger than the band-gap energy of the semiconductor and that a space-charge layer (SCL), either a depletion layer or an accumulation layer, should be formed at the surface (Figure 9.4(a)). The role of the SCL is to facilitate charge separation of photogenerated excitons because the forces are applied to the electron and the hole in opposite directions along the potential gradient in the SCL. In the depletion layer, therefore, the electrons and holes are drifted to the bottom of the SCL and the semiconductor surface, respectively, and vice versa in the accumulation layer, as shown in Figure 9.4(c). Charge separation induces the potential which cancels the intrinsic potential of the SCL. The SPV is defined as a light-induced change of the surface potential, i.e., the magnitude of surface band bending that is diminished as the light is irradiated on the semiconductor surface (Figure 9.4(c)). The SPV depends on the number of separated carriers, and the maximum SPV equals the magnitude of intrinsic band bending, i.e., the band becomes flat. A comprehensive review of the SPV is given by Kronik and Shapira [29].

The SPV diminishes gradually with time since the carriers that are drifted to the bottom of the SCL diffuse thermally to the surface against the surface potential gradient and recombine with the counter carriers at the surface (Figure 9.4(d)). This SPV decay process usually takes nanoseconds to microseconds irrespective of the materials [30–34]. Since the carriers should overcome the potential

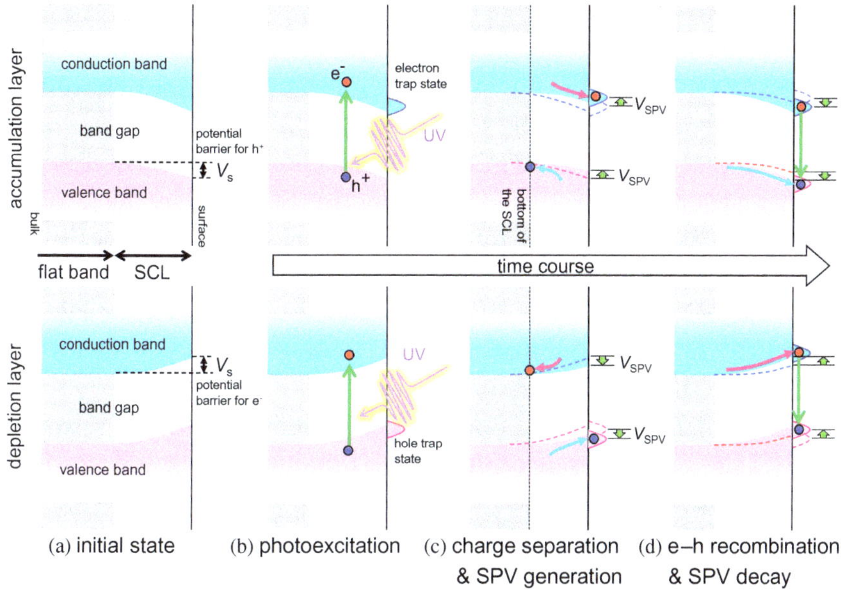

Fig. 9.4. (a) Energy level diagrams of accumulation- and depletion-type space charge layers (SCLs). (b–d) Changes of the band bending structures from the photoexcitation by the UV pulse irradiation to charge separation and to the e–h recombination. The surface photovoltage (SPV) is generated upon charge separation, which induces the shift of the energy levels by V_{SCL}. The SPV is, then, diminishes gradually as the electrons and the holes recombine on the surface. Band bending acts as a surface potential barrier for the carriers diffusing from the bulk to the surface.

barrier in the SCL when they come out on the surface, the decay time depends strongly on the barrier height, i.e., the magnitude of band bending at the surface. In the TR-XPS measurements, the decay process of the SPV is monitored via a temporal energy shift of core-level or valence band photoemission signals from the semiconductor surfaces.

The generation process of the SPV is much faster than the decay process [31]. In some cases, the onset of the SPV takes less than 50 ps because it has not been detected by the TR-XPS experiments with the time resolution of ~50 ps [32–34]. The fast generation process suggests that the forces that are applied to the charged particles of the exciton in the SCL are large enough to swiftly dissociate into the free carriers and to bring them to the surface and to the bottom of the SCL. This behavior of the photoexcited excitons should be different from those in the bulk beneath the SCL, where the band is flat so that the potential gradient to facilitate the exciton dissociation is absent. Thus, the excitons in the SCL should have more chance to be dissociated into the free carriers than those in the bulk. Considering that the free carriers are involved in photocatalytic reactions, the SCL on the photocatalyst surfaces, either a depletion type or an accumulation type, is one of the key factors to determine the photocatalytic activity.

It is worth mentioning that the SPV generation and the resulting shift of the energy levels of the semiconductor surface tell us which type of SCL (either an accumulation or a depletion type) is formed on the surface. This is because the SPV always diminishes the magnitude of intrinsic band bending (Figure 9.4(c)). Namely, if the energy levels move to the lower (higher) binding energies by the SPV, we can say that the intrinsic SCL is an accumulation (a depletion) type.

9.2.3 *Carrier lifetime on TiO_2 surfaces*

The pump-probe TR-XPS technique allows us to monitor the decay process of the SPV in real time via an observation of the transient binding-energy shift of the XPS peaks by changing the delay time. Figure 9.5(a) shows a result of the TR-XPS measurement for the rutile TiO_2(110) surface. The surface was irradiated by UV laser pulses of 3.06 eV to generate the free photocarriers, and probe SR pulses of 600 eV were illuminated after desired delay times to acquire a Ti $2p_{3/2}$ spectrum. An obvious energy shift of the Ti $2p_{3/2}$ peak toward the lower binding energies is seen, especially at short delay times; the shift is ~0.12 eV at the delay time of 0.2 ns, while it diminishes to 0.08 and 0.02 eV at 10 and 100 ns, respectively.

The observed shift is directly converted to the SPV. The direction of the SPV-induced shift, i.e., the lower binding-energy side, indicates that an accumulation-type SCL is formed on the rutile

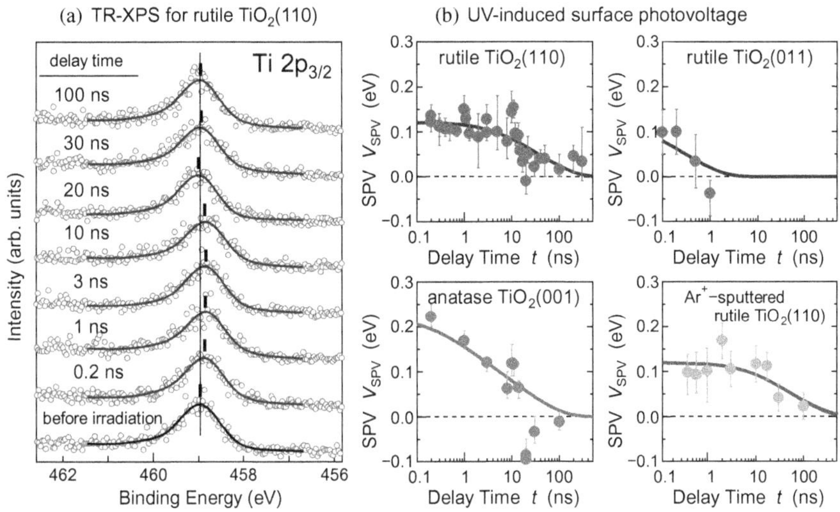

Fig. 9.5. (a) Change in the Ti $2p_{3/2}$ spectra of rutile $TiO_2(110)$ as a function of the delay time. The pump and probe pulse energies were 3.06 and 600 eV, respectively. Curves drawn by open circles are the raw data, and each solid line through the raw spectrum is a result of least-square fitting using a sum function of a Gaussian function and a Shirley-type curve. A thin vertical line and bold tick marks indicate the peak position of the spectra before and after the UV-laser irradiation, respectively. The UV-induced peak shift is cause by the generation of the SPV. (b) Plots of the SPV against the delay time for four TiO_2 surfaces. The power densities of the UV laser pulse were 12, 40, 12, and 48 mJ cm^{-2} $pulse^{-1}$ for rutile $TiO_2(110)$, rutile $TiO_2(011)$, anatase $TiO_2(001)$, and Ar^+-sputtered rutile $TiO_2(110)$, respectively. Solid circles with error bars are the experimental data, and solid lines are obtained by least-square fitting using Eq. (9.1).

$TiO_2(110)$ surface, which was subjected to annealing and Ar^+ sputtering in an ultrahigh vacuum chamber to obtain a carbo-free and well-ordered surface before the measurement. The accumulation layers are usually formed on the TiO_2 surfaces irrespective of the crystal structure and the surface orientation when the surfaces are cleaned in the ultrahigh vacuum conditions.

The decrease in the magnitude of the peak shift (the SPV) as a function of the delay time means the decreasing number of the free carriers as a result of the e–h recombination. Therefore, the SPV decay rate is a good measure of the carrier lifetime. In a framework of the thermionic emission model [30,31,35], the delay-time dependent SPV $V_{SPV}(t)$ and the carrier lifetime τ_∞ are known to be related as

$$V_{SPV}(t) = -\eta k_B T \ln\left\{1 - \left[1 - \exp\left(-\frac{V_0}{\eta k_B T}\right)\right] e^{-t/\tau_\infty}\right\}. \quad (9.1)$$

Here, t is the delay time and V_0 is the SPV at $t = 0$, i.e., $V_{SPV}(0) = V_0$. η is the ideality factor [36], which is an indicator of an e–h recombination rate before the carriers overpass the surface potential barrier and has generally a value between 1 and 2. k_B and T are the Boltzmann constant and the temperature, respectively.

The upper-left panel in Figure 9.5(b) shows a delay-time dependent change of the SPV on rutile $TiO_2(110)$. Filled circles with error bars represent the experimental data and a solid line is a result of least-square fitting with Eq. (9.1), in which τ_∞ and V_0 were treated as variables and η was fixed at 1.4 [37]. The obtained values are $\tau_\infty = 150$ ns and $V_0 = 0.12$ eV. The same TR-XPS measurements and data analyses were carried out for rutile $TiO_2(011)$, anatase $TiO_2(001)$, and Ar^+-sputtered rutile $TiO_2(110)$, and the results are also shown in Figure 9.5(b). The evaluated V_0 values are 0.12, 0.22, and 0.12 eV and the τ_∞ values are 1.3, 90, and 230 ns for rutile $TiO_2(011)$, anatase $TiO_2(001)$, and sputtered rutile $TiO_2(110)$, respectively.

V_0 depends on the power density of the UV laser pulse as well as the e–h recombination rate below 50 ps, a time region where the TR-XPS study was not accessible due to the time resolution of the measurement system. Since the different laser power was used for four TiO_2 surfaces, it is not straightforward to make a parallel comparison of V_0 among the surfaces. Nevertheless, it is inferred that a higher fraction of the photocarriers might be quenched by a fast e–h recombination within picoseconds (below 50 ps) on rutile $TiO_2(011)$ than the other three surfaces because the SPV magnitude at 0.1 ns is the lowest (Figure 9.5(b)) despite a relatively high power density (40 mJ cm^{-2} $pulse^{-1}$) to stimulate the SPV. This may explain the very short carrier lifetime on rutile $TiO_2(011)$ (τ_∞ = 1.3 ns) in comparison with the other three surfaces, namely, the tail of the picosecond-order decay of the SPV could be observed on rutile $TiO_2(011)$ in the nanosecond time window. A possible origin of the fast e–h recombination on this surface has been discussed in Ref. [37], where an existence of characteristic deep hole-trap states, whose capture cross-section is so large that the hole capture and a subsequent e–h recombination are facilitated, may be responsible for the fast recombination and, thus, the short lifetime.

It is interesting to note that the different lifetimes between rutile $TiO_2(110)$ and sputtered rutile $TiO_2(110)$ (τ_∞ = 150 and 230 ns) must originate purely from the surface condition of the crystal. On the sputtered rutile $TiO_2(110)$ surface, the clean surface was subjected to mild Ar^+ sputtering to produce oxygen vacancies with a density of 6×10^{14} cm^{-2} [37]. These vacancies are usually considered to be e–h recombination centers so that the lifetime is expected to be shortened on the sputtered surface. However, the prolonged lifetime from 150 to 230 ns means that there is another role for the vacancies other than the recombination centers; it is an enlargement of the surface potential barrier height. The surface oxygen vacancies on the TiO_2 surfaces are known to push the valence band position downwards [35,37] so that the higher density of the vacancies on the surface leads to larger downward band bending and, thus, a higher potential barrier height. As will be shown in Section 9.4.1, the barrier height is the most influential factor to determine the carrier lifetime among other factors that affect the lifetime, i.e., a carrier-capture cross-section, a thermal velocity of the carriers, and a density of the carrier trap states. Therefore, the photocarriers on the sputtered $TiO_2(110)$ surface should have a longer lifetime on the non-sputtered surface.

9.3 XPS for Evaluation of Photocatalytic Activity

9.3.1 *Surface science approach*

The use of bulk single crystals or single crystalline films with well-defined surfaces is an example of surface science approaches to clarify physics underlaying surface physicochemical phenomena including photocatalysis. This approach can avoid possible complexity arising from morphology of the material surface. One of the pioneering photocatalytic studies by the surface science approach is a comparative study of $Ag^+ \rightarrow Ag^0$ photoreduction on different crystal facets of rutile TiO_2 in a $AgNO_3$ solution [38,39]. A clear surface dependence of the photoreduction activity was revealed; the rutile TiO_2(101) surface exhibited the highest activity, while the activity was poorer on the (100) and (110) surfaces. In a more recent work by Ahmed *et al.* [40], photocatalytic methanol oxidation and terephthalic acid hydroxylation were compared on anatase and rutile TiO_2 surfaces. The anatase (101) surface exhibits the higher activity than the rutile TiO_2 surfaces with (001), (100), and (101) orientations. For photocatalytic decomposition of methyl orange in an aqueous solution [41], the anatase surface also shows a higher activity than the rutile surface. The trend that the photocatalytic activity of anatase TiO_2 is superior to that of rutile one is in good agreement with the already known trend deduced from studies employing catalytic approaches using TiO_2 powders [42–44]. Thus, the surface science approach is valid for the photocatalytic study.

Figure 9.6(a) shows low-energy electron diffraction (LEED) patterns of single-crystal rutile and anatase TiO_2 surfaces, which were cleaned by cycles of Ar^+ sputtering (1–2 kV, 1×10^{-4} Pa Ar, 10 min) and annealing in O_2 atmosphere (900–1000 K, (1–2.5) $\times 10^{-4}$ Pa O_2, 10 min) until no carbon contaminant was observed. The (110), (100), and (001) surfaces of rutile TiO_2 and the anatase TiO_2(101) surface exhibited (1 × 1) patterns, whereas the rutile TiO_2(011) and anatase TiO_2(001) surfaces showed (2 × 1) and (4 × 1) reconstructions, respectively. Although there still exist some controversial issues, widely accepted models of the surface atomic structures are shown in Figure 9.6(b) [45]. For the XPS and TR-XPS experiments presented in this and previous sections, these surfaces with long-range ordered structures were used for a precise characterization of the TiO_2 surfaces as well as for a guarantee of reproducibility of the measurement results.

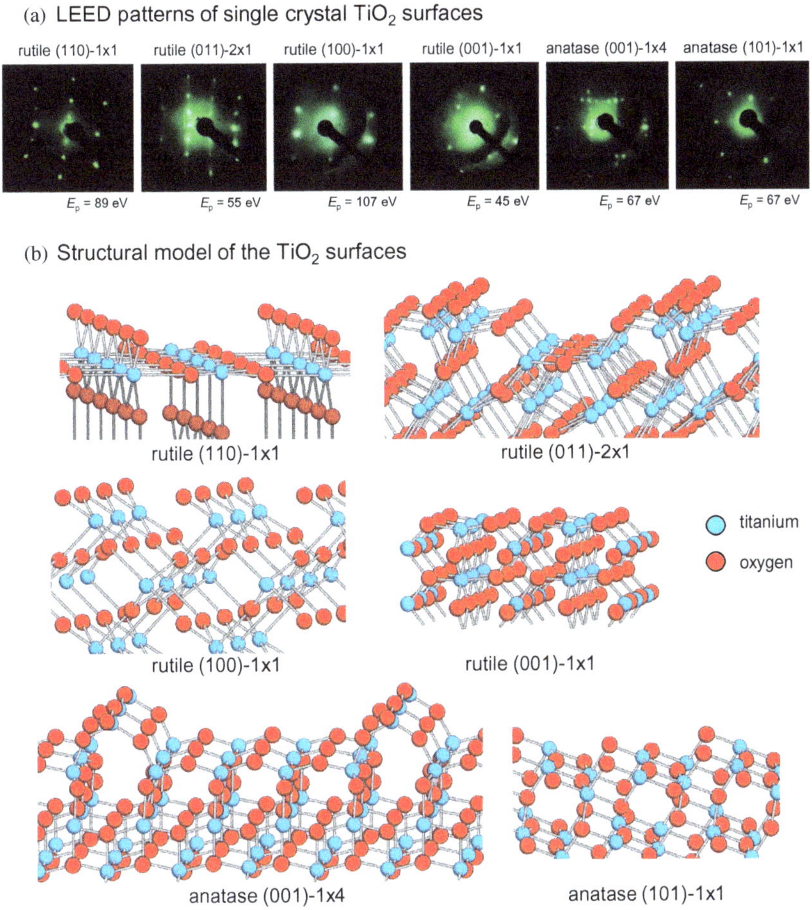

Fig. 9.6. (a) LEED patterns of single-crystal rutile and anatase TiO_2 surfaces, which were cleaned in ultrahigh vacuum conditions. The images were acquired by a LEED optics with a flat microchannel plate (BLD800IR, OCI Vacuum Microengineering). E_p is a primary energy of the electron beam. (b) Structural models of the TiO_2 surfaces [45].

9.3.2 *Experimental setup*

In photocatalytic studies, XPS is usually used for the characterization of photocatalyst surfaces and/or of the adsorption state of target molecules, and the XPS data are utilized to interpret how the photocatalysts work during photocatalytic reactions. Reaction rates are usually examined by, e.g., UV-visible absorption spectroscopy [41,44] or gas/liquid chromatography [40,43]. These techniques assess amounts of reactant or product molecules in ambient gases or solutions surrounding the photocatalysts. On the other hand, XPS can only examine the molecules which adsorb on the photocatalyst surface. Furthermore, a measurement environment is limited to ultrahigh and high vacuum or near-ambient pressure (up to several pascals [46]) conditions. Therefore, the rate determination by XPS is quite rare [47,48]. However, as will be demonstrated in the following, to clarify the role of the photoexcited carriers in the photocatalytic reaction, it is indispensable to measure both the photocatalytic reaction rate and the surface potential barrier height in the same environment. For this purpose, XPS is an ideal technique.

Figure 9.7 shows an experimental apparatus at beamline BL-13B of the Photon Factory [49], an SR facility in Tsukuba, Japan. A hemispherical electron energy analyzer (SES200, Gamma Data/Scienta) was used for the XPS study. The TiO_2 surfaces were cleaned in a preparation chamber and were transferred to an analysis chamber without exposing the samples to the air. The TiO_2 samples were

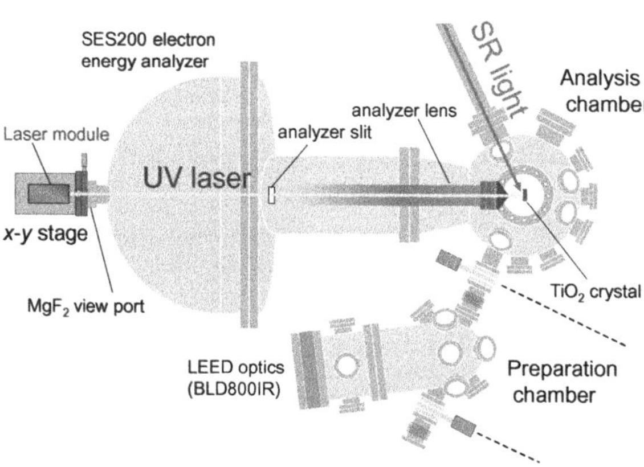

Fig. 9.7. Schematic illustration of the XPS measurement system at BL-13B of the Photon Factory. A sample loadlock chamber (not shown) was attached to the preparation chamber.

placed so that the surfaces faced the analyzer lens. The SR X-ray was incident on the sample surface at 65° from the surface normal direction. A continuous-wave UV laser with a wavelength of 375 nm (3.31 eV) and a power density of 2.4 W cm^{-2} was delivered from a diode laser module (Obis375LX, Coherent Inc.) to induce photocatalytic reactions. The module was mounted on an *x-y* stage, which was attached to a MgF_2 viewport on the back side of the analyzer. The laser was illuminated on the surface through a slit and the analyzer lens.

A spatial overlap between the UV laser and the SR light was confirmed by a laser-induced SPV shift of a Si 2p XPS peak. As the UV laser was irradiated on Si(111)-7 × 7, the Si 2p peak moved to the higher-binding energy side by an SPV shift [48]. The direction of the shift suggests that a depletion layer was developed on the surface. The magnitude of the shift depended on the *x* and *y* positions of the micrometers of the *x-y* stage on which the laser module was mounted. The laser-SR overlap was guaranteed when the maximum SPV shift was obtained. It should be noted that the UV laser with a power density of 2.4 W cm^{-2} was insufficient to induce an observable SPV on the TiO_2 surfaces. This is the reason why the Si(111) surface was used.

9.3.3 *Acetic acid as a probe molecule*

Figure 9.8(a) shows C 1s spectra of acetic acid adsorbed on various TiO_2 surfaces. Acetic acid is known to absorb on the TiO_2 surfaces in the form of acetate at room temperature by breaking the O–H bond ($CH_3COOH \rightarrow CH_3COO^- + H^+$). Two C 1s peaks at ~286 and ~289 eV are assigned to methyl carbon and carboxylate carbon, respectively. Such a two-peak structure is observed irrespective of the TiO_2 surfaces. A slight difference in the peak energy may be due to a different adsorption state of acetate and/or a different magnitude of band bending on the TiO_2 surfaces.

An interesting point to note is that the intensity ratio of the 286 and 289 eV peaks varies depending on the surface. This should reflect the adsorption structure of acetate on the surface, namely, an

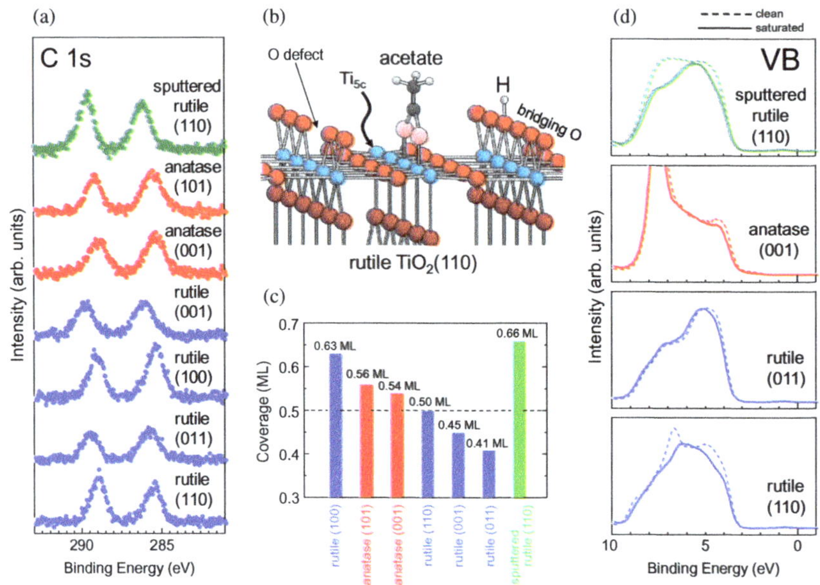

Fig. 9.8. (a) C 1s spectra of acetic-acid-saturated TiO_2 surfaces. Adsorption was carried out at room temperature by backfilling the loadlock chamber with the vapor of acetic acid. The photon energy used was 753 eV. (b) Adsorption structure of acetic acid on rutile TiO_2(110). The molecule adsorbs dissociatively and resultant acetate bounds to the surface Ti_{5c} atoms via its two O atoms. A dissociated H atom forms an OH group. If a bridging O atom is removed by, for instance, Ar^+ sputtering and an O defect is formed, two Ti_{5c} atoms are newly provided as an adsorption site. (c) Comparison of the saturation coverage of acetic acid on seven surfaces at room temperature. The coverages were evaluated from the relative C 1s peak intensities to that on rutile TiO_2(110), where the saturation coverage is known to be 0.5 ML [50]. (d) Valence-band (VB) spectra of four TiO_2 surfaces acquired before (dashed lines) and after (solid lines) acetic-acid adsorption. The photon energy was 100 eV.

upright or tilted orientation, and a bidentate or monodentate coordination. On the rutile TiO_2(110) surface, acetate is adsorbed with the upright orientation via its O atoms that are bonded to two five-coordinated Ti (Ti_{5c}) atoms, as illustrated in Figure 9.8(b).

Another point to be noted is that the integrated intensity of the C 1s peaks, which reflects the coverage of acetic acid, depends on the surfaces. It is established that a saturation coverage of acetic acid on rutile $TiO_2(110)$ is half of the density of the surface Ti_{5c} atoms (5.2×10^{14} cm^{-2}) [50] and is defined as 0.5 monolayers (ML). The coverage on the defective (sputtered) rutile $TiO_2(110)$ surface is 1.3 times larger than that on rutile $TiO_2(110)$. This suggests that the surface defects, mainly the O defects, are adsorption sites other than the Ti_{5c} atoms. Among the defect-free (or less defective) surfaces, rutile $TiO_2(100)$ can accommodate the largest amounts of acetic acid, followed by anatase (101), anatase (001), rutile (110), rutile (001), and rutile (011) (Figure 9.8(c)). This order is viewed as the order of the surface chemical reactivity toward acetic-acid adsorption.

9.3.4 *UV irradiation effect*

UV laser irradiation on the acetic-acid-covered TiO_2 surfaces leads to a gradual intensity decrease of the C 1s XPS peak. On rutile $TiO_2(110)$, for example, the intensity was decreased to 80% for the first ~1 h (Figure 9.9(a)). The decrease should be caused by photocatalytic decomposition and desorption of acetate. A possible reaction process is that a photogenerated hole induces the decomposition of acetate into a methyl group and carbon dioxide, i.e., $CH_3COO^- + h^+ \rightarrow CH_3 + CO_2$, and the methyl group and dissociated H^+ forms methane upon accepting a photogenerated electron, i.e., $CH_3 + H^+ + e^- \rightarrow CH_4$. Although this is the simplest reaction path, alternative reaction paths to produce H_2 and C_2H_6 have also been proposed [51]. Nevertheless, the reaction is always initiated by the hole-acetate interaction. It should be noted that photocatalytic reactions are affected by coexisting molecules, and different photocatalytic decomposition paths of acetic acid are possible under the presence of O_2 and H_2O [51,52].

Although the UV irradiation induces the C 1s peak intensity to decrease, the peak intensity also decreases even when the XPS spectra are measured without UV irradiation. This means that elimination of adsorbed acetic acid is caused by the X-rays. A recent study by Yang *et al.* [53] has suggested that X-ray irradiation generates hot electrons and holes in the conduction and valence bands,

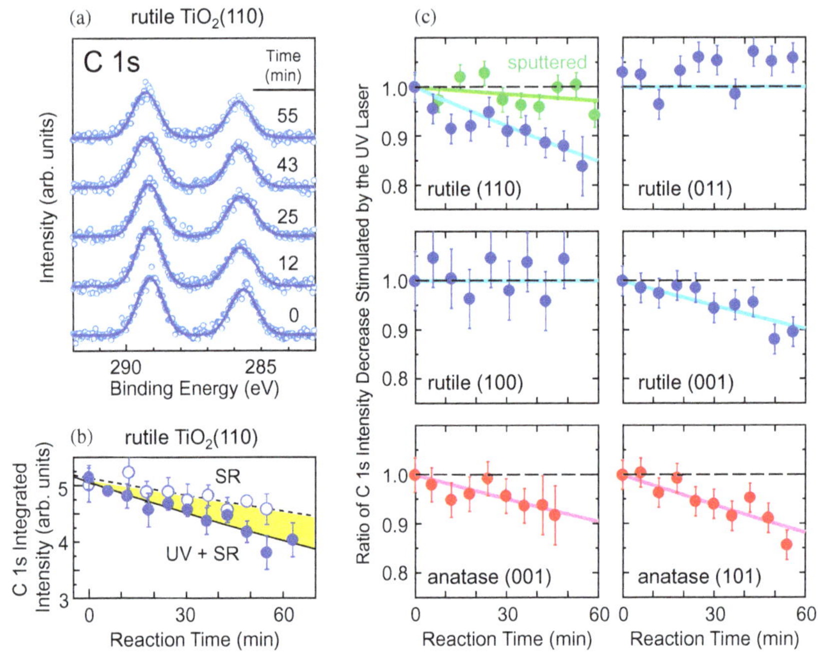

Fig. 9.9. (a) Change of the C 1s spectrum of acetic-acid-saturated rutile TiO_2(110) as a function of the irradiation time of the UV laser (3.31 eV). The spectra were acquired by the SR light at 753 eV. Curves drawn by the symbols are the experimental data, whose background was subtracted by polynomial functions. Solid lines are the results of least-square fitting using two Gaussian functions. (b) Plots of the C 1s integrated intensities against the UV-irradiation time, i.e., the reaction time. Filled and open symbols are the results of the measurements with and without the UV-laser irradiation. A yellow hutched area represents a net effect of the UV laser on the C 1s intensity decrease. (c) Intensity decreasing ratios of the C 1s peak by photocatalytic decomposition desorption of acetic acids on the TiO_2 surfaces. Thick solid lines are the results of exponential fitting.

respectively, and these carriers induce reactions similar to photocatalytic reactions. On the other hand, since SR-induced decomposition of adsorbates is widely seen irrespective of substrate materials, direct photochemical decomposition [54], decomposition by secondary electrons emitted from the substrate [55], and/or photon-stimulated desorption [56] could be more plausible causes. Thus, to extract the net amount of acetic acid eliminated by photocatalytic decomposition desorption, the C 1s intensity decrease by the SR irradiation should be excluded from the overall C 1s peak intensity decrease by simultaneous irradiation of both UV laser and SR X-ray. The yellow hatched area in Figure 9.9(b) corresponds to the amount of acetic acid eliminated from rutile $TiO_2(110)$ by the photocatalytic reaction.

9.3.5 *Photocatalytic reaction rate*

Figure 9.9(c) shows time-dependent changes of the acetate C 1s peak intensities induced only by the UV laser on seven TiO_2 surfaces. The decreasing rate on each surface is evaluated by fitting the experimental data with an exponential function under the assumption that photocatalytic decomposition and desorption can be approximated by a first-order reaction; $I = \exp(-t/\tau_{ph})$, where I, t, and τ_{ph} are the normalized C 1s peak intensity, the UV-irradiation time, and the decay constant, respectively. A smaller decay constant means a faster elimination of acetic acid from the surface and, thus, a higher photocatalytic activity. τ_{ph} on rutile $TiO_2(110)$ is the smallest among all the surfaces investigated with τ_{ph} = 370 min, whereas rutile $TiO_2(011)$ and (100) surfaces do not exhibit any activity with an infinite value of τ_{ph}.

A reciprocal of τ_{ph} is a rate constant k_{ph} and is a direct measure of the photocatalytic activity. Table 9.1 summarizes τ_{ph} and k_{ph} for seven TiO_2 surfaces. Rutile $TiO_2(110)$ exhibits the highest activity with $k_{ph} = 2.7 \times 10^{-3}$ min^{-1}. The value is even larger than those of two anatase TiO_2 surfaces. This indicates that anatase TiO_2 is not always more active than rutile TiO_2, but the crystallographic orientation of the surface is more important to determine the activity.

An interesting point is that the photocatalytic activity on the sputtered rutile $TiO_2(110)$ surface ($k_{ph} = 0.5 \times 10^{-3}$ min^{-1}) is much lower than that on rutile $TiO_2(110)$ (2.7×10^{-3} min^{-1}). This suppression is induced even though the surface O defects enhance the surface chemical reactivity towards acetic-acid adsorption (Figure 9.8(c)). Therefore, it is speculated that the photocatalytic activity should be determined by a factor other than amounts of reactants on the photocatalyst surface.

Table 9.1. Decay time constant of the C 1s peak intensity (τ_{ph}) and the rate constant of photocatalytic decomposition desorption of acetic acid (k_{ph}).

Surface	τ_{ph}(min)	k_{ph} (min^{-1})
Rutile TiO_2(110)	370	2.7×10^{-3}
Anatase TiO_2(101)	480	2.1×10^{-3}
Rutile TiO_2(001)	580	1.7×10^{-3}
Anatase TiO_2(001)	600	1.7×10^{-3}
Rutile TiO_2(110)[a]	2100	0.5×10^{-3}
Rutile TiO_2(011)	∞	0
Rutile TiO_2(100)	∞	0

Note: [a]Ar^+ sputtered surface.

9.4 Correlation Between Photocatalytic Activity and Carrier Lifetime

9.4.1 *Carrier lifetime on adsorbate-covered surface*

The TR-XPS study has clarified the lifetime of the UV-excited photocarriers τ_∞ on four TiO_2 surfaces (Figure 9.5). It is known that the carrier dynamics is sensitive to the SCL formed in the near-surface region of the photocatalyst crystal. In the case of the TiO_2 surfaces where the accumulation layers are formed and the holes have a longer diffusion length (from the bottom of the SCL to the surface) than the electrons, τ_∞ should depend on the width of the SCL W_{SCL}, the effective surface recombination velocity of the holes S_p^*, and the hole diffusion constant D_p as follows [57]:

$$\tau_\infty = \frac{W_{SCL}}{S_p^*} + \frac{4}{D_p}\left(\frac{W_{SCL}}{\pi}\right)^2. \quad (9.2)$$

The first and second terms on the right-hand side of Eq. (9.2) are called a surface-recombination-velocity term and a carrier diffusion term, respectively. The diffusion term is calculated to be less than 2 ps on the TiO_2 surfaces because of $W_{SCL} \simeq 3$ nm and $D_p = 1.34 \times 10^{-2}$ cm^2 s^{-1} for the hole diffusion in the SCL of TiO_2 [58].

This value is much smaller than the carrier lifetimes we are now considering (Figure 9.5). Thus, the lifetime is approximated by the surface-recombination-velocity term:

$$\tau_{\infty} \simeq \frac{W_{\mathrm{SCL}}}{S_{\mathrm{p}}^{*}} = \frac{W_{\mathrm{SCL}}}{S_{\mathrm{p}}} \exp\left(\frac{V_{\mathrm{s}}}{\eta k_{\mathrm{B}} T}\right). \tag{9.3}$$

Here, S_p is the surface recombination velocity and V_s is the surface potential barrier height (see Figure 9.4(a) for definition). Comparing the three parameters that determine τ_{∞}, i.e., W_{SCL}, S_p, and V_s, a small change in V_s leads to a large change in τ_{∞} since τ_{∞} depends on V_s exponentially. On the other hand, changes in the other two parameters are less significant. Thus, molecular adsorption and the resultant modification of the TiO_2 surfaces affect the carrier lifetime mainly via the change in V_s.

Acetic-acid adsorption leads to a slight variation of V_s on the TiO_2 surfaces. Adsorption pushes the valence-band-maximum (VBM) positions down toward the higher-binding-energy side on anatase TiO_2(001), rutile TiO_2(011), and rutile TiO_2(110) by 0.05, 0.06, and 0.04 eV, respectively (Figure 9.8(d)). These values correspond to the change in V_s (ΔV_s) induced by adsorption.

Regarding the defective rutile TiO_2(110) surface, it is difficult to precisely control the O defect densities on the surface. The defective rutile TiO_2 surface used for the TR-XPS study (Figure 9.5(b)) had the O defects with 3×10^{14} cm^{-2}, while the defect density was 2×10^{14} cm^{-2} on the surface prepared for the photocatalytic activity measurements (Figure 9.9(c)). An energetic difference of the VBMs on these surfaces was 0.05 eV with a shallower positioning of the VBM on the surface with a lower defect density. As the defective surface with 2×10^{14} cm^{-2} was covered with a saturation amount of acetic acid, the VBM was found to be hardly affected (the top panel in Figure 9.8(d)). Therefore, the ΔV_s value for defective rutile TiO_2(110) was estimated to be −0.05 eV.

The carrier lifetime on the clean TiO_2 surface τ_{∞} determined by TR-XPS and the acetic-acid-induced change of the surface barrier height ΔV_s determined by XPS allow us to estimate the carrier lifetime on the acetic-acid-adsorbed TiO_2 surfaces $\tau_{\infty,AC}$ by the following equation:

$$\tau_{\infty,Ac} = \tau_{\infty} \exp\left(\frac{\Delta V_s}{\eta k_B T}\right), \tag{9.4}$$

which is derived from Eq. (9.3) under the assumption that the changes of W_{SCL} and S_p by acetic-acid adsorption are smaller enough than the change of $\exp[V_s/(\eta k_B T)]$. By inserting the experimentally determined τ_{∞} and ΔV_s as well as $\eta = 1.4$, $\tau_{\infty,Ac}$ are 360, 7, 450, and 60 ns for the acetic-acid-adsorbed surfaces of anatase $TiO_2(001)$, rutile $TiO_2(011)$, rutile $TiO_2(110)$, and defective rutile $TiO_2(110)$, respectively. Table 9.2 summarizes τ_{∞}, ΔV_s, and $\tau_{\infty,Ac}$ for four surfaces.

9.4.2 *Activity-lifetime correlation*

A relation between the rate constant of photocatalytic decomposition and desorption of acetic acid (k_{ph}) and the carrier lifetime on the acetic-acid-saturated TiO_2 surface ($\tau_{\infty,Ac}$) is shown in Figure 9.10(a). Although there are only four plotted points, we see a linear and positive correlation between k_{ph} and $\tau_{\infty,Ac}$. This result demonstrates that the photoexcited carriers surely contribute to the photocatalytic reaction and, more importantly, that they contribute to the rate-determining step. In the acetic-acid-adsorption system,

Table 9.2. Carrier lifetime on adsorbate-free surfaces (τ_{∞}), acetic-acid-induced change in the surface potential barrier height (ΔV_s), and the expected carrier lifetime on the acetic-acid-adsorbed surfaces ($\tau_{\infty,Ac}$).

Surfaces	τ_{∞} (ns)	ΔV_s (eV)	$\tau_{\infty,Ac}$ (ns)
Rutile $TiO_2(110)$	150	−0.04	450
Anatase $TiO_2(001)$	90	−0.05	360
Rutile $TiO_2(110)$[a]	230	−0.05	60
Rutile $TiO_2(011)$	1.3	−0.06	7

Note: [a]Ar^+ sputtered surface.

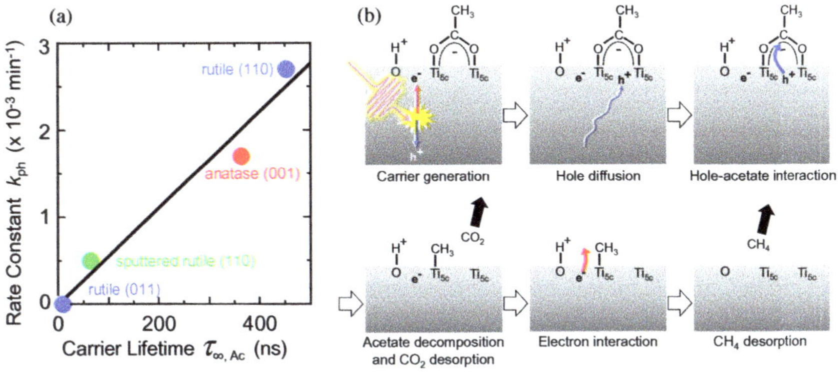

Fig. 9.10. (a) Plot of the photocatalytic activity (k_{ph}) against the carrier lifetime ($\tau_{\infty,Ac}$). A solid line is a least-square fitting result. (b) One of the possible photocatalytic decomposition desorption processes of acetic acid in the absence of O_2 and H_2O. The linear k_{ph}–$\tau_{\infty,Ac}$ correlation suggests that the hole diffusion to the surface is a rate-determining step.

the holes rather than the electrons should play a role to determine the overall photocatalytic activity since the hole-acetate reaction ($CH_3COO^- + h^+ \rightarrow CH_3 + CO_2$) is considered to initiate the reaction (Figure 9.10(b)).

After dissociation of the photogenerated excitons in the SCL on the TiO_2 surface, the holes are drifted to the bottom of the SCL. These holes thermally diffuse to the surface against the potential gradient and recombine with the electrons on the surface. These processes determine the carrier lifetime, as described in Section 9.2.2. Remember that the carrier lifetime τ_∞ is expressed by Eq. (9.3), it is natural to consider that k_{ph} is proportional to τ_∞ because Eq. (9.3) can be converted to the Arrhenius equation by replacing W_{SCL}/S_p and V_s/η with a pre-exponential factor (A) and a negative of an activation energy ($-E_a$), respectively. Therefore, the hole diffusion process should govern the reaction rate. If k_{ph} is not proportional to τ_∞, the rate-determining step should not be the hole diffusion but a carrier transfer from the surface to acetate or other later reaction steps (Figure 9.10(b)).

There is a general consensus that anatase TiO_2 is photocatalytically more active than the rutile one. However, the k_{ph} values listed in Table 9.1 tell us that this is not always the case. A more important factor is the formation of the SCL and the resultant surface potential barrier height V_s, as inferred from Eq. (9.3). Thus, a TiO_2 surface having an SCL with a large V_s value in "working" conditions should exhibit a high photocatalytic activity regardless of the crystal structure of TiO_2. This means that an artificial modification of the TiO_2 surface by, for instance, an introduction of the O vacancies on the surface or surface doping of metal atoms to enlarge V_s is a good guideline to develop highly active photocatalysts.

On the other hand, such a V_s guideline is not appreciated for nanometer-sized crystals because the SCL is not developed in the nanocrystals with the sizes less than 10 nm [59], where band bending, if exists, is negligibly small. In the flat-band crystals, carrier trapping by vacancies, impurities, or any kinds of crystal dislocations is one of influential factors to determine the carrier diffusion and, thus, the carrier lifetime. Theoretical studies have predicted that the formation of O defects, which act as carrier trap sites, is more favorable in anatase TiO_2 than in rutile TiO_2 [12,60]. Thus, the carrier lifetime may be longer in anatase TiO_2 than in the rutile one even when the surface potential barrier is absent, as has already been suggested by the TR-XPS study [33].

9.5 Concluding Remarks

TR-XPS utilizing a pump-probe technique is unique in that the photoexcited charged carrier dynamics can be pursued in real time via a temporal energy shift of a core-level peak of a constituent element. Since the core electrons are localized at the atoms, a local view of the excited state can be obtained. This fact makes TR-XPS an ideal tool to investigate the carrier dynamics, especially in complex systems which are composed of multiple components and/or elements. Another advantage of TR-XPS is that very small energy shifts of a few meV can be detected because a sharp spectral feature of the core-level peaks is available. By taking these advantages of TR-XPS, the lifetime of the photogenerated carriers on the TiO_2 surfaces has been evaluated from the SPV-induced energy shift of the Ti $2p_{3/2}$ peak.

Among the several factors that determine the carrier lifetime, a surface potential barrier height, that is equivalent to the magnitude of band bending in the SCL, is the most influential one. This means that the lifetime is sensitive to the surface modification which affects band bending. Regarding the TiO_2 surfaces, acetic-acid adsorption enlarged the barrier height by 0.04–0.06 eV. Although the change is very small, more than three times longer carrier lifetime is achieved.

On the acetic-acid/TiO_2 surfaces, where the SCL with an accumulation type is formed, the photogenerated holes rather than the electrons play dual roles: one is to determine the carrier lifetime and the other is to initiate photocatalytic decomposition desorption of acetic acid. A linear correlation between the carrier lifetime and the reaction rate constant found in our study indicates that diffusion of the holes from the bottom of the SCL to the surface should be a more time-consuming process than the hole–acetate interaction so that the hole diffusion should be a rate-limiting step.

Utilizing the linear activity-lifetime correlation, the photocatalytic activity of TiO_2 can be predicted from the carrier lifetime, namely, the TiO_2 surface with a longer carrier lifetime may exhibit a higher photocatalytic activity. Even more importantly, this correlation suggests that the photocatalytic activity should be, in principle, controllable by adjusting the carrier lifetime. Since the lifetime depends strongly on the surface potential barrier height, surface modification to enlarge band bending is an easy but practical way to enhance the lifetime and, thus, the activity.

For future studies, it is important to confirm how universal the linear activity-lifetime correlation is, especially in ambient conditions where photocatalytic reactions proceed under the presence of O_2 and H_2O. XPS and/or TR-XPS measurements in the ambient conditions are challenging because they usually require a high or ultrahigh vacuum to avoid photoelectron scattering by gaseous molecules. However, the recent progress of the measurement apparatus enables us to conduct the XPS measurements in ambient and near-ambient conditions. Therefore, *in situ* photocatalytic studies utilizing ambient-pressure XPS, like the one pioneered by Lampimäki *et al.* [61], and ambient-pressure TR-XPS will be flourished in the coming decade and will clarify the relation between the carrier dynamics and the photocatalytic activity.

References

[1] Fujishima A, Honda K. Electrochemical evidence for the mechanism of the primary stage of photosynthesis. Bull Chem Soc Jpn. 1971; 44(4):1148–1150.

[2] Fujishima A, Honda K. Electrochemical photolysis of water at a semiconductor electrode. Nature. 1972;238:37–38.

[3] Schrauzer GN, Guth TD. Photolysis of water and photoreduction of nitrogen on titanium dioxide. J Am Chem Soc. 1977;99(22): 7189–7193.

[4] Frank SN, Bard AJ. Heterogeneous photocatalytic oxidation of cyanide ion in aqueous solutions at titanium dioxide powder. J Am Chem Soc. 1977;99(1):303–304.

[5] Ibusuki T, Takeuchi K. Removal of low concentration nitrogen oxides through photoassisted heterogeneous catalysis. J Mol Catal. 1994; 88(1):93–102.

[6] Matsunaga T, Tomoda R, Nakajima T, Wake H. Photoelectrochemical sterilization of microbial cells by semiconductor powders. FEMS Microbiol Lett. 1985;29(1–2):211–214.

[7] Wang R, Hashimoto K, Fujishima A, Chikuni M, Kojima E, Kitamura A, Shimohigoshi M, Watanabe T. Light-induced amphiphilic surfaces. Nature. 1997;388:431–432.

[8] Tang H, Levy F, Berger H, Schmid PE. Urbach tail of anatase TiO_2. PhysRevB.1995; 52(11):7771–7774.

[9] Hashimoto K, Irie H, Fujishima A. TiO_2 photocatalysis: A historical overview and future prospects. Jpn J Appl Phys. 2005;44(12R): 8269–8285.

[10] Schneider J, Matsuoka M, Takeuchi M, Zhang J, Horiuchi Y, Anpo M, Bahnemann DW. Understanding TiO_2 photocatalysis: Mechanisms and materials. Chem Rev. 2014;114(19):9919–9986.

[11] Hoffmann MR, Martin ST, Choi W, Bahnemann DW. Environmental applications of semiconductor photocatalysis. Chem Rev. 1995; 95(1):69–96.

[12] Mattioli G, Alippi P, Filippone F, Caminiti R, Banapasta AA. Deep versus shallow behavior of intrinsic defects in rutile and anatase TiO_2 polymorphs. J Phys Chem C. 2010;114(49):21694–21704.

[13] de Jongh PE, Vanmaekelbergh D. Trap-limited electronic transport in assemblies of nanometer-size TiO_2 particles. Phys Rev Lett. 1996;77(16):3427–3430.

[14] van de Lagemaat J, Frank AJ. Nonthermalized electron transport in dye-sensitized nanocrystalline TiO_2 films: Transient photocurrent and random-walk modeling studies. J Phys Chem B. 2001;105(45): 11194–11205.

[15] Shkrob IA, Sauer Jr MC. Hole scavenging and photo-stimulated recombination of electron-hole pairs in aqueous TiO_2 nanoparticles. J Phys Chem B. 2004;108(33):12497–12511.

[16] Willmott P. *An Introduction to Synchrotron Radiation: Techniques and Applications*. Chichester, UK: John Wiley & Sons, 2011.

[17] Fauster T. Two-photon photoemission. Prog Surf Sci. 1994;46(2–3): 177–186.

[18] Wang YH, Hsieh D, Sie EJ, Steinberg H, Gardner DR, Lee YS, Jarillo-Herrero P, Gedik N. Measurement of intrinsic Dirac fermion cooling on the surface of the topological insulator Bi_2Se_3 using time-resolved and angle-resolved photoemission spectroscopy. Phys Rev Lett. 2012;109:127401-1–5.

[19] Ito H, Otaki Y, Tomohiro Y, Ishida Y, Akiyama R, Kimura A, Shin S, Kuroda S. Observation of unoccupied states of SnTe(111) using pump-probe ARPES measurement. Phys Rev Res. 2020;2: 043120-1–9.

[20] Siefermann KR, Pemmaraju CD, Neppl S, Shavorskiy A, Cordones AA, Vura-Weis J, Slaughter DS, Sturm FP, Weise F, Bluhm H, Strader ML, Cho H, Lin M-F, Bacellar C, Khurmi C, Guo J, Coslovich G, Robinson JS, Kaindl RA, Schoenlein RW, Belkacem A, Neumark DM, Leone SR, Nordlund D, Ogasawara H, Krupin O, Turner JJ, Schlotter WF, Holmes MR, Messerschmidt M, Minitti MP, Gul S, Zhang JZ, Huse N, Prendergast D, Gessner O. Atomic-scale

perspective of ultrafast charge transfer at a dye-semiconductor interface. J Phys Chem Lett. 2014;5(15):2753–2759.

[21] Roth F, Neppl S, Shavorskiy A, Arion T, Mahl J, Seo HO, Bluhm H, Hussain Z, Gessner O, Eberhardt W. Efficient charge generation from triplet excitons in metal-organic heterojunctions. Phys Rev B. 2019;99:020303-1–6.

[22] Ozawa K, Yamamoto S, D'angelo M, Natsui Y, Terashima N, Mase K, Matsuda I. Enhanced photoresponsivity of fullerene in the presence of phthalocyanine: A time-resolved X-ray photoelectron spectroscopy study of phthalocyanine/C_{60}/TiO_2(110). J Phys Chem C. 2019; 123(7):4388–4395.

[23] Ogawa M, Yamamoto S, Kousa Y, Nakamura F, Yukawa R, Fukushima A, Harasawa A, Kondoh H, Tanaka Y, Kakizaki A, Matsuda I. Development of soft X-ray time-resolved photoemission spectroscopy system with a two-dimensional angle-resolved time-of-flight analyzer at SPring-8 BL07LSU. Rev Sci Instrum. 2012;83:023109-1–7.

[24] Yamamoto S, Matsuda I. Time-resolved photoelectron spectroscopies using synchrotron radiation: Past, present, and future. J Phys Soc Jpn. 2013;82:021003-1–18.

[25] Yamamoto S, Senba Y, Tanaka T, Ohashi H, Hirono T, Kimura H, Fujisawa M, Miyawaki J, Harasawa A, Seike T, Takahashi S, Nariyama N, Matsushita T, Takeuchi M, Ohata T, Furukawa Y, Takeshita K, Goto S, Harada Y, Shin S, Kitamura H, Kakizaki A, Oshima M, Matsuda I. New soft X-ray beamline BL07LSU at SPring-8. J Synchrotron Radiat. 2014;21:352–365.

[26] Glownia JM, Cryan J, Andreasson J, Belkacem A, Berrah N, Blaga CI, Bostedt C, Bozek J, DiMauro LF, Fang L, Frisch J, Gessner O, Guhr M, Hajdu J, Hertlein MP, Hoener M, Huang G, Kornilov O, Marangos JP, March AM, McFarland BK, Merdji H, Petrovic VS, Raman C, Ray D, Reis DA, Trigo M, White JL, White W, Wilcox R, Young L, Coffee RN, Bucksbaum PH. Time-resolved pump-probe experiments at the LCLS. Opt Express. 2010;18:17620–17630.

[27] Bergeard N, Silly MG, Krizmancic D, Chauvet C, Guzzo M, Ricaud JP, Izquierdo M, Stebel L, Pittana P, Sergo R, Cautero G, Dufour G, Rochet F, Sirotti F. Time-resolved photoelectron spectroscopy using synchrotron radiation time structure. J Synchrotron Radiat. 2011; 18:245–250.

[28] Neppl S, Gessner O. Time-resolved X-ray photoelectron spectroscopy techniques for the study of interfacial charge dynamics. J Electron Spectrosc Relat Phenom. 2015;200:64–77.

[29] Kronik L, Shapira Y. Surface photovoltage phenomena: Theory, experiment, and applications. Surf Sci Rep. 1999;37(1–5):1–206.

[30] Bröcker D, Gieβel T, Widdra W. Charge carrier dynamics at the SiO_2/Si(100) surface: A time-resolved photoemission study with combined laser and synchrotron radiation. Chem Phys. 2004;299(2–3):247–251.

[31] Spencer BF, Graham DM, Hardman SJO, Seddon EA, Cliffe MJ, Syres KL, Thomas AG, Stubbs SK, Sirotti F, Silly MG, Kirkham PF, Kumarasinghe AR, Hirst GJ, Moss AJ, Hill SF, Shaw DA, Chattopadhyay S, Flavell WR. Time-resolved surface photovoltage measurements at n-type photovoltaic surfaces: Si(111) and ZnO(10$\bar{1}$0). Phys Rev B. 2013;88:195301-1–16.

[32] Ogawa M, Yamamoto S, Fujikawa K, Hobara R, Yukawa R, Yamamoto Sh, Kitagawa S, Pierucci D, Silly MG, Lin C-H, Liu R-Y, Daimon H, Sirotti F, Tang S-J, Matsuda I. Relaxations of the surface photovoltage effect on the atomically controlled semiconductor surfaces studied by time-resolved photoemission spectroscopy. Phys Rev B. 2013;88:165313-1–9.

[33] Ozawa K, Emori M, Yamamoto S, Yukawa R, Yamamoto Sh, Hobara R, Fujikawa K, Sakama H, Matsuda I. Electron-hole recombination time at TiO_2 single-crystal surfaces: Influence of surface band bending. J Phys Chem Lett. 2014;5(11):1953–1957.

[34] Liu R-Y, Ozawa K, Terashima N, Natsui Y, Feng B, Ito S, Chen W-C, Cheng C-M, Yamamoto S, Kato H, Chiang T-C, Matsuda I. Controlling the surface photovoltage on WSe_2 by surface chemical modification. Appl Phys Lett. 2018;112:211603-1–5.

[35] Ozawa K, Yamamoto S, Mase K, Matsuda I. A surface science approach to unveiling the TiO_2 photocatalytic mechanism: Correlation between photocatalytic activity and carrier lifetime. e-J Surf Sci Nanotechnol. 2019;17:130–147.

[36] Sah C-T, Noyce RN, Shockley W. Carrier generation and recombination in p-n junctions and p-n junction characteristics. Proc IRE. 1957;45(9):1228–1243.

[37] Ozawa K, Yamamoto S, Yukawa R, Liu R, Emori M, Inoue K, Higuchi T, Sakama H, Mase K, Matsuda I. What determines the lifetime of

photoexcited carriers on TiO_2 surfaces? J Phys Chem C. 2016;120(51): 29283–29289.

[38] Morris Hotsenpiller PA, Bolt JD, Farneth WE, Lowekamp JB, Rohrer GS. Orientation dependence of photochemical reactions on TiO_2 surfaces. J Phys Chem B. 1998;102(17):3216–3226.

[39] Lowekamp JB, Rohrer GS, Morris Hotsenpiller PA, Bolt JD, Farneth WE. Anisotropic photochemical reactivity of bulk TiO_2 crystals. J Phys Chem B. 1998;102(38):7323–7327.

[40] Ahmed AY, Kandiel TA, Oekermann T, Bahnemann D. Photocatalytic activities of different well-defined single crystal TiO_2 surfaces: Anatase versus rutile. J Phys Chem Lett. 2011;2(19):2461–2465.

[41] Luttrell T, Halpegamage S, Tao J, Kramer A, Sutter E, Batzill M. Why is anatase a better photocatalyst than rutile? Model studies on epitaxial TiO_2 films. Sci Rep. 2014;4:4043-1–8.

[42] Tanaka K, Capule MFV, Hisanaga T. Effect of crystallinity of TiO_2 on its photocatalytic action. Chem Phys Lett. 1991;187(1–2):73–76.

[43] Sclafani A, Herrmann JM. Comparison of the photoelectronic and photocatalytic activities of various anatase and rutile forms of titania in pure liquid organic phases and in aqueous solutions. J Phys Chem. 1996;100(32):13655–13661.

[44] Su R, Bechstein R, Sø L, Vang RT, Sillassen M, Esbjörnsson B, Palmqvist A, Besenbacher F. How the anatase-to-rutile ratio influences the photoreactivity of TiO_2. J Phys Chem C. 2011;115(49): 24287–24292.

[45] Bourikas K, Kordulis C, Lycourghiotis A. Titanium dioxide (anatase and rutile): Surface chemistry, liquid-solid interface chemistry, and scientific synthesis of supported catalysts. Chem Rev. 2011;114(19): 9754–9823.

[46] Schnadt J, Knudsen J, Johansson N. Present and new frontiers in materials research by ambient pressure X-ray photoelectron spectroscopy. J Phys Condens Matter. 2020;32(41):413003-1–29.

[47] Sandell A, Ragazzon D, Schaefer A, Farstad MH, Borg A. Photochemistry of carboxylate on TiO_2(110) studied with synchrotron radiation photoelectron spectroscopy. Langmuir. 2016;32(44): 11456–11464.

[48] Ozawa K, Yamamoto S, Yukawa R, Liu R-Y, Terashima N, Natsui Y, Kato H, Mase K, Matsuda I. Correlation between photocatalytic

activity and carrier lifetime: Acetic acid on single-crystal surfaces of anatase and rutile TiO_2. J Phys Chem C. 2018;122(17):9562–9569.

[49] Ozawa K, Aiura Y, Wakabayashi D, Tanaka H, Kikuchi T, Toyoshima A, Mase K. Beamline commissioning for microscopic measurements with ultraviolet and soft X-ray beam at the upgraded beamline BL-13B of the Photon Factory. J Synchrotron Radiat. 2022;29:400–408.

[50] Guo Q, Cocks I, Williamsa EM. The orientation of acetate on a TiO_2(110) surface. J Chem Phys. 1997;106:2924–2931.

[51] Kraeutler B, Bard AJ. Heterojeneous photocatalytic decomposition of saturated carboxylic acids on TiO_2 powder. Decarboxylative route to alkanes. J Am Chem Soc. 1978;100(9):5985–5992.

[52] Liao L-F, Lien C-F, Lin J-L. FTIR study of adsorption and photoreactions of acetic acid on TiO_2. Phys Chem Chem Phys. 2001;3(17): 3831–3837.

[53] Yang C, Meng Y, Xia S, Pan G. Hot carrier transfer-induced photodegradation at a thiolated Au/TiO_2 interface under X-ray irradiation. J Phys Chem C. 2020;124(40):22212–22220.

[54] O'Shea JN, Ben Taylor J, Mayor LC, Swarbrick JC, Schnadt J. Molecular damage in bi-isonicotinic acid adsorbed on rutile TiO_2(110). Surf Sci. 2008;602(9):1693–1698.

[55] Heister K, Zharnikov M, Grunze M, Johansson LSO, Ulman A. Characterization of X-ray induced damage in alkanethiolate monolayers by high-resolution photoelectron spectroscopy. Langmuir. 2001;17(1):8–11.

[56] Carbone M, Piancastelli MN, Zanoni R, Comtet G, Dujardin G, Hellner L. Photon-stimulated desorption of methanol on Si(111) 7×7 and Si(100) 2×1 at the C 1s and O 1s thresholds. Surf Sci. 1997;390(1–3):219–223.

[57] Sproul AB. Dimensionless solution of the equation describing the effect of surface recombination on carrier decay in semiconductors. J Appl Phys.1994;76:2851–2854.

[58] Nakabayashi S, Komuro S, Aoyagi Y, Kira A. Transient grating method applied to electron-transfer dynamics at a semiconductor/liquid interface. J Phys Chem. 1987;91(7):1696–1698.

[59] Zhang Z, Yates Jr JT. Band bending in semiconductors: Chemical and physical consequences at surfaces and interfaces. Chem Rev. 2012;112(10):5520–5551.

[60] Morgan BJ, Watson GW. Intrinsic n-type defect formation in TiO_2: A comparison of rutile and anatase from GGA + *U* calculations. J Phys Chem C. 2010;114(5):2321–2328.

[61] Lampimäki M, Schreiber S, Zelenay V, Křepelová A, Birrer M, Axnanda S, Mao B, Liu Z, Bluhm H, Ammann M. Exploring the environmental photochemistry on the TiO_2(110) surface in situ by near ambient pressure X-ray photoelectron spectroscopy. J Phys Chem C. 2015;119(13):7076–7085.

https://doi.org/10.1142/9781800613294_0010

Chapter 10

Catalysts at Work by Near-Ambient Pressure X-ray Photoelectron Spectroscopy

Emilia A. Carbonio[*,†,¶] and Detre Teschner[*,‡,§]

[*]*Department of Inorganic Chemistry, Fritz-Haber-Institut der Max-Planck-Gesellschaft, Berlin 14195, Germany*

[†]*Helmholtz-Zentrum Berlin für Materialien und Energie GmbH, BESSY II Catalysis for Energy, CE-GKAT, Berlin 12489, Germany*

[‡]*Department of Heterogeneous Reactions, Max-Planck-Institute for Chemical Energy Conversion, Mülheim an der Ruhr 45470, Germany*

[¶]*carbonio@fhi-berlin.mpg.de*

[§]*teschner@fhi-berlin.mpg.de*

Abstract

Near-ambient pressure X-ray photoelectron spectroscopy (NAP-XPS) is becoming a popular characterization tool of surfaces. Developments in the last 20–30 years brought photoelectron spectroscopy into the realm of techniques that can be used under realistic use case scenarios in practical environments. Catalysis and surface science are still the dominant scientific fields that employ NAP-XPS most frequently. Heterogeneous catalysts are typically solid materials that convert reactants into products without being consumed in the reaction. This however does not mean that catalysts are rigid entities under reaction conditions. Just to the contrary: the as-synthesized catalyst can be best seen as only a precursor to the active-phase/active-site ensemble, which is created under operational conditions. Phase transition, segregation, change of the nanoparticle morphology, surface reconstruction, surface melting, deposition of reactant fragments on and dissolution into the catalyst, and other phenomena may occur under the chemical potential of the reactants. NAP-XPS is a suitable experimental technique to study the electronic structure of a catalyst surface under reaction conditions. It provides a variety of potential information ranging from the actual surface phase/state of the catalyst under reaction conditions, the surface species present, and their coverage, up to more subtle information like work function changes, band bending, and more. Here, in this chapter, after some introductory paragraphs on technical aspects, sample contamination, and beam damage, we discuss various selected examples of NAP-XPS in heterogeneous gas-phase catalysis research. The examples are grouped together under the themes of (i) quantification of surface coverages, (ii) selected catalytic applications, and (iii) selected catalysts for more than one application. Due to the vast literature, it was not our intention to provide a comprehensive overview of all reactions pursued and scrutinized by NAP-XPS rather the examples were selected based on our own interests and with the goal to highlight many of the facets of catalyst dynamics. Finally, despite the broad adaptation of the technique, we also point out current shortcomings and limitations.

10.1 Introduction

Catalysis is the science and technology in which the rates of chemical reactions are manipulated. A heterogeneous catalyst is (typically) a solid material that, without being consumed, modifies the reaction path and increases the rate of the target chemical transformation. This tells us that if several competing pathways and products are feasible, the catalyst may change the distribution of these products compared to the non-catalytic reaction. Hence, the selectivity of a chemical reaction can be tuned. Frequently, the target product is not thermodynamically favored. The catalyst must then activate the reactants and favor the path to the target product without further activating the latter.

In a catalytic reaction, reactants move to the surface and by chemisorption undergo initial activation. Due to the undercoordination of surface atoms, as compared to atoms in the bulk, surface atoms can be considered as high-energy sites with an intrinsic property to chemically interact with adsorbates. This interaction depends on the electronic and geometric structure of the surface site. Often, chemisorbed species need further activation, which may require reorientation or diffusion to another surface site. Surface reactions including dissociation and associative reactions by new chemical bond formation between adsorbed species follow. Again, these elementary steps are strongly influenced by the local electronic and geometric properties of the surface. In the end, products are formed and leave the surface by desorption. The surface elementary steps that occur consecutively and in parallel build the surface reaction network. This is however only one side of the story. Our model of heterogeneous catalysis evolved in the last 100 years and now we understand that the catalyst itself does not remain completely unchanged under reaction conditions. In fact, the as-synthesized catalyst can be best seen as only a precursor to the active-phase/active-site ensemble, which is created under operational conditions. Phase transition, segregation, change of the nanoparticle morphology, surface reconstruction, surface melting, deposition of reactant fragments on and dissolution into the catalyst, and other phenomena may occur under the chemical potential of the reactants. These processes constitute the catalyst dynamics on the timescale of milliseconds to minutes and these alterations may only be stable under the set chemical potential of the applied condition. The extent to which these processes occur depends also on the solid-state kinetics, which, in turn, may depend on the dimensions of the particles and transport mechanisms. A set of shorter timescale dynamics, such as adsorbate vibrations, lattice distortion upon surface elementary steps, and charge-carrier dynamics of semiconductor catalysts, complement the dynamics in heterogeneous catalysis observable in various time and length scales. As mass and energy transport, the local chemical potential, and the geometric and electronic structure of the catalyst are strongly interconnected through the catalyst's dynamics, these create feedback loops that make it necessary to study catalytic phenomena under reaction conditions [1].

In situ or *operando* experiments combine the analysis of the geometric or electric structure of the materials with the real-time proof of catalytic action. Unfortunately, many powerful techniques to evaluate the geometric and electronic structure of catalysts are not very sensitive to the surface layer and its dynamics. On the other hand, those methods that are extremely surface sensitive, e.g., chemical reactions and adsorption with probe molecules, have only limited capability to be utilized *in situ*.

10.2 From UHV to Near-Ambient Pressure X-ray Photoelectron Spectroscopy (NAP-XPS)

Conventional UHV XPS instruments commercially available for several decades are powerful tools of catalytic research, on their own. However, as we detailed in our Introduction, there is a strong driving force to study catalysts under more realistic conditions and thus researchers implemented certain modifications to the conventional setup to make NAP-XPS possible. Since the X-ray source and detector require UHV conditions, the minimum requirement for *in situ* operations is the separation of these units from the gas atmosphere. Furthermore, the short mean free path of electrons in a gaseous environment necessitates the minimization of the electron path in the gas medium. These constraints can be met by differentially pumped and gas-tight X-ray sources and introducing differential pumping stages between the sample environment and the electron analyzer. For *in situ* XPS experiments of (electro)catalysts in liquid, the reader is referred to Chapter 12.

After the first application of NAP-XPS for liquids by Hans and Kai Siegbahn [2] in 1973, Roberts [3] and colleagues developed the first NAP-XPS setup for studying solid surfaces in the presence of gas at up to 1 Torr pressure. In 1990, Ruppender *et al.* [4] described a NAP-XPS instrument with three differential pumping stages that was designed to allow applications of high pressures (up to 1 mbar) at the same sample position where UHV data were recorded and where the sample could be cleaned and characterized by He^+ ion scattering. The first NAP-XPS instrument to be used at a synchrotron facility was developed at the Lawrence Berkeley National Laboratory (LBNL) and installed at the Advanced Light Source (ALS), beamline 9.3.2 [5]. This instrument featured a differentially pumped electrostatic lens system, which increased the collection efficiency for electrons, therefore offering a significant performance improvement relative to passive differential pumping approaches. The second generation of this instrument was co-developed by the Fritz Haber Institute, LBNL, and SPECS GmbH, and two such NAP-XPS setups were installed at BESSY II and ALS [6]. The combination of high-brilliant synchrotron light source and the differentially pumped electrostatic lens system brought about the required breakthrough, with a strong increase in the number of publications utilizing NAP-XPS since 2005 (Figure 10.1). Besides other applications,

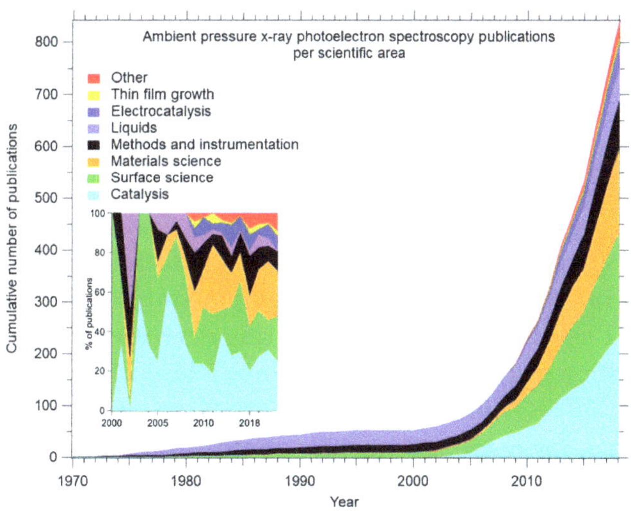

Fig. 10.1. Categorization of the NAP-XPS papers published between 1970 and 2018 in scientific areas. The inset shows the share of publications of respective fields during each year between 2000 and 2018. Adapted with permission from Ref. [7] CC BY 4.0.

up to the day of writing this book chapter, over 400 publications deal with some sort of (electro)catalytic systems or some related aspects.

Several excellent reviews appeared on NAP-XPS; some of them are listed in the following: Schnadt *et al.* [7], Knop-Gericke *et al.* [8], Salmeron and Schlögl [9], Bluhm [10], Starr *et al.* [11], Shavorskiy *et al.* [12], Trotochaud *et al.* [13], Head and Schnadt [14], Salmeron [15], Roy *et al.* [16], Weatherup [17], Nguyen *et al.* [18], and Zhong *et al.* [19]. Since 2014, the NAP-XPS community has had the opportunity to gather and exchange ideas in an annually organized international workshop, the last installment of which was held in December 2021 by the Brookhaven National Laboratory, USA [20]. Schnadt and colleagues have recently set up a NAP-XPS publications database, which currently contains over 700 publications [7,21].

10.3 Some Technical Aspects of NAP-XPS

Neglecting beam effects in the first approximation, the time of acquisition and peak intensities define the quality of a spectrum with a given sample. Peak intensities of a particular spectrum at a given excitation energy depend on the brightness of the X-ray beam, the transmission efficiency of the photoelectrons, and the sensitivity of the detector. With NAP-XPS, however, additional parameters must be taken into account.

(1) X-rays have to pass through an X-ray transparent window to enter the *in situ* cell/chamber, in which transmission plays an important role. Photoabsorption and scattering decrease transmission and thus the photon flux reaching the sample. Typical window materials are Si_3N_4 and SiC with a thickness of 50–100 nm. For example, Si_3N_4 in the soft X-ray range will strongly photoabsorb at the Si $L_{2,3}$ (~100 eV), N K (~400 eV), and Si K (~1840 eV) edges; the attenuation decreases with increasing energies. Obviously, the thicker the window, the stronger the attenuation but the mechanical stability of the window to withstand a certain pressure difference improves with thickness and with a smaller window area. Thus, the thickness of the

window is a compromise between mechanical stability and transmission. An alternative approach is to add differential pumping between the X-ray source (synchrotron) and the *in situ* cell with a small (~0.3 mm) diameter aperture close (~20 mm) to the sample [22].

(2) Just like with the absorption of X-rays by the window material, the gas phase will also attenuate the incoming photon flux. The transmission of photons (T) through a certain path length z and above the absorption edge can be described in a Beer–Lambert-type fashion, as

$$T = \exp(-n\mu_a(E)z), \tag{10.1}$$

with n being the number of molecules or atoms per unit volume and $\mu_a(E)$ being the atomic photoabsorption cross-section.

Figure 10.2 depicts the attenuation of X-ray photons on the example of water vapor for a path length of 3 cm at different gas

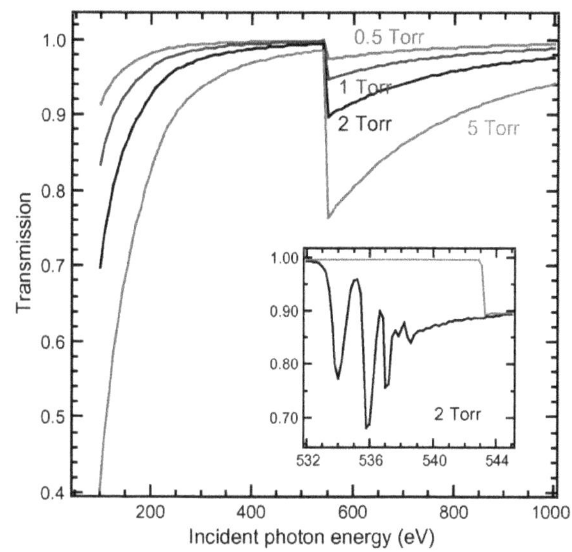

Fig. 10.2. Absorption of X-rays by water vapor for a path length of 3 cm at pressures of 0.5, 1.0, 2.0, and 5.0 Torr. The inset shows the absorption fine structure for water vapor for the case of 2 Torr. Adapted with permission from Ref. [10]. Copyright (2010) Elsevier B.V.

pressures, indicating that attenuation by the gas phase can be a substantial factor at high pressures (and large path lengths), and for incident photon energies close to the absorption edge of elements contributing to the gas phase. Technically, the X-ray window is placed close to the sample to minimize attenuation but not too close to avoid thermal damage by a heated sample.

(3) The strongest limiting factor of signal intensity in NAP-XPS is the scattering of photoelectrons by the gas-phase molecules. Figure 10.3 shows experimental data of signal attenuation as a function of water vapor pressure and electron kinetic energy [10]. As with the exponential dependence of photon transmission, the attenuation of the photoelectron signal in a gaseous ambient is

$$I_p/I_0(KE,p) = \exp(-z\sigma(KE)p/kT), \qquad (10.2)$$

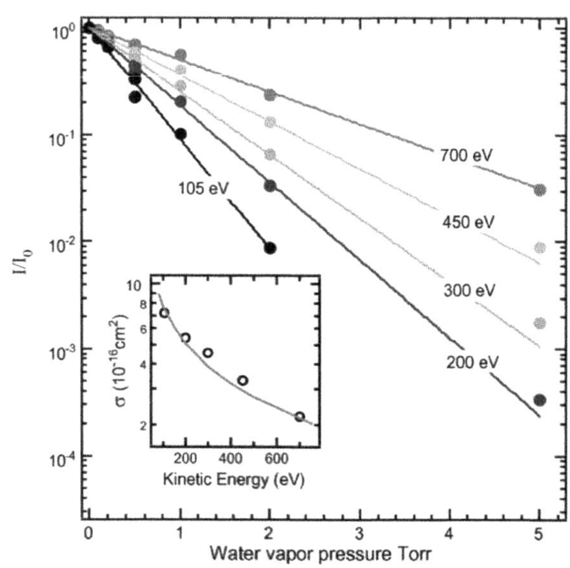

Fig. 10.3. Experimentally determined attenuation of the photoelectron signal as a function of water vapor pressure and electron kinetic energy. Adapted with permission from Ref. [10]. Copyright (2010) Elsevier B.V.

with I_p being the signal at pressure p, I_0 being the signal in vacuum, $\sigma(KE)$ being the electron scattering cross-section, and z being the path length that the electrons travel in gas at pressure p and temperature T.

Figure 10.3 also clearly indicates that lower kinetic energy photoelectrons, those that provide more surface-sensitive XPS measurements, are more prone to scattering. Since the photoelectron signal decreases exponentially with the pressure and path length, it is paramount to minimize the influence of this contribution. To achieve this goal, the technical solution is to introduce differential pumping stages with small slit/aperture sizes between the sample and the analyzer/detector unit and move the sample as close as feasible to the first aperture. However, the first differential stage will induce the pressure to drop in the proximity of the front aperture thus making it not ideal to place the sample "too close" to the aperture. Simulations of pressure drop show that it is already 5 % at a sample aperture distance of one aperture diameter. Hence, it is recommended to place the sample at a distance to the first aperture that is 1.5–2 times the aperture diameter, i.e., for a 1 mm aperture, the distance is 1.5–2 mm. In Figure 10.3, with a sample aperture distance of 0.9 mm (0.3 mm diameter aperture), 2 Torrs of water induced an approximately hundred-fold reduction of signal intensity of 105 eV photoelectrons. To minimize the attenuation of photoelectrons, one obvious solution is to use small aperture sizes that enable the placement of the sample at a short distance to the aperture. This, in turn, requires a focused beam with a small spot size, which in combination with high-brilliance synchrotron light sources poses a requirement of high structural stability of the investigated catalyst with respect to beam damage. A very short distance with a small aperture diameter induces high thermal stress on the nozzle and its coating when catalysts are heated to high temperatures in the mbar pressure range. Also, small grains of catalyst powders may fly and attach to the orifice of the nozzle, which can seriously diminish electron transmission into the lens system. Nowadays, the typical diameter of the first aperture to the pumping stage is 0.3 mm [22–25], but NAP-XPS experiments have been carried out up to 100 Torr with soft X-rays using a 50 μm aperture [26]. After the first aperture, three to four differential pumping stages ensure a pressure reduction on the order of 10^8 towards the analyzer, which enables stable detector operation. Additionally, electrostatic focusing inside the differential pumping stages needs to maximize the photoelectron acceptance angle, as collection efficiency scales with the collection plane half-angle squared.

To perform a NAP-XPS experiment, the sample environment plays a crucial role. Three basic layouts of *in situ* cells are proposed (see Figure 10.4) [11,27]. The standard design developed first and still in operation is a vacuum chamber, which is backfilled with the gas of interest. As this is the traditional design, these spectrometers are often equipped with other vacuum components, such as sample preparation and transfer chambers, cleaning, and other analytic options.

Gas analysis is possible by mass spectrometers (MS) connected to the main chamber or placed in the first differential pumping stage or by gas chromatography (GC) once the exhaust gas is compressed to 1 bar. As in this design the whole chamber is exposed to the gas atmosphere, it is difficult to switch fast from *in situ* to UHV experiments when the latter requires high cleanliness. When the chamber has a volume of several liters, the achievable conversion of the studied reaction, sampled as an average of the whole chamber, will be low. (We note here that the reaction might still become diffusion limited by one of the reactants. This can be shown when sampled locally in front of the sample.) Also, different *in situ* experiments pose the potential problem of cross-contamination. To overcome some of these limitations, another design [23] (Figure 10.4(b)) utilizes an *in situ* cell placed inside the main vacuum chamber. In this particular design, the AP cell is docked to the analyzer. As the AP cell volume is sealed off from the surrounding UHV chamber, the gaseous sample environment is limited to the volume of the AP cell and the UHV chamber is retained under vacuum conditions. Sample heating in this particular design is done by electron bombardment of the wall behind the sample (the filament being in vacuum behind the sample). Mass spectrometric gas analysis is

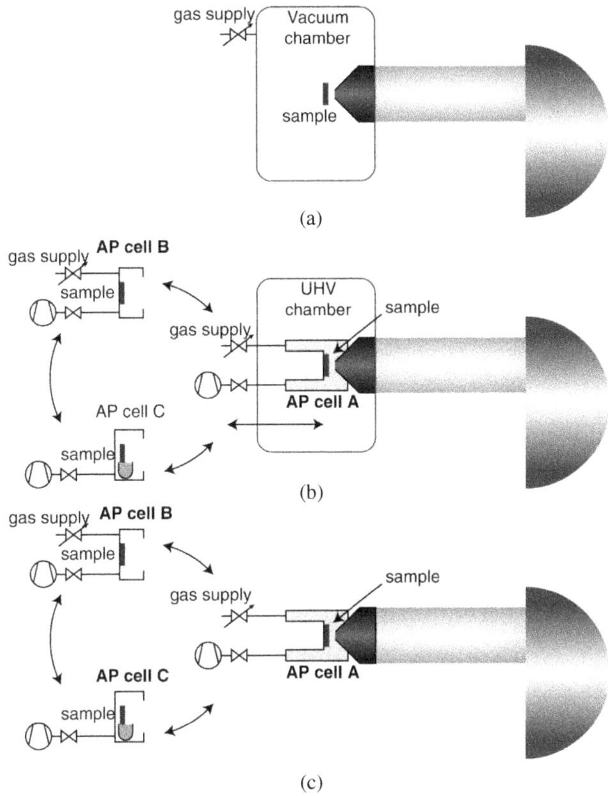

Fig. 10.4. Three basic concepts of sample environment: (a) a standard vacuum chamber is backfilled with the gas or gas mixture, (b) a standard UHV setup in combination with a detachable and exchangeable ambient pressure (AP) cell, and (c) an AP cell without the standard vacuum chamber. Adapted with permission from Ref. [27]. Copyright (2016) Elsevier B.V.

connected to the gas exhaust line and due to the lower cell volume, gas exchange is reportedly possible within 15 seconds. Similar setups have been described by others [28–30]. A special version of this cell design uses a membrane approach, [17,31] that is described elsewhere in this book. In the third cell design (Figure 10.4(c)), the NAP-XPS system incorporates an exchangeable, small volume, tailor-made sample cell without any UHV chamber. An example is a small volume analysis chamber that can be removed from a Scienta

HiPP-2 analyzer, as reported by Newberg *et al.* [32]. This concept may hold promise for complex sample environments, such as in electrocatalytic research or when UHV experiments are not important and cross-contamination is an issue.

In a recent cell design, a gas inlet was integrated into the front cone of a spectrometer as an outer ring channel, and in the middle, a 70 μm thin Ti foil with a laser-drilled set of micrometer-sized holes arranged in line served as the first aperture [33]. Synchrotron-based hard X-ray excitation was used to increase the electron mean free path in the gas region between the sample and analyzer while grazing incidence <5° close to total external refection conditions enhanced surface sensitivity (of a flat sample). With this approach, the authors reported measurements at up to 2.5 bar local pressure. More details about this instrument can be found in Chapter 6.

For some time, the main drivers of NAP-XPS studies were the advantages (increased photon flux, smaller spot size, and tunable photon energy) brought about by the high-brilliance synchrotron light sources, with many beamlines nowadays operating NAP-XPS setups [5,6,22,23,25,34–39]. The availability of commercial setups [40,41] and their further development made NAP-XPS highly attractive for research institutions. Hence, in recent years, based on these commercial products, a large number of laboratory-based instruments have been installed. These instruments greatly benefit from the small-spot, monochromated Al Kα X-ray source.

10.4 Miscellaneous Other Aspects

10.4.1 *Detection of gas-phase spectra*

In NAP-XPS experiments, the incident X-ray beam irradiates not only the sample surface but also a part of the gas phase, a section of which is within the acceptance angle of the analyzer. Gas-phase signals at typical mbar pressure conditions can be comparable in intensity to surface signals. Unlike surface species of metallic samples, of which binding energy is referenced to the Fermi level of the sample, gas-phase peaks are referenced to the vacuum level. The vacuum level of the gas is coupled to the vacuum level of the surrounding surfaces, the sample, and the nozzle to the differentially pumped lens system being the relevant surfaces. Since these vacuum levels are usually not identical and the gas phase signal is the sum of the irradiated gas phase volume along the surface normal, this induces a small broadening of the gas phase peaks. In case gas-phase peaks are not desired, this effect can be used, in which a positive bias is applied to the nozzle. This smears out the gas-phase peaks making their spectral intensity disappear in the spectrum background. Detection of gas-phase molecules, however, provides an additional means of determining the gas-phase composition directly in front of the sample. This is often more representative of the chemical potential at the sample surface than the gas detection as a mean average of the whole cell or chamber. Another useful aspect of detecting gas-phase peaks is that their apparent binding energies may provide information about the magnitude of changes in the work function of the sample (Figure 10.5) [42].

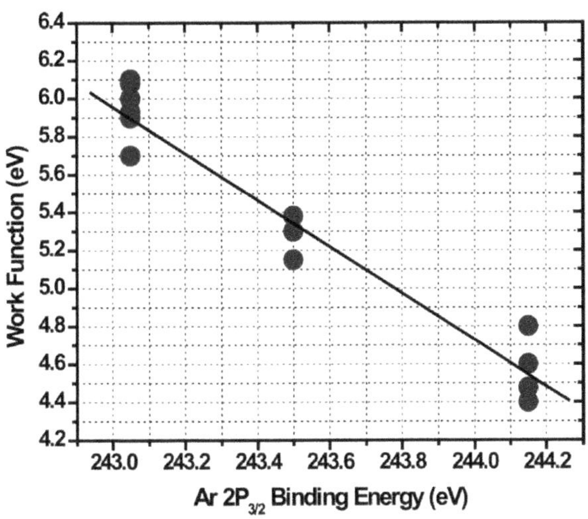

Fig. 10.5. Work function versus measured Ar $2p_{3/2}$ gas-phase binding energy values. The work function values are of HOPG, Au(111), and Pt(111) from the listed references. The Ar $2p_{3/2}$ BEs are measured from near surface of HOPG, Au(111), and Pt(111) at 250 mTorr of Ar gas for each measurement. Adapted with permission from Ref. [42]. Copyright (2013) American Chemical Society.

10.4.2 *Sample contamination*

One of the advantages of XPS for catalysis research is its sensitivity to detect surface-related species and phases. This brings about the side effect that elements not thought to be part of the sample can be detected. This "contamination" can be intrinsic to the sample or be a real contamination from the NAP-XPS setup or the feed gas. If extrinsic, this raises the question of whether we look for, find, and properly identify the contaminants. This is crucial to avoid misinterpretation of other spectral peaks (e.g., an O 1s peak). Furthermore, if we properly assign the contamination, another question is, whether or not it is relevant. Does the contaminant modify the function of the catalyst or is it merely an annoyance, a spectator? These questions should concern us if a NAP-XPS study aims to correlate spectral observations with catalyst performance, that is, if we are in pursuit of structure-function correlations. Nevertheless, we should not forget that in any other catalytic experiments, in a standard reactor or with other not surface or element-specific *in situ* spectroscopy, where we do not scrutinize the surface, contamination may occur. Although some contaminants can safely be traced back to previous experiments in the NAP-XPS setup, some may come from the dosed feed gas. At the typical mbar pressure condition, an impurity in the gas supply on the order of 1 ppm but with a high sticking probability may fast show up as a contaminant at the catalyst surface. Moreover, gases introduced into the sample cell/chamber can displace species already adsorbed on the walls, which can adsorb onto the sample surface. Also, as catalysts are meta-stable under operation, they can partially produce mobile species contaminating the sample holder, manipulator, or any solid surfaces nearby. Once these surfaces get hot in a follow-up experiment with a new catalyst, the mobile species may move to the sample surface and induce contamination. One particular case of contamination is when the X-ray beam strongly interacts with organic molecules in a reducing environment facilitating the decomposition of the organic molecule and leaving carbon deposits on the surface. This can be checked by moving the beam spot along the sample surface.

10.4.3 *Radiation damage and beam effects*

Beam damage is well known in the XPS community even for classical UHV experiments. As just mentioned, organic molecules are particularly prone to decomposition by X-rays and by secondary electrons produced in the sample. Inorganic materials are also susceptible to beam damage [43]. NAP-XPS experiments create however additional complications.

It is well known that small metallic nanoparticles can be synthesized by reducing metal ions through traditional chemical reactants but also by radiolysis [44,45]. Not only gamma rays but also X-rays are capable and often efficient ionizing photon sources. Arble *et al.* demonstrated the influence of beam-induced radiolytic reactions on electrochemical processes at the electrode/aqueous electrolyte interfaces using SPEM imaging and X-ray photoelectron spectroscopy [46]. They also find that radiolitic processes often overshadow local electrometry.

The radiolysis of water due to ionizing radiation results in the production of electrons, H atoms, OH radicals, H_3O^+ ions, and molecules (H_2 and hydrogen peroxide). Radiolytic H_2 production is strongly modified at water/oxide interfaces [47]. Hydroxyl radical quantification through OH scavenging using benzoate was performed by Huart *et al.* upon irradiation of liquid water by soft X-rays, in the 200–2000 eV range [48]. They find that sub-keV electrons behave as high linear energy transfer particles since they are able to deposit tens to hundreds of electronvolts in nanometric volumes. The highest OH yield (Figure 10.6) was obtained at the lowest studied energy (215 eV).

Using soft X-ray absorption measurements, Weatherup *et al.* have shown that under the conditions at which atmospheric pressure X-ray spectroscopies are performed, the effects of radiolysis at the interface between a solid electrode and a high-pressure gas or liquid environment can be significant [49]. X-ray illumination led to the generation of reactive species by radiolysis, which can strongly influence the behavior observed in these systems. This was found to depend not only on the electrode material under consideration but

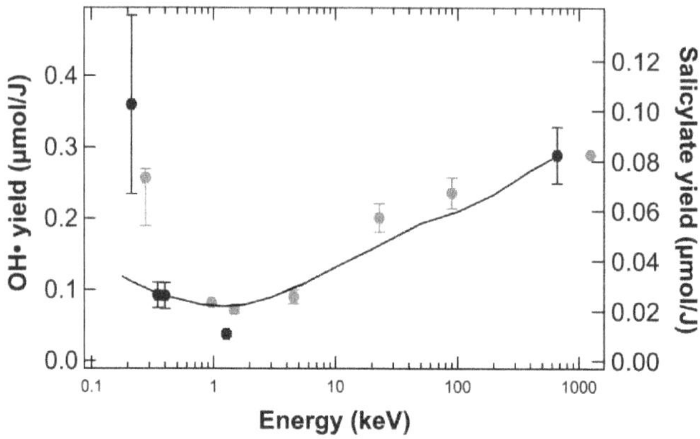

Fig. 10.6. Yield of HO• radicals as a function of photon energy (black symbols). The benzoate sodium concentration is 0.01 mol/L. Gray symbols are measurements by Fulford *et al.* [50]. The experimental results are compared to MC simulations for monoenergetic electrons (solid line). Adapted with permission from Ref. [48]. Copyright (2020) American Chemical Society.

also critically on the environment in which it is being measured. In air, and aqueous hydroxide solutions, the predominantly oxidizing products of radiolysis led Cu electrodes to oxidize to CuO. The addition of only small amounts of CH_3OH to the aqueous hydroxide solutions changed the balance of radiolysis products toward reducing species that caused the reduction of the electrode to Cu_2O. On the other hand, the oxidation state of Ni films in $NaHCO_3$ solution was found to remain stable under X-ray illumination and could be electrochemically cycled between a reduced and oxidized state.

NAP-XPS was used to examine the effect of water and photon flux on the stability of $CH_3NH_3PbI_3$ perovskite films and solar cells [51]. The authors found that the used amount of water vapor alone (up to 1 mbar) had a negligible impact on the perovskite film. However, the higher the photon flux was, the more prominent were the changes in the NAP-XPS spectra and concomitantly a faster decline in power conversion efficiency (PCE) and a more substantial hysteresis in the J-V characteristics were observed.

The soft X-ray heterogeneous radiolysis of pyridine was studied by NAP-XPS in the presence of hydrated clay (Strontium-Hydroxyhectorite) by Boucly *et al.* [52]. They find that under high flux (~10^{16} photons × cm^{-2} × s^{-1}) soft X-rays, the beam can induce the heterogeneous radiolysis of pyridine, leading to the steady production of N_2. Another, catalytically relevant observation was made by Salmeron and colleagues who have demonstrated in a NAP-XPS study that gold can activate oxygen in the mbar pressure range in the presence of X-ray irradiation to produce oxidic gold [53]. A similar study by Dumbuya *et al.* confirmed the beam-induced oxidation of gold in oxygen [54]. Although radiolytic beam damage is rarely investigated in catalytic NAP-XPS studies, all these examples emphasize that care has to be taken to rule out beam effects and, if present, one has to carefully minimize its influence on the spectroscopic observation. The quest for higher and higher pressures in NAP-XPS studies is plastered with complications rooted in beam effects.

10.5 NAP-XPS Examples for Heterogeneous Catalytic Applications

Catalysis, and particularly heterogeneous catalysis, was and still is the most dominant application of NAP-XPS [7]. From the technological perspective, real catalysts are mostly supported systems with multiple promoters and additives. Support is very often an insulator, which makes the utilization of NAP-XPS problematic due to sample charging. A promoter package frequently applied to the main active component makes industrial catalysts typically too complex to enable structure-function correlations to be drawn from the data. Thus, more often than not, model catalysts are prepared with varying degrees of simplification. This involves, for example, leaving aside the additives and promoters, resorting to less-problematic supports or removing the support completely, and using surface science type model samples as nanoparticles deposited on thin-film supports (and their inverse), single crystals, and foils. Such simplifications are useful when the catalytic function is not hampered. Here, we will review numerous catalytic applications and catalyst systems, a significant portion of which are from our own research or studies performed in collaboration with partner research institutions. The examples are grouped together under the themes, as follows:

- quantification of surface coverages,
- selected catalytic applications,
- selected catalysts for more than one application.

As we will see, some of the studies are lumped and discussed together with others, whereas some systems are explained separately in more detail. Due to the vast literature, it was not our intention to provide a comprehensive overview of all reactions pursued and scrutinized by NAP-XPS rather the examples were selected based on our own interests and with the goal to highlight many of the facets of catalyst dynamics explained in the Introduction.

10.5.1 *NAP-XPS studies with quantification of the surface coverages*

Typically, the kinetics of surface reactions can be described by a complex function of rate constants and coverages. The activation energy of a surface reaction step is influenced by the local environment around the reaction site, and the chemisorption energy of adsorbates is often a function of coverage. These have the consequences that in the scheme of a microkinetic model, the critical rate constant will be a function of the coverage of a certain species (or empty site), and adsorbate–adsorbate interactions can have a significant influence on the rate of every reaction step. In even more complex cases, the rate-limiting coverage contributions also relate to the entropic part (configurational entropy) of active surface sites [55]. Hence, the surface coverage of adsorbates plays a critical role in heterogeneous catalysis. Unfortunately, many different surface intermediates and spectator species can form, and due to the heterogeneity of surface sites, the reactivity over these different sites can and most likely will vary. XPS being quantitative and specific to the chemical environment may provide quantitative information on surface adsorbed species. That is, NAP-XPS studies can be utilized to evaluate the surface coverage of adsorbates under various reaction conditions in the hope to relate surface coverage information to the reaction rate. Also, the selectivity of a reaction can be governed by modulating surface coverages. To calculate coverages on a technical catalyst based on NAP-XPS measurements is, however, bound to be a difficult task because of the heterogeneity of the surface, the lack of a good coverage reference point to relate coverages under reaction conditions, and the general issue of XPS quantification in a non-flat geometry. Taken all these together, surface coverage evaluation can be useful, but it is very rarely done in NAP-XPS studies. Those studies that report the evaluation of surface coverages are done with single crystals and essentially only in CO oxidation. Our chapter is devoted to more complex catalytic systems, and therefore here we only briefly review two of these studies.

The spectral area of an adsorbed species must be related to the available surface sites. This might be relatively simple if the core-level spectrum of the surface holds both of these information. For example, Toyoshima *et al.* [56] adsorbed CO on Pt(111) at RT from UHV conditions up to 0.1 Torr and followed the evolution of the high-resolution Pt 4f spectrum (hν: 150 eV). The CO coverage was estimated from the peak components of the unoccupied surface (clean surface core-level shift) and the surface Pt atoms binding to linear and bridged CO (adsorbate-induced core-level shifts). Essentially, the same method was utilized by Johansson *et al.* [57] in calculating the O coverage on Pt(111) and Ir(111) under CO oxidation.

Another approach is to use a reference spectrum of a sample with known coverage prepared, for example, in UHV condition (or calculated theoretically) and calibrate the *in situ* reaction spectrum to the reference coverage. This approach was used by Blomberg *et al.* [58] to estimate the active O coverage on Pd(100) or by Kondoh *et al.* [59] for three different Pd single crystalline surfaces under CO oxidation. Johansson *et al.* [57] investigated CO oxidation over Pt(111) and Ir(111) in an oxygen-rich feed of O_2:CO = 9:1.

The Pt(111), below 530 K, was CO covered (CO coverage of 0.33 ± 0.06 ML at 450–500 K and 0.25 ± 0.05 ML at 517 K) and largely inactive. We want to point out here that despite the roughly factor of two reduction of the CO coverage found here, as compared to the dense $(\sqrt{19} \times \sqrt{19})R23.4°$-13 phase with 0.68 ML coverage observed by Toyoshima *et al.* [56] at RT, no signature of adsorbed O was visible in the spectra. At about 530 K, there was a large increase in the CO_2 and a simultaneous decrease in the CO signal in the mass spectrometer accompanied by a shift of the gas phase O 1s peaks indicating a change in the sample work function, disappearing the top and bridged CO_{ads} signals, and the appearance of O_{ads} with a coverage of 0.21 ± 0.02 ML. The sudden ignition of CO oxidation and the parallel change of the adsorbate structure from a CO to O-covered surface are in line with similar experiments over Pd single crystals [58,59]. Analogous experiments on Ir(111), however, indicated that adsorbed CO and O coexisted on the surface and that the transition from a CO to O-covered surface occurred in a larger 450–550 K temperature window (Figure 10.7). Based on DFT calculations, the difference was explained in terms of the adsorption energies

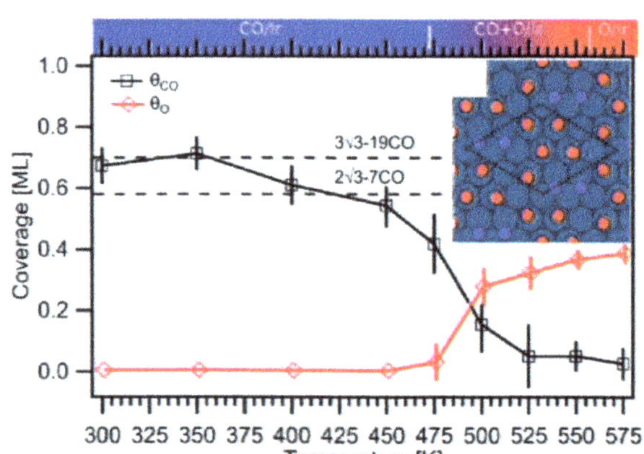

Fig. 10.7. Coverage of CO and oxygen for all measured temperatures on Ir(111). The inset in the top right corner shows the DFT-optimized structure that forms once adsorbed oxygen starts to deplete CO on the surface. Adapted with permission from Ref. [7] CC BY 4.0.

of O and CO on the Pt(111) and Ir(111) surfaces: on the latter, the O and CO adsorption energies are larger, large enough to overcome the CO–O repulsive interaction and enable the coexistence of both types of adsorbed species.

Ortega and coworkers [60] utilized a curved Pt(111) crystal to probe the chemical composition of the surface with varied step densities under CO oxidation. The sample center was the (111) surface and vicinal surfaces with increasing density of close-packed steps with {100} (A-type) and {111} (B-type) microfacets appeared as one moves away from the center. Determination of CO coverage of terraces and steps was based on a reference spectrum upon 10 L CO dosage in UHV and RT, which was claimed to develop a c(4 × 2) structure, yielding a coverage of 0.5 ML CO, with 0.25 ML terrace-top, and 0.25 ML terrace-bridge population.

In Figure 10.8, the C 1s and O 1s spectra are plotted in the temperature range of 520–570 K for three sections of the sample: the (111) plane and the two A- and B-type stepped edges. The low-active CO-covered (Chem-CO) surface vanished and an oxygen-covered active surface (Chem-O and oxides) developed suddenly over all flat and stepped crystal planes at T= 562.5 K. Although this temperature is ~30 K higher, the abrupt change of the surface and the entering of the mass-transfer limited regime of CO oxidation are both identical to that found by Johansson *et al.* [57]. Surprisingly, despite the different chemisorption energy of CO on steps and terrace sites, the removal of CO occurred at the same temperature.

Figure 10.9 compares the C 1s spectra of the sample at two selected temperatures to the reference spectrum in UHV at 0.5 ML CO coverage. As compared to the (111) spectrum, in the vicinal (223) and (553) planes, additionally step-top and step-bridge species appear (see sketch), with a small decrease in total coverage. The same four Chem-CO species were found in the UHV and NAP-XPS experiments, but the terrace/step and the terrace bridge/top ratios are notably lower in the $CO+O_2$ feed. The CO coverage at terraces decreases from 0.50 ML at 300 K to about 0.3 ML at 350 K and to about 0.2 ML at 550 K. Additionally, the coverage of the step CO remained almost constant up to 550 K, implying that when

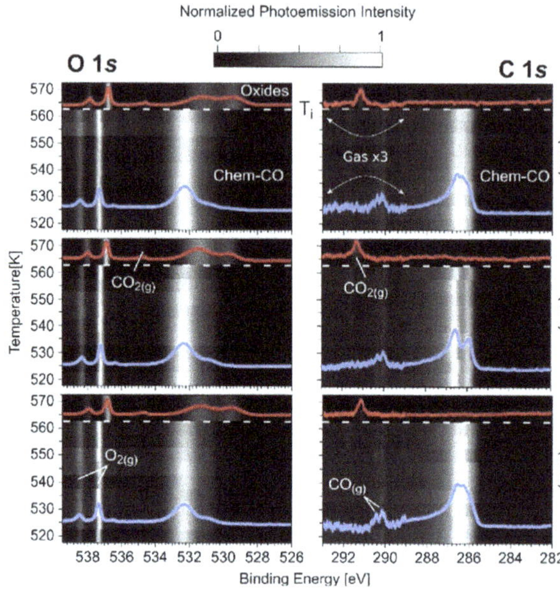

Fig. 10.8. O 1s (left) and C 1s (right) photoemission intensity images measured at three different points on the c-Pt(111) sample exposed to about 1 mbar, CO:O_2 of 1:6 during a heating cycle. Individual spectra for active and poisoned phases appear overlaid. The abrupt ignition of the reaction is marked with the dotted horizontal line at 562.5 K. Adapted with permission from Ref. [60]. Copyright (2020) Wiley-VCH GmbH.

approaching the ignition T, CO stays anchored to steps while terraces become partially emptied. In Figure 10.10, the parallel evolutions of C 1s and O 1s spectra are compared before and above ignition. First, three different temperatures are contrasted: far from the ignition (350 K), right before (560 K), and immediately thereafter (570 K). The 350 K and 560 K spectra, though they belong to the same CO-poisoned phase, look markedly different. The total O 1s to C 1s intensity ratio during the thermal activation slowly increases up to 555 K and overshoots near the ignition temperature. This can be explained by an additional O species besides the adsorbed CO phase. According to the fit, this transient-O peak emerges at about 532.1 eV and builds up below the reaction light-off. Based on the activity profile, the transient O species exhibits low catalytic activity

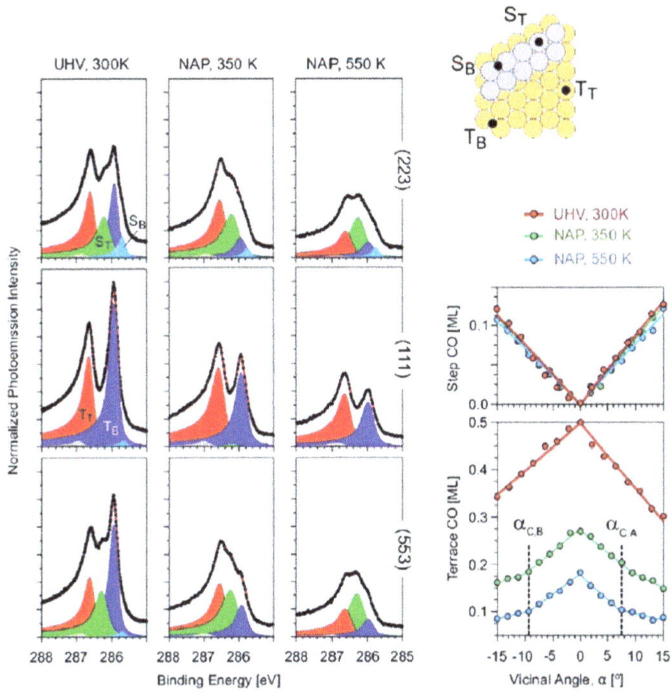

Fig. 10.9. Line fitting of individual spectra at the (223), (111), and (553) planes on the curved surface. TT, TB, ST, and SB stand for emission from CO molecules, respectively, adsorbed in Terrace-Top, Terrace-Bridge, Step-Top, and Step-Bridge sites, sketched on the right-hand side of the panel. Step (top) and Terrace (bottom) CO coverage across the curved surface. Adapted with permission from Ref. [60]. Copyright (2020) Wiley-VCH GmbH.

and is thermally stable. The high binding energy of the transient-O species indicates a different chemical environment compared to the hollow-adsorbed Chem-O of the active state, whose O 1s peak appears below 530 eV. Based on DFT calculations and considering subsurface oxygen adsorption below the hcp surface site, two stable surface-CO/subsurface-O complexes were found. The more stable one is depicted in Figure 10.10 and corresponds to a geometry where the subsurface oxygen and the CO adsorbed molecule share one Pt surface atom. The DFT predicted binding energy corresponds well to the experimentally derived fit value. The calculated

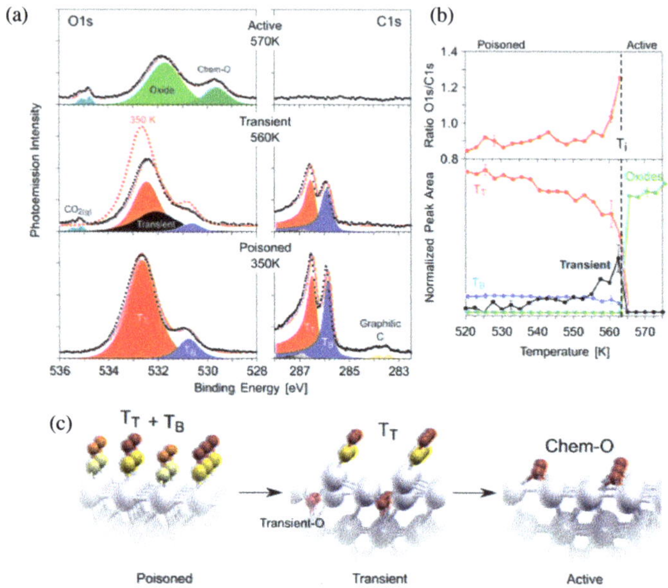

Fig. 10.10. Presence of oxygen before light-off. (a) O 1s and C 1s spectra at the (111) surface exposed to 1 mbar, $CO:O_2$ of 1:7 in a slow heating cycle (0.25 $Kmin^{-1}$, 2.5 K steps), acquired far from the ignition (poisoned, 350 K), right below (transient, 560 K), and immediately after (active, 570 K). In the O 1s panel for the transient stage, the spectrum of the 350 K poisoned phase appears overlaid (red dashed line). Differences between both are attributed to the Transient-O oxygen peak at about 532.1 eV. (b) Top: evolution of the O 1s/C 1s intensity ratio before light-off. Bottom: temperature variation of the O 1s intensity for each surface species around the ignition. (c) From left to right, model structures for the poisoned (Chem-CO), transient (CO-Pt-O complex), and active (only Chem-O) stages. Adapted with permission from Ref. [60]. Copyright (2020) Wiley-VCH GmbH.

adsorption energy for the CO molecule in the CO-Pt-O complex increases by about 0.3 eV with respect to that in the bare (2×2) CO layer at the terrace site, roughly equaling the adsorption energy difference between terraces and steps' sites. Therefore, this can explain the simultaneous desorption of CO in the whole curved Pt surface at the ignition point.

These two and other studies [58,59,61] demonstrate a potential problem surrounding the "active" surface state and coverages. At the conditions where adsorbed CO is the sole surface species observed, essentially no or minimal reaction occurred. Though such correlation was not explicitly established, if any, the reaction rate in this case will correlate with the coverage of empty sites. When no CO_{ads} was observed at the surface, the reaction rate is high but does not increase with temperature, and the reaction is mass transfer limited by the diffusion of CO through a CO-depleted gas segment over the surface. In this case, the rate is by definition not controlled by the surface event and surface coverage. Hence, again we are not able to correlate the surface coverage to the reaction rate. That is, in neither regime could we derive a microkinetic rate law from the NAP-XPS observations. Let us consider the following thought experiment: if we would be able to get rid of the CO-depleted region, the reaction rate would be much higher. Then, by increasing the mass flow of the feed gas, the rate would further increase until it goes to saturation. At the point of rate maximum, the chemical potential of gas-phase CO would be much higher than in the mass transfer limited regime and thus the chemical state of the surface is expected to be different than these studies indicate. Hence, the O-covered surface, as identified in these studies, is more active than the CO-poisoned surface but less active than the surface in our thought experiment at the rate maximum. As pointed out by Matera and Reuter using a first-principles-based multiscale modeling approach that integrated first-principles kinetic Monte Carlo simulations of the surface reaction chemistry into a fluid dynamical treatment of the macroscale flow structures in the reactor, in the high catalytic activity stage, both reactants (CO and O) have to be adsorbed on the surface, and a range of heat and mass transport effects have to be considered and properly accounted for to prevent wrong mechanistic conclusions from *in situ* experiments [62].

10.5.2 *Selected catalytic applications to understand the active surface state/phase*

In this section, we review selected examples of NAP-XPS application in heterogeneous catalysis with the overarching theme of catalyst dynamics.

10.5.2.1 *Methanol steam reforming (MSR)*

MSR is one of the promising routes to produce high purity hydrogen for fuel cell applications. The goal is to run the reaction selectively towards H_2 and CO_2 without producing CO. Methanol might also be used directly in fuel cells (direct methanol fuel cells) when the anode does not produce CO or it is resistant to CO poisoning. Zafeiratos *et al.* [63] investigated a ternary PtRuCo catalyst, a catalyst candidate for direct methanol fuel cells, in MSR and water–gas shift (WGS) reaction. The catalyst produced mostly CO (CO:CO2 = 4:1) at 570 K under NAP-XPS conditions at ~35% methanol conversion. At the same temperature, the CO conversion in WGS was solely 2% suggesting that methanol decomposition is a significant pathway under MSR conditions and that the catalyst, at least under WGS, was not particularly active to convert the produced CO. However, the surface state of the pre-reduced catalysts under these two reaction feeds was significantly different. Whereas Pt and Ru were mostly metallic in both feeds, with a small ionic Pt under WGS, Co was primarily metallic under MSR but almost entirely oxidized to Co^{2+} in WGS. Co oxidation gave rise to significant Co segregation to the surface. The authors concluded that the ternary system was too complex to afford structure–function correlations. Thus, a polycrystalline PtCo foil was also studied in MSR. Comparing the Pt/Co ratio as a function of the photoelectron kinetic energy, with a reference material recorded in UHV, it was suggested that by surface segregation a Pt-skin layer formed over the Co-oxide layer.

Pd/ZnO has been shown to be an excellent MSR catalyst, however, there was some controversy about whether the high CO_2 selectivity was related to the $PdZn_\alpha$ phase (structurally similar to the fcc Pd) or related to the tetragonal $PdZn_{\beta 1}$. Halevi *et al.* [64] synthesized the $PdZn_\alpha$ phase with 12 at.% Zn and determined the surface chemical state of the sample under MSR and CO-oxidation conditions. Except for a slight change in the Pd 3d peak shape, indicating a small degree of oxidation under CO-oxidation conditions, the chemical state remained constant. Furthermore, while ZnO was not observed, surface Zn content increased under oxidative conditions and decreased under reductive MSR conditions. $PdZn_\alpha$ was not selective to CO_2 under MSR.

Rameshan *et al.* [65] prepared two types of PdZn (near-)surface alloys by Zn deposition in ultrahigh vacuum on a Pd(111) single crystal and annealing: both had a Pd:Zn of 1:1 alloy composition at the surface but differed regarding their Pd:Zn subsurface composition. The samples were denoted as "multilayer" and "monolayer" alloy states. They have shown that the CO_2 selectivity of a PdZn 1:1 surface alloy is governed by the subsurface layers. Whereas the multilayer PdZn near-surface alloy exhibited high CO_2 selectivity, a predominantly palladium-coordinated monolayer PdZn surface only produced CO and H_2 despite its identical initial surface composition. Under NAP-XPS conditions (Figure 10.11) besides the difference in the Pd3d core-level shift, the valence band revealed another difference. While the multilayer alloy showed a "Cu-like" valence band, that is, a significantly reduced density of states near the Fermi level and downshifted d-band, the monolayer alloy resembled a modified Pd surface. Probably the most important difference was, however, observed in the Zn 3d spectra: an oxidized Zn species (most likely OH species) formed on the multilayer alloy was not present in any significant quantities with the monolayer alloy, indicating a difference in the capability of the samples towards water activation. That is, the monolayer alloy does not seem to be able to activate water, and the CH_2O formed from methanol is thus not converted into CO_2 but rather dehydrogenated to CO.

Friedrich *et al.* [66] synthesized intermetallic $PdZn_{\beta 1}$ (tetragonal structure) in a broad Pd:Zn ratio (Pd content: 47–59 at.%) by heating elemental Zn and Pd in the targeted atomic ratio in an evacuated and sealed quartz glass ampoule up to 900°C and annealing for six days. Samples were ground and pressed to pellets. NAP-XPS investigations were performed on as-synthesized and reduced PdZn samples as well as under MSR conditions, displaying clear differences in the observed Pd 3d binding energy, the Zn/Pd ratio, and the amount of oxidized Zn species on the surface, depending on the bulk Pd content. All $Zn_{100-x}Pd_x$ samples preserved their intermetallic character in the near-surface region under MSR conditions.

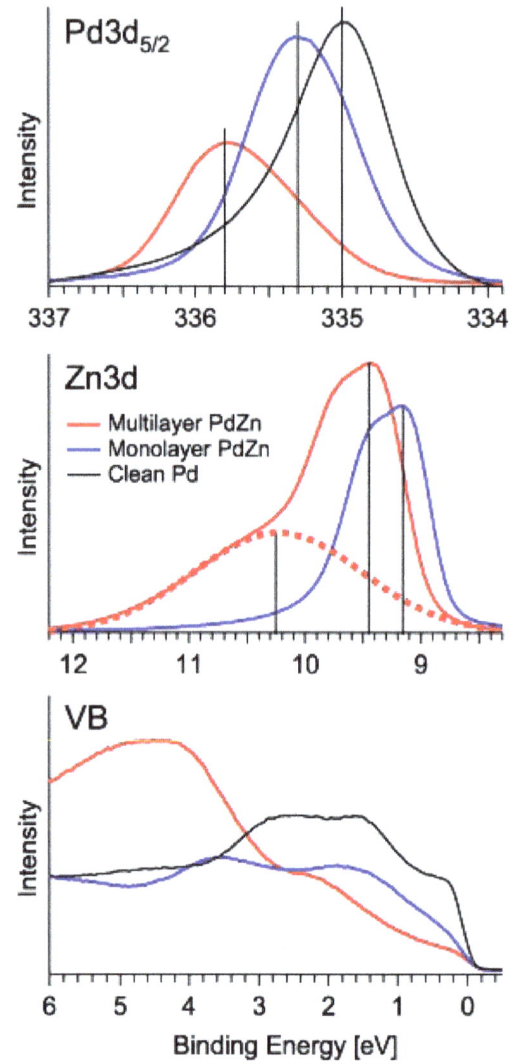

Fig. 10.11. AP-XPS spectra (Pd 3d, Zn 3d, and valence-band (VB) regions) acquired *in situ* during MSR on the PdZn 1:1 multilayer (red curves) and monolayer alloy (blue curves). For comparison, the respective clean Pd spectra are added (black lines). The oxidized ZnOH component is highlighted by the dashed red line (middle panel). The Pd 3d spectra were recorded with 650 eV photon energy, the Zn 3d, and valence band regions with 120 eV. Reaction conditions: 0.12 mbar methanol, 0.24 mbar water, 553 K. Adapted with permission from Ref. [65]. Copyright (2010) Wiley-VCH Verlag GmbH & Co.

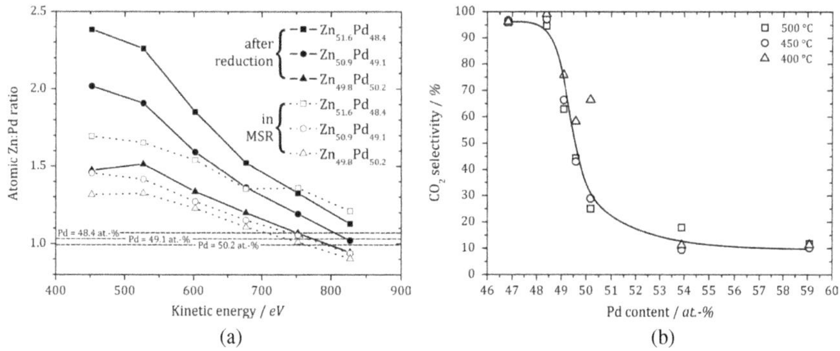

Fig. 10.12. (a) Depth-depending atomic Zn/Pd ratio of $Zn_{51.6}Pd_{48.4}$, $Zn_{50.9}Pd_{49.1}$, and $Zn_{49.8}Pd_{50.2}$ after hydrogen reduction (420°C, 0.5 mbar) and during methanol steam reforming (360°C, 0.2 mbar, H_2O:MeOH = 2:1). Horizontal dashed lines represent the Pd bulk content of the investigated samples. (b) CO_2 selectivity of $Zn_{100\text{-}x}Pd_x$ samples in a flow reactor under methanol steam reforming at three reaction temperatures (atmospheric pressure, H_2O:MeOH = 1:1). The line is a guide to the eye. Adapted with permission from Ref. [66]. Copyright (2011) Elsevier Inc.

Compounds that were Zn-rich in the bulk possessed the Zn-richest surfaces (Figure 10.12(a)) and showed the strongest core-level shift in Pd 3d binding energy compared to elemental Pd. Oxidized Zn species on the reduced samples could only be observed on Zn-rich samples. These findings strongly correlate with the methanol steam reforming properties under atmospheric pressure; samples rich in Zn provided the highest activities and high selectivity to CO_2 (Figure 10.12(b)), whereas samples rich in Pd possessed poor catalytic activity and mainly formed CO during MSR. A balance between oxidized Zn species and intermetallic Pd on the surface of a Zn-rich sample resulted in CO_2 selectivity up to 99.4%.

10.5.2.2 *Ethanol steam reforming (ESR)*

Just as MSR, ethanol steam reforming (ESR) is an industrially relevant reaction as ethanol has become more readily available from renewable sources. Due to the endothermic reaction, often a high reaction temperature is required to achieve good ethanol conversion. The high temperature combined with the reducing conditions holds a high risk of fast catalyst deactivation by coking. Recently, several ceria-supported transition metal catalysts were proposed as a good catalyst for ESR, and some of these were investigated by NAP-XPS. Liu *et al.* prepared a model Ni/CeO_2(111) catalyst by evaporating Ce onto a Ru (0001) single crystal in the presence of 5×10^{-7} Torr O_2 and then annealed it to 800 K for 10 min at the same O_2 pressure [67]. Then, Ni with an approximately 0.15 ML coverage was vacuum deposited on the thus prepared CeO_2(111) surface. Their experiments using NAP-XPS revealed that the surface of the ceria substrate was highly reduced and hydroxylated under ESR, whereas the small supported Ni nanoparticles were present as metallic Ni and Ni carbide. Their results, in combination with *in situ* Infrared Reflection Absorption Spectroscopy, suggested that the catalyst components have a cooperative dual function. Reduced ceria assists in the deprotonation of ethanol and water with subsequent generation of hydroxyls, and metallic Ni was the actual active site of both C–C and C–H cleavages of the chemisorbed ethanol. The authors proposed that coke formation/removal may occur through a Ni carbide intermediate phase and hydroxyl/oxygen transfer can play an important role in mitigating carbon accumulation.

Sohn *et al.* studied the surface reduction characteristics of bare ceria and cobalt nanoparticles supported on ceria under ESR conditions using NAP-XPS, either with pre-calcined [68] or with pre-reduced [69] catalysts. Ceria particles were prepared in two different particle sizes, one in nano and the other in micron size (CeO_2-NP and CeO_2-MP), with average particle sizes of 4 and 120 nm, respectively. It was found (Figure 10.13) that the particle size of ceria, and thus the microstructure and defect density, affects surface reducibility. Smaller particles led to a higher extent of surface reduction. Supported cobalt nanoparticles have a significant effect on the surface reducibility of both CeO_2-NP and CeO_2-MP. Compared to bare ceria particles, the presence of fully oxidized cobalt nanoparticles on the surface of ceria support retards surface reduction of ceria, leading to a higher average oxidation state of cerium

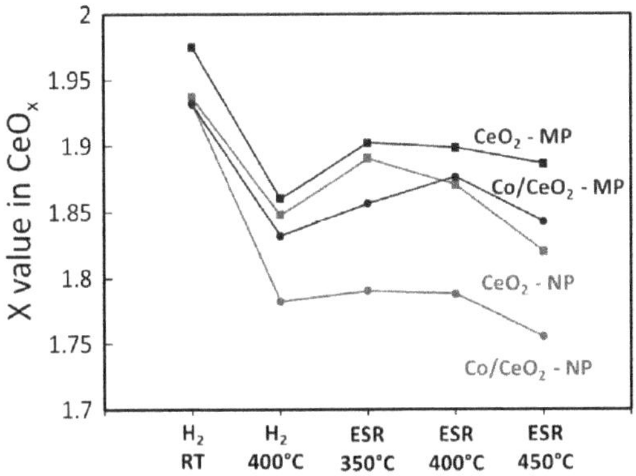

Fig. 10.13. Extent of reduction of ceria for CeO_2-NP, -MP and Co/CeO_2-NP, -MP samples under reducing and ethanol steam reforming conditions (ethanol-to-water ratio = 1:10). Adapted with permission from Ref. [69]. Copyright (2017) Springer.

compared to the bare ceria supports in ESR. The presence of fully reduced cobalt particles (Co°) facilitated ceria reduction through hydrogen spillover.

The comparison of cerium oxidation state between pre-reduced CeO_2 and pre-reduced Co/CeO_2 indicated a higher extent of reduction of cerium in the case of Co/CeO_2 for ethanol steam reforming. These results suggest that both the initial state of cobalt as well as the microstructure of the ceria play an important role in the actual surface state of the catalysts. Co/CeO_2-NP was the most active sample with a high H_2 yield. The oxidation state of cobalt in various forms (single crystal, unsupported nanoparticles, and supported nanoparticles on CeO_2 and ZnO) and the type of chemisorbed species were also investigated by Turczyniak *et al.* under ESR condition by NAP-XPS [70]. They have found a profound effect of the support on the reducibility of cobalt oxide. ZnO interacted strongly with CoO and probably a mixed $Zn_{1-x}Co_xO$ spinel oxide formed that retarded further reduction. Also, the amount of carbon deposition was significantly higher on unsupported samples, in agreement with the observations of Liu *et al.* [67]. The influence of the support was also explicitly investigated when unsupported and ceria-supported bimetallic $Rh_{0.5}Pd_{0.5}$ catalysts were compared in H_2, in O_2, and under ESR [71]. The authors used three different photoelectron kinetic energies when recording their spectra, which enabled them to estimate the depth distribution of the elements in the samples. The results indicated that supported nanoparticles were less strongly reduced under identical conditions. Furthermore, the elemental distribution changed as a function of the applied condition: in H_2, Pd segregated to the surface for both supported and unsupported samples. On the other hand, Rh surface segregation was found under ESR with the unsupported sample; the Rh migration towards the surface was absent in $Rh_{0.5}Pd_{0.5}/CeO_2$. In addition, Pd was oxidized at the topmost surface in $Rh_{0.5}Pd_{0.5}/CeO_2$. The authors suggested that this was due to water activation on ceria sites, providing the bimetallic particles with active oxygen, which re-oxidizes the surface Pd sites. In a follow-up publication, the same group (Soler *et al.* [72]) investigated the influence of the support morphology by comparing nanocube- and nanorod-shaped ceria-supported $Rh_{0.5}Pd_{0.5}$ samples under ESR. They found that $Rh_{0.5}Pd_{0.5}$ supported on ceria nanocubes were drastically more oxidized, which they attributed to the lower vacancy formation energy of the {100} planes of the nanocube morphology giving rise to easier oxygen transfer to the $Rh_{0.5}Pd_{0.5}$ nanoparticles. This difference was associated with the strongly different product distribution over the two catalysts.

10.5.2.3 *Selective alkyne hydrogenation over Pd catalysts*

Hydrogenation is one of the oldest chemical reactions performed in the chemical industry and researched at the fundamental level. The petrochemical product streams based on steam cracking and catalytic cracking of naphtha fractions require the removal of byproducts with higher unsaturation to avoid problems in subsequent downstream technological steps. Removal of traces of alkynes or multiple unsaturated hydrocarbons from ethylene or propylene feed to produce polymer-grade alkenes is achieved by selective hydrogenation. The target here is to produce selectively the partial hydrogenation product (Figure 10.14) and prevent the conversion of the alkene. Another important application of selective alkyne

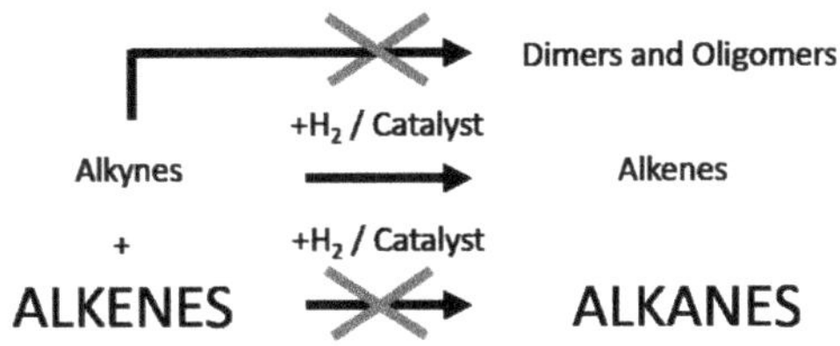

Fig. 10.14. Scheme of selective hydrogenation of alkynes in a typical mixed alkyne/alkene stream.

hydrogenation is in the production of fine and specialty chemicals. Palladium and its derivatives were found to be highly active and selective catalysts for these reactions.

For any practical hydrogenation reaction to occur, adsorption of both unsaturated hydrocarbon and hydrogen is mandatory. Since on palladium surfaces molecularly adsorbed H_2 is held very weakly (10–20 kJ mol^{-1}), and thus desorbs at temperatures far below any relevant hydrogenation reaction may occur, hydrogen needs to dissociate at the surface. Mitsui *et al.* found that dissociative adsorption on Pd(111) requires aggregates of three or more vacant sites [73]. Thus, at high hydrocarbon coverages, H_2 activation might be limiting. Palladium absorbs large amounts of hydrogen in its bulk, with hydrogen atoms occupying octahedral interstitial sites. The bulk and subsurface H atoms are less strongly held than surface-adsorbed H. Surface diffusion of adsorbed H is facile, at least without the presence of large quantities of other adsorbates, with the activation energy for diffusion typically in the range of 10–20 kJ mol^{-1} [74,75]. This enables rapid diffusion of H over the surface.

A characteristic difference of alkynes compared to alkenes is their stronger adsorption to any of the transition metals. For example, acetylene binding, at 25% coverage, was computed over Pd(111) to be 172 kJ mol^{-1} for hollow, 137 kJ mol^{-1} for bridge, and 55 kJ mol^{-1} for atop position [76], whereas the hollow site destabilizes to 136 kJ mol^{-1} at 33% coverage [77]. Noteworthy, for the same 25% coverage, ethylene alone adsorbs by only 59 kJ mol^{-1} to bridge sites. Thus, the good selectivity of Pd, and some other transition metals, is frequently explained by the differences in the heat of adsorption. As long as alkyne is not entirely removed from the reactant feed, owing to its higher heat of adsorption, alkyne should occupy surface sites displacing ethylene from the surface and will not allow its re-adsorption. However, we need to consider a few observations. Ethylene hydrogenation is faster than acetylene, it is typically first (or near first) order in H_2, and the reaction was shown to prefer π-bonded ethylene with the most weakly bound hydrogen. Furthermore, taking Pd(111) as an example, acetylene binds most strongly to three-fold hollow sites, as mentioned above, whereas ethylene prefers bridge sites. In fact, it was shown experimentally, that ethylene can adsorb on Pd catalysts in the presence of acetylene [78]. Hence, it is not directly evident that strong adsorption is beneficial for high reactivity. As convincing as the heat of adsorption argumentation is, it was not clear whether this alone would afford for the high selectivity of Pd catalysts. As it turned out, NAP-XPS was instrumental to challenge the prevailing view at that time on selective hydrogenation over Pd.

We have investigated the near surface of palladium catalysts under alkyne selective hydrogenation by NAP-XPS in conjunction with *in situ* prompt gamma activation analysis (PGAA) assessing the bulk hydrogen concentration of Pd and DFT calculations [79–82]. First [79], the hydrogenation of 1-pentyne was studied over supported and unsupported Pd materials and we found that, during XPS experiments, partial hydrogenation was the dominant reaction path at the applied mbar pressure conditions. Moreover, the reaction was accompanied by the build-up of a surface Pd–C phase (Figure 10.15) by fragmentation of reactant molecules and carbon dissolving in the top few Pd layers. This was identified by the ~335.6 eV Pd 3d component; its contribution to the overall Pd 3d envelopes strongly exceeding the limit of the surface core-level shift component of hydrocarbonaceous adsorbates.

Additionally, the low binding energy C 1s component (283.4 eV), being close to those of carbides, substantiated the assignment of some carbon atoms occupying octahedral subsurface positions. By performing non-destructive depth profiling via photon energy

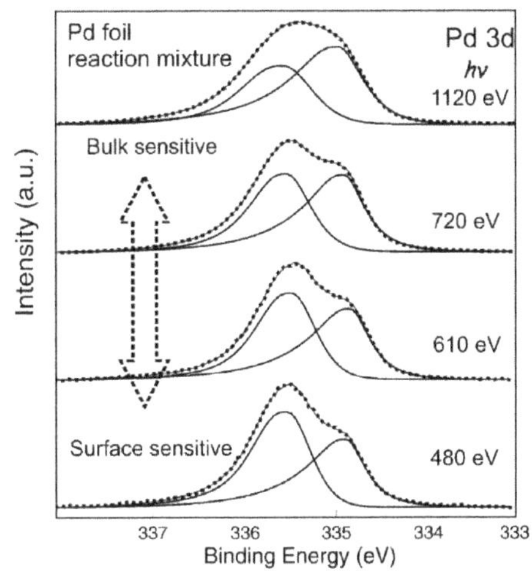

Fig. 10.15. Non-destructive depth profiling experiment on Pd 3d of Pd foil during 1-pentyne hydrogenation. All the spectra were recorded at 358 K. Adapted with permission from [79] Copyright (2006) Elsevier Inc.

variation, we concluded that the Pd–C phase was 2–3 atomic layers thick. Thus, the Pd–C surface phase defines the top few Pd layers with subsurface carbon participation. The spectral features were further investigated upon switching off the reactants individually. Curve 1 in Figures 10.16(a) and 10.16(b) represents the Pd 3d and C 1s spectra under 1-pentyne selective hydrogenation. As 1-pentyne was removed from the inlet gas (curve 2), the low-BE C 1s component and a significant intensity from the high-BE side disappeared. The latter disappeared instantaneously, whereas the former took several minutes to do so. The difference in time constant of the two components indicates that these components are not directly related and are not a single species.

The slight asymmetry of the peak at ~285 eV in the difference curve (1–2) suggests that it contains two components (284.8 and 285.2 eV). Since the gas phase is not contributing to the 285 eV peak (see spectrum 4; only negligible intensity below 285.7 eV), the peak was attributed to chemisorbed pentyne molecules, which desorbed as the 1-pentyne feed was stopped. After switching back 1-pentyne,

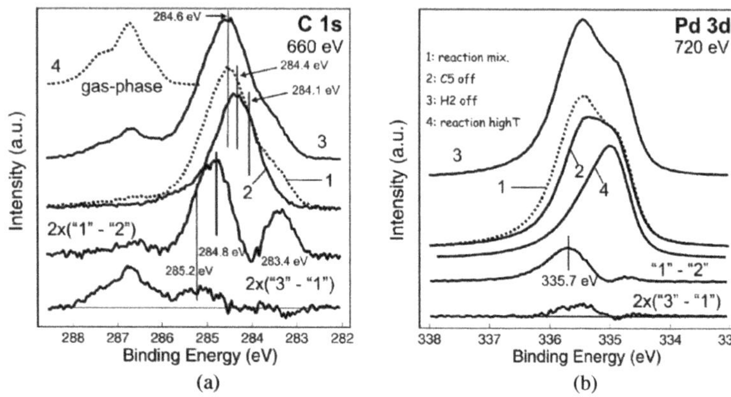

Fig. 10.16. C 1s (a) and Pd 3d (b) regions of Pd foil at different conditions. 1: In the reaction mixture (1-pentyne hydrogenation); 2: after switching off 1-pentyne (only H_2); 3: after switching off H_2 (only 1-pentyne). All the spectra were recorded at 358 K. Some different spectra are also shown. In (b), Pd 3d of Pd foil at 523 K is also included. Adapted with permission from Ref. [79]. Copyright (2006) Elsevier Inc.

the spectrum (not shown) immediately developed as in the reaction mixture (curve 1). Moreover, a pentene signal appeared in the MS spectrum. Calculation of Pd-to-C ratios indicated a <0.5% increase of carbon content in the information depth between states before switching off and after switching back to the 1-pentyne feed. Therefore, beam-induced carbon accumulation was negligible in this experiment. Spectrum 3 was recorded after the hydrogen supply was turned off. The lack of visible spectral changes (except the increased gas-phase components) suggests the stability of Pd–C without H_2. Nevertheless, the difference spectrum (3–1) suggested a tiny increase at 285.2 eV, that is, the 1-pentyne surface coverage increased slightly in the absence of H_2. As hydrogen was reintroduced, the reaction continued, and again no beam-induced carbon accumulation was observed from the investigation of the spectra. In Pd 3d spectrum, as 1-pentyne was switched off, the Pd–C component partially decreased. The difference spectrum indicates a somewhat higher (335.7 eV) BE; thus, the 335.6 eV component is the sum of more than one component (Pd–C and adsorbate-induced surface core-level shift). The original Pd 3d spectrum was observed as pentyne was reintroduced in the feed (not shown separately). Moreover, hardly any modification was obtained with just 1-pentyne in the feed. The slight increase at ~335.6 might be correlated with higher pentyne coverage, also seen in C 1s. As the palladium samples were heated further (up to 523 K), their catalytic activity in this mbar pressure condition diminished. Parallel to this, the 335.6 eV Pd–C component also strongly decreased, indicating its metastable nature with respect to segregation into carbon and pure palladium.

Later [80,81], we generalized these observations (Figure 10.17), as other alkynes were also shown to selectively hydrogenate on a carbon-modified Pd surface, whereas alkenes did not induce such a surface modification. Notably, lower chain alkynes gave rise to more carbon dissolving in Pd, thickening the Pd–C surface phase. We also showed that by applying a high hydrogen/alkyne ratio, the total hydrogenation starts to be the prevailing reaction path. This can be correlated to the vanishing Pd–C component at higher H_2/C_5 ratios [80]. Therefore, we concluded that alkynes were selectively

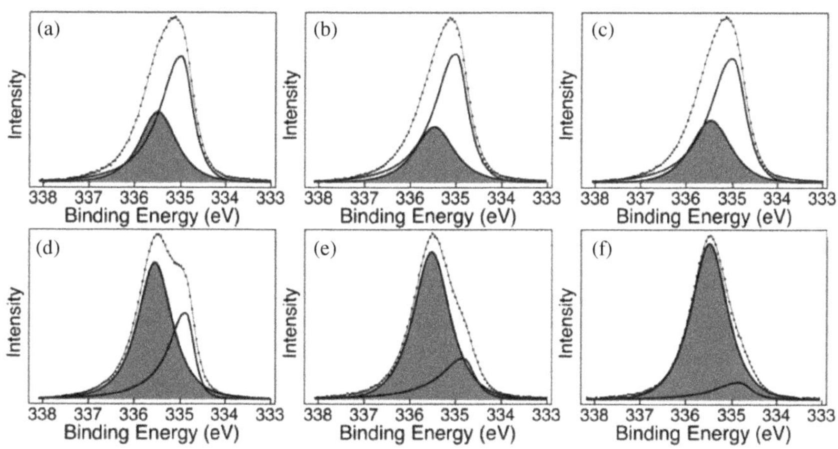

Fig. 10.17. Comparison of *in situ* Pd $3d_{5/2}$ spectra of Pd foil under alkene and alkyne hydrogenation at 1 mbar (H_2/C_xH_y: 9/1) and 343–353 K. (a) 1-pentene; (b) propene; (c) ethylene; (d) 1-pentyne; (e) propyne; (f) acetylene. Adapted with permission from Ref. [81]. Copyright (2008) Wiley-VCH Verlag GmbH & Co.

hydrogenated on Pd–C, whereas unselective alkyne hydrogenation, as well as alkene hydrogenation, proceeded on Pd surfaces without subsurface C modification. Recently, using a graphene membrane cell approach, similar observations were found at 1 bar by NAP-XPS: selective propyne hydrogenation proceeded on Pd–C whereas at high H_2:C_3H_4 ratio hydrogen reduces the population of Pd–C with a concomitant increase in unselective hydrogenation [83]. These observations are in line with DFT predictions considering surface free energies and the effective carbon chemical potentials showing that alkynes on palladium are prone to decomposition, whereas alkenes are not, and that subsurface carbon can be removed as methane at high hydrogen and low carbon chemical potentials [81,84]. It is interesting to note that not only alkynes were shown to be able to produce the Pd–C surface phase. Reversible Pd–C formation was found by NAP-XPS with ethylene oxidation at lower pressure [85] and when exposing Pd nanoparticles to CO [86].

Complementary *in situ* PGAA experiments revealed no correlation between activity in the regime of selective hydrogenation and the amount of hydrogen dissolved in palladium. Both high (palladium β-hydride) and low bulk hydrogen concentrations were compatible with selective alkyne hydrogenation despite the first-order dependence on $p(H_2)$. Additional hydrogenation experiments at total alkyne hydrogenation, as well as with alkenes, indicated that the bulk exists exclusively in the β-hydride form. These observations are in line with the NAP-XPS results emphasizing the role of Pd–C to decouple the surface and bulk H concentration.

As mentioned above, palladium is used in industry to hydrogenate alkynes in a mixed alkyne/alkene feed, and hence we have investigated the temporal evolution of the surface using time-resolved NAP-XPS when co-dosing propyne to a propene/H_2 feed [81]. In line with Figure 10.17(b), the Pd 3d core level was dominated by the bulk component in the propene+H_2 feed. Then, as propyne was introduced, we observed a two-step process: a fast initial C uptake and rise of the Pd–C component (within the first two minutes), likely related to the occupation of the first subsurface sites, and a second slow process of further growth of Pd–C with time on stream.

Since this second process is much slower, the buildup of additional mixed Pd–C layers should be energetically less favorable. The calculated growth rate of the Pd–C layer was approximately 7 Åh^{-1}, which is equivalent to a diffusion coefficient of carbon of approximately 10^{-22} m^{2}s^{-1}, in good agreement with the results of Yokoyama *et al.* (10^{-23} m^{2}s^{-1}) extrapolated to our temperature.

To complement the NAP-XPS observations, the energetics of H adsorption and diffusion of hydrogen from the surface toward the bulk was analyzed at low 0.33 ML surface coverage [82]. It was found (see Figure 10.18) that (i) H binds the strongest on the surface and binding is progressively weaker in the first and second interlayers, then approaching the limit of –0.1 eV bulk absorption value, and (ii) the energy barrier to move from surface to subsurface is ~0.4 eV and to deeper interlayers is smaller (0.2–0.1 eV), and the extraction

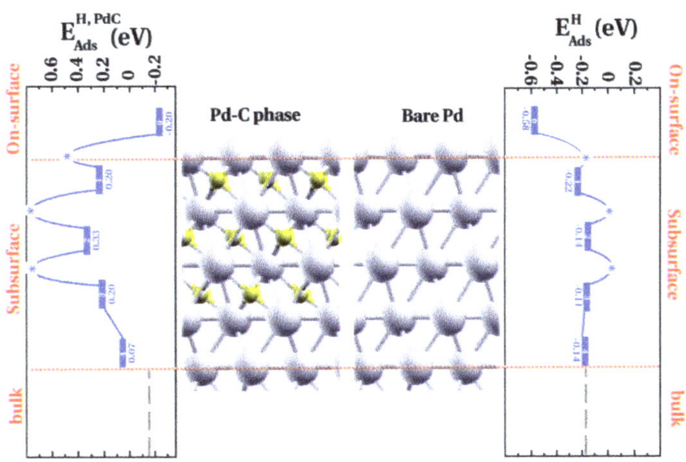

Fig. 10.18. (Right panel) Energy profile for the adsorption and migration of H toward the bulk for pure Pd(111). (Left panel) Energy profile for the adsorption and migration of H toward the bulk in the presence of a three-layer Pd-C phase. First, second, and third interlayers are populated by C atoms (concentration 1 C for 3 Pd). Horizontal dotted lines are used to differentiate the three different binding zones of Pd; the vertical dashed lines indicate hydrogen binding energy in the bulk of Pd. Adapted with permission from Ref. [82]. Copyright (2010) American Chemical Society.

barrier is solely 0.1 eV. Once the subsurface is populated with C, (iii) surface H adsorption energy is reduced and an exclusion zone is formed in the subsurface, and (iv) the barrier is only moderately higher for the penetration, 0.9 eV, but considerably increased for the extraction, 0.8 eV. Hence, the presence of hydrogen atoms in the subsurface for the Pd–C phase is thermodynamically disfavored; nevertheless, bulk H is not captured, even in the presence of the Pd-C phase, but the dynamic exchange with the surface is slowed down. There seem to be multiple roles of subsurface carbon: (i) it shifts the d-band center, heightens the barrier for alkene hydrogenation, and facilitates alkene desorption [87]; (ii) it lowers the binding energy of surface H; (iii) it forms an exclusion zone for subsurface H; and (iv) it increases the extraction barrier of subsurface H. These factors ultimately lead to low-surface H concentration necessary to avoid total hydrogenation and facilitate selective partial hydrogenation of alkynes.

Inspired by this knowledge, several research teams developed hydrothermal (glucose- or Na_2EDTA-assisted) synthesis methods for liquid-phase hydrogenation systems, which led to Pd catalysts with subsurface and interstitial carbon occupation [88–91]. These modifications turned the palladium catalysts to be ultra-selective in various liquid-phase alkyne, dienes, and imine reduction reactions.

Unfortunately, subsurface carbon as a modifier of gas-phase hydrogenation is only metastable [84]. Once the chemical potential of surface carbon does not stabilize this form and that of hydrogen reaches the critical level at which surface H will diffuse in high concentration into the subsurface, the subsurface carbon will be removed. When this occurs, usually at near completion of alkyne conversion, the reaction will accelerate, and alkene hydrogenation sets in. This leads to a rapid loss of selectivity. Palladium alloys may overcome some of these problems, but a potential issue of using alloys is their tendency to undergo segregation and island formation [92]. In an attempt to find a replacement for carbon that acts similarly, but to circumvent the stability problem, palladium intermetallic compounds were investigated. They have a unique lattice structure and covalent bonding between Pd and the second component, the latter factor hinders the transformation back to metallic Pd. Moreover, Pd intermetallic compounds allow for the separation of Pd sites, which are beneficial for diminishing oligomerization side reactions. Guided by these principles, Pd–Ga intermetallic compounds were introduced as a potential alternative catalyst for selective alkyne transformation [93–97]. The intermetallic compounds Pd_3Ga_7, PdGa, and Pd_2Ga exhibit different well-ordered crystal structures with mainly heteroatomic Pd coordination and pronounced covalent atomic interactions, as also evidenced by the strongly modified electronic structure (see Figure 10.19 valence band): with a reduction of the density of states at the Fermi level and a shift of the d-band center away from the Fermi level. Concomitantly, and following the shift in the d-band onset, the Pd 3d of the intermetallic compounds experience a shift to higher binding energy (Figure 10.19 Pd 3d). The ordering and the similar magnitude of the d-band and Pd 3d shifts suggest that the Pd 3d shift has predominantly a final state origin. NAP-XPS experiments at around 1 mbar under acetylene hydrogenation conditions evidenced neither a shift nor any additional components formed in Pd 3d for any of the stoichiometric intermetallic samples (Figure 10.20 on the example of PdGa) [93–95].

The intermetallic compounds are highly selective and stable catalysts but with low activities compared to Pd/Al_2O_3 due to the low specific surface area (Figure 10.21). Increasing the surface with harsher top-down treatments like milling or even etching

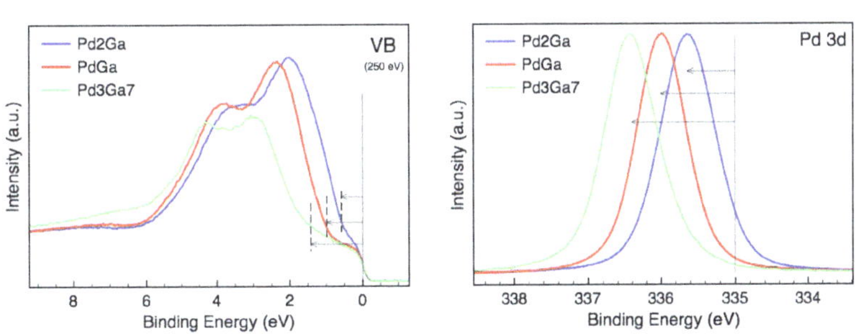

Fig. 10.19. Valence band and Pd $3d_{5/2}$ of Pd-Ga intermetallic compounds.

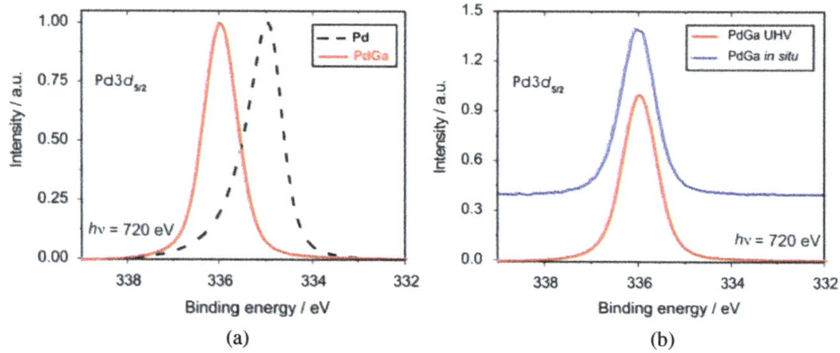

Fig. 10.20. (a) Comparison of the Pd $3d_{5/2}$ UHV XPS spectra of metallic Pd and PdGa. The shift to higher binding energy for PdGa is clearly visible. (b) Identical Pd $3d_{5/2}$ spectra of PdGa in UHV and *in situ* acetylene hydrogenation conditions. Adapted with permission from Ref. [93]. Copyright (2007) Elsevier Science Ltd.

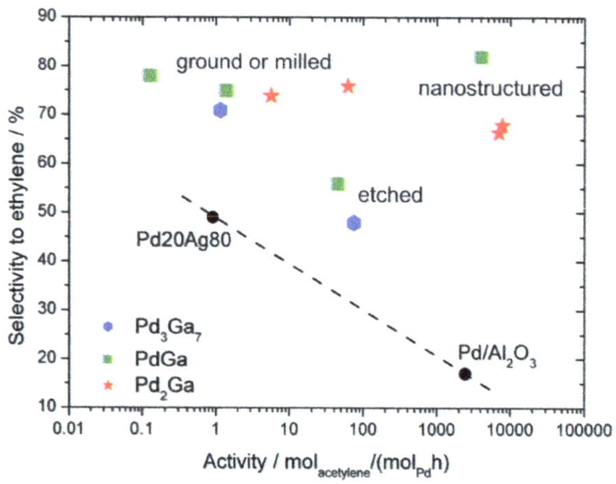

Fig. 10.21. Selectivity to ethylene in % over the rate of acetylene hydrogenation of model systems and technical catalysts (dashed line is a guide to the eye). All data after 20 h time on stream at 200°C in 0.5 % C_2H_2, 5 % H_2, and 50 % C_2H_4 in He with a total flow of 30 mL min^{-1}. Activity data adapted with permission from Ref. [96]. Copyright (2010) American Chemical Society.

with ammonia solution enhances the activity, especially the etching results in Pd-enriched surfaces by partly decomposing the intermetallic compounds, which leads to lower selectivity. Pd_2Ga was also successfully synthesized as nanoparticles supported on oxygen-functionalized CNTs by impregnation of the CNTs with palladium nitrate and gallium nitrate hydrate dissolved in ethanol followed by calcination and reduction. Pd_2Ga/CNT was active and highly selective in acetylene and propyne hydrogenation and, just as with the other intermetallic compounds, NAP-XPS did not detect any modification of the sample [97,98]. Later, Pd_2In, which is isostructural to Pd_2Ga, was also successfully prepared [99]. Pd_2In turned out to possess similar structural and catalytic properties as Pd_2Ga and NAP-XPS confirmed its stability under acetylene hydrogenation conditions at 1 mbar.

10.5.2.4 *Self-sustained kinetic oscillations in the catalytic oxidation of propane by NAP-XPS*

Self-sustained reaction-rate oscillations, though not typically the desired operation, are often observed in heterogeneous catalysis; they may appear in a wide pressure range from ultrahigh vacuum up to atmospheric pressure over different types of catalysts including single crystals, foils, and supported catalysts. Kaichev and coworkers studied the reactions of propane with oxygen over a Ni foil by NAP-XPS and observed self-sustained oscillations in propane conversion [100–102]. After cleaning the Ni foil by Ar-ion sputtering, the sample was heated in a C_3H_8:O_2 = 9:1 feed to 680°C and the heating power of the near-IR laser was fixed, that is, the operation was adiabatic, instead of isothermal. Oscillations can be observed in a broad range of oxygen-deficient conditions (between C_3H_8:O_2 = 15:1 to 1:1). As the sample was heated in the reactant mixture up to 680°C, the conversion of propane first decreased from 0.6% to 0.1% and then slowly increased back to ~0.5–0.7%, as first smaller and later larger rate oscillations appeared (Figure 10.22). At the applied condition, the conversion of propane oscillated in the range of 1–23% and propene, CO, CO_2, H_2, and H_2O were the products detected by gas chromatography. That is, at least three catalytic reactions occurred: the partial oxidation of propane to CO and H_2, the total oxidation of propane to CO_2 and H_2O, and the direct dehydrogenation of propane to propene and H_2.

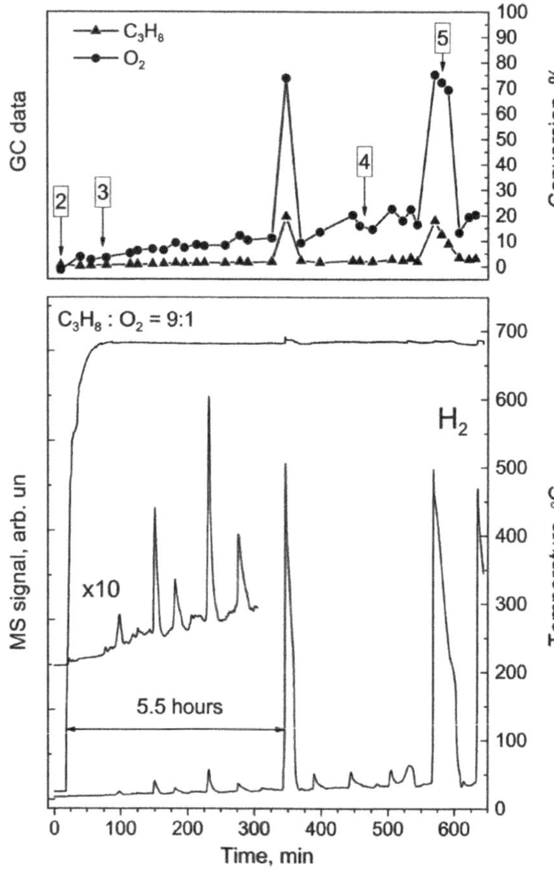

Fig. 10.22. Evolution of self-sustained rate oscillations in the oxidation of propane over nickel: conversions of propane and oxygen measured by GC are presented in the top panel; the evolution of the H_2 signal detected by MS is presented in the low panel. The oxygen flow is 0.5 sccm and the propane flow is 4.5 sccm at a total pressure of 0.5 mbar. Labels 2, 3, 4, and 5 refer to the times when the corresponding core-level spectra depicted in Figure 10.24 were measured. Adapted with permission from Ref. [101]. Copyright (2015) Elsevier Inc.

The rate oscillations were accompanied by synchronous changes in the color of the Ni foil (Figure 10.23). In the first picture, corresponding to the low-active state of the sample, the color of the entire Ni surface was dark gray. In the second picture, a lighter spot

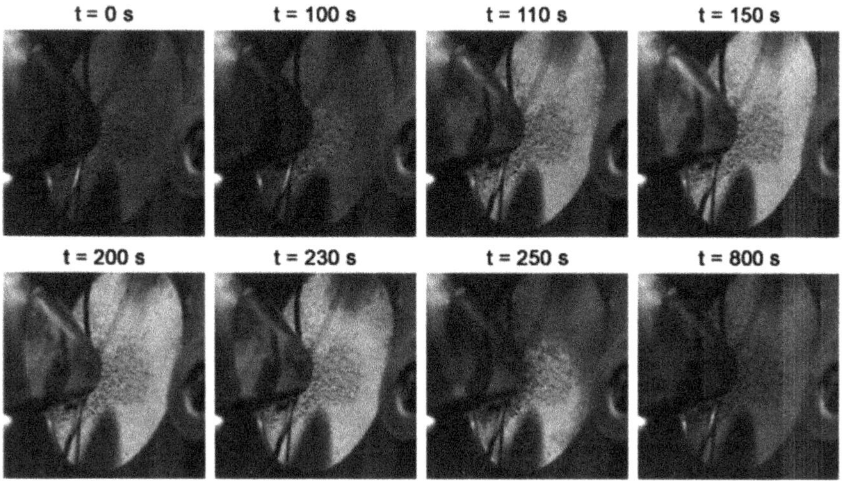

Fig. 10.23. Photos taken during the self-sustained oscillations. The molar ratio of propane to oxygen is 9:1. The circle in the center of each photo, changing its color, is the surface of Ni foil attached to a sample holder by a plate with a round hole of 8 mm diameter. Adapted with permission from Ref. [102]. Copyright (2017) American Physical Society.

appeared at the center of the foil. The boundary of this light spot moved rapidly to the perimeter and, as a result, the entire surface of the Ni foil got the bright metallic color, as shown in the next two pictures. This state was the highly active phase of the sample. In the reverse process, the color changed to dark gray and the boundary of the light spot moved from the perimeter of the foil to its center. The last picture with dark gray color shows again the sample at its low active state. This process repeated itself regularly in time. Thus, the rate oscillation in this case could be observed even with the naked eye as chemical waves propagated through the catalyst surface.

To understand the origin of these oscillations, the chemical state of the Ni surface was followed by NAP-XPS, simultaneously to MS and GC measurements. The Ni $2p_{3/2}$ and O 1s core-level spectra are depicted in Figure 10.24; the label numbers correspond to the time marked in Figure 10.22.

The Ni $2p_{3/2}$ spectrum obtained in vacuum after Ar-ions sputtering was clearly metallic. The corresponding O 1s spectrum

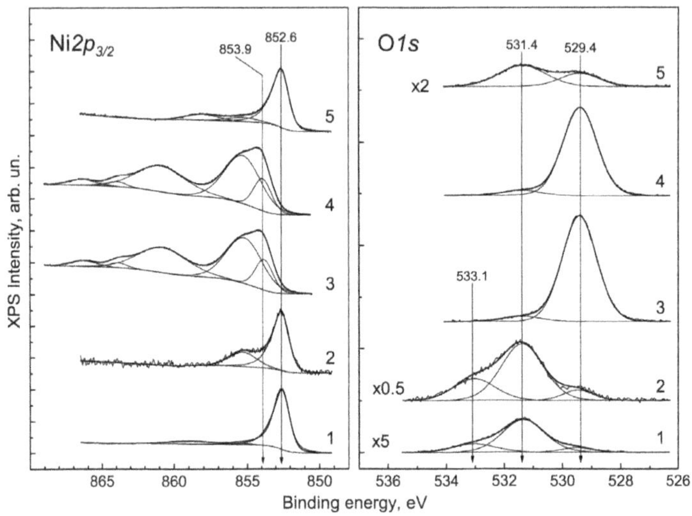

Fig. 10.24. Normalized Ni $2p_{3/2}$ and O 1s core-level spectra obtained in vacuum after the Ar-ions sputtering (1), under the reactant mixture at room temperature (2), after heating to 680°C in the reactant mixture (3), in moments of low (near 2%) conversion of propane (4), and high (near 18%) conversion of propane (5) after the oscillations commence (see Figure 10.22). Adapted with permission from Ref. [101]. Copyright (2015) Elsevier Inc.

contained three smaller peaks at 529.4, 531.4, and 533.1 eV. These could be attributed to chemisorbed oxygen species, some C–O-containing species, and molecular water adsorbed to the surface from the residual gases of the XPS chamber. After introducing the reactant mixture, the intensity of the O 1s spectrum increased strongly; however, the spectrum contained the same three peaks with similar relative intensities. The Ni 2p changed slightly, undergoing some weak oxidation as the surface interacted with the adsorbing gases. The spectra changed completely during heating the sample in the reactant mixture and once the temperature of 680°C was reached, the chemical state of the sample could be best described as defect-rich NiO. The main peak in O 1s is very close to the position of adsorbed O and corresponds to the lattice O in nickel oxide. The small peak at higher binding energy is either some oxygen-carbon functional group or surface OH. The NiO is thick enough that the metallic Ni is not visible in the Ni 2p spectrum. Thus, NiO is characteristic of the low active surface state. As the sample activated and a high activity oscillation commenced, the surface state changed dramatically back to metallic (spectra number 5). Comparing Figures 10.22 and 10.24, metallic Ni catalyzes mostly the partial oxidation of propane, whereas in the low active state, all the above-mentioned reactions occur. Based on the C 1s spectrum, nickel carbide could be ruled out as the active state of the sample. As adsorbed O and NiO have a similar O 1s signature, differentiation of these two in the high active state is not possible.

Nevertheless, the Ni 2p spectrum clearly suggests that, if at all, any oxide contribution is negligible. In a follow-up experiment, the periodic oxidation and reduction of nickel were followed during the regular self-sustained reaction rate oscillations (Figure 10.25). As observed, the rate of reaction is inversely correlated to the amount of oxygen on the surface, completely in line with the observations in

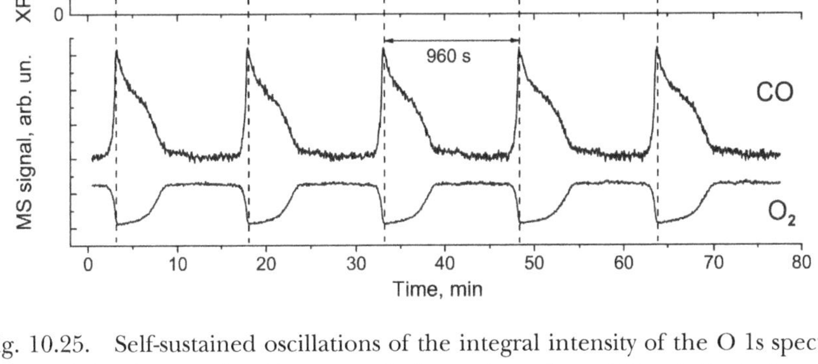

Fig. 10.25. Self-sustained oscillations of the integral intensity of the O 1s spectra observed simultaneously with the oscillations of gas-phase CO and O_2 during the oxidation of propane over a Ni foil. The propane/oxygen ratio is 9/1, the total pressure is 0.5 mbar, and the average temperature is approximately 610°C. Adapted with permission from Ref. [101]. Copyright (2015) Elsevier Inc.

Figure 10.24. *Ex situ* SEM investigation of the sample revealed a strong reconstruction of the catalyst surface affecting not only the surface but also several μm thicknesses. Based on the observations, the authors proposed a switch in the mechanism between Langmuir–Hinshelwood mechanism in the active phase to the Mars–van Krevelen mechanism when the sample is oxidized and less active. The reasons for the spontaneous transition between the highly active and less active states, that is, between metallic Ni and NiO, were not clear, though. Nevertheless, oscillations require a feedback mechanism most probably related to some structural changes of the sample in parallel with the adiabatic operation that enables modulation of the sample temperature either from the reaction heat (note the quite different heat of reaction for dehydrogenation up to total oxidation) or from the oxidation/reduction of the sample itself.

10.5.2.5 *Selective oxidation of alkanes over complex oxides*

Selective oxidation of hydrocarbons is of key importance in producing chemical intermediates and consumer products. It is always a multistep process with the consecutive abstraction of hydrogen and addition of oxygen atoms. The challenge of this reaction type is that the potential reaction network is vast and numerous side products are possible, and even more importantly, the target product is not the thermodynamically favored product (CO_2+H_2O). These stringent requirements can be met in certain cases by complex mixed metal oxides, such as vanadium phosphorous oxides (VPO) and MoVTeNb-oxide. Here, we review some of the NAP-XPS works done on complex metal oxides used in selective oxidation.

VPO are industrially applied in the direct synthesis of maleic anhydride by selective oxidation of n-butane. Teschner *et al.* reviewed the relevant literature as they discussed NAP-XPS data of VPO in n-butane oxidation [103]. These we will not recite; here, we solely look at one recent publication briefly, where Heenemann *et al.* [104] used vanadyl pyrophosphate (VPP) for propane oxidation in the presence and absence of steam. Steam reduced the reaction rate but improved selectivity to the target product, here to acrylic acid.

The authors used NAP-XPS in conjunction with V L-edge absorption measurements and *operando* Microwave Conductivity experiments. VPP is a p-type semiconductor, with electron holes as a major charge carrier. The conductivity of the sample, after an initial decline, stabilized at around 0.015 Sm^{-1}. In wet/dry feed cycles, the conductivity decreased (σ = 0.0146 Sm^{-1}) under wet feed conditions and increased under dry feed conditions (σ = 0.0155 Sm^{-1}). NAP-XPS experiments in the dry feed indicated an initial average V oxidation state of 4.1/4.3 depending on the used photoelectron kinetic energy (750/150 eV, respectively), that is, above the expected V^{4+} in stoichiometric VPP. Also, the P/V ratio was higher than 1, though the surface sensitive data (KE: 150 eV) showed lower values than the more bulk sensitive one. Upon changing to wet feed, steam irreversibly depleted phosphorus in the topmost surface layer of VPP but only in the first treatment with steam. Simultaneously, the V^{5+} population increased at the surface to yield an average oxidation state of $V^{4.4+}$. In the less surface sensitive data, changes were at the limit of detection. After the first treatment in steam, the concentration of V^{5+} surface species remained constant in further dry/wet cycles. The surface concentration of phosphorus in these cycles kept constant, as well. Hydrolysis of VPO upon high steam-containing feed is well known, though the influence on the actual P/V ratio can depend on various factors. For example, Richter *et al.* [105] have found that exposure to water vapor resulted in an increase in the V oxidation state and a rise in the P/V ratio of the surface layers. During water vapor treatment, a migration and a loss of phosphorus were found using thermogravimetric experiments, which is also what one finds under n-butane oxidation in the technical system. The enrichment in phosphorus is driven by the gain in energy of the system due to the hydration and hydrolysis of P–O–P or P–O–V bonds. They also found that the P/V ratio decreased with increasing temperature of water vapor treatment, that is, with decreasing amounts of adsorbed water molecules or with more efficient P_xO_y removal by the enhanced kinetics. That is, water drives the surface hydrolysis and P surface enrichment, but in parallel when the conditions allow, P will be lost, and eventually the surface may end up P deficient, just as

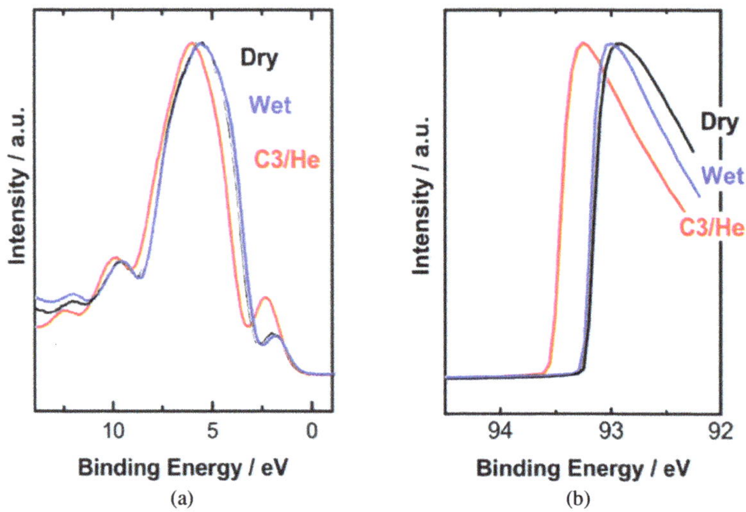

Fig. 10.26. Valence band spectra (normalized to their maximum) recorded at E_{hv} of 100 eV with constant photon flux (a) and secondary electron cutoff (b) of VPP in different gas mixtures: dry and wet propane oxidation feed and in propane/He. Adapted with permission from Ref. [104]. Copyright (2017) American Chemical Society.

found by Heenemann *et al.* During the NAP-XPS experiments, the valence band position and work function of the VPP sample were also assessed by recording the valence band and the secondary electron cutoff (Figure 10.26). The valence band edge position did not shift in any detectable manner by adding steam. (In contrast, a clear shift of the valence band to higher binding energies was obtained in a reducing feed.) On the other hand, the work function decreased in wet feeds and increased in dry feeds, by oscillating around 6.82 eV. The upward band bending was very similar under dry and wet conditions. But, when adding water to the reactive propane feed, the change of the work function without changes in the valence band position can be caused by a modified surface dipole. The authors concluded that steam did not change the space charge layer in VPP significantly and thus the hole concentration was not affected by steam addition. In turn, the reduced conductivity under wet conditions was assigned to reduced carrier mobility.

The mixed metal MoVTeNb-oxide with the orthorhombic M1 phase has attracted much interest as a promising catalyst for the (amm)-oxidation of propane. Recently, the tellurium- and niobium-free MoV-oxide in the same M1 phase was shown to be able to produce acrylic acid as well, though with much lower selectivity compared to MoVTeNb-oxide. Trunschke *et al.* [106] investigated the surface of MoV-oxide by NAP-XPS and compared these to their previously published results on MoVTeNb-oxide to understand the clues of the lower acrylic acid selectivity. Under working conditions, the MoV-oxide surface is reconstructed and differs from the bulk in terms of composition and oxidation state of vanadium. They observed a substantial enrichment of V^{5+} species at the surface, and their accumulation was facilitated by the wet reaction feed (Figure 10.27).

Just as we have seen above using VPP, acrylic acid production was strongly enhanced in wet feed over MoV-oxide, but based on the temporal evolution, the increase of the V^{5+} species seemed to

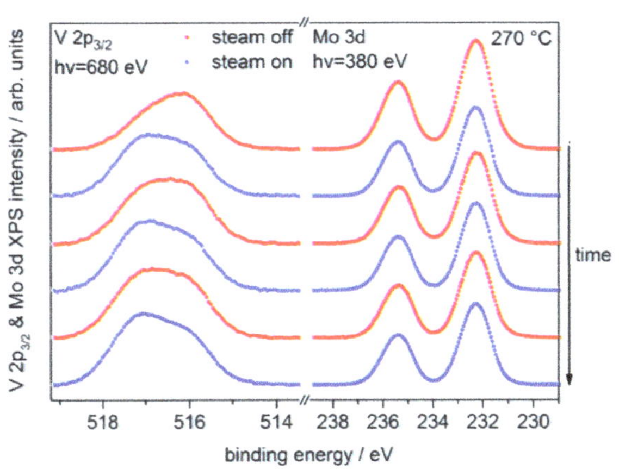

Fig. 10.27. Long-term changes (0–1200 min time on stream) in the profile of the V $2p_{3/2}$ (left) and the Mo 3d signal (right) for MoV-oxide during switching from dry (red spectra) to wet (blue spectra) feed. Reaction conditions: $O_2/C_3H_8/He/H_2O$ = 2/0.5/1.7/0 sccm (steam off) and 2/0.5/0/1.7 sccm (steam on), at T = 270°C. Adapted with permission from Ref. [106]. Copyright (2017) American Chemical Society.

correlate with the decreasing acrylic acid production under NAP-XPS conditions. Although the spectral resolution was not enough to differentiate two V^{5+} species, based on various arguments, the authors identify two electronically different V^{5+} species, assigned to M1 lattice and extra-framework vanadium species (Figure 10.28). The enrichment of extra-framework vanadium species was attributed to the migration of vanadium from M12 sites in the hexagonal channel to the surface under reaction conditions. This process was enhanced when the oxidation was carried out in the presence of steam. Accumulation of vanadium in the highest oxidation state is thought to be detrimental as it would induce a further reaction of the intermediate or consecutive combustion of the desired acrylic acid. On the other hand, the selective MoVTeNb-oxide features dynamic site isolation due to segregation of Te [107]. Te in MoVTeNb-oxide occupies the hexagonal channel, and when segregating to the surface, it serves as an uncritical diluent of surface V^{5+}

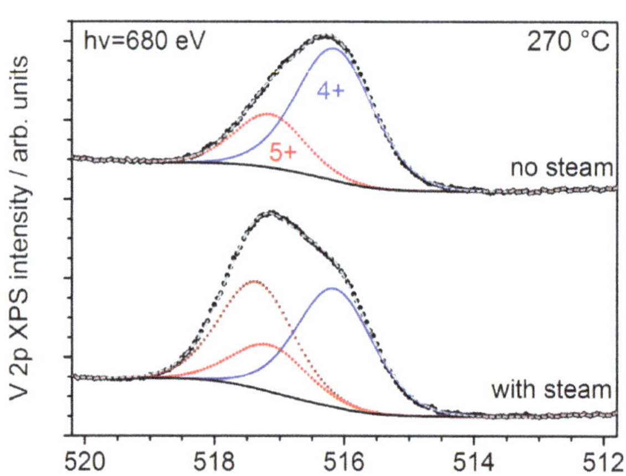

Fig. 10.28. Least squares fit of V $2p_{3/2}$ XP core-level spectra of MoV-oxide without and with steam in the propane oxidation feed. Black dots represent the experimental data, and blue, red, and brown dots represent vanadium 4+, 5+ (I), and 5+ (II) species, respectively. Gray dots are the sum of fit profiles and a Shirley background. Adapted with permission from Ref. [106]. Copyright (2017) American Chemical Society.

species that are proposed to be involved in propane activation and oxygen insertion. In MoVTeNb-oxide, V^{5+} accounted for less than 15% of the surface cations, whereas in MoV-oxide, V^{5+} increased up to 30%.

Later, Trunschke and co-workers compared the MoV-oxide in the oxidation of ethane, propane, and n-butane by NAP-XPS and *operando* Microwave Conductivity experiments [108]. In contrast to the n-type MoVTeNb-oxide, MoV-oxide showed p-type semiconducting behavior. While the conductivity of the sample depended sensitively on the surrounding atmosphere, not only on alkane chain lengths but also on reactant conversion levels, the different alkanes exhibited practically no influence on the surface composition as indicated by the NAP-XPS data. The work function of MoV-oxide was 6.68 ± 0.02 eV in the dry alkane oxidation gas feeds. The work function, however, decreased by ~200 meV with small changes in the valence band onset upon wet propane oxidation feed, suggesting a change in band bending. Similarly, the impact of steam on the electronic structure of MoVTeNb-oxide was studied by the same group using the same methodology [109]. Steam increased both the conversion of propane and the selectivity to the acrylic acid. In line with the above observations, they concluded that the increased catalytic performance was associated with a decreased conductivity, a modification of the surface chemical and electronic structure with an enrichment with V and Te compared to Mo^{6+}, an enrichment of covalently bonded V^{5+} species (though to a lesser degree as with MoV-oxide), a decreased work function and hence polarity of the surface, and a modified valence band structure.

Perovskites are known combustion catalysts, but some of them might be applicable for less demanding selective oxidation reactions, such as the oxidative dehydrogenation of alkanes to alkenes. Based on the above segregation processes, Koch *et al.* [110] hypothesized that the selectivity of $AMnO_3$ (A = La, Sm) in the oxidative dehydrogenation of propane to propene can be changed by surface segregation of manganese oxide species. The authors synthesized perovskite catalysts via combustion and hydrothermal synthesis to explore the potential of this total oxidation catalyst in selective oxidation of propane by varying the surface composition. The surface-near region was investigated by NAP-XPS and NEXAFS at the Mn L-edge in the same machine. The catalysts (especially $SmMnO_3$-c, prepared with combustion synthesis) exhibited high activity with an onset temperature of around 200°C. The catalysts with combustion synthesis, $SmMnO_3$-c and $LaMnO_3$-c, showed approximately 20% propene selectivity in a dry (propane/O_2/N_2/H_2O = 5/10/85/0) feed, but when a wet feed was used (propane/O_2/N_2/H_2O = 5/10/45/40), the selectivity increased to 40–50%, with a factor x3-5 drop of reaction rate and an increase of apparent activation energy by 23–29 kJ mol^{-1}.

The $SmMnO_3$-c catalyst exhibited a slightly higher Mn concentration on the surface of the fresh catalysts. La-containing catalysts, in particular, $LaMnO_3$-h prepared by hydrothermal synthesis, was lean of Mn at the surface. No significant changes were observed in the Mn/(Mn+A) surface ratio under the different reaction conditions, but the distribution of surface oxygen species as well as the oxidation state of manganese underwent considerable changes. The nature of the oxygen surface species was analyzed by fitting the O 1s core-level spectra. Differentiation and assignment of the components, as shown in Figure 10.29, were derived from the development of the O 1s spectrum during variation of temperature, feed composition, and excitation energy. During heating to 270°C in O_2, most of the adsorbates were removed including the major part of the surface carbonate, the process of which was identical for both $LaMnO_3$-c and $SmMnO_3$-c. Thus, the stability of adsorbates and carbonates on the surface was independent of the A element. The reaction feed-induced structural changes in the near surface were also similar for both catalysts. The O 1s spectra changed upon the introduction of the propane oxidation feed and the relative concentration of $O_{lattice}$ was reduced, while the contributions of O_{defect} and $O_{surface}$ increased. Hydroxyl and $O_{surface}$ accounted for about 10–20% of the detected oxygen species. When 40% of H_2O was added to the feed, a further decrease in $O_{lattice}$ was observed indicating structural changes in the near-surface region. Remarkably, not only OH but also $O_{surface}$ increased. As mentioned above, these surface changes in wet feed

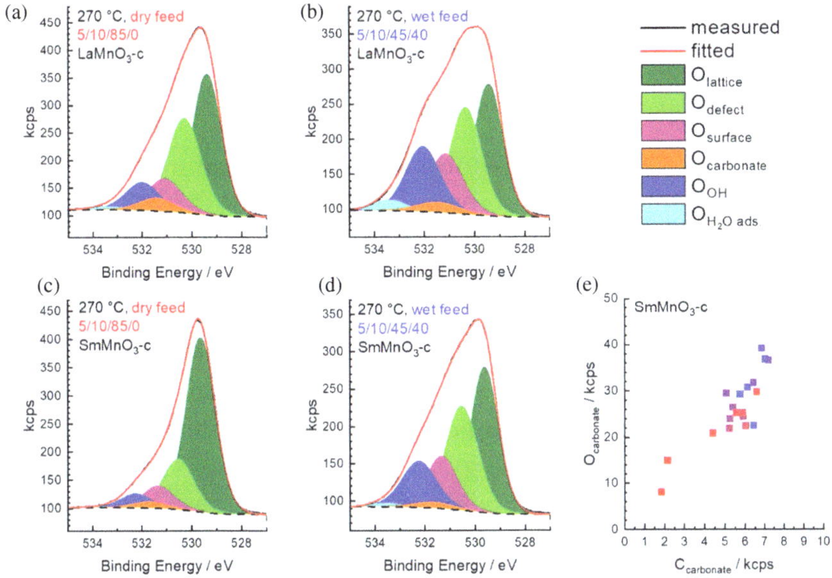

Fig. 10.29. Measured O 1s spectra (black) collected at 270°C for $LaMnO_3$-c in a dry (a) and wet propane oxidation feed (b) as well as $SmMnO_3$-c in a dry (c) and wet (d) feed (feed composition indicated in the Figures as propane/O_2/He/H_2O ratios) and the corresponding fits (red) using seven single contributions (colored peaks with the assignment provided in the legend). The broken line represents the background. (e) Correlation of the intensity of the O 1s peak due to $O_{carbonate}$ as a function of the C 1s peak of $C_{carbonate}$ during heating from room temperature to 270°C (the temperature is illustrated by using a color code from blue (room temperature) to red (270°C)) in 20% O_2. Adapted with permission from Ref. [110]. Copyright (2020) American Chemical Society.

were accompanied by an increase in selectivity to propene. In addition to the changes of the O 1s spectra, *in situ* NEXAFS difference spectra at the Mn L-edge indicated the formation of Mn^{2+} species upon switching to the wet feed. The work concluded that in presence of a two-dimensional MnO_x surface phase, the addition of steam to the feed resulted in an increase in selectivity to the partial oxidation product propene. The selectivity increase was associated with the presence of Mn in a low oxidation state (2+), an increased concentration of hydroxyl groups, and a higher abundance of adsorbed activated oxygen surface species.

10.5.3 *Selected catalysts for more than one application*

10.5.3.1 *Cobalt oxides and their reactivity in CO oxidation with or without the presence of hydrogen*

In 2009, Xie *et al.* [111] reported that Co_3O_4 nanorods, predominately exposing (110) planes, and enriched in Co^{3+} cations possess very high activity for low-temperature CO oxidation. This fueled NAP-XPS investigations of cobalt oxides in CO oxidation and in the preferential oxidation of CO in hydrogen (PROX) [112–117]. As we will see, some of these reports are reasonably consistent, but some of the claims are clearly contradictory.

Lukashuk *et al.* [115] studied CO oxidation over Co_3O_4 utilizing a multi *operando* approach, combining NAP-XPS, FTIR, and XRD. Before any of the experiments, the sample was treated in O_2 at 400°C to remove carbon and carbonate contaminations. Additionally, they also used switching experiments between CO and O_2. Under NAP-XPS, CO/O_2 switching (each 0.15 mbar for ~50 min) was carried out at RT, 100°C, 150°C, and 200°C, simultaneously recording surface-sensitive Co 2p and C 1s spectra (KE = 200 eV; Figure 10.30) and the signature of CO_2 in mass spectrometry. A significant increase in CO_2 evolution at 200°C and continuous CO_2 production for ~30 min upon switching to CO indicated that not only adsorbed and surface lattice oxygen might take part in CO_2 production but also oxygen from the bulk. Upon switching from O_2 to CO, the Co_3O_4 surface oxidation state did not change at RT but some Co reduction was observed at higher temperatures, especially at 200°C. Nevertheless, the dominant surface phase was still the spinel Co_3O_4. The amount of CO_2 produced at 200°C during steady-state CO oxidation was 20–40 times higher than that in the switching experiment. Without gas-phase oxygen, CO reduced part of the surface, while switching to O_2 re-oxidizes it. Concerning the C 1s region, elementary carbon increased with temperature in CO, whereas switching to O_2 decreased the carbon signal. Thus, a (minor) reaction pathway toward CO_2 may proceed via CO dissociation followed by carbon oxidation to CO_2. In contrast, the carbonate

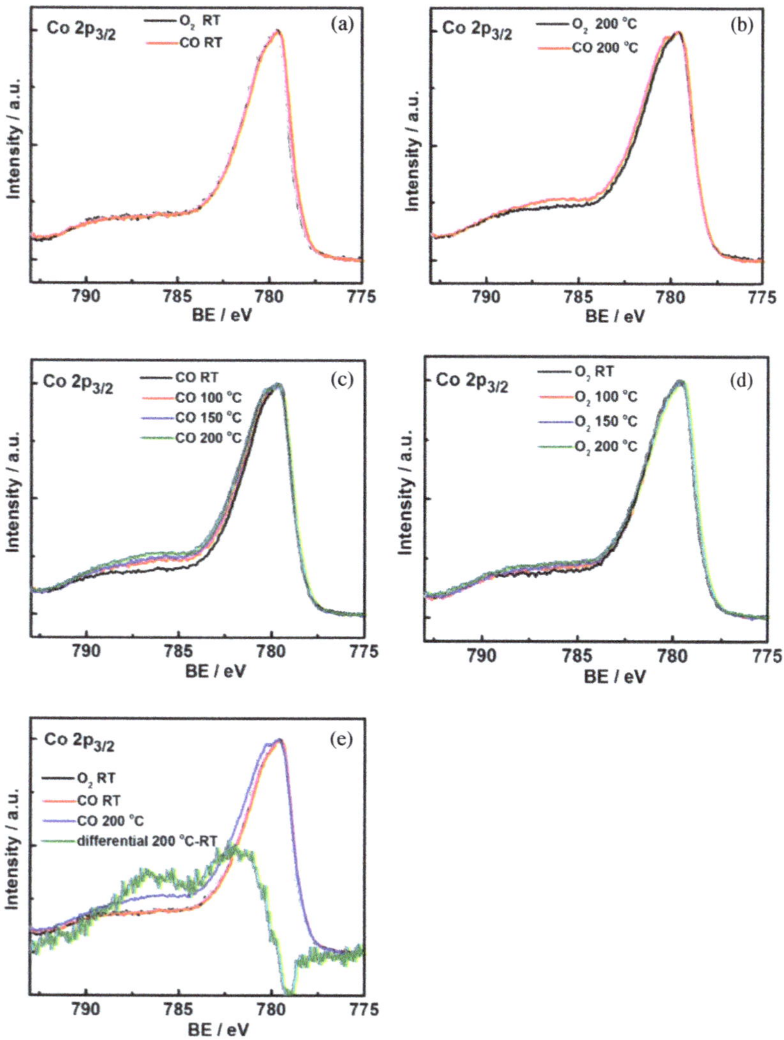

Fig. 10.30. *Operando* NAP-XPS on Co_3O_4 during CO/O_2 switching (0.15 mbar CO vs 0.15 mbar O_2), Co $2p_{3/2}$ region ($h\nu = 1015$ eV; KE = 200 eV): comparison in CO and O_2 at (a) RT and (b) 200°C; the Co $2p_{3/2}$ region at various temperatures in (c) CO and (d) O_2; (e) difference spectrum (Co $2p_{3/2}$ 200°C in CO; Co $2p_{3/2}$ RT in O_2). Adapted with permission from Ref. [115]. Copyright (2018) American Chemical Society.

signal decreased less at higher temperatures, indicating that carbonate decomposition was more difficult than carbon oxidation. Also switching experiments in FTIR suggested that surface carbonates were rather stable spectators, with their intensity decreasing rather slowly and mostly at high temperatures, which precludes carbonates from being reaction intermediates of CO oxidation. Under steady-state CO oxidation, Co 2p did not indicate any significant surface reduction, but a close inspection with difference spectra suggested some minor reduction at around 150°C and above. This indicates a rapid and dynamic reduction/re-oxidation cycling of the cobalt oxide surface in the $CO+O_2$ mixture under steady state, the equilibrium being shifted to the side of re-oxidation. The absolute amount of elementary carbon, carbonates, and weakly adsorbed CO increased from RT to 100°C but decreased upon further heating. Since the C concentration was much higher in pure CO, the oxygen vacancies are likely the reaction sites of CO dissociation, or the equilibrium coverage of C is effectively moderated by the chemisorbed oxygen from the gas phase.

Taken these together, *in situ* NAP-XPS and FTIR spectroscopy were in line with a Mars–van Krevelen-type CO oxidation mechanism, via continuous redox cycles of the Co_3O_4 surface by CO and O_2. That is, CO adsorbed to cobalt cations reacted with lattice surface oxygen, and the created oxygen vacancies were replenished by O_2 from the gas phase, leaving an active O species on the surface. This species can recombine with another CO, though this can be seen as a classical Langmuir–Hinshelwood mechanism. Alternatively, O_{ads} can find and heal another vacancy site. Also, adsorbed CO (likely near an oxygen vacancy) may undergo dissociation, filling an oxygen vacancy and depositing elementary carbon on the surface, which can be re-oxidized by O_2. This process seems to occur already at RT. Overall, CO dissociation and carbon re-oxidation are likely additional pathways of CO oxidation on Co_3O_4. The authors studied the effect of reductive pre-treatment, too. When the sample was reduced in H_2 at 300 or 400°C, activity was lower and CO conversion started at a higher temperature.

As opposed to Lukashuk, Tang *et al.* [113] reduced the Co_3O_4 nanoparticles in H_2 at 250°C, which gave rise to CoO formation, though only in the surface-near region. TEM indicated that the bulk of the nanoparticles was still in the Co_3O_4 state. Then, the surface state was followed by NAP-XPS in CO oxidation at various temperatures under an oxygen-rich (0.4 mbar CO + 1 mbar O_2) feed. The Co_3O_4@CoO core–shell structure was stable during CO oxidation in the temperature range of 60–140°C. At 140°C and above, the CoO surface phase was gradually oxidized to Co_3O_4. The gas-phase evolution, as sampled by mass spectrometry, indicated that the sample was certainly active at 100°C (probably already at 60°C) and the CO conversion reached ~60% at 140°C. The estimated CO conversion was ~80% at 220°C. Due to the small increase in CO conversion from 140°C to 220°C, we anticipate that the reaction rate was strongly limited in this temperature range. If we consider our earlier discussion on CO oxidation in the diffusion-limited regime, all CO molecules that could diffuse through the CO-depleted zone in front of the sample react immediately on the surface and produce CO_2. That is, the surface essentially did not experience the chemical potential of CO in this case and sensed only O_2 and CO_2. Since CoO can be re-oxidized in O_2 at and above 140°C, surface oxidation to Co_3O_4 is not surprising. In fact, it is the expected process. In a kinetic investigation, limiting the CO conversion below 20%, it was found that the apparent activation energy was 49.3 kJ mol^{-1} in the temperature range of 80–140°C and 36.9 kJ mol^{-1} in the temperature range of 160–200°C. The authors thus argued that the lower apparent activation energy found in the higher T range was linked to the re-oxidized surface state and this should be beneficial.

Lukashuk and colleagues investigated Co_3O_4 in PROX reaction, too [116]. They observed approximately the same reaction temperature onset (60–70°C), as without H_2 in the feed, with peak CO_2 formation at 175°C. Water formation was negligible below 130–150°C, but increased at higher T and was essentially the only product from O_2 above 275°C. Above this temperature, CO/CO_2 transforms into methane. TPR and *in situ* XANES at the Co K-edge in H_2 indicated partial reduction to CoO at ~250°C, in agreement with Tang *et al.* [113]. At and above 300°C, reduction to metallic Co commenced. Under PROX conditions, the bulk was investigated by *in situ* XANES and the surface by NAP-XPS. As opposed to pure H_2, XANES did not indicate any reduction at 250°C and below. At 300°C, besides Co_3O_4, both CoO and metallic Co were found by XANES but not by NAP-XPS. In fact, the authors claimed no surface reduction of Co_3O_4 under PROX, at all, though a closer inspection of their Co 2p spectra in comparison to the spectrum of the as-prepared Co_3O_4 in O_2 at RT indicates some minor differences pointing to a tiny extra Co^{2+} contribution. The mass spectrometry data recorded during NAP-XPS was in line with the reactor data, except for the lack of CH_4 formation above 250°C. The reduced rate enhancement of CO_2 production once significant amounts of water are formed is in line with the water-mediated deactivation of Co_3O_4 nanorods in CO oxidation proposed by Jain *et al.* [114] based on their NAP-XPS study. The lack of CH_4 production at low (mbar) pressure can be linked to the absence of metallic Co, not observed under PROX in NAP-XPS conditions. Similar carbonaceous adsorbates, elemental carbon, carbonates, and minor adsorbed CO (or C-OH) were found as in CO oxidation. A similar Mars–van Krevelen-type mechanism was suggested, as we detailed above, with the difference that once O_2 conversion is complete and the temperature is high enough to reduce the catalyst to metallic Co, methanation kicks in.

Nguyen *et al.* [112] took Co_3O_4 nanorods and used these after various reductive pre-treatments in PROX. They find that H_2 pre-treatment at 200°C increased but at 300°C decreased the PROX activity of the sample as compared to the untreated as-synthesized Co_3O_4. The 200°C treated sample was capable of 100% CO conversion at around 120°C. Interestingly, the 300°C treated sample was highly unselective below ~140°C. From NAP-XPS experiments, the authors could show that the sample reduced at 300°C was CoO upon the pre-treatment and stayed CoO under PROX conditions (at RT and 140°C). Both untreated and 200°C samples were predominantly Co_3O_4 but the mild reductive treatment induced the formation of surface defects. Hence, they concluded that a defective Co_3O_4 surface was key for an efficient Co-based PROX catalyst.

Zafeiratos and coworkers [117] based on the presumption that reduced sites or oxygen vacancies are beneficial for the PROX reactivity of cobalt oxides compared a Co_3O_4 calcined at 350°C in synthetic air ($PreO_2$-Co) with the same sample pre-treated in H_2 at 400°C ($PreH_2$-Co). They find that the H_2 pre-treated sample was more active, though, to be fair, their calcined sample was very inactive below 200°C. We must note here that both samples were really low surface area samples; BET surface area was around 1 $m^2\ g^{-1}$. In line with the observation of Lukashuk [116], $PreO_2$-Co in PROX was, based on the Co 2p spectra, in the form of Co_3O_4 from RT to 300°C. Their Co 2p spectra of $PreH_2$-Co are depicted in Figure 10.31. At 50°C, the Co 2p spectrum contains the contribution of both

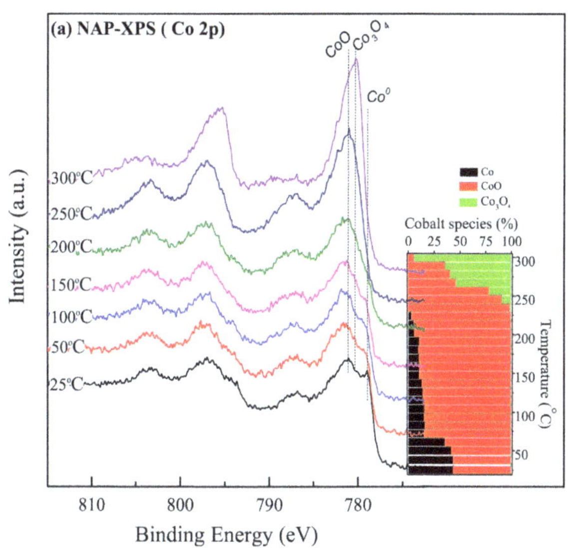

Fig. 10.31. NAP-XPS spectra of Co 2p on pre-reduced Co_3O_4 in PROX at various temperatures. The inset shows the distribution of cobalt phases obtained by Co 2p peak deconvolution. Reaction conditions: 1% CO, 2% O_2, and 97% H_2 at 0.5 mbar with a temperature ramping rate of 5°C min^{-1}; prior to the reaction, cobalt was reduced in 1 mbar of H_2 at 400°C for 30 min. Adapted with permission from Ref. [117]. Copyright (2019) American Chemical Society.

metallic Co at 778.8 eV and Co^{2+} is also clearly visible by the characteristic peak around 781 eV and its intense satellite at +6 eV higher binding energy. As the reaction temperature was increased, the metallic Co feature gradually disappeared and CoO dominated the spectrum up to 250°C.

At 300°C, the Co 2p spectrum of $PreH_2$-Co resembles that of $PreO_2$-Co, clearly indicating further oxidation to Co_3O_4. A peak deconvolution procedure using Co 2p profiles recorded over reference Co, CoO, and Co_3O_4 samples was used to assess quantitatively the evolution of oxidation states as a function of temperature. As shown in the inset, in the temperature window of 75–250°C, Co^{2+} was the dominant oxidation state of $PreH_2$-Co, while metallic Co and Co_3O_4 appeared at low and high reaction temperatures, respectively. The depth distribution assessed by non-destructive depth profiling suggested that metallic cobalt was buried under a CoO surface layer in a CoO-shell/Co-core-like arrangement. The C 1s spectra indicated very low carbon and carbonate coverage for $PreH_2$-Co, while the typical three component C 1s spectra were observed on $PreO_2$-Co, as noted above, with high carbon and carbonate coverage. Additional DFT calculations suggested that a possible origin of the activity differences was due to blocking of surface sites by carbonates which should be favored on Co_3O_4(001) but not on the CoO (001) surfaces.

As this section on cobalt oxides indicates, some of these observations apparently contradict each other. These disagreements can be reconciled if we remind ourselves that the microstructure, surface orientation, and the as-synthesized defect density of the samples can all play an important part in both the catalytic properties and the solid-state dynamics. As all these studies were carried out with different catalysts, it gets clear that the bulk phase identity of Co_3O_4 in their as-synthesized form tells little about the potential complexity of the system and the catalytic outcome.

10.5.3.2 *Pt/CeO$_2$ in water–gas shift and PROX reactions*

The water–gas shift (WGS) followed by the PROX reactions are important processes for the production of high-purity H_2 for fuel cell applications. Pt/ceria is capable of catalyzing both of these reactions at considerably low temperatures (WGS: 200–300°C; PROX: 50–120°C), due to a cooperative mechanism. In the most likely scenario, ceria provides oxidation agents to Pt nanoparticles to oxidize CO to CO_2. We studied the very same Pt/CeO_2 catalyst in both reactions. Here, we start with unpublished WGS data followed by our results on PROX.

In a fuel processor for fuel cell applications, the WGS feed contains not only CO and (over-stoichiometric) H_2O but also significant CO_2 concentration and high excess of H_2. Under these conditions, the WGS rate is close to zero order for CO, half order for H_2O (stronger for low H_2O concentrations and less for high H_2O excess), and inhibited by both CO_2 and especially H_2 [118,119]. Under model feed conditions, without the addition of products, both H_2O and CO are close to first order at low concentrations approaching zero order in excess of either reactant. Two major types of mechanisms were proposed. According to the redox mechanism, CO adsorbs on metal sites and reacts with oxygen, supplied by ceria to the metal–ceria interface, while water dissociation takes place on an oxygen-vacancy site of ceria to form hydrogen and atomic oxygen, which re-oxidizes ceria. The associative mechanism involves interaction between CO molecules and surface hydroxyl groups and the formation of surface intermediates (formate or carboxyl species were proposed), which further decompose to the reaction products, CO_2 and H_2.

NAP-XPS experiments using a pre-oxidized 5% Pt/CeO_2 under model feed conditions show (Figure 10.32(a)) that the surface of ceria was reduced. Here, we studied the capability of water (or its lack thereof) to re-oxidize the ceria surface, without the presence of large H_2 excess. 1035 eV excitation energy was used and hence this was a surface-sensitive measurement: KE: ~150 eV; IMFP: ~6 Å. The Ce 3d spectrum (only a section of it is depicted) indicated a mixture of Ce^{3+} and Ce^{4+}. The peak fitting provided a Ce^{3+}/Ce^{4+} ratio of 0.84 ±0.04 and, in the range of the estimated uncertainty, independent of the WGS feed composition. For orientation, hydrogen treatment in 0.48 mbar at 573 K induced a stronger reduction with a Ce^{3+}/Ce^{4+} ratio of 1.3. Importantly, no re-oxidation of Ce^{3+} was observed when

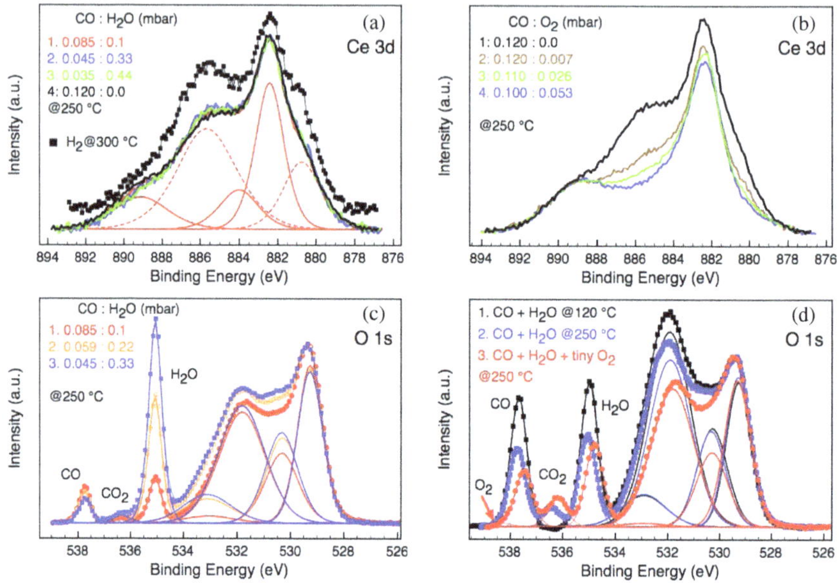

Fig. 10.32. NAP-XPS Ce 3d (a,b) and O 1s (c,d) spectra of 5% Pt/CeO_2 in the order of experiments and at the conditions indicated in the legends. In (a), a Ce 3d spectrum in 0.48 mbar H_2 at 300°C and in (d) an O1s spectrum in 0.2 mbar CO + 0.23 mbar H_2O at 120°C are overlayed. Otherwise, all other spectra are at 250°C. In (a), the red spectrum is curve fitted with five components, the solid lines correspond to Ce^{4+}, and the dashed lines to Ce^{3+}. Spectra are intensity scaled. Note that we adopted the view that the line shape of the *v* component (882.4 eV) is asymmetric and thus fitted with an additional component at 884.0 eV [120–123].

the water/CO ratio of the feed increased. Rather, fitting indicated a minor further reduction, but only within the estimated uncertainty, and hence we consider the reduction degree to be constant. Even pure CO at 523 K did not reduce the surface of ceria further (Ce^{3+}/Ce^{4+}: 0.8). We interpret the observed Ce 3d spectra under WGS as the ceria support being principally reduced at the topmost layer. Unlike Ce 3d, O 1s of ceria, however, varied somewhat when increasing the feed water content (Figure 10.32(c)). In a near-stoichiometric WGS feed, a conservative fit provided four catalyst components and three gas-phase peaks: 529.25 eV: lattice O, 530.3–530.4 eV surface oxygenates, 531.8 eV: surface OH, 533.0–533.1 eV surface water, as well as H_2O, CO_2, and CO gas-phase contributions. The intensity of the lattice O and OH groups remained almost constant, whereas the amount of oxygenates (530.3 eV) and especially of adsorbed water (533 eV) increased as the feed H_2O/CO ratio was increased up to six-fold. Hence, the amount of adsorbed water can be increased without re-oxidizing the ceria surface. As for the platinum state (Figure 10.33), interpretation of Pt 4f is not unambiguous. The supported nanoparticulate nature of Pt (in combination with the applied experimental settings of the analyzer and beamline) did not allow us to resolve well the various contributions. Thus, we restrict ourselves to a qualitative description. Besides the metallic bulk Pt 4f at 71 eV, three other peaks were found. The distortion of the otherwise asymmetric line shape of metallic Pt suggests a peak at ~+0.4 eV to the main peak that can be related to bridged-bonded CO. Linear CO appears at ~72 eV and an oxidized Pt (probably due to small Pt clusters or interface sites to ceria) at 72.7–72.8 eV. The low concentration of linearly bonded CO is consistent with Pt 4f spectra in PROX at the same temperature and low pressure but at odds with PROX data at 85°C (see later) or *in situ* DRIFTS data at higher pressure. This we attribute to the overall low partial pressure of CO.

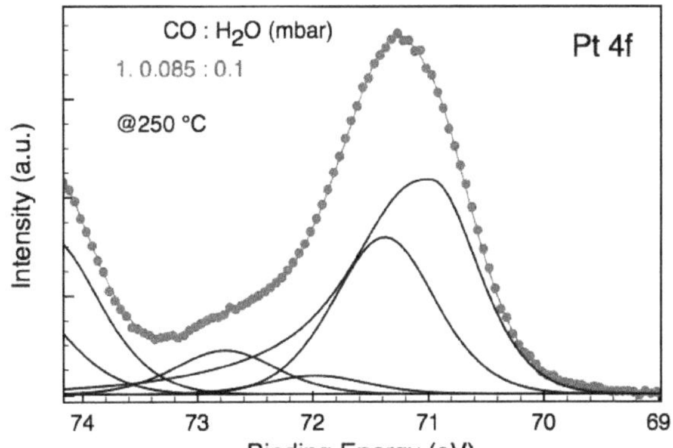

Fig. 10.33. NAP-XPS Pt 4f spectrum of 5% Pt/CeO_2 in a 0.085 mbar CO + 0.1 mbar H_2O reaction mixture at 250°C.

As opposed to water, oxygen readily re-oxidized the ceria (Figure 10.32(b)): even when minute quantities of O_2 were co-dosed to CO (6% of O_2 in CO), the Ce^{3+}/Ce^{4+} dropped to 0.5, and in a CO:O_2 feed of ~1:0.5, the Ce^{3+}/Ce^{4+} ratio was below 0.3. This suggests that pure water is insufficient to induce any measurable re-oxidation of ceria under WGS conditions, whereas O_2, even in the smallest quantities, can. Additionally, as we will see later, the ceria of our Pt/ceria catalyst at the same temperature, however in a PROX reaction feed, was much more oxidized, and thus we can exclude the influence of X-rays.

In a separate experiment, the effect of oxygen addition was investigated in the O 1s (Figure 10.32(d)). In this case, a CO:H_2O feed of ~0.2:0.23 mbar was introduced (first H_2O followed by CO) at 50°C and the sample was heated to 120°C where we took the first spectrum. The concentration of surface OH groups is significantly higher in this experiment, which most likely relates to the lower temperature, the dosing sequence, and the pre-history of the sample. As the sample was heated to 250°C, approximately 10% of the dosed CO was converted to CO_2, as deduced from the gas-phase peak integrals. Apart from an approximately 10% reduction of OH intensity, the O 1s spectrum remains quite similar. Upon addition of 7% O_2, most of which was consumed to raise the CO conversion to ~20%, both the oxygenate and OH peaks reduced in intensity, whereas chemisorbed water was largely eliminated. The gas-phase peaks shift to lower apparent binding energy, signaling a change in the sample work function due to the re-oxidation of the ceria surface, as we have seen in the Ce 3d.

The zeroth reaction order for water in water-rich model feed suggests that the surface is not deficient in an oxidizing agent. The outer surface of ceria is identically reduced both in pure CO and in the presence of H_2O-rich feed; in other words, ceria re-oxidation (detectable by XPS) does not occur under our reaction conditions. WGS, however, clearly took place, as indicated by the CO_2 signal in O 1s (and also the mass spectrometry analysis, not shown). Since the classical redox mechanism was considered to occur in the sense that H_2O adsorption should take place directly on reduced ceria, that is, not in an indirect fashion via the first adsorption on Pt, re-oxidation of ceria should not be restricted to ceria sites adjacent to Pt. Hence, the whole ceria surface is expected to react to the changing feed composition and re-equilibrate in a more oxidized state. Therefore, with the classical redox mechanism, one should expect to see a certain degree of surface re-oxidation of ceria when adding and increasing H_2O into CO, and this should correlate with the WGS activity. This is however not what we observed at 250°C under our model reaction conditions. We note here that oxygen-deficient ceria is known to re-oxidize when contacted with water vapor [124] but in that case at a much higher temperature. Also, note that Heyden and coworkers [125,126] analyzed various reaction pathways by DFT calculations (including one with water activation on Pt next to an O vacancy) and proposed an associative carboxyl pathway with redox regeneration as the most likely scenario. This mechanism involves a carbonyl bound to Pt at the interface and an oxygen vacancy formation next to it at the ceria surface in the step of CO_2 desorption. Heyden and coworkers [125] noted that only a few interface atoms are supposed to be active due to strong adsorption of CO. Our unpublished *in situ* DRIFTS data indicate that CO bounds very strongly at the interface, and thus, we think it is unlikely a reaction intermediate.

Since the ceria surface in H_2 at 300°C was more strongly reduced, we cannot rule out that with real reformate feeds, the ceria experiences a stronger reduction, and then water may moderate the over-reduction of ceria. Nevertheless, the species thus produced will unlikely provide the active oxygen for the reaction, as it did not in the model H_2O+CO feed. We also point out that the reaction rate is higher without co-dosing H_2 (and CO_2). The stronger reduction in H_2-containing feed however fits Pt/CeO_2 deactivation under real WGS conditions: one of the proposed mechanisms is that the high H_2 content creates an environment that facilitates the irreversible over-reduction of the support, unfavorable for WGS [127].

The CO content of the reformer feed after WGS is still too high for a proton-exchange membrane fuel cell, but CO can be removed by the PROX reaction. We investigated ceria-supported precious metals, especially Pt/CeO_2, in PROX using various characterization methods, including *in situ* DRIFTS and NAP-XPS [128–132]. Pt/CeO_2 can effectively remove CO from PROX feed at 323–373 K: the lower the T, the better the selectivity, thus less hydrogen is lost. On the other hand, at higher temperatures, selectivity drops fast and most CO cannot be removed from the feed because all oxygen gets converted to H_2O.

The surface-near chemical state of 5% Pt/CeO_2 catalyst was investigated by NAP-XPS. The sample was first activated in 0.5 mbar O_2 at 573 K and then cooled in O_2. At ~340 K, oxygen was evacuated and replaced by hydrogen (0.48 mbar). As room temperature was reached, first CO (0.032 mbar) and then O_2 (0.015 mbar) were added to the hydrogen and the sample was heated to two different temperatures. Ceria was largely oxidized at 573 K in O_2, whereas hydrogen induced the formation of Ce^{3+} species (Figure 10.34). Fitting the spectrum in hydrogen indicated a Ce^{3+}/Ce^{4+} ratio of 0.87, thus similar to the reduction degree of ceria under WGS conditions. In the PROX mixture, however, the surface was clearly re-oxidized (Ce^{3+}/Ce^{4+} ratio of 0.4–0.45). Note that although the gas feed contained only 3% oxygen, with 91% of H_2, the ceria surface was half as reduced in PROX than in WGS at identical temperatures. When hydrogen was removed from the gas feed, further ceria oxidation occurred, with a Ce^{3+}/Ce^{4+} ratio of 0.25, reproducing the observed reduction degree of the support as reported above in the oxygen addition experiment to CO in the WGS section.

The O 1s spectra acquired next to Ce 3d are depicted in Figure 10.35. We applied the same fitting strategy and essentially the same four surface components were found: lattice O, oxygenates, OH, and surface water (with increasing binding energy). Data in H_2 suggest a high degree of surface hydroxylation and surface water formation. Despite the surface partially re-oxidizing in PROX, at low temperatures, most of the oxygen species remained approximately constant in intensity, except for oxygenates, those of which doubled in intensity. On the other hand, when the sample temperature in the PROX feed was raised to 523 K, a significant portion of

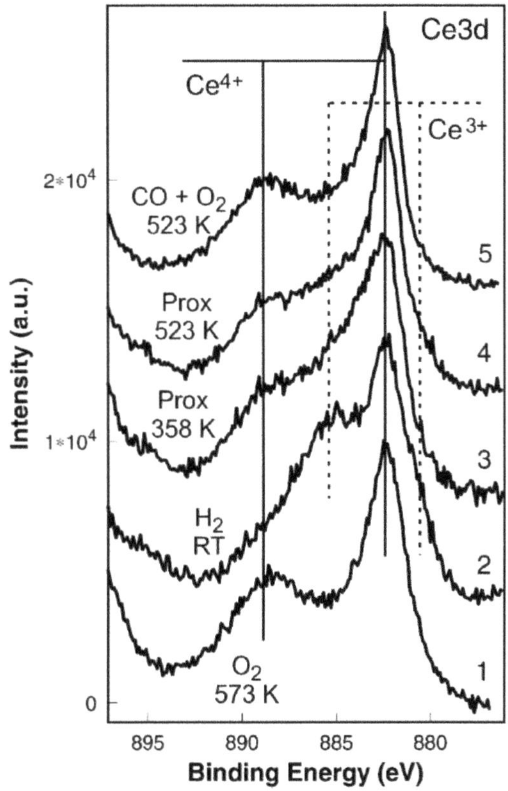

Fig. 10.34. Part of the Ce 3d region of the 5% Pt/CeO_2 at different conditions: 1: in 0.5 mbar O_2 at 573 K; 2: in 0.48 mbar H_2 at RT; 3: in a PROX mixture (0.48 mbar H_2, 0.032 mbar CO, and 0.015 mbar O_2) at 358 K; 4: in the same PROX mixture at 523 K; 5: in a mixture of 0.032 mbar CO and 0.015 mbar O_2 (as hydrogen was switched off after 4). The measurements were carried out in sequence as indicated by the numbers. Adapted with permission from Ref. [130]. Copyright (2005) Elsevier Inc.

oxygenates and OH were converted and removed from the surface. Furthermore, surface water has almost completely desorbed.

Figure 10.36 shows the platinum 4f core level in hydrogen and in the reaction mixture at two different temperatures. In H_2 at RT, after O_2 activation at 573 K, Pt exists in two distinct species: one is the zero-valent state and the other with a BE shift of +1.6 eV

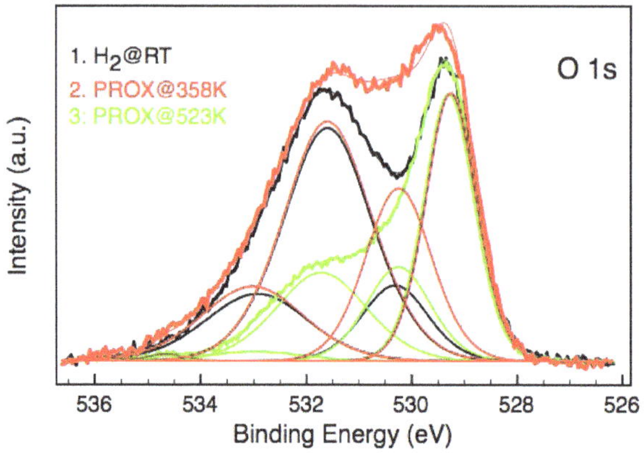

Fig. 10.35. O 1s region of 5% Pt/CeO_2 at different conditions: 1: in 0.48 mbar H_2 at RT (after O_2 activation at 573 K; H_2 introduced at ~340 K during cooling); 2: in a PROX mixture (0.48 mbar H_2, 0.032 mbar CO, and 0.015 mbar O_2) at 358 K.; 3: in the same PROX mixture at 523 K. The measurements were carried out in sequence as indicated by the numbers. The PROX@358K spectrum is scaled to provide the same lattice O component as with the other spectra. Adapted (spectra reanalyzed) with permission from Ref. [131]. Copyright (2005) Elsevier Inc.

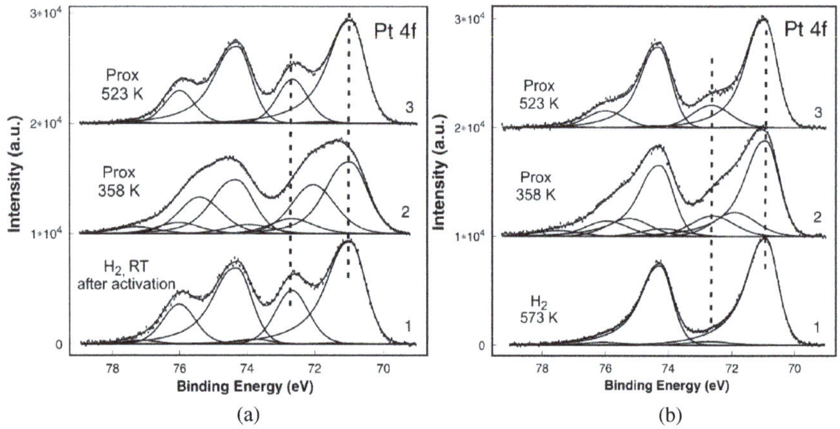

Fig. 10.36. (a) Pt 4f region of 5% Pt/CeO_2 at different conditions: 1: in 0.48 mbar H_2 at RT (after O_2 activation at 573 K); 2: in ~0.5 mbar PROX mixture at 358 K; 3: in ~0.5 mbar PROX mixture at 523 K; (b) 1: in 0.48 mbar H_2 at 573 K; 2: in ~0.5 mbar PROX mixture at 358 K; 3: in ~0.5 mbar PROX mixture at 523 K. The measurements were carried out in sequence as indicated by the numbers. Adapted with permission from Ref. [130]. Copyright (2005) Elsevier Inc.

representing small, oxidized clusters or interface sites to ceria. At 358 K under PROX conditions, the spectrum changed appreciably; the 72.6 eV component was no longer resolved, and a shoulder at 72.0 eV appeared. Furthermore, a tail at around 77.2 eV is evident, indicating a non-resolvable 7/2 component at ~73.9 eV. As we have discussed above, the surface core-level shift of adsorbed bridged and linear CO appears at +0.4 and +1.0 eV from the main metallic bulk peak, respectively. Thus, the PROX spectrum at 358 K contains linear bound CO. Bridged CO, though may exist, cannot be resolved from our spectrum. On further heating of the sample to 523 K, the new peaks disappeared, and the spectrum closely resembled that measured after activation. Accordingly, in this low-pressure NAP-XPS experiment, the coverage of CO was appreciable at 358 K but low at 523 K. In the second experiment after activation in oxygen, the sample was reduced in hydrogen at 573 K and tested in the PROX reaction. Catalytically, the sample behaves analogous to the non-reduced sample, with similar activity at low T and selectivity decay at 523 K. During the 573 K hydrogen treatment, platinum was totally reduced, however, in the PROX reaction mixture at low T, all the afore-mentioned components appeared, albeit with smaller relative intensities.

Combining the above NAP-XPS data with *in situ* DRIFTS and the other characterization, the following picture of the PROX mechanism appeared. Based on *in situ* DRIFTS analysis, the coverage of none of the adsorbed CO species (in form of carbonyl, carboxyl, formate, or carbonate) could be directly correlated with the CO_2 production rate. We remind our readers that in Section 10.5.1, we have seen that CO was stabilized on Pt up to 500–550 K so oxygen activation on Pt was not possible. This holds true on Pt(111) under PROX conditions, too [7]. Thus, the low-temperature activity of Pt/CeO_2 necessitates a non-competitive mechanism, that is, CO adsorption on metal, oxygen activation on the support, and reaction at/near the metal/oxide interface. The PROX performance improved significantly at higher p(CO), which increased the number of oxygen vacancies on the ceria surface. Water, created by hydrogen spillover from Pt sites and OH recombination on ceria, stabilizes on such vacancy ceria sites. At low temperatures, this slows down further water production by poisoning H oxidation sites as water desorption is hindered. Since O_2 activation on ceria is possible (seen, for example, by the re-oxidation in Ce 3d upon PROX), O_2 activation and water retention improved CO oxidation selectivity. The positive correlation between CO_2 production and surface water concentration (seen by *in situ* DRIFTS) is in line with this proposition. At higher temperatures, however, surface water desorption is facile and thus much of the oxygen activated on the ceria surface is involved in the unwanted side reaction of H oxidation, as H spills over from Pt to the support.

Now, when we compare the WGS and PROX data, we see that the major difference is in the degree of ceria reduction and thus in the retention of OH and adsorbed water on the surface. Retention of water on the surface is facilitated by oxygen vacancies in ceria, and when O_2 is dosed, as in PROX, part of these vacancies is quickly eliminated. We know that PROX is facile below 373 K, a temperature at which WGS does not occur. We have also seen that no re-oxidation of ceria by water was discernable in NAP-XPS. *Operando* DRIFTS-SSITKA experiments under WGS have shown that formates (as one of the proposed intermediates in an associate mechanism) cannot be a true intermediate as its surface temporal evolution upon isotope exchange was significantly slower than the appearance of isotope CO_2 [133]. Therefore, we can propose that under WGS either H_2O directly spills over to the Pt and gets activated or the carbonyl at the Pt/ceria interface reacts with a surface OH group, though the latter is less likely due to the high stability of interface carbonyls. Based on pulsed isotope experiments, Penkala *et al.* [134] have found strong indications for a Langmuir–Hinshelwood-type mechanism of CO oxidation on Pt/ceria; that is, surface O from gas phase and not the lattice oxygen of ceria was involved in the reaction. Therefore, in PROX, oxygen activation on ceria can lead to $(O)_{ads}$ which by reverse spillover may diffuse to the Pt/ceria interface and react with surface carbonyl on Pt in forming CO_2. A nice and comprehensive review on the use of NAP-XPS for the investigations of ceria-based catalysts was recently published by Garcia *et al.* [135].

10.5.3.3 *Silver in ethylene epoxidation and high-temperature applications*

Ethylene epoxidation over silver is one of the simplest kinetically controlled heterogeneously catalyzed reactions [136–140]. In a simplified picture, a silver catalyst under a mixture of C_2H_4 and O_2 at ca. 500 K will convert ca. 50% of ethylene molecules into ethylene oxide, C_2H_4O (EO), instead of the thermodynamically favored total combustion product, CO_2 and water (Figure 10.37) [136,137,140,141].

In epoxidation, a single oxygen atom is added across the carbon–carbon double bond of ethylene to produce EO. With the precise addition of promoters, the selectivity to EO can reach ca. 90% [140,142]. Why and how silver has the unique ability to catalyze this reaction with such ease and efficiency is intriguing and has been a subject of studies for many decades [136–138,140,142–144]. The first mechanisms to account for the different reported selectivities had been proposed in the 1940s, suggesting that a special form of oxygen would be active in epoxidation [145]. We will refer to this special species (as it became later known) as electrophilic oxygen [146–149]. Several surface science studies carried out during the 1980s showed important evidence of electrophilic oxygen [150–155]. Ag (111) and Ag(110) which could produce EO under a mixture of ca. 27 mbar C_2H_4 with 200 mbar O_2 were shown, after 2–10 minutes of reaction, to have only an adsorbed atomic oxygen surface reconstruction (the term refers to atomic oxygen on the reconstructed surface) when transferred to a UHV chamber

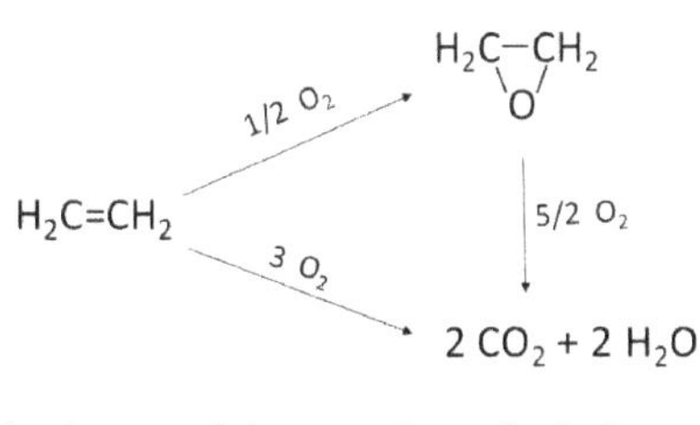

Fig. 10.37. Simplified scheme of the reaction of ethylene (C_2H_4) with O_2 to give the partial oxidation product ethylene oxide (C_2H_4O) and CO_2 (+H_2O).

equipped with low energy electron diffraction (LEED) and XPS analysis [150,151]. This atomic oxygen on the reconstructed surface was shown to lead only to ethylene combustion, as the oxygen pre-covered surfaces exposed to ethylene would result in the detection of only CO_2 during temperature-programmed reaction (TPR) [150,151]. We will refer to this atomic oxygen species (as it became later known) as nucleophilic oxygen [146–149]. These results suggested a mechanism where the electrophilic oxygen species responsible for the formation of EO was gas-phase O_2, and nucleophilic oxygen, which is active in total combustion, was thought to create sites for ethylene adsorption [151]. However, other experiments using Ag(111) [154,155] and silver powders [152,153] showed that EO could be produced through TPR with ethylene in the absence of gas-phase or adsorbed O_2 suggesting that electrophilic oxygen would not be molecular oxygen. The sample pre-treatments that would lead to a surface able to produce EO during TPR had to consist of either long time (ca. 30 min) exposure to O_2 at ca. 2–30 mbar at 475 K [152] or reaction conditions [154–157]. Both strategies resulted in the formation of reconstructed and dissolved oxygen (whose presence was determined by TPD), thus the combination of these two species was considered essential to obtain EO with TPR [152–154]. This was first explained by assuming that dissolved oxygen pulls electron density from the nucleophilic oxygen, making the adsorbed species more electrophilic [154].

Electrophilic oxygen was found to be a species with a characteristic O 1s BE of ca. 530.5 eV [156,158]. When first observed in XPS in 1978 [159], this species was assigned to subsurface oxygen and later to atomic oxygen adsorbed on the unreconstructed silver surface [160]. During the 1980s and 1990s publications, combining UHV-XPS with TPD and TPR measurements revealed two important points. First, that nucleophilic oxygen, the surface reconstruction characterized by an O 1s binding energy of ca. 528.4 eV, participates in the total combustion of ethylene even when dissolved oxygen is present [157,158]. Second, while TPD experiments show the presence of dissolved oxygen, the UHV-XPS measurements would only detect a single O 1s contribution with a BE of ca. 528.5 eV. This form of dissolved or subsurface oxygen was proposed then to be similar in nature to nucleophilic oxygen. This made the assignment of electrophilic oxygen, with an O 1s BE of ca. 530.5 eV, to dissolved or subsurface oxygen inconsistent. These contradictions made the role of dissolved or subsurface oxygen in the reaction and its relation to electrophilic oxygen to be unclear. Using a mixture of CO + O_2 as a pre-treatment to activate the silver surface (instead of C_2H_4 + O_2) showed to hinder oxygen dissolution, yet electrophilic oxygen did form [156]. Angle-resolved XPS showed that the electrophilic oxygen was located on the surface of the silver catalyst [156] and subsurface oxygen was thought to be necessary and responsible for making the adsorbed species more electrophilic [154]. Electrophilic oxygen was thus thought to be oxygen more weakly or "covalently" bond to silver [156,157]. By using isotope labeling, a combined XPS and TPR study demonstrated that the species with an O 1s BE of ca. 530.5 eV is the species responsible for EO formation (at least under UHV) [157]. By treating a silver surface with a 1 mbar mixture of C_2H_4 (5%) + $^{16}O_2$ at 470 K, followed by exposure to 10^{-3} mbar of $^{18}O_2$, a surface containing ^{16}O-electrophilic oxygen and ^{18}O-nucleophilic oxygen was obtained, which was confirmed by the combined results of XPS and TPD (Figures 10.38(a) and 10.38(b)). Nucleophilic oxygen, which desorbs below 600 K, was desorbed as purely $^{18}O_2$, while electrophilic oxygen, which desorbs slightly above 700 K, was desorbed as $^{16}O_2$. TPR results (Figure 10.38(c)) show that after adsorption of ethylene on the surface containing both nucleophilic and electrophilic oxygen, the produced EO contained only ^{16}O, while both $C^{16}O_2$ and $C^{18}O_2$ were obtained. This suggests that when both nucleophilic and electrophilic oxygens are present, the former participates only in combustion, while the latter is the only species to incorporate an oxygen atom into ethylene to form EO.

Following the assignment of electrophilic oxygen to unreconstructed adsorbed atomic oxygen (the term refers to oxygen adsorbed on the unreconstructed silver surface) [160], an oxametallacycle (OMC) mechanism was developed, wherein ethylene reacts with atomic oxygen on the unreconstructed surface to make an oxametallacycle intermediate that can decompose to either

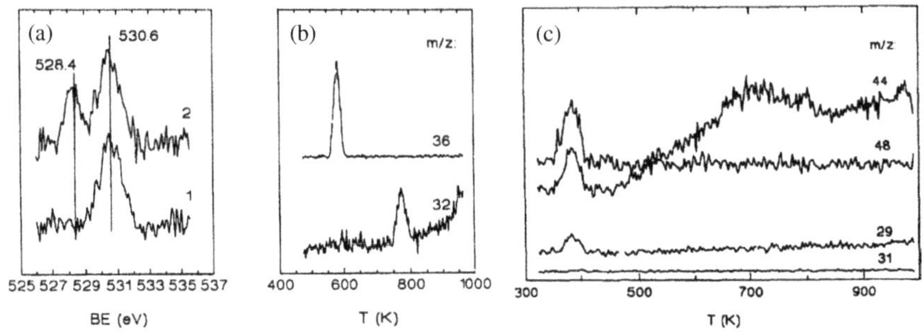

Fig. 10.38. (a) O ls spectra and (b) TPD spectra of O_2 recorded after treatment of Ag surface with C_2H_4 (5%) + $^{16}O_2$ at P = 1 mbar and T= 470 K (spectra 1) followed by $^{18}O_2$ adsorption at P = 10^{-3} mbar and T = 470 K (spectra 2). (c) TPR spectra of CO_2 and C_2H_4O recorded after ethylene adsorption at the Ag surface containing both nucleophilic and electrophilic oxygen. Adapted with permission from Ref. [157]. Copyright (1994) Elsevier Science B.V.

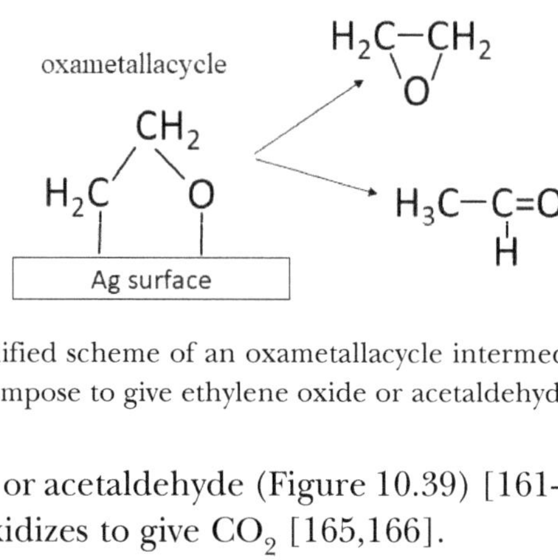

Fig. 10.39. Simplified scheme of an oxametallacycle intermediate on a silver surface that can decompose to give ethylene oxide or acetaldehyde.

ethylene oxide or acetaldehyde (Figure 10.39) [161–164], where the later rapidly oxidizes to give CO_2 [165,166].

Thus, unreconstructed atomic oxygen has remained a focus in understanding the mechanism of ethylene epoxidation over silver [162,167–171]. Experimentally, the OMC intermediate in this mechanism has been synthesized on silver surfaces through various precursors [172,173], including EO [143,168]. This intermediate was shown to decompose partially into EO during TPD [143,168]. Although such result showed that EO can form an OMC that can again decompose to EO, it does not clearly demonstrate that ethylene reacts with oxygen to form the OMC intermediate.

Nevertheless, many authors still consider unreconstructed atomic oxygen as the active species in EO formation.

Post-reaction analysis, which is often used in conjunction with surface science techniques in UHV, can lead to the loss of the active sites/species on the catalyst surface during the evacuation of the reaction feed and sample transfer [147]. By *in situ* measurements using NAP-XPS, the investigation of the oxygen species present on the silver surface under O_2 [147,149,159,174,175] or reaction conditions [146–148,176] has been possible. Particularly important has been the use of on-line product analysis along with NAP-XPS for ethylene epoxidation [147,148,176]. Using NAP-XPS combined with proton transfer mass spectrometry (PTRMS) to measure EO production under different reaction conditions, it was shown that there is a correlation between the surface concentration of electrophilic oxygen and the amount of EO produced [147]. Figure 10.40 shows how changing the temperature of the Ag sample under reaction conditions changed EO production, and at each temperature, the EO production can be correlated with electrophilic oxygen formation quantified by XPS (Figure 10.40(a)). And while EO production at each temperature showed a linear correlation with electrophilic oxygen, nucleophilic oxygen showed no correlation to EO production (Figure 10.40(b)). This was also observed by changing the reactant feed ratio (Figures 10.40(b) and 10.40(c)). For a clean Ag surface under ca. 0.3 mbar O_2 at 420 K, only nucleophilic oxygen was present on the surface, while electrophilic oxygen appeared upon the addition of ethylene to the gas-phase atmosphere [147]. Even a small partial pressure of ethylene ($p(C_2H_4)/p(O_2) = 1{:}10$) was shown to induce a decrease in nucleophilic oxygen coverage and led to the appearance of electrophilic oxygen. The increase in the partial pressure of ethylene up to $p(C_2H_4)/p(O_2) = 1{:}3$ and further to 1:1 led to a further reduction of nucleophilic oxygen and an increased surface concentration of electrophilic oxygen (Figure 10.40(c)) [147]. This came to show that electrophilic oxygen and nucleophilic oxygen compete for the same surface site. The role of nucleophilic oxygen was then still thought to be to facilitate ethylene adsorption, while participating only in

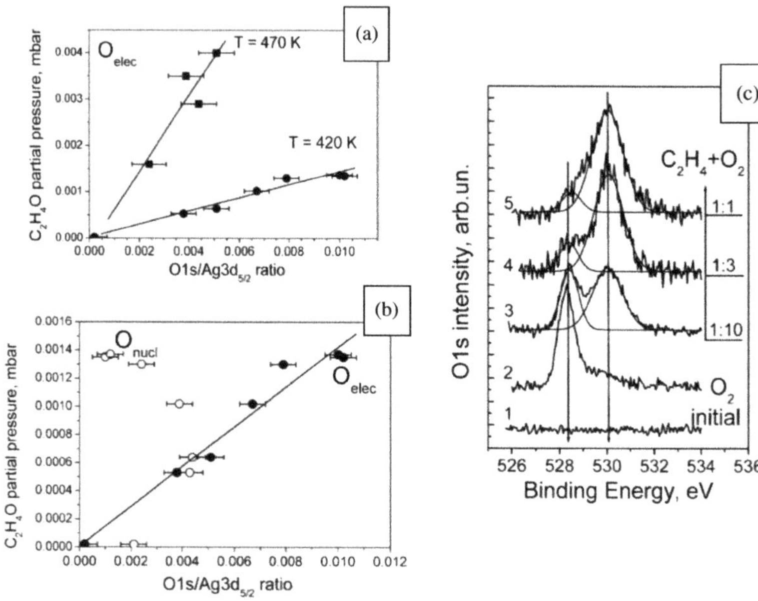

Fig. 10.40. (a) Yield of ethylene oxide as a function of the abundance of the electrophilic oxygen measured at different temperatures: 420 K and 470 K. (b) Yield of ethylene oxide as a function of the abundance of the nucleophilic (open symbols) and the electrophilic (full symbols) oxygen. The measurements were performed at 420 K and at various partial pressures of oxygen and ethylene (c) O 1s spectra measured *in situ* at 420 K, $p(O_2) = 0.33$ mbar and various $p(C_2H_4)$: (2) 0 mbar (pure O_2); (3) 0.03 mbar (1:10); (4) 0.1 mbar (1:3); (5) 0.24 mbar (1:1). The spectra from initial, "clean" silver surface measured in UHV (1) is also shown for comparison. Adapted with permission from Ref. [147]. Copyright (2005) Elsevier Inc.

ethylene combustion [147], in agreement with previous interpretations [151]. Here, one should note that these results agree with the previous UHV results where electrophilic oxygen was more easily obtained when the sample pre-treatment was done under reaction conditions [154–157] while O_2 pre-treatments needed to be at high pressures and longer time [152].

PTRMS can only detect EO but not CO_2. Thus, the selectivity of the reaction cannot be determined. When combining NAP-XPS measurements with on-line gas chromatography analysis, using a

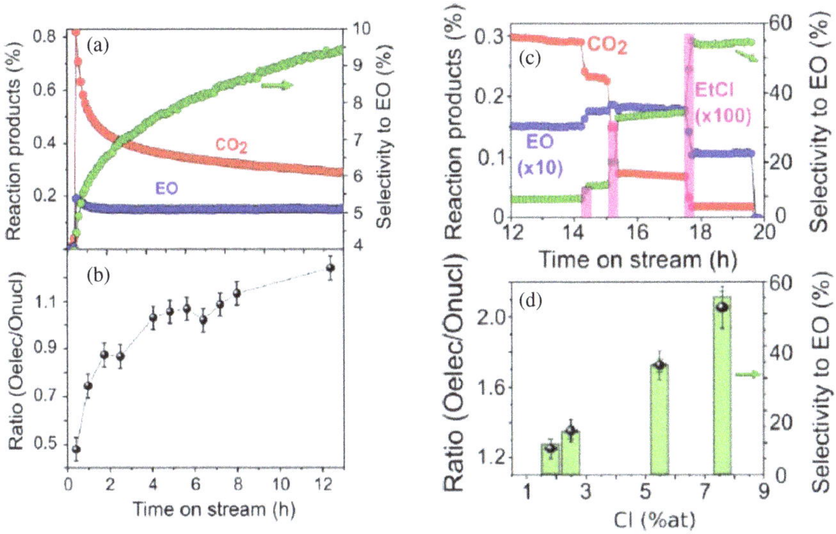

Fig. 10.41. Dynamics of non-promoted (a,b) and Cl-promoted (c,d) silver catalyst under ethylene epoxidation at 0.3 mbar, 230°C, $C_2H_4:O_2 = 1:2$. (a) Mole fraction of reaction products of CO_2 (red) and EO (blue) together with selectivity to EO (green) as a function of time. (b) Ratio of electrophilic to nucleophilic oxygen species (Oelec/Onucl) as a function of time. (c) Mole fraction of reaction products of CO_2 (red) and EO (blue) together with selectivity to EO (green) as a function of time. *In situ* CH_3CH_2Cl pulses are represented by the purple bars. (d) Ratio of electrophilic to nucleophilic oxygen species Oelec/Onucl (black dots) and selectivity (green bars) as a function of the Cl atomic concentration. Adapted with permission from Ref. [148]. Copyright (2014) Elsevier Inc.

silver powder as a catalyst, the ratio of electrophilic to nucleophilic oxygen was shown to correlate with the selectivity of EO [148].

Figures 10.41(a) and 10.41(b) show how the ratio of these oxygen species mimics the selectivity evolution over time under reaction conditions. Here, the electrophilic oxygen surface concentration was shown to increase gradually with time accompanied by a simultaneous decrease in nucleophilic oxygen. The same trend was observed for Cl-promoted silver powder, where the *in situ* addition of Cl to the Ag catalyst surface was shown to decrease the amount of nucleophilic oxygen yet not affecting the surface concentration of electrophilic oxygen. The selectivity, which increased upon Cl addition, correlated once again with the ratio between the electrophilic and nucleophilic oxygen species (Figures 10.41(c) and 10.41(d)) [148].

NAP-XPS broadened our understanding of silver catalysts' surfaces under ethylene epoxidation conditions. And though the investigation of the atomic structure of nucleophilic oxygen has been extensive, the structure of electrophilic oxygen remained elusive. This has been mainly because, as mentioned before, the preparation of electrophilic oxygen requires high pressures and long-time exposure of the silver surface to the gas atmosphere, which makes it difficult to study under UHV without critical contamination. And while it has been commonly accepted that electrophilic oxygen would be subsurface or unreconstructed atomic oxygen, there was not any strong experimental evidence for these species being present under steady-state epoxidation. Thus, the continued interest in the Ag/O system remained to be a central topic of investigation.

The first publications to report on atomic oxygen species with an O 1s BE different from that of nucleophilic O (> 528.3 eV) and electrophilic oxygen (ca. 530.5 eV) species were on experiments using UHV-XPS on an Ag(210) surface exposed to O_2 at low temperatures [177,178]. The authors assigned the observed low BE of the O 1s spectra component (BE < 528 eV) to two O species: adsorbed O atoms in four-fold hollows and O–Ag rows at the steps. It was shown later that, because of the ionic nature of the Ag/O interaction, weakly bound unreconstructed atomic O should have a BE < 528 eV [174]. The O 1s measured for Ag(111) and Ag(110) under 10^{-4} mbar O_2 at 423 K, at 20 and 40 minutes of exposure, has a main BE of 528.15 eV and 528.3 eV, respectively, which is the characteristic of nucleophilic oxygen [174,179]. And while DFT calculations have shown that the introduction of subsurface oxygen can increase the O 1s BE of unreconstructed atomic oxygen from 527–528 eV up to 528.9 eV [179], all the calculated BEs for the different unreconstructed atomic oxygen species were shown to be well below that of electrophilic oxygen [174,179]. Significant experimental evidence that unreconstructed atomic oxygen has an O 1s BE < 528 eV and cannot explain the high surface concentration of electrophilic oxygen species observed in XPS (with O1s BE of ca. 530.5 eV) came from NAP-XPS studies in combination with DFT calculations [174,175,179]. Figure 10.42 illustrates the main results.

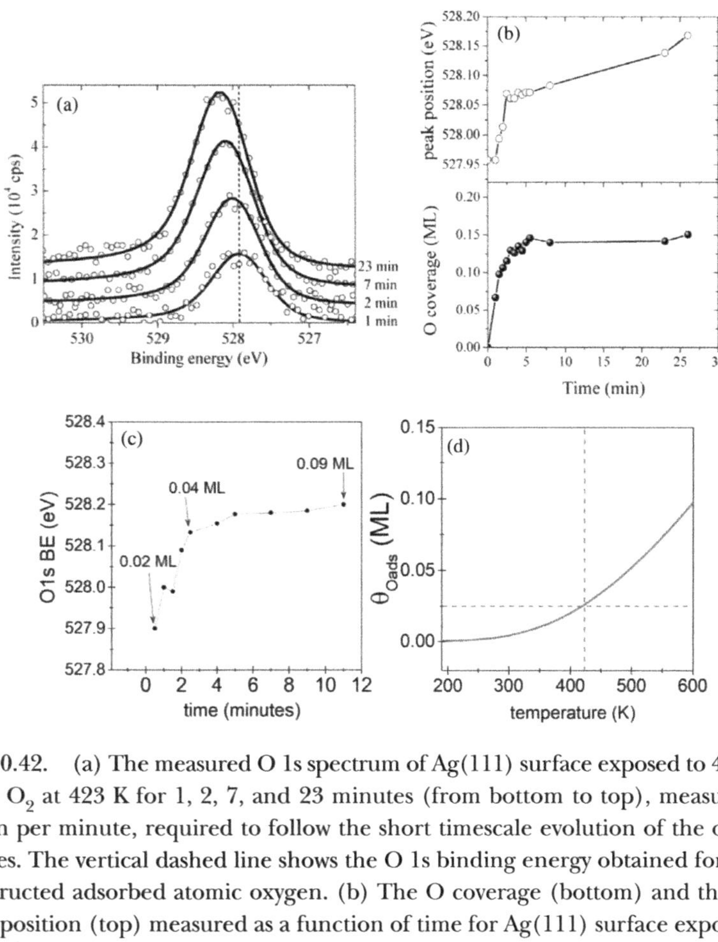

Fig. 10.42. (a) The measured O 1s spectrum of Ag(111) surface exposed to 4×10^{-4} mbar O_2 at 423 K for 1, 2, 7, and 23 minutes (from bottom to top), measured at 1 scan per minute, required to follow the short timescale evolution of the oxygen species. The vertical dashed line shows the O 1s binding energy obtained for unreconstructed adsorbed atomic oxygen. (b) The O coverage (bottom) and the O 1s peak position (top) measured as a function of time for Ag(111) surface exposed to 4×10^{-4} mbar O_2 at 423 K. (c) O 1s peak position as a function of time for Ag(110) exposed to 10^{-6} mbar O_2 at 423 K, (d) predicted maximum coverage of unreconstructed atomic oxygen as a function of temperature, assuming the added-row Ag–O chains in the p(N × 1) contribute no configurational entropy. The dashed lines show the predicted coverage at the temperature of the experiment. (a) and (b) Adapted with permission from Ref. [174] from the PCCP Owner Societies and (c) and (d) Adapted with permission from Ref. [175] from the Royal Society of Chemistry. CC BY 3.0.

It was shown that unreconstructed atomic oxygen appears as a transient species when a clean silver surface is exposed continuously to O_2 at low pressures (ca. 10^{-6}–10^{-4} mbar) showing a BE of 527.95 eV [174] and 527.9 eV [175] on Ag(111) and Ag(110) respectively (Figures 10.42(a)–10.42(c)), in agreement with the BE obtained by DFT calculations [174,175,179] and with the previous assignment on Ag(210) [177,178]. Confirmed by theory (Figure 10.42(d)), NAP-XPS experiments show that it is only possible to obtain a maximum coverage of ca. 0.02–0.03 ML of unreconstructed atomic oxygen on Ag(110) [175] and ca. 0.05–0.06 ML on Ag(111) at 423 K [174] in agreement with previous STM results [180]. Higher coverages (higher oxygen pressure) lead immediately to the thermodynamically favored surface reconstruction, i.e., nucleophilic oxygen, shown by the characteristic O 1s BE > 528 eV (Figure 10.42). Then, it was suggested that possibly more than one species could be active in epoxidation. (1) Unreconstructed atomic oxygen with a characteristic BE ≤ 528 eV [174,175,179], which according to DFT calculations, would be partially selective to EO [162,167,169–171,175]. (2) Electrophilic oxygen with a characteristic BE of ca. 530.5 eV [146–149,159], the only species which had been demonstrated experimentally to be active in epoxidation [157]. Its identity and structure were, however, still unknown. And while studies of the OMC mechanism have offered significant insights into how ethylene epoxidation proceeds through the reaction of ethylene with unreconstructed atomic oxygen, this species with an O 1s BE < 528 eV [174,175,177–179] has never been observed on active surfaces [147,148,157]. Instead, it forms surface reconstructions (nucleophilic oxygen) already at the oxygen chemical potentials needed for ethylene epoxidation, suggesting only low coverages (< 0.05 ML) might be present under epoxidation reaction conditions [147–149,174,175]. The combination of NAP-XPS and DFT has shown that unreconstructed atomic oxygen cannot account for the electrophilic oxygen species (with BE ca. 530.5 eV). DFT calculations suggested that the types of weakly bound oxygen that would give an O 1s BE > 530 eV are covalently bound, though to an element other than Ag, such as C or H [179]. Interestingly, following a group of patents from the

1980s that claimed that the addition of ppm NO_x to the reaction feed led to high selectivity to EO, a publication from the group of Lambert in 2000 reported on the addition of ppm NO to the reactant feed in ethylene epoxidation [181]. They also showed an increase in EO selectivity upon NO addition and proposed that the surface oxyanion (NO_3) formed on the catalyst surface upon NO addition might have been directly involved in alkene epoxidation. They suggested a cycle where oxygen transfer occurred from the oxyanion to the alkene, and the NO_2 generated would re-oxidize to NO_3 by atomic oxygen, as illustrated in Figure 10.43 [181].

An oxidative preparation method using NO_2 as an efficient source of atomic oxygen that had been previously used to prepare the surface oxygen reconstruction on Ag(111) and Ag(110) [182] was successfully adapted and used in 2013 to prepare an oxygen species with O 1s BEs consistent with electrophilic oxygen on an Ag(111) surface [183,184]. The use of NO_2 rather than O_2 circumvents the problematic low sticking probability of O_2, allowing sample preparation at much lower pressures [182] making this procedure a (clean) UHV-compatible recipe. Using NO_2, an Ag(111) surface with three types of oxygen species was prepared: nucleophilic oxygen with its characteristic O 1s BE of 528.4 eV [174,183,184] and other two oxygen species with O 1s BE of 530.7 eV and 530.2 eV. The latter appears gradually with nucleophilic oxygen [183]. The oxygen species with BE of 530.7 eV was found to form a surface reconstruction characterized by a $(7\times\sqrt{3})$*rect* LEED pattern [184]. It was shown (Figure 10.44(a)) later that the species with a BE of 530.2 eV is the oxygen in adsorbed SO_4, $SO_{4,ads}$, on the unreconstructed silver

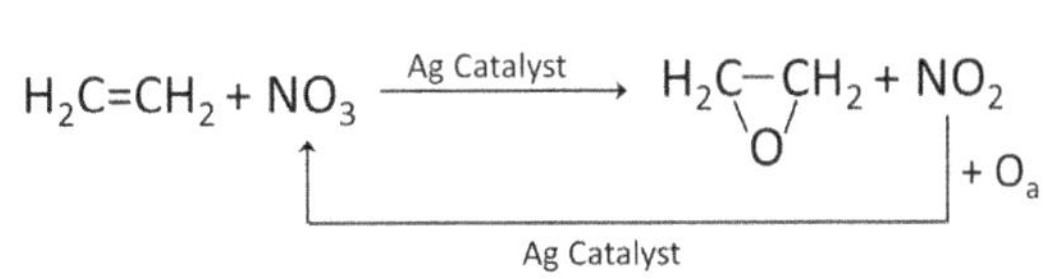

Fig. 10.43. Simplified scheme of the cycle proposed to participate in the production of EO from the reaction of ethylene with the NO_3 oxyanion on silver and the subsequent reoxidation of NO_2 to regenerate NO_3.

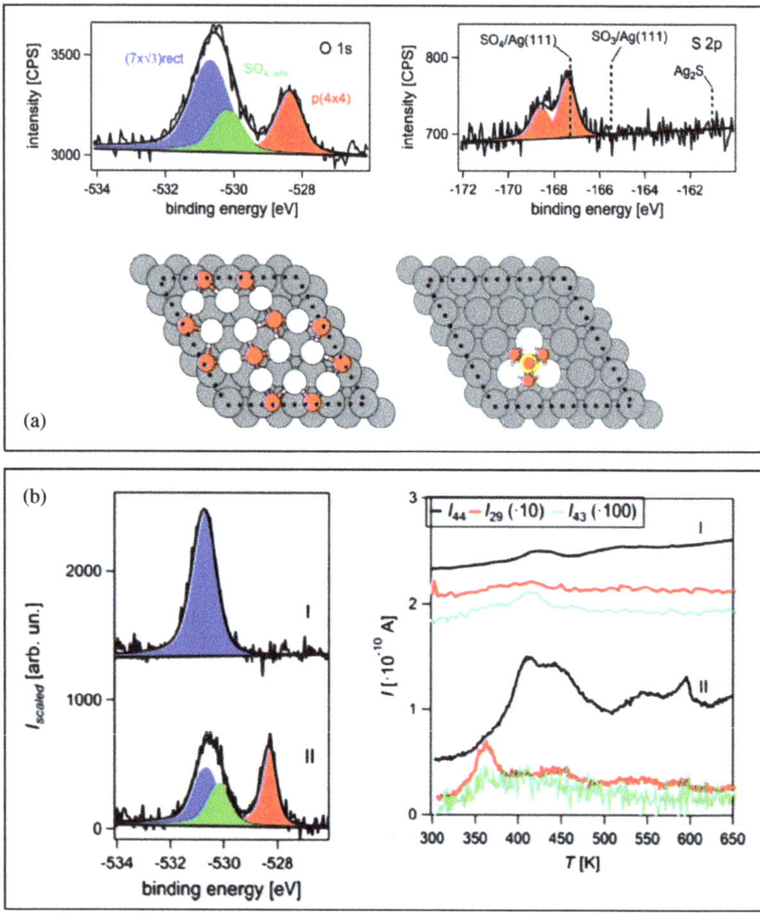

Fig. 10.44. (a) O 1s and S 2p spectra (upper left and right) [S $2p_{3/2}$ reference BEs (dashed lines)] of the reconstructed SO_4 phase, after oxidation with 270 L at 500 K. Structures of the p(4 × 4)-O phase (bottom left) and $SO_{4,ads}$ (bottom right). (b) O 1s (left) and TPR (right) spectra of the (7 × √3)rect reconstruction (I) and the surface active in ethylene epoxidation (II). CO_2 (EO) production leads to m/z = 44 amu (m/z = 29 amu and m/z = 43 amu). Adapted with permission from Ref. [176]. Copyright (2018) American Chemical Society.

surface, whereas the species with the BE of 530.7 eV is the oxygen of SO_4 that forms the $(7\times\sqrt{3})$*rect* surface reconstruction [176]. (As for the source of sulfur, we will come to this point soon.) Oxygen treatment of the SO_4-covered surface with this reconstruction was found to lead to the formation of reconstructed atomic oxygen and unreconstructed $SO_{4,ads}$ [176]. It was found that when oxygen-induced surface reconstructions are produced, it leads to a local competition for silver adatoms and this partially lifts the $(7\times\sqrt{3})rect$ reconstruction, generating the unreconstructed $SO_{4,ads}$.

Figure 10.44(b) depicts TPR results showing that only the surface containing all three O-species is active in EO production. This resembles the previous results on electrophilic oxygen [152–158] as can be seen by comparing Figure 10.44(b), where only surfaces with unreconstructed $SO_{4,ads}$ are active in EO formation during TPR, with Figure 10.38, where the surfaces actively forming EO during TPR were shown to require both electrophilic and nucleophilic atomic oxygen [157]. By noting that $SO_{4,ads}$ forms under UHV conditions only when both the SO_4-$(7\times\sqrt{3})rect$ and the oxygen surface reconstructions are present, electrophilic oxygen observed previously in UHV and shown to participate in EO formation during TPR [157] was suggested to be the oxygen in unreconstructed $SO_{4,ads}$. The O 1s and S 2p BEs of these species were confirmed by DFT calculations [176]. Moreover, it was shown by DFT calculations that EO formation on the $(7\times\sqrt{3})rect$ surface reconstruction is highly activated (1.8 eV relative to adsorbed C_2H_4), while the activation barrier to EO formation from O in $SO_{4,ads}$ is only 0.96 eV relative to adsorbed ethylene. This explains why a surface containing only the $(7\times\sqrt{3})rect$ reconstruction, as in Figure 10.44 (or only electrophilic oxygen in Figure 10.38), cannot produce EO, and only a surface with both electrophilic and nucleophilic oxygen is able to do so. It was proposed that the role of the oxygen reconstruction during TPR appears to be in stabilizing electrophilic oxygen by partially transforming the inactive $(7\times\sqrt{3})rect$ reconstruction into the active unreconstructed, $SO_{4,ads}$ [176].

NAP-XPS was used then to confirm the presence of this species under steady-state epoxidation and its relation to EO formation. Figure 10.45 shows the PTRMS detection of EO production and the S 2p spectra for an Ag powder under 0.3 mbar C_2H_4 + O_2 mixtures. In Figure 10.45(a), silver under the reaction mixture at room temperature shows no EO production, as expected at this temperature [147]. After a temperature ramp, at 523 K, EO is produced and S 2p shows the presence of SO_4/Ag. To verify its influence on EO production, the sample was treated in pure ethylene at 723 K. After this treatment, the surface concentration of SO_4 and the production of EO decrease. To corroborate this trend, a different experiment was performed (Figure 10.45(b)). Ag powder was placed under C_2H_4 at 503 K and no SO_4 was observed. After introducing O_2 to the stream, SO_4 immediately appeared on the silver surface while EO was produced. After some time, the sample was treated in pure O_2 at 673 K that increased the surface

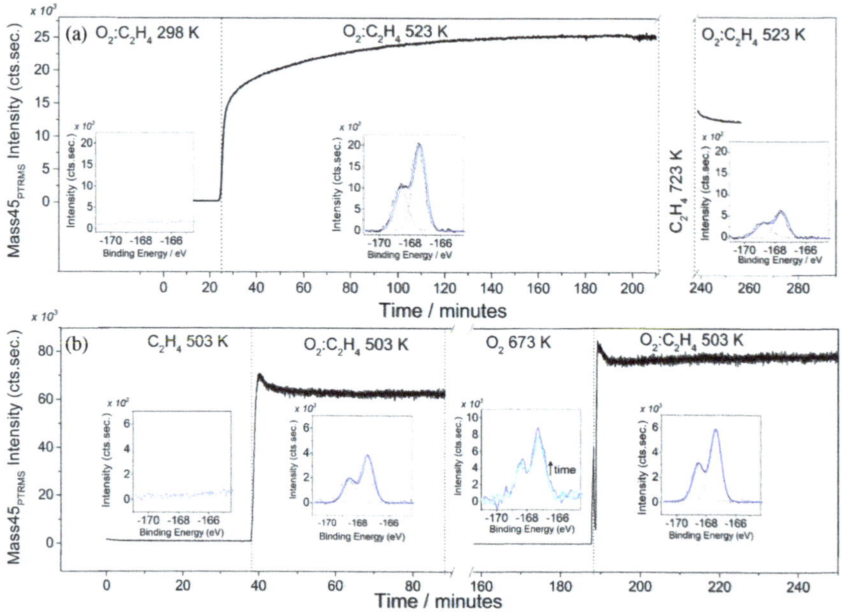

Fig. 10.45. (a) PTRMS (mass 45) showing the EO production from an Ag powder catalyst under various conditions (total pressure of 0.3 mbar) with corresponding S 2p spectra inset. (a) The catalyst was exposed to the reaction gas mixture at 298 K and then heated to 523 K to initiate the reaction. After 210 min on stream, the adsorbed SO_4 formed during epoxidation was cleaned by ethylene titration before returning to the reaction gas mixture. S 2p spectra measured with hν = 355 eV. (b) An Ag powder catalyst was held under ethylene at 503 K before introducing O_2 to initiate the reaction. After 90 min on stream, more S was segregated to the surface under pure O_2 before returning to the reaction gas mixture. S 2p spectra were measured with hν = 390 eV. Adapted with permission from Ref. [176]. Copyright (2018) American Chemical Society.

concentration of SO_4. After this treatment, the sample was cooled down to 503 K again and C_2H_4 was introduced into the stream. Higher EO production was detected for the higher surface concentration of SO_4. These results resemble those from Figure 10.40, where higher EO production was correlated with a higher surface concentration of electrophilic oxygen [147].

The correlation of EO and the SO_4 coverage was also followed by NAP-XPS and on-line GC. It was shown that at higher coverages, the EO production increased linearly with SO_4 coverage while CO_2 production decreased (hence, the selectivity increased) [176]. Moreover, the *in situ* measurement of S 2p under pure ethylene at 503 K (after exposing the Ag powder to the reaction mixture at the same temperature) showed that SO_3 appears as a transient on the silver surface while SO_4 is titrated by ethylene. This indicates that SO_4 can transfer an O atom to ethylene. However, under reaction conditions, the SO_4 coverage reaches a steady state, which suggests that the SO_3 formed is in equilibrium with SO_4. This is remarkably similar to the reaction cycle proposed by Lambert for NO_x-promoted silver catalysts: NO_3 reacting with ethylene on silver to form EO and NO_2, the latter subsequently re-oxidizing to NO_3 (Figure 10.43) [181]. This also agrees with the idea that electrophilic oxygen should be an oxygen covalently bound to an element other than silver. According to DFT, these would have BEs above 530 eV [179], which agrees with the experimentally observed BE of the reactive electrophilic oxygen species [147,148,157]. These new findings are in line with the early correlations associated with the preparation of electrophilic oxygen. While an oxygen surface reconstruction, i.e., nucleophilic oxygen, is almost immediately formed under O_2 [174,175] or NO_2 [183] when the oxygen coverage is higher than ca. 0.03–0.06 ML [174,175], electrophilic oxygen formation has been linked to the dissolution of oxygen into the silver bulk [152–155,183,184]. In many publications, an O_2 atmosphere was used to generate electrophilic oxygen and required high pressure and long-time exposures [149,159,183,185]. The recent findings show that in such cases, the silver bulk is the likely source of sulfur [176], a known impurity of silver, thus dissolved oxygen is required to segregate electrophilic oxygen (SO_4) to the surface [176,183].

This could explain why under an O_2 atmosphere the formation of electrophilic oxygen is a slow process depending on the O_2 pressure and temperature [149,158]. On the other hand, electrophilic oxygen seems to be generated faster in reaction mixtures [148,154–157]. This can also be explained by the faster formation of SO_4 on the silver surface by sulfur trace impurities present in ethylene [186], since this does not require significant dissolution of oxygen into silver.

Apart from ethylene epoxidation, silver-based catalysts are used in several alcohol partial oxidation reactions, many of which are carried out at high temperatures (650–1000 K) [187]. Among these, formaldehyde synthesis is an important industrial reaction that has been the subject of numerous investigations. Formaldehyde can be produced from methanol mainly through two major reactions: one is methanol direct dehydrogenation which produces formaldehyde and hydrogen. The other is the partial oxidation of methanol, which produces formaldehyde and water. Other secondary reactions may also occur where byproducts, such as CO and CO_2, are formed and to a lesser extent methyl formate, methane, and formic acid [188]. Nowadays, formaldehyde is produced at an industrial scale by two major processes generally carried out at atmospheric pressure and at 500–720°C [188,189]. An air-deficient process using methanol and air, with or without the addition of steam, is carried out using silver catalysts [188]. It is believed that formaldehyde is produced from both direct dehydrogenation and partial oxidation reactions over silver. The extent to which these reactions occur is largely determined by the temperature, water content, and $CH_3OH:O_2$ ratio [189–191]. A complete description of the different industrial processes can be found in Refs. [188,192].

In general, two main species are thought to participate in the different reaction paths: (1) adsorbed atomic oxygen (named unreconstructed atomic oxygen in the previous section and frequently called O_α in the literature related to methanol oxidation) and (2) an O species observed by TPD to desorb at higher temperatures (>850 K), known as O_γ. [193,194]. The latter has been proposed to be a thermally stable suboxide or oxide-like layer and to be strongly bound [193]. The catalytic activity of silver for methanol oxidation is believed to be related to the ratio between these two oxygen species, and this is thought to dictate which reaction path is prevailing [190,191]. The direct dehydrogenation path is favored at higher temperatures (>773 K) resulting in a higher selectivity for formaldehyde [190]. At this temperature (which is above the atomic oxygen desorption temperature), O_γ is believed to play a major role in the methanol dehydrogenation path [190,193]. Several XPS studies done after exposing silver to different gas atmospheres (O_2, CH_3OH, and different ratios of $CH_3OH:O_2$) have identified O_γ with an O 1s BE of 529.0–529.8 eV [149,191,193]. Understanding the oxygen–silver interaction is essential to understand the catalytic activity of silver in partial oxidation reactions. NAP-XPS has been used to study the oxygen species present at high temperatures under O_2 and the evolution of the surface population of these species with time [149]. It was shown that two oxygen species form at 773 K under 0.25 mbar O_2 (Figure 10.46): one with a BE of 529.0–529.3 eV (peak 1) and a

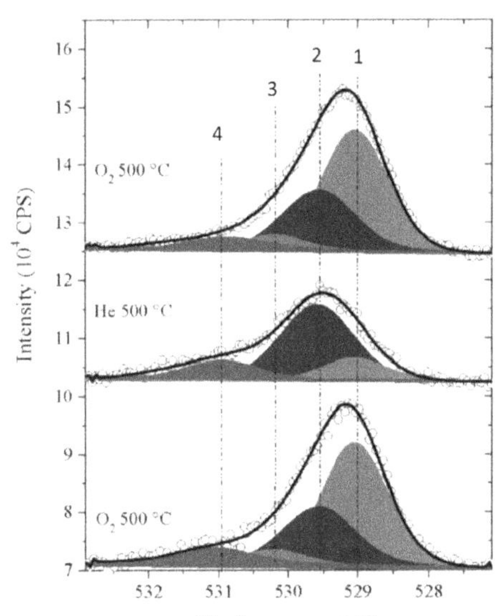

Fig. 10.46. O 1s spectra for silver foil at 773 K under different gas environments, measured at the ISISS beamline at BESSY II with a NAP-XPS endstation. Figure adapted with permission from Ref. [149]. Copyright (2012) Royal Society of Chemistry.

second species with a BE of 529.3–529.8 eV (peak 2). The species of peak 1, which appears after only a few minutes of exposure of Ag to 0.25 mbar O_2 at 423 K, is not stable without an O_2 atmosphere at 773 K. The species from peak 2, which appears only after many hours of exposure to O_2 at 773 K, is stable without O_2 in the gas phase at the same temperature. Due to its high-temperature stability, peak 2 was assigned to O_γ species and the species of peak 1 was proposed to be oxide-like species located at steps [149]. This latter species can only be characterized by *in situ* methods and might not be present (in large amounts) in post-reaction analysis, as it is not stable at high temperatures in the absence of an O_2 atmosphere. Peaks 3 and 4 (named electrophilic oxygen) have been discussed in the previous section.

It has been proposed that the strongly bound oxygen (O_γ) species stable at high temperatures under UHV could be (in some cases) a metal oxide (metal = Mo, Ta, W) on the silver surface [195]. The authors proposed that these metal oxide traces could arise from small parts of some setups' sample holders, as these elements are typically present in parts of sample holders made for UHV setups. The authors reported that the metal oxide peak observed in their experiment was ca. 1% of the Ag signal and that this might have been easily overlooked in some cases. It remains yet to be established if metal trace impurities might be present at the silver surface under *in situ* conditions, as is the case for sulfur trace impurities on silver [176].

Figure 10.47 shows the *in situ* O 1s spectra for an Ag pellet under 0.5 mbar O_2 and 1 mbar O_2:CH_3OH:H_2O:N_2 reaction mixture at 773 K. Under O_2 (spectra A), there are four different oxygen species present on the silver surface. Species 1 with a BE of 529 eV can be assigned to O_γ if we consider the UHV post-analysis literature [191,193] or as the same as species 1 from Figure 10.46 [149]. Unlike the results from Rocha *et al.* [149], only one species with BE in the range of 529–529.8 eV is observed in Figure 10.47 (peak 1). Species 2 and 3 with BE of 530.3 eV and 530.9 eV (named electrophilic oxygen species, previously discussed) are thought not to participate in the methanol oxidation reaction [190,191] and will not

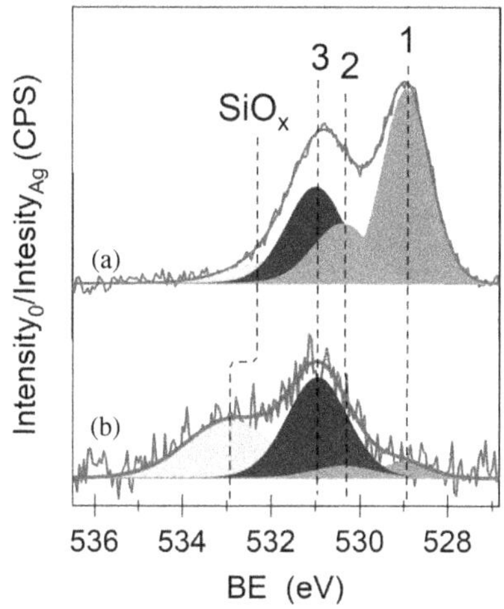

Fig. 10.47. O 1s spectra for a silver pellet (a) under 0.5 mbar O_2 at 773 K and (b) under 1 mbar $CH_3OH/O_2/H_2O/N_2$ = 2/1/1.3/75.7 at 773 K (after four hours). The O 1s spectra were normalized to the Ag 3d intensity. The pellet was pre-treated under 1 bar O_2 at 923 K for 24 hours. The spectra were measured at the ISISS beamline at BESSY II with a NAP-XPS endstation. We thank Frederic Sulzmann for providing the data.

be discussed further in this section. Species 4 with a BE of 531–532 eV is from SiO_x trace impurities (as we will address this in the following). A different species distribution is observed under methanol oxidation. The notorious decrease observed for species 1 under methanol oxidation conditions has been attributed to the consumption of this species by its direct reaction with methanol.[186] Further investigation of silver under methanol oxidation using NAP-XPS is still required to determine the nature and role of the different species, which remains an open question.

The presence of the SiO_x peak can be explained by silver bulk trace impurities segregation. Rocha *et al.* showed using NAP-XPS that when silver is exposed to 0.2 mbar O_2 at 773 K for long periods of time, thermal segregation of bulk impurities can occur and be enhanced by the presence of a gas phase and that it seems to depend on the sample history [149]. They observed that a high binding energy feature between 531.5 and 532.5 eV evolved occasionally for

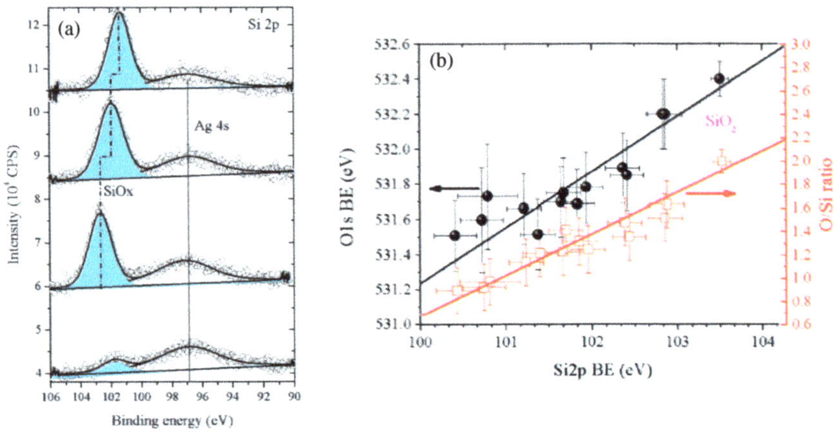

Fig. 10.48. (a) Si 2p of silicon contamination in Ag foils exposed to 0.25 mbar O_2 at 773 K. (b) Correlation of the Si 2p peak position with the BE of the SiO_x component in the O 1s spectra (filled dots) and with the ratio of atomic concentrations of oxygen and silicon (open squares). The spectra were measured at the ISISS beamline at BESSY II with a NAP-XPS endstation. Figure adapted with permission from Ref. [149]. Copyright (2012) Royal Society of Chemistry.

silver foils exposed to O_2 at high temperatures, and at the same time, they detected SiO_x on the silver surface (Figure 10.48(a)). SiO_x was shown to be present in a distribution of Si oxidation states with different BEs, and the changes in O 1s BE of the SiO_x species correlate with the changes in BE of Si 2p and the O/Si ratio (Figure 10.48(b)). The difference in sensitivity factors between silicon and oxygen for XPS analysis based on Al/Mg K_α sources might easily hinder the detection of the Si 2p signal while the corresponding O 1s feature is still observed. In a survey scan taken with a photon energy of 800 eV, Si 2p is buried within the noise due to its proximity to the Ag 2s peak. In contrast, surface-sensitive spectra taken at 350 eV clearly showed the Si 2p peak on the silver surface [149]. This goes to show how the segregation of bulk trace impurity can be easily overlooked and not be considered for the O 1s spectral interpretation, as some authors occasionally observe such high BE peak after high-pressure treatments [191]. Thus, the use of synchrotron radiation for NAP-XPS experiments is of fundamental importance, not only because one can measure with high surface sensitivity, using photon energies where the cross-section for different elements is maximized, but also due to the higher photon flux (higher intensity and signal/noise ratio). This and a critical eye are important when evaluating the O species on silver, mainly in the case of partial oxidation reactions, where the oxygen chemical potential at high pressure and temperature might drive surface segregation of bulk impurities.

10.6 Concluding Remarks

NAP-XPS has the potential to provide a variety of information on heterogeneous catalysis. This ranges from the actual surface phase/state of the catalyst under reaction conditions, the surface species present, and their coverage, up to more subtle information like work function changes, band bending, and more. Based on the selected examples, we have shown how NAP-XPS is a viable experimental technique to study the electronic structure of a catalyst surface under reaction conditions and to identify active surface species. It was shown that catalysis needs to be understood as a co-operation of reactants, products, and the catalytic material. The catalyst itself is not left unchanged under high-performance catalysis. Combined with on-line product analysis and DFT calculations, NAP-XPS is a powerful tool to identify how the surface and subsurface species might correlate with the catalyst activity and selectivity, and it brings important knowledge to help elucidate the reaction mechanisms. Although the technique has been used in catalysis research for almost two decades, the examples also clearly illustrate some of the current shortcomings and limitations.

1. Most studies are restricted to one or a low number of catalysts and/or one or a few catalytic conditions. Although the observations are often claimed to have a causal relation to the catalytic action, due to the limited number of variables explored, the importance of the observation is not cross-validated. Furthermore, spectral observations and catalytic function may not have a trivial mathematic correlation.

2. As a certain electrical conductivity is required to record interpretable and useful NAP-XPS spectra, the classical supported catalysts based on insulator supports are not feasible for this technique. More often than not, model catalysts are studied, where promoters, additives, and supports are neglected or exchanged. These simplifications are useful to enable NAP-XPS experiments, but it necessitates the validation that the catalytic function is not hampered.
3. Following point 2, the importance of microstructure, synthesis details, etc. are sometimes overlooked, and the observed dynamics (or the lack thereof) may relate to such parameters. Therefore, results from one study to the next may not be transferable, which can explain sometimes contradictory observations.
4. Most studies are still constrained to a few mbar pressure, which can raise the question of relevance to one bar or high-pressure catalysis. Thus, it is imperative to prove that the catalytic function is similar in these two different pressure conditions.
5. Although the technique is intrinsically capable of deriving quantitative coverage information, it is difficult to achieve for samples other than foils and single crystals.

To allow a deeper mechanistic understanding of the catalytic function, future studies should move beyond the current practice of low-dimensional experimentation with respect to variables and using purely steady-state experiments. Properly designed experiments with a larger sample size can open up avenues for utilizing machine learning and thus deriving complex two or multivariable descriptor function correlations. These in turn can feed back to catalyst synthesis to predict new catalyst formulations. Further instrumental developments, for example, implementing fast switching dosing valves in combination with low-volume flow cells, can enable time-resolved transient measurements that may distinguish reactive and spectator species. Sub-nanosecond time-resolved pump-probe NAP-XPS experiments [196] may prove useful for studying charge carrier dynamics or other elementary processes in the available time domain. An interesting current development for sub-second time-resolved NAP-XPS is the event-averaging pulse-probe spectroscopy, in which various kinds of repetitive perturbations (gas phase composition, pressure, and temperature) can be used to trigger and generate event-averaged spectra. The only requirement is that the time-resolved data contain features that can be identified by image recognition. Knudsen and coworkers [197] utilized cycling gas pulse experiments using a triggering signal obtained by image recognition and originating from true changes on the surface. They applied this method in a feasibility study on CO oxidation over Pd(100) to investigate the forced transition between the CO-covered and oxidized surfaces.

The quest for higher and higher pressures to be used in NAP-XPS experiments will certainly continue. Special flow cells with focused beams will enable us to close the pressure gap to one bar catalysis, even without resorting to tender X-ray excitation. However, the writing is on the wall that radiolysis can influence and change the surface state during one bar XPS experimentation. Standing waves NAP-XPS [198] and measuring charge transfer between adsorbates and metal surfaces [199] are niche applications of NAP-XPS that may help solve particular questions. Combining NAP-XPS in one experiment with other spectroscopic techniques, such as vibrational spectroscopy, has the potential to further augment our understanding of heterogeneous catalysis. Here, vibrational spectroscopy can provide surface coverage information not realistically available for supported catalysts by XPS. NAP-XPS has the potential to become an indispensable tool for catalysis research, but we are not yet there.

References

[1] Schlögl R. Heterogeneous catalysis. Angew Chemie Int Ed. 2015;54:3465–3520. Doi:10.1002/anie.201410738.

[2] Siegbahn H, Siegbahn K. ESCA applied to liquids. J Electron Spectros Relat Phenomena. 1973;2:319–325. Doi:10.1016/0368-2048(73)80023-4.

[3] Joyner RW, Roberts MW, Yates K. A "high-pressure" electron spectrometer for surface studies. Surf Sci. 1979;87:501–509. Doi:10.1016/0039-6028(79)90544-2.
[4] Ruppender HJ, Grunze M, Kong CW, Wilmers M. *In situ* X-ray photoelectron spectroscopy of surfaces at pressures up to 1 mbar. Surf Interface Anal. 1990;15:245–253. Doi:10.1002/sia.740150403.
[5] Ogletree DF, Bluhm H, Lebedev G, Fadley CS, Hussain Z, Salmeron M. A differentially pumped electrostatic lens system for photoemission studies in the millibar range. Rev Sci Instrum. 2002;73:3872–3877. Doi:10.1063/1.1512336.
[6] Bluhm H, Hävecker M, Knop-Gericke A, Kleimenov E, Schlögl R, Teschner D, Bukhtiyarov VI, Ogletree DF, Salmeron M. Methanol oxidation on a copper catalyst investigated using *in situ* X-ray photoelectron spectroscopy. J Phys Chem B. 2004;108:14340–14347. Doi:10.1021/jp040080j.
[7] Schnadt J, Knudsen J, Johansson N. Present and new frontiers in materials research by ambient pressure X-ray photoelectron spectroscopy. J Phys Condens Matter. 2020;32:413003. Doi:10.1088/1361-648x/ab9565.
[8] Knop-Gericke A, Kleimenov E, Hävecker M, Blume R, Teschner D, Zafeiratos S, Schlögl R, Bukhtiyarov VI, Kaichev VV, Prosvirin IP, Nizovskii AI, Bluhm H, Barinov A, Dudin P, Kiskinova M. Chapter 4: X-ray photoelectron spectroscopy for investigation of heterogeneous catalytic processes. In BC Gates, H Knözinger (Eds.), *Advanced Catalysis* (pp. 213–272). Academic Press, 2009. Doi:10.1016/S0360-0564(08)00004-7.
[9] Salmeron M, Schlögl R. Ambient pressure photoelectron spectroscopy: A new tool for surface science and nanotechnology. Surf Sci Rep. 2008;63:169–199. Doi:10.1016/j.surfrep.2008.01.001.
[10] Bluhm H. Photoelectron spectroscopy of surfaces under humid conditions. J Electron Spectros Relat Phenomena. 2010;177:71–84. Doi:10.1016/j.elspec.2009.08.006.
[11] Starr DE, Liu Z, Hävecker M, Knop-Gericke A, Bluhm H. Investigation of solid/vapor interfaces using ambient pressure X-ray photoelectron spectroscopy. Chem Soc Rev. 2013;42:5833. Doi:10.1039/c3cs60057b.

[12] Shavorskiy A, Karslioglu O, Zegkinoglou I, Bluhm H. Synchrotron-based ambient pressure X-ray photoelectron spectroscopy. Synchrotron Radiat News. 2014;27:14–23. Doi:10.1080/08940886.2014.889547.

[13] Trotochaud L, Head AR, Karslıoğlu O, Kyhl L, Bluhm H. Ambient pressure photoelectron spectroscopy: Practical considerations and experimental frontiers. J Phys Condens Matter. 2017;29:53002. Doi:10.1088/1361-648x/29/5/053002.

[14] Head AR, Schnadt J. UHV and ambient pressure XPS: Potentials for Mg, MgO, and $Mg(OH)_2$ surface analysis. JOM. 2016;68:3070–3077. Doi:10.1007/s11837-016-2112-x.

[15] Salmeron M. From surfaces to interfaces: Ambient pressure XPS and beyond. Top Catal. 2018;61:2044–2051. Doi:10.1007/s11244-018-1069-0.

[16] Roy K, Artiglia L, van Bokhoven JA. Ambient pressure photoelectron spectroscopy: Opportunities in catalysis from solids to liquids and introducing time resolution. ChemCatChem. 2018;10:666–682. Doi:10.1002/cctc.201701522.

[17] Weatherup RS. 2D material membranes for operando atmospheric pressure photoelectron spectroscopy. Top Catal. 2018;61:2085–2102. Doi:10.1007/s11244-018-1075-2.

[18] Nguyen L, Tao FF, Tang Y, Dou J, Bao X-J. Understanding catalyst surfaces during catalysis through near ambient pressure X-ray photoelectron spectroscopy. Chem Rev. 2019;119:6822–6905. Doi:10.1021/acs.chemrev.8b00114.

[19] Zhong L, Chen D, Zafeiratos S. A mini review of *in situ* near-ambient pressure XPS studies on non-noble, late transition metal catalysts. Catal Sci Technol. 2019;9:3851–3867. Doi:10.1039/C9CY00632J.

[20] 8th International APXPS Workshop. (n.d.). https://www.bnl.gov/apxps2021/index.php.

[21] APXPS publications database. (n.d.). https://apxpspubs.maxiv.lu.se/.

[22] Held G, Venturini F, Grinter DC, Ferrer P, Arrigo R, Deacon L, Quevedo Garzon W, Roy K, Large A, Stephens C, Watts A, Larkin P, Hand M, Wang H, Pratt L, Mudd JJ, Richardson T, Patel S, Hillman M, Scott S. Ambient-pressure endstation of the Versatile Soft X-ray

(VerSoX) beamline at Diamond Light Source. J Synchrotron Radiat. 2020;27:1153–1166. Doi:10.1107/S1600577520009157.

[23] Schnadt J, Knudsen J, Andersen JN, Siegbahn H, Pietzsch A, Hennies F, Johansson N, Mårtensson N, Öhrwall G, Bahr S, Mähl S, Schaff O. The new ambient-pressure X-ray photoelectron spectroscopy instrument at MAX-lab. J Synchrotron Radiat. 2012;19:701–704. Doi:10.1107/S0909049512032700.

[24] Palomino RM, Hamlyn R, Liu Z, Grinter DC, Waluyo I, Rodriguez JA, Senanayake SD. Interfaces in heterogeneous catalytic reactions: Ambient pressure XPS as a tool to unravel surface chemistry. J Electron Spectros Relat Phenomena. 2017;221:28–43. Doi:10.1016/j.elspec.2017.04.006.

[25] Pérez-Dieste V, Aballe L, Ferrer S, Nicolàs J, Escudero C, Milán A, Pellegrin E. Near ambient pressure XPS at ALBA. J Phys Conf Ser. 2013;425:72023. Doi:10.1088/1742-6596/425/7/072023.

[26] Kaya S, Ogasawara H, Näslund L-Å, Forsell JO, Casalongue HS, Miller DJ, Nilsson A. Ambient-pressure photoelectron spectroscopy for heterogeneous catalysis and electrochemistry. Catal Today. 2013;205:101–105. Doi:10.1016/j.cattod.2012.08.005.

[27] Knudsen J, Andersen JN, Schnadt J. A versatile instrument for ambient pressure X-ray photoelectron spectroscopy: The Lund cell approach. Surf Sci. 2016;646:160–169. Doi:10.1016/j.susc.2015.10.038.

[28] Nguyen L, (Feng) Tao F. Development of a reaction cell for *in-situ*/operando studies of surface of a catalyst under a reaction condition and during catalysis. Rev Sci Instrum. 2016;87:64101. Doi:10.1063/1.4946877.

[29] Kerherve G, Regoutz A, Bentley D, Hood C, Feeley K, Knight S, Robson A, Turner C, Singh N, Pontefract J, Åhlund J, Kahk JM, Villar-Garcia IJ, Payne DJ. Laboratory-based high pressure X-ray photoelectron spectroscopy: A novel and flexible reaction cell approach. Rev Sci Instrum. 2017;88:33102. Doi:10.1063/1.4975096.

[30] Orlando F, Waldner A, Bartels-Rausch T, Birrer M, Kato S, Lee M-T, Proff C, Huthwelker T, Kleibert A, van Bokhoven J, Ammann M. The environmental photochemistry of oxide surfaces and the

nature of frozen salt solutions: A new *in situ* XPS approach. Top Catal. 2016;59:591–604. Doi:10.1007/s11244-015-0515-5.

[31] Velasco-Vélez JJ, Pfeifer V, Hävecker M, Wang R, Centeno A, Zurutuza A, Algara-Siller G, Stotz E, Skorupska K, Teschner D, Kube P, Braeuninger-Weimer P, Hofmann S, Schlögl R, Knop-Gericke A. Atmospheric pressure X-ray photoelectron spectroscopy apparatus: Bridging the pressure gap. Rev Sci Instrum. 2016;87:053121. Doi:10.1063/1.4951724.

[32] Newberg JT, Åhlund J, Arble C, Goodwin C, Khalifa Y, Broderick A. A lab-based ambient pressure X-ray photoelectron spectrometer with exchangeable analysis chambers. Rev Sci Instrum. 2015;86: 85113. Doi:10.1063/1.4928498.

[33] Amann P, Degerman D, Lee M-T, Alexander JD, Shipilin M, Wang H-Y, Cavalca F, Weston M, Gladh J, Blom M, Björkhage M, Löfgren P, Schlueter C, Loemker P, Ederer K, Drube W, Noei H, Zehetner J, Wentzel H, Åhlund J, Nilsson A. A high-pressure X-ray photoelectron spectroscopy instrument for studies of industrially relevant catalytic reactions at pressures of several bars. Rev Sci Instrum. 2019;90:103102. Doi:10.1063/1.5109321.

[34] Olivieri G, Giorgi JB, Green RG, Brown MA. Five years of ambient pressure photoelectron spectroscopy (APPES) at the Swiss Light Source (SLS). J Electron Spectros Relat Phenomena. 2017;216: 1–16. Doi:10.1016/j.elspec.2017.01.003.

[35] Ogletree FD, Bluhm H, Hebenstreit ED, Salmeron M. Photoelectron spectroscopy under ambient pressure and temperature conditions. Nucl Instrum Methods Phys Res Sect A. 2009;601:151–160. Doi:10.1016/j.nima.2008.12.155.

[36] Knop-Gericke A, Pfeifer V, Velasco-Velez J-J, Jones T, Arrigo R, Hävecker M, Schlögl R. *In situ* X-ray photoelectron spectroscopy of electrochemically active solid-gas and solid-liquid interfaces. J Electron Spectros Relat Phenomena. 2017;221:10–17. Doi:10.1016/j.elspec.2017.03.010.

[37] Grass ME, Karlsson PG, Aksoy F, Lundqvist M, Wannberg B, Mun BS, Hussain Z, Liu Z. New ambient pressure photoemission endstation at Advanced Light Source beamline 9.3.2. Rev Sci Instrum. 2010;81:53106. Doi:10.1063/1.3427218.

[38] Urpelainen S, Såthe C, Grizolli W, Agåker M, Head AR, Andersson M, Huang S-W, Jensen BN, Wallén E, Tarawneh H, Sankari R, Nyholm R, Lindberg M, Sjöblom P, Johansson N, Reinecke BN, Arman MA, Merte LR, Knudsen J, Schnadt J, Andersen JN, Hennies F. The SPECIES beamline at the MAX IV laboratory: A facility for soft X-ray RIXS and APXPS. J Synchrotron Radiat. 2017;24:344–353. Doi:10.1107/S1600577516019056.

[39] Yoshida M, Kondoh H. *In situ* observation of model catalysts under reaction conditions using X-ray core-level spectroscopy. Chem Rec. 2014;14:806–818. Doi:10.1002/tcr.201402025.

[40] SPECS GmbH, NAP-XPS products. (n.d.). https://www.specs-group.com/.

[41] ScientaOmicron NAP-XPS products. (n.d.). https://scientaomicron.com/en/system-solutions/electron-spectroscopy.

[42] Axnanda S, Scheele M, Crumlin E, Mao B, Chang R, Rani S, Faiz M, Wang S, Alivisatos AP, Liu Z. Direct work function measurement by gas phase photoelectron spectroscopy and its application on PbS nanoparticles. Nano Lett. 2013;13:6176–6182. Doi:10.1021/nl403524a.

[43] Copperthwaite RG, Lloyd J. X-ray photoelectron spectroscopic evidence for trapped oxygen species in irradiated $NaClO_3$. Nature. 1978;271:141–142. Doi:10.1038/271141a0.

[44] Remita H, Remita S. Metal clusters and nanomaterials: Contribution of radiation chemistry. Recent Trends Radiat Chem. n.d.;347–383. Doi:10.1142/9789814282093_0013.

[45] Bondaz L, Fontaine P, Muller F, Pantoustier N, Perrin P, Morfin I, Goldmann M, Cousin F. Controlled synthesis of gold nanoparticles in copolymers nanomolds by X-ray radiolysis. Langmuir. 2020;36:6132–6144. Doi:10.1021/acs.langmuir.0c00554.

[46] Arble C, Guo H, Strelcov E, Hoskins B, Zeller P, Amati M, Gregoratti L, Kolmakov A. Radiation damage of liquid electrolyte during focused X-ray beam photoelectron spectroscopy. Surf Sci. 2020;697:121608. Doi:10.1016/j.susc.2020.121608.

[47] Le Caër S. Water radiolysis: Influence of oxide surfaces on H_2 production under ionizing radiation. Water. 2011;3:235–253. Doi:10.3390/w3010235.

[48] Huart L, Nicolas C, Kaddissy JA, Guigner J-M, Touati A, Politis M-F, Mercere P, Gervais B, Renault J-P, Hervé du Penhoat M-A. Soft X-ray radiation and Monte Carlo simulations: Good tools to describe the radiation chemistry of Sub-keV electrons. J Phys Chem A. 2020;124:1896–1902. Doi:10.1021/acs.jpca.9b10539.

[49] Weatherup RS, Wu CH, Escudero C, Pérez-Dieste V, Salmeron MB. Environment-dependent radiation damage in atmospheric pressure X-ray spectroscopy. J Phys Chem B. 2018;122:737–744. Doi:10.1021/acs.jpcb.7b06397.

[50] Fulford J, Bonner P, Goodhead DT, Hill MA, O'Neill. Experimental determination of the dependence of OH radical yield on photon energy: A comparison with theoretical simulations. J Phys Chem A. 1999;103:11345–11349. Doi:10.1021/jp993087n.

[51] Kot M, Kegelmann L, Köbler H, Vorokhta M, Escudero C, Kúš P, Šmíd B, Tallarida M, Albrecht S, Abate A, Matolínová I, Schmeißer D, Flege JI. *In situ* near-ambient pressure X-ray photoelectron spectroscopy reveals the influence of photon flux and water on the stability of halide perovskite. ChemSusChem. 2020;13:5722–5730. Doi:10.1002/cssc.202001527.

[52] Boucly A, Rochet F, Arnoux Q, Gallet J-J, Bournel F, Tissot H, Marry V, Dubois E, Michot L. Soft X-ray heterogeneous radiolysis of pyridine in the presence of hydrated strontium-hydroxyhectorite and its monitoring by near-ambient pressure photoelectron spectroscopy. Sci Rep. 2018;8:6164. Doi:10.1038/s41598-018-24329-8.

[53] Jiang P, Porsgaard S, Borondics F, Köber M, Caballero A, Bluhm H, Besenbacher F, Salmeron M. Room-temperature reaction of oxygen with gold: An *in situ* ambient-pressure X-ray photoelectron spectroscopy investigation. J Am Chem Soc. 2010;132:2858–2859. Doi:10.1021/ja909987j.

[54] Dumbuya K, Cabailh G, Lazzari R, Jupille J, Ringel L, Pistor M, Lytken O, Steinrück H-P, Gottfried JM. Evidence for an active oxygen species on Au/TiO_2(110) model catalysts during investigation with *in situ* X-ray photoelectron spectroscopy. Catal Today. 2012;181:20–25. Doi:10.1016/j.cattod.2011.09.035.

[55] Teschner D, Novell-Leruth G, Farra R, Knop-Gericke A, Schlögl R, Szentmiklósi L, Hevia MG, Soerijanto H, Schomäcker R, Pérez-

Ramírez J, López N. *In situ* surface coverage analysis of RuO_2-catalysed HCl oxidation reveals the entropic origin of compensation in heterogeneous catalysis. Nat Chem. 2012;4:739–745. Doi:10.1038/nchem.1411.

[56] Toyoshima R, Yoshida M, Monya Y, Suzuki K, Amemiya K, Mase K, Mun BS, Kondoh H. A high-pressure-induced dense CO overlayer on a Pt(111) surface: A chemical analysis using *in situ* near ambient pressure XPS. Phys Chem Chem Phys. 2014;16:23564–23567. Doi:10.1039/C4CP04318A.

[57] Johansson N, Andersen M, Monya Y, Andersen JN, Kondoh H, Schnadt J, Knudsen J. Ambient pressure phase transitions over Ir(1 1 1): At the onset of CO oxidation. J Phys Condens Matter. 2017;29:444002. Doi:10.1088/1361-648x/aa8a44.

[58] Blomberg S, Hoffmann MJ, Gustafson J, Martin NM, Fernandes VR, Borg A, Liu Z, Chang R, Matera S, Reuter K, Lundgren E. *In situ* X-ray photoelectron spectroscopy of model catalysts: At the edge of the gap. Phys Rev Lett. 2013;110:117601. Doi:10.1103/PhysRevLett.110.117601.

[59] Kondoh H, Toyoshima R, Monya Y, Yoshida M, Mase K, Amemiya K, Mun BS. *In situ* analysis of catalytically active Pd surfaces for CO oxidation with near ambient pressure XPS. Catal Today. 2016;260:14–20. Doi:10.1016/j.cattod.2015.05.016.

[60] Garcia-Martinez F, García-Fernández C, Simonovis JP, Hunt A, Walter A, Waluyo I, Bertram F, Merte LR, Shipilin M, Pfaff S, Blomberg S, Zetterberg J, Gustafson J, Lundgren E, Sánchez-Portal D, Schiller F, Ortega JE. Catalytic oxidation of CO on a curved Pt(111) surface: Simultaneous ignition at all facets through a transient CO-O complex**. Angew Chemie Int Ed. 2020;59:20037–20043. Doi:10.1002/anie.202007195.

[61] Krick Calderón S, Grabau M, Yoo JE, Killian MS, Schmuki P, Steinrück H-P, Papp C. Reactivity of TiO_2 nanotube-supported platinum particles in the CO oxidation reaction. ChemCatChem. 2017;9:564–572. Doi:10.1002/cctc.201600913.

[62] Matera S, Reuter K. Transport limitations and bistability for *in situ* CO oxidation at RuO_2(110): First-principles based multiscale modeling. Phys Rev B. 2010;82:85446. Doi:10.1103/PhysRevB.82.085446.

[63] Zafeiratos S, Paloukis F, Papakonstantinou G, Teschner D, Hävecker M, Vass E, Schnörch P, Knop-Gericke A, Schlögl R, Moreno B. A comparative *in situ* XPS study of PtRuCo catalyst in methanol steam reforming and water gas shift reactions. Catal Today. 2010;157:250–256. Doi:10.1016/j.cattod.2010.03.030.

[64] Halevi B, Peterson EJ, Roy A, DeLariva A, Jeroro E, Gao F, Wang Y, Vohs JM, Kiefer B, Kunkes E, Hävecker M, Behrens M, Schlögl R, Datye AK. Catalytic reactivity of face centered cubic $PdZn_{\alpha}$ for the steam reforming of methanol. J Catal. 2012;291:44–54. Doi:10.1016/j.jcat.2012.04.002.

[65] Rameshan C, Stadlmayr W, Weilach C, Penner S, Lorenz H, Hävecker M, Blume R, Rocha T, Teschner D, Knop-Gericke A, Schlögl R, Memmel N, Zemlyanov D, Rupprechter G, Klötzer B. Subsurface-controlled CO_2 selectivity of PdZn near-surface alloys in H_2 generation by methanol steam reforming. Angew Chemie Int Ed. 2010;49:3224–3227. Doi:10.1002/anie.200905815.

[66] Friedrich M, Teschner D, Knop-Gericke A, Armbrüster M. Influence of bulk composition of the intermetallic compound ZnPd on surface composition and methanol steam reforming properties. J Catal. 2012;285:41–47. Doi:10.1016/j.jcat.2011.09.013.

[67] Liu Z, Duchoň T, Wang H, Grinter DC, Waluyo I, Zhou J, Liu Q, Jeong B, Crumlin EJ, Matolín V, Stacchiola DJ, Rodriguez JA, Senanayake SD. Ambient pressure XPS and IRRAS investigation of ethanol steam reforming on Ni–CeO_2(111) catalysts: An *in situ* study of C–C and O–H bond scission. Phys Chem Chem Phys. 2016;18:16621–16628. Doi:10.1039/C6CP01212D.

[68] Sohn H, Soykal II, Zhang S, Shan J, Tao F, Miller JT, Ozkan US. Effect of cobalt on reduction characteristics of ceria under ethanol steam reforming conditions: AP-XPS and XANES studies. J Phys Chem C. 2016;120:14631–14642. Doi:10.1021/acs.jpcc.6b02490.

[69] Sohn H, Celik G, Gunduz S, Dogu D, Zhang S, Shan J, Tao FF, Ozkan US. Oxygen mobility in pre-reduced nano- and macro-ceria with co loading: An AP-XPS, in-situ DRIFTS and TPR study. Catal Lett. 2017;147:2863–2876. Doi:10.1007/s10562-017-2176-4.

[70] Turczyniak S, Luo W, Papaefthimiou V, Ramgir NS, Haevecker M, MacHocki A, Zafeiratos S. A comparative ambient pressure X-ray

photoelectron and absorption spectroscopy study of various cobalt-based catalysts in reactive atmospheres. Top Catal. 2016;59:532–542. Doi:10.1007/s11244-015-0531-5.

[71] Divins NJ, Angurell I, Escudero C, Pérez-Dieste V, Llorca J. Influence of the support on surface rearrangements of bimetallic nanoparticles in real catalysts. Science (80-.). 2014;346:620 LP–623. Doi:10.1126/science.1258106.

[72] Soler L, Casanovas A, Ryan J, Angurell I, Escudero C, Pérez-Dieste V, Llorca J. Dynamic reorganization of bimetallic nanoparticles under reaction depending on the support nanoshape: The case of RhPd over ceria nanocubes and nanorods under ethanol steam reforming. ACS Catal. 2019;9:3641–3647. Doi:10.1021/acscatal.9b00463.

[73] Mitsui T, Rose MK, Fomin E, Ogletree DF, Salmeron M. Dissociative hydrogen adsorption on palladium requires aggregates of three or more vacancies. Nature. 2003;422:705–707. Doi:10.1038/nature01557.

[74] Greeley J, Mavrikakis M. Surface and subsurface hydrogen: Adsorption properties on transition metals and near-surface alloys. J Phys Chem B. 2005;109:3460–3471. Doi:10.1021/jp046540q.

[75] Christmann K. Interaction of hydrogen with solid surfaces. Surf Sci Rep. 1988;9:1–163. Doi:10.1016/0167-5729(88)90009-X.

[76] Mei D, Sheth PA, Neurock M, Smith CM. First-principles-based kinetic Monte Carlo simulation of the selective hydrogenation of acetylene over Pd(111). J Catal. 2006;242:1–15. Doi:10.1016/j.jcat.2006.05.009.

[77] Sheth PA, Neurock M, Smith CM. A first-principles analysis of acetylene hydrogenation over Pd(111). J Phys Chem B. 2003;107:2009–2017. Doi:10.1021/jp021342p.

[78] Al-Ammar AS, Webb G. Hydrogenation of acetylene over supported metal catalysts. Part 1. — Adsorption of [14C]acetylene and [14C] ethylene on silica supported rhodium, iridium and palladium and alumina supported palladium. J Chem Soc Faraday Trans 1. 1978;74:195–205. Doi:10.1039/F19787400195.

[79] Teschner D, Vass E, Hävecker M, Zafeiratos S, Schnörch P, Sauer H, Knop-Gericke A, Schlögl R, Chamam M, Wootsch A, Canning AS,

Gamman JJ, Jackson SD, McGregor J, Gladden LF. Alkyne hydrogenation over Pd catalysts: A new paradigm. J Catal. 2006;242:26–37. Doi:10.1016/j.jcat.2006.05.030.

[80] Teschner D, Borsodi J, Wootsch A, Révay Z, Hävecker M, Knop-Gericke A, Jackson SD, Schlögl R. The roles of subsurface carbon and hydrogen in palladium-catalyzed alkyne hydrogenation. Science (80-.). 2008;320:86–89. Doi:10.1126/science.1155200.

[81] Teschner D, Révay Z, Borsodi J, Hävecker M, Knop-Gericke A, Schlögl R, Milroy D, Jackson SD, Torres D, Sautet P. Understanding palladium hydrogenation catalysts: When the nature of the reactive molecule controls the nature of the catalyst active phase. Angew Chemie Int Ed. 2008;47:9274–9278. Doi:10.1002/anie.200802134.

[82] Teschner D, Borsodi J, Kis Z, Szentmiklósi L, Révay Z, Knop-Gericke A, Schlögl R, Torres D, Sautet P. Role of hydrogen species in palladium-catalyzed alkyne hydrogenation. J Phys Chem C. 2010;114: 2293–2299. Doi:10.1021/jp9103799.

[83] Velasco-Vélez JJ, Teschner D, Girgsdies F, Hävecker M, Streibel V, Willinger MG, Cao J, Lamoth M, Frei E, Wang R, Centeno A, Zurutuza A, Hofmann S, Schlögl R, Knop-Gericke A. The role of adsorbed and subsurface carbon species for the selective alkyne hydrogenation over a Pd-black catalyst: An operando study of bulk and surface. Top Catal. 2018;61:2052–2061. Doi:10.1007/s11244-018-1071-6.

[84] Torres D, Cinquini F, Sautet P. Pressure and temperature effects on the formation of a Pd/C surface carbide: Insights into the role of Pd/C as a selective catalytic state for the partial hydrogenation of acetylene. J Phys Chem C. 2013;117:11059–11065. Doi:10.1021/jp400059m.

[85] Gabasch H, Kleimenov E, Teschner D, Zafeiratos S, Hävecker M, Knop-Gericke A, Schlögl R, Zemlyanov D, Aszalos-Kiss B, Hayek K, Klötzer B. Carbon incorporation during ethene oxidation on Pd(111) studied by *in situ* X-ray photoelectron spectroscopy at 2×10^{-3} mbar. J Catal. 2006;242:340–348. Doi:10.1016/j.jcat.2006.06.022.

[86] Balmes O, Resta A, Wermeille D, Felici R, Messing ME, Deppert K, Liu Z, Grass ME, Bluhm H, van Rijn R, Frenken JWM, Westerström R, Blomberg S, Gustafson J, Andersen JN, Lundgren E. Reversible

formation of a PdC_x phase in Pd nanoparticles upon CO and O_2 exposure. Phys Chem Chem Phys. 2012;14:4796–4801. Doi:10.1039/C2CP22873D.

[87] Studt F, Abild-Pedersen F, Bligaard T, Sørensen RZ, Christensen CH, Nørskov JK. On the role of surface modifications of palladium catalysts in the selective hydrogenation of acetylene. Angew Chemie Int Ed. 2008;47:9299–9302. Doi:10.1002/anie.200802844.

[88] Chan CWA, Tam KY, Cookson J, Bishop P, Tsang SC. Palladium with interstitial carbon atoms as a catalyst for ultraselective hydrogenation in the liquid phase. Catal Sci Technol. 2011;1:1584–1592. Doi:10.1039/C1CY00257K.

[89] Chan CWA, Xie Y, Cailuo N, Yu KMK, Cookson J, Bishop P, Tsang SC. New environmentally friendly catalysts containing Pd–interstitial carbon made from Pd–glucose precursors for ultraselective hydrogenations in the liquid phase. Chem Commun. 2011;47:7971–7973. Doi:10.1039/C1CC12681D.

[90] Garcia-Ortiz A, Vidal JD, Iborra S, Climent MJ, Cored J, Ruano D, Pérez-Dieste V, Concepción P, Corma A. Synthesis of a hybrid Pd^0/Pd-carbide/carbon catalyst material with high selectivity for hydrogenation reactions. J Catal. 2020;389:706–713. Doi:10.1016/j.jcat.2020.06.036.

[91] Guo R, Chen Q, Li X, Liu Y, Wang C, Bi W, Zhao C, Guo Y, Jin M. PdC_x nanocrystals with tunable compositions for alkyne semihydrogenation. J Mater Chem A. 2019;7:4714–4720. Doi:10.1039/c8ta12002a.

[92] Zafeiratos S, Piccinin S, Teschner D. Alloys in catalysis: Phase separation and surface segregation phenomena in response to the reactive environment. Catal Sci Technol. 2012;2:1787–1801. Doi:10.1039/c2cy00487a.

[93] Kovnir K, Armbrüster M, Teschner D, Venkov TV, Jentoft FC, Knop-Gericke A, Grin Y, Schlögl R. A new approach to well-defined, stable and site-isolated catalysts. Sci Technol Adv Mater. 2007;8:420–427. Doi:10.1016/j.stam.2007.05.004.

[94] Kovnir K, Armbrüster M, Teschner D, Venkov TV, Szentmiklósi L, Jentoft FC, Knop-Gericke A, Grin Y, Schlögl R. *In situ* surface characterization of the intermetallic compound PdGa — A highly

selective hydrogenation catalyst. Surf Sci. 2009;603:1784–1792. Doi:10.1016/j.susc.2008.09.058.

[95] Kovnir K, Osswald J, Armbrüster M, Teschner D, Weinberg G, Wild U, Knop-Gericke A, Ressler T, Grin Y, Schlögl R. Etching of the intermetallic compounds PdGa and Pd_3Ga_7: An effective way to increase catalytic activity? J Catal. 2009;264:93–103. Doi:10.1016/j.jcat.2009.03.007.

[96] Armbrüster M, Kovnir K, Behrens M, Teschner D, Grin Y, Schlögl R. Pd-Ga intermetallic compounds as highly selective semihydrogenation catalysts. J Am Chem Soc. 2010;132:14745–14747. Doi:10.1021/ja106568t.

[97] Shao L, Zhang W, Armbrüster M, Teschner D, Girgsdies F, Zhang B, Timpe O, Friedrich M, Schlögl R, Su DS. Nanosizing intermetallic compounds onto carbon nanotubes: Active and selective hydrogenation catalysts. Angew Chemie Int Ed. 2011;50:10231–10235. Doi:10.1002/anie.201008013.

[98] Bauer M, Schoch R, Shao L, Zhang B, Knop-Gericke A, Willinger M, Schlögl R, Teschner D. Structure-activity studies on highly active palladium hydrogenation catalysts by X-ray absorption spectroscopy. J Phys Chem C. 2012;116:22375–22385. Doi:10.1021/jp306962v.

[99] Luo Y, Alarcón Villaseca S, Friedrich M, Teschner D, Knop-Gericke A, Armbrüster M. Addressing electronic effects in the semi-hydrogenation of ethyne by $InPd_2$ and intermetallic Ga-Pd compounds. J Catal. 2016;338:265–272. Doi:10.1016/j.jcat.2016.03.025.

[100] Kaichev VV, Saraev AA, Gladky AY, Prosvirin IP, Knop-Gericke A, Bukhtiyarov VI. *In situ* study of self-sustained oscillations in propane oxidation and propane steam reforming with oxygen over nickel. Catal Lett. 2019;149:313–321. Doi:10.1007/s10562-018-2604-0.

[101] Kaichev VV, Teschner D, Saraev AA, Kosolobov SS, Gladky AY, Prosvirin IP, Rudina NA, Ayupov AB, Blume R, Hävecker M, Knop-Gericke A, Schlögl R, Latyshev AV, Bukhtiyarov VI. Evolution of self-sustained kinetic oscillations in the catalytic oxidation of propane over a nickel foil. J Catal. 2016;334:23–33. Doi:10.1016/j.jcat.2015.11.009.

[102] Kaichev VV, Saraev AA, Gladky AY, Prosvirin IP, Blume R, Teschner D, Hävecker M, Knop-Gericke A, Schlögl R, Bukhtiyarov VI. Reversible bulk oxidation of Ni foil during oscillatory catalytic

oxidation of propane: A novel type of spatiotemporal self-organization. Phys Rev Lett. 2017;119:026001. Doi:10.1103/PhysRevLett.119.026001.

[103] Teschner D, Vass EM, Schlögl R. Photoelectron spectroscopy of catalytic oxide materials. In D Jackson, JS Hargreaves (Eds.). *Metal Oxide Catalysts* (pp. 243–298). Weinheim: WILEY-VCH Verlag GmbH & Co, 2008. Doi:10.1002/9783527626113.ch6.

[104] Heenemann M, Heine C, Hävecker M, Trunschke A, Schlögl R. Influence of steam on a vanadyl pyrophosphate catalyst during propane oxidation. J Phys Chem B. 2018;122:695–704. Doi:10.1021/acs.jpcb.7b06314.

[105] Richter F, Papp H, Götze T, Wolf GU, Kubias B. Investigation of the surface of vanadyl pyrophosphate catalysts. Surf Interface Anal. 1998;26:736–741. Doi:10.1002/(SICI)1096-9918(199809)26:10<736::AID-SIA410>3.0.CO;2-Z.

[106] Trunschke A, Noack J, Trojanov S, Girgsdies F, Lunkenbein T, Pfeifer V, Hävecker M, Kube P, Sprung C, Rosowski F, Schlögl R. The impact of the bulk structure on surface dynamics of complex Mo–V-based oxide catalysts. ACS Catal. 2017;7:3061–3071. Doi:10.1021/acscatal.7b00130.

[107] Sanfiz AC, Hansen TW, Teschner D, Schnörch P, Girgsdies F, Trunschke A, Schlögl R, Looi MH, Hamid SBA. Dynamics of the MoVTeNb oxide M1 phase in propane oxidation. J Phys Chem C. 2010;114:1912–1921. Doi:10.1021/jp909352u.

[108] Wernbacher AM, Kube P, Hävecker M, Schlögl R, Trunschke A. Electronic and dielectric properties of MoV-oxide (M1 Phase) under alkane oxidation conditions. J Phys Chem C. 2019;123:13269–13282. Doi:10.1021/acs.jpcc.9b01273.

[109] Heine C, Hävecker M, Trunschke A, Schlögl R, Eichelbaum M. The impact of steam on the electronic structure of the selective propane oxidation catalyst MoVTeNb oxide (orthorhombic M1 phase). Phys Chem Chem Phys. 2015;17:8983–8993. Doi:10.1039/C5CP00289C.

[110] Koch G, Hävecker M, Teschner D, Carey SJ, Wang Y, Kube P, Hetaba W, Lunkenbein T, Auffermann G, Timpe O, Rosowski F, Schlögl R, Trunschke A. Surface conditions that constrain alkane oxidation on

perovskites. ACS Catal. 2020;10:7007–7020. Doi:10.1021/acscatal.0c01289.

[111] Xie X, Li Y, Liu Z-Q, Haruta M, Shen W. Low-temperature oxidation of CO catalysed by Co_3O_4 nanorods. Nature. 2009;458:746–749. Doi:10.1038/nature07877.

[112] Nguyen L, Zhang S, Yoon SJ, (Feng) Tao F. Preferential oxidation of CO in H_2 on pure Co_3O_{4-x} and Pt/Co_3O_{4-x}. ChemCatChem. 2015;7:2346–2353. Doi:10.1002/cctc.201500320.

[113] Tang Y, Ma L, Dou J, Andolina CM, Li Y, Ma H, House SD, Zhang X, Yang J, (Feng) Tao F. Transition of surface phase of cobalt oxide during CO oxidation. Phys Chem Chem Phys. 2018;20:6440–6449. Doi:10.1039/C7CP07407G.

[114] Jain R, Reddy KP, Ghosalya MK, Gopinath CS. Water mediated deactivation of Co_3O_4 naonrods catalyst for CO oxidation and resumption of activity at and above 373 K: Electronic structural aspects by NAPPES. J Phys Chem C. 2017;121:20296–20305. Doi:10.1021/acs.jpcc.7b05480.

[115] Lukashuk L, Yigit N, Rameshan R, Kolar E, Teschner D, Hävecker M, Knop-Gericke A, Schlögl R, Föttinger K, Rupprechter G. Operando insights into CO oxidation on cobalt oxide catalysts by NAP-XPS, FTIR, and XRD. ACS Catal. 2018;8:8630–8641. Doi:10.1021/acscatal.8b01237.

[116] Lukashuk L, Föttinger K, Kolar E, Rameshan C, Teschner D, Hävecker M, Knop-Gericke A, Yigit N, Li H, McDermott E, Stöger-Pollach M, Rupprechter G. Operando XAS and NAP-XPS studies of preferential CO oxidation on Co_3O_4 and CeO_2-Co_3O_4 catalysts. J Catal. 2016;344:1–15. Doi:10.1016/j.jcat.2016.09.002.

[117] Zhong L, Kropp T, Baaziz W, Ersen O, Teschner D, Schlögl R, Mavrikakis M, Zafeiratos S. Correlation between reactivity and oxidation state of cobalt oxide catalysts for CO preferential oxidation. ACS Catal. 2019;9:8325–8336. Doi:10.1021/acscatal.9b02582.

[118] Hilaire S, Wang X, Luo T, Gorte RJ, Wagner J. A comparative study of water-gas-shift reaction over ceria-supported metallic catalysts. Appl Catal A Gen. 2004;258:271–276. Doi:10.1016/j.apcata.2003.09.026.

[119] Phatak AA, Koryabkina N, Rai S, Ratts JL, Ruettinger W, Farrauto RJ, Blau GE, Delgass WN, Ribeiro FH. Kinetics of the water–gas shift

reaction on Pt catalysts supported on alumina and ceria. Catal Today. 2007;123:224–234. Doi:10.1016/j.cattod.2007.02.031.
[120] Trudeau ML, Tschöpe A, Ying JY. XPS investigation of surface oxidation and reduction in nanocrystalline $Ce_xLa_{1-x}O_{2-y}$. Surf Interface Anal. 1995;23:219–226. Doi:10.1002/sia.740230405.
[121] Skála T, Šutara F, Prince KC, Matolín V. Cerium oxide stoichiometry alteration via Sn deposition: Influence of temperature. J Electron Spectros Relat Phenomena. 2009;169:20–25. Doi:10.1016/j.elspec.2008.10.003.
[122] Kato S, Ammann M, Huthwelker T, Paun C, Lampimäki M, Lee M-T, Rothensteiner M, van Bokhoven JA. Quantitative depth profiling of Ce^{3+} in Pt/CeO_2 by *in situ* high-energy XPS in a hydrogen atmosphere. Phys Chem Chem Phys. 2015;17:5078–5083. Doi:10.1039/C4CP05643D.
[123] Nelin CJ, Bagus PS, Ilton ES, Chambers SA, Kuhlenbeck H, Freund H-J. Relationships between complex core level spectra and materials properties. Int J Quantum Chem. 2010;110:2752–2764. Doi:10.1002/qua.22807.
[124] Padeste C, Cant NW, Trimm DL. The influence of water on the reduction and reoxidation of ceria. Catal Lett. 1993;18:305–316. Doi:10.1007/BF00769451.
[125] Aranifard S, Ammal SC, Heyden A. On the importance of metal–oxide interface sites for the water–gas shift reaction over Pt/CeO_2 catalysts. J Catal. 2014;309:314–324. Doi:10.1016/j.jcat.2013.10.012.
[126] Aranifard S, Ammal SC, Heyden A. On the importance of the associative carboxyl mechanism for the water-gas shift reaction at Pt/CeO_2 interface sites. J Phys Chem C. 2014;118:6314–6323. Doi:10.1021/jp5000649.
[127] Zalc JM, Sokolovskii V, Löffler DG. Are noble metal-based water–gas shift catalysts practical for automotive fuel processing? J Catal. 2002;206:169–171. Doi:10.1006/jcat.2001.3465.
[128] Teschner D, Wootsch A, Pozdnyakova O, Sauer H, Knop-Gericke A, Schlögl R. Surface and structural properties of Pt/CeO_2 catalyst under preferential CO oxidation in hydrogen (PROX). React Kinet Catal Lett. 2006;87:235–247. Doi:10.1007/s11144-006-0030-6.

[129] Teschner D, Wootsch A, Pozdnyakova-Tellinger O, Kröhnert J, Vass EM, Hävecker M, Zafeiratos S, Schnörch P, Jentoft PC, Knop-Gericke A, Schlögl R. Partial pressure dependent *in situ* spectroscopic study on the preferential CO oxidation in hydrogen (PROX) over Pt/ceria catalysts. J Catal. 2007;249:318–327. Doi:10.1016/j.jcat.2007.05.010.

[130] Pozdnyakova O, Teschner D, Wootsch A, Kröhnert J, Steinhauer B, Sauer H, Toth L, Jentoft FC, Knop-Gericke A, Paál Z, Schlögl R. Preferential CO oxidation in hydrogen (PROX) on ceria-supported catalysts, part I: Oxidation state and surface species on Pt/CeO_2 under reaction conditions. J Catal. 2006;237:1–16. Doi:10.1016/j.jcat.2005.10.014.

[131] Pozdnyakova O, Teschner D, Wootsch A, Kröhnert J, Steinhauer B, Sauer H, Toth L, Jentoft FC, Knop-Gericke A, Paál Z, Schlögl R. Preferential CO oxidation in hydrogen (PROX) on ceria-supported catalysts, part II: Oxidation states and surface species on Pd/CeO_2 under reaction conditions, suggested reaction mechanism. J Catal. 2006;237:17–28. Doi:10.1016/j.jcat.2005.10.015.

[132] Pozdnyakova-Tellinger O, Teschner D, Kröhnert J, Jentoft FC, Knop-Gericke A, Schlögl R, Wootsch A. Surface water-assisted preferential CO oxidation on Pt/CeO_2 catalyst. J Phys Chem C. 2007;111:5426–5431. Doi:10.1021/jp0669862.

[133] Meunier FC. The power of quantitative kinetic studies of adsorbate reactivity by operando FTIR spectroscopy carried out at chemical potential steady-state. Catal Today. 2010;155:164–171. Doi: 10.1016/j.cattod.2009.11.017.

[134] Penkala B, Aubert D, Kaper H, Tardivat C, Conder K, Paulus W. The role of lattice oxygen in CO oxidation over Ce(18)O_2-based catalysts revealed under operando conditions. Catal Sci Technol. 2015;5: 4839–4848. Doi:10.1039/C5CY00842E.

[135] Garcia X, Soler L, Divins NJ, Vendrell X, Serrano I, Lucentini I, Prat J, Solano E, Tallarida M, Escudero C, Llorca J. Ceria-based catalysts studied by near ambient pressure X-ray photoelectron spectroscopy: A review. Catalysts. 2020;10:286. Doi:10.3390/catal10030286.

[136] van Santen RA, Kuipers HPCE. The mechanism of ethylene epoxidation. Adv Catal. 1987;35:265–321. Doi:10.1016/S0360-0564(08)60095-4.

[137] Sachtler WMH, Backx C, van Santen RA. On the mechanism of ethylene epoxidation. Catal Rev Eng. 1981;23:127–149. Doi:10.1080/03602458108068072.

[138] Bukhtiyarov VI, Knop-Gericke A. Ethylene epoxidation over silver catalysts. Nanostructured Catal Sel Oxidations. 2011;214–247. Doi:10.1039/9781847559876-00214.

[139] Serafin JG, Liu AC, Seyedmonir SR. Surface science and the silver-catalyzed epoxidation of ethylene: An industrial perspective. J Mol Catal A Chem. 1998;131:157–168. Doi:10.1016/S1381-1169(97)00263-X.

[140] Rebsdat S, Mayer D. Ethylene oxide. In *Ullmann's Encyclopedia of Industrial Chemistry*. Weinheim: Wiley-VCH Verlag GmbH & Co, 2012. Doi:10.1002/14356007.a10_117.

[141] Stegelmann C, Schiødt NC, Campbell CT, Stoltze P. Microkinetic modeling of ethylene oxidation over silver. J Catal. 2004;221:630–649. Doi:10.1016/j.jcat.2003.10.004.

[142] Pu T, Tian H, Ford M, Rangarajan S, Wachs IE. Overview of selective oxidation of ethylene to ethylene oxide by Ag catalysts. ACS Catal. 2019;9:10727–10750. Doi:10.1021/acscatal.9b03443.

[143] Lukaski A, Barteau M. Investigation of ethylene oxide on clean and oxygen-covered Ag(110) surfaces. Catal Lett. 2009;128:9–17. Doi:10.1007/s10562-008-9802-0.

[144] Barteau MA. Surface science and the advancement of direct olefin epoxidation — A perspective on the article, "Partial oxidation of higher olefins on Ag(111) and Ag(110): Conversion of styrene to styrene oxide, benzene, and benzoic acid", by Andreas Must and Robert J. Surf Sci. 2006;600:5021–5023. Doi:10.1016/j.susc.2006.09.024.

[145] Twigg GH. The catalytic oxidation of ethylene. Trans Faraday Soc. 1946;42:284–291. Doi:10.1039/tf9464200284.

[146] Bukhtiyarov VI, Havecker M, Kaichev VV, Knop-Gericke A, Mayer RW, Schlogl R. Atomic oxygen species on silver: Photoelectron spectroscopy and X-ray absorption studies. Phys Rev B. 2003;67:235422. Doi:10.1103/PhysRevB.67.235422.

[147] Bukhtiyarov VI, Nizovskii AI, Bluhm H, Hävecker M, Kleimenov E, Knop-Gericke A, Schlögl R. Combined *in situ* XPS and PTRMS study of ethylene epoxidation over silver. J Catal. 2006;238:260–269. Doi:10.1016/j.jcat.2005.11.043.

[148] Rocha TCR, Hävecker M, Knop-Gericke A, Schlögl R. Promoters in heterogeneous catalysis: The role of Cl on ethylene epoxidation over Ag. J Catal. 2014;312:12–16. Doi:10.1016/j.jcat.2014.01.002.

[149] Rocha TCR, Oestereich A, Demidov DV, Havecker M, Zafeiratos S, Weinberg G, Bukhtiyarov VI, Knop-Gericke A, Schlogl R. The silver-oxygen system in catalysis: New insights by near ambient pressure X-ray photoelectron spectroscopy. Phys Chem Chem Phys. 2012;14:4554–4564.

[150] Campbell CT. The selective epoxidation of ethylene catalyzed by Ag(111): A comparison with Ag(110). J Catal. 1985;94:436–444. Doi:10.1016/0021-9517(85)90208-8.

[151] Campbell CT, Paffett MT. Model studies of ethylene epoxidation catalyzed by the Ag(110) surface. Surf Sci. 1984;139:396–416. Doi:10.1016/0039-6028(84)90059-1.

[152] van Santen RA, de Groot CPM. The mechanism of ethylene epoxidation. J Catal. 1986;98:530–539. Doi:10.1016/0021-9517(86)90341-6.

[153] Backx C, Moolhuysen J, Geenen P, van Santen RA. Reactivity of oxygen adsorbed on silver powder in the epoxidation of ethylene. J Catal. 1981;72:364–368. Doi:10.1016/0021-9517(81)90019-1.

[154] Grant RB, Lambert RM. Mechanism of the silver-catalyzed heterogeneous epoxidation of ethylene. J Chem Soc Commun. 1983;12:662–663. Doi:10.1039/c39830000662.

[155] Grant RB, Lambert RM. A single crystal study of the silver-catalysed selective oxidation and total oxidation of ethylene. J Catal. 1985;92:364–375. Doi:10.1016/0021-9517(85)90270-2.

[156] Bukhtiyarov VI, Boronin AI, Prosvirin IP, Savchenko VI. Stages in the modification of a silver surface for catalysis of the partial oxidation of ethylene: II. Action of the reaction medium. J Catal. 1994;150:268–273. Doi:10.1006/jcat.1994.1345.

[157] Bukhtiyarov VI, Prosvirin IP, Kvon RI. Study of reactivity of oxygen states adsorbed at a silver surface towards C_2H_4 by XPS, TPD and TPR. Surf Sci. 1994;320:L47–L50. Doi:10.1016/0039-6028(94)00562-1.

[158] Bukhtiyarov VI, Boronin AI, Savchenko VI. Stages in the modification of a silver surface for catalysis of the partial oxidation of ethylene. J Catal. 1994;150:262–267. Doi:10.1006/jcat.1994.1344.

[159] Joyner RW, Roberts MW. Study of the adsorption of oxygen on silver at high-pressure by electron-spectroscopy. Chem Phys Lett. 1979;60:459–462. Doi:10.1016/0009-2614(79)80612-0.
[160] Au CT, Singhboparai S, Roberts MW, Joyner RW. Chemisorption of oxygen at Ag(110) surfaces and its role in adsorbate activation. J Chem Soc Trans I. 1983;79:1779–1791. Doi:10.1039/f19837901779.
[161] Özbek MO, van Santen RA. The mechanism of ethylene epoxidation catalysis. Catal Lett. 2013;143:131–141. Doi:10.1007/s10562-012-0957-3.
[162] Ozbek MO, Onal I, van Santen RA. Effect of surface and oxygen coverage on ethylene epoxidation. Top Catal. 2012;55:710–717. Doi:10.1007/s11244-012-9870-7.
[163] Linic S, Barteau MA. Construction of a reaction coordinate and a microkinetic model for ethylene epoxidation on silver from DFT calculations and surface science experiments. J Catal. 2003;214: 200–212. Doi:10.1016/S0021-9517(02)00156-2.
[164] Bocquet ML, Loffreda D. Ethene epoxidation selectivity inhibited by twisted oxametallacycle: A DFT study on Ag surface-oxide. J Am Chem Soc. 2005;127:17207–17215. Doi:10.1021/ja051397f.
[165] Barteau MA, Bowker M, Madix RJ. Formation and decomposition of acetate intermediates on the Ag(110) surface. J Catal. 1981;67: 118–128. Doi:10.1016/0021-9517(81)90266-9.
[166] Sault AG, Madix RJ. The mechanism of acetate oxidation on Ag(110). Surf Sci. 1986;172:598–614. Doi:10.1016/0039-6028(86)90503-0.
[167] Linic S, Barteau MA. Control of ethylene epoxidation selectivity by surface oxametallacycles. J Am Chem Soc. 2003;125:4034–4035. Doi:10.1021/ja029076g.
[168] Linic S, Barteau MA. Formation of a stable surface oxametallacycle that produces ethylene oxide. J Am Chem Soc. 2002;124:310–317. Doi:10.1021/ja0118136.
[169] Christopher P, Linic S. Engineering selectivity in heterogeneous catalysis: Ag nanowires as selective ethylene epoxidation catalysts. J Am Chem Soc. 2008;130:11264–11265. Doi:10.1021/ja803818k.
[170] Kokalj A, Gava P, de Gironcoli S, Baroni S. What determines the catalyst's selectivity in the ethylene epoxidation reaction. J Catal. 2008;254:304–309. Doi:10.1016/j.jcat.2008.01.008.

[171] Greeley J, Mavrikakis M. On the role of subsurface oxygen and ethylenedioxy in ethylene epoxidation on silver. J Phys Chem C. 2007;111:7992–7999. Doi:10.1021/jp070490i.

[172] Linic S, Medlin JW, Barteau MA. Synthesis of oxametallacycles from 2-iodoethanol on Ag(111) and the structure dependence of their reactivity. Langmuir. 2002;18:5197–5204. Doi:10.1021/la011783k.

[173] Jones GS, Mavrikakis M, Barteau MA, Vohs JM. First synthesis, experimental and theoretical vibrational spectra of an oxametallacycle on a metal surface. J Am Chem Soc. 1998;120:3196–3204. Doi:10.1021/ja973609h.

[174] Jones TE, Rocha TCR, Knop-Gericke A, Stampfl C, Schlogl R, Piccinin S. Thermodynamic and spectroscopic properties of oxygen on silver under an oxygen atmosphere. Phys Chem Chem Phys. 2015;17:9288–9312. Doi:10.1039/C5CP00342C.

[175] Carbonio EA, Rocha TCR, Klyushin AY, Píš I, Magnano E, Nappini S, Piccinin S, Knop-Gericke A, Schlögl R, Jones TE. Are multiple oxygen species selective in ethylene epoxidation on silver? Chem Sci. 2018;9:990–998. Doi:10.1039/C7SC04728B.

[176] Jones TE, Wyrwich R, Bocklein S, Carbonio EA, Greiner MT, Klyushin AY, Moritz W, Locatelli A, Mentes TO, Nino MA, Knop-Gericke A, Schlogl R, Gunther S, Wintterlin J, Piccinin S. The selective species in ethylene epoxidation on silver. ACS Catal. 2018;8:3844–3852.

[177] Savio L, Gerbi A, Vattuone L, Baraldi A, Comelli G, Rocca M. Monitoring super- and subsurface oxygen on Ag(210) by high energy resolution X-ray photoelectron spectroscopy: Subsurface diffusion and segregation. J Phys Chem B. 2006;110:942–947. Doi:10.1021/jp0533168.

[178] Savio L, Gerbi A, Vattuone L, Pushpa R, Bonini N, de Gironcoli S, Rocca M. Subsurface oxygen stabilization by a third species: Carbonates on Ag(210). J Phys Chem C. 2007;111:10923–10930. Doi:10.1021/jp0714421.

[179] Jones TE, Rocha TCR, Knop-Gericke A, Stampfl C, Schlogl R, Piccinin S. Insights into the electronic structure of the oxygen species active in alkene epoxidation on silver. ACS Catal. 2015;5:5846–5850. Doi:10.1021/acscatal.5b01543.

[180] Carlisle CI, Fujimoto T, Sim WS, King DA. Atomic imaging of the transition between oxygen chemisorption and oxide film growth on Ag{111}. Surf Sci. 2000;470:15–31. Doi:10.1016/S0039-6028(00)00831-1.

[181] Palermo A, Husain A, Lambert RM. Triply-promoted ethene epoxidation: NO_x promotion of the Ag-catalysed reaction in the presence of alkali and chlorine under electrochemical control. Catal Lett. 2000;69:175–179. Doi:10.1023/A:1019026124235.

[182] Bare SR, Griffiths K, Lennard WN, Tang HT. Generation of atomic oxygen on Ag(111) and Ag(110) using NO_2: A TPD, LEED, HREELS, XPS and NRA study. Surf Sci. 1995;342:185–198. Doi:10.1016/0039-6028(95)00670-2.

[183] Günther S, Böcklein S, Wintterlin J, Niño MA, Menteş TO, Locatelli A. Locating catalytically active oxygen on Ag(1 1 1)-A spectromicroscopy study. ChemCatChem. 2013;5:3342–3350. Doi:10.1002/cctc.201300355.

[184] Bocklein S, Gunther S, Wintterlin J. High-pressure scanning tunneling microscopy of a silver surface during catalytic formation of ethylene oxide. Angew Chemie Int Ed. 2013;52:5518–5521. Doi:10.1002/anie.201210209.

[185] Grant RB, Lambert RM. Ethylene-oxide isomerization on single-crystal Ag(111) in atomically clean and Cs-moderated conditions. J Catal. 1985;93:92–99. Doi:10.1016/0021-9517(85)90154-X.

[186] Zimmermann H, Walzl R. Ethylene. In *Ullmann's Encyclopedia of Industrial Chemistry*. n.d. Doi:10.1002/14356007.a10_045.pub3.

[187] Torbina VV, Vodyankin AA, Ten S, Mamontov GV, Salaev MA, Sobolev VI, Vodyankina OV. Ag-based catalysts in heterogeneous selective oxidation of alcohols: A review. Catalysts. 2018;8:447. Doi:10.3390/catal8100447.

[188] Franz AW, Kronemayer H, Pfeiffer D, Pilz RD, Reuss G, Disteldorf W, Gamer AO, Hilt A. Formaldehyde. In *Ullmann's Encyclopedia of Industrial Chemistry* (pp. 1–34). Weinheim: Wiley-VCH Verlag GmbH, 2016, Doi:10.1002/14356007.a11_619.pub2.

[189] Bahmanpour AM, Hoadley A, Tanksale A. Critical review and exergy analysis of formaldehyde production processes. Rev Chem Eng. 2014;30:583–604. Doi:10.1515/revce-2014-0022.

[190] Nagy A, Mestl G. High temperature partial oxidation reactions over silver catalysts. Appl Catal A Gen. 1999;188:337–353. Doi:10.1016/S0926-860X(99)00246-X.

[191] Waterhouse GIN, Bowmaker GA, Metson JB. Mechanism and active sites for the partial oxidation of methanol to formaldehyde over an electrolytic silver catalyst. Appl Catal A Gen. 2004;265:85–101. Doi:10.1016/j.apcata.2004.01.016.

[192] Millar GJ, Collins M. Industrial production of formaldehyde using polycrystalline silver catalyst. Ind Eng Chem Res. 2017;56:9247–9265. Doi:10.1021/acs.iecr.7b02388.

[193] Bao X, Muhler M, Schedel-Niedrig T, Schlögl R. Interaction of oxygen with silver at high temperature and atmospheric pressure: A spectroscopic and structural analysis of a strongly bound surface species. Phys Rev B. 1996;54:2249–2262. Doi:10.1103/PhysRevB.54.2249.

[194] Rehren C, Isaac G, Schlögl R, Ertl G. Surface and subsurface products of the interaction of O_2 with Ag under catalytic conditions. Catal Lett. 1991;11:253–265. Doi:10.1007/BF00764316.

[195] Reichelt R, Günther S, Wintterlin J. Strongly-bound oxygen on silver surfaces: A molybdenum oxide contamination? J Phys Chem C. 2011;115:17417–17428. Doi:10.1021/jp203434s.

[196] Shavorskiy A, Neppl S, Slaughter DS, Cryan JP, Siefermann KR, Weise F, Lin M-F, Bacellar C, Ziemkiewicz MP, Zegkinoglou I, Fraund MW, Khurmi C, Hertlein MP, Wright TW, Huse N, Schoenlein RW, Tyliszczak T, Coslovich G, Robinson J, Kaindl RA, Rude BS, Ölsner A, Mähl S, Bluhm H, Gessner O. Sub-nanosecond time-resolved ambient-pressure X-ray photoelectron spectroscopy setup for pulsed and constant wave X-ray light sources. Rev Sci Instrum. 2014;85:93102. Doi:10.1063/1.4894208.

[197] Knudsen J, Gallo T, Boix V, Strømsheim M, D'Acunto G, Goodwin C, Wallander H, Zhu S, Soldemo M, Lömker P, Cavalca F, Scardamaglia M, Degerman D, Nilsson A, Amann P, Shavo A, Schnadt J. Stroboscopic operando spectroscopy of the dynamics in heterogeneous catalysis by event-averaging. Nat Commun. 2021;12:6117. Doi: 10.1038/s41467-021-26372-y.

[198] Nemšák S, Shavorskiy A, Karslioglu O, Zegkinoglou I, Rattanachata A, Conlon CS, Keqi A, Greene PK, Burks EC, Salmassi F, Gullikson

EM, Yang S-H, Liu K, Bluhm H, Fadley CS. Concentration and chemical-state profiles at heterogeneous interfaces with sub-nm accuracy from standing-wave ambient-pressure photoemission. Nat Commun. 2014;5:5441. Doi:10.1038/ncomms6441.

[199] Wang J, Lu D, Li C, Zhu Y, Boscoboinik JA, Zhou G. Measuring charge transfer between adsorbate and metal surfaces. J Phys Chem Lett. 2020;11:6827–6834. Doi:10.1021/acs.jpclett.0c02002.

https://doi.org/10.1142/9781800613294_0011

Chapter 11

Scanning and Full-Field Imaging Photoelectron Microscopy Studies Relevant to Heterogeneous Catalysis

Luca Gregoratti*, Andrea Locatelli† and Maya Kiskinova‡

Elettra-Sincrotrone Trieste S.C.p.A, Area Science Park, Basovizza, Trieste 34149, Italy

*luca.gregoratti@elettra.eu

†andrea.locatelli@elettra.eu

‡maya.kiskinova@elettra.eu

Abstract

The complexity of catalytic systems is multilevel and needs profound knowledge of their properties and their chemical-structural evolution under reaction conditions. In this respect, the complementary capabilities of synchrotron-based photoelectron microscopy methods in terms of imaging, spectroscopy, and spatial resolution have opened unique opportunities to explore the structure and chemical composition of catalyst systems at relevant length and time scales and correlate them to the fabrication or operating conditions. Using selected results, this chapter illustrates the potential of scanning and full-field imaging photoelectron microscopes operated at synchrotron facilities in characterization of metal catalyst and electrocatalyst systems.

11.1 Introduction

Heterogeneous catalysis is one of the most extensively studied topics in the field of modern chemistry and chemical engineering, since it is in the heart of the chemical industry, fuel and energy production, conversion, and storage, while it is also crucial for environmental protection. The complexity of catalytic systems is multilevel both in space and time [1], and to attain and preserve the desired functionality, we need to acquire profound knowledge of the catalysts' properties and their evolution under operation conditions. Among the most critical issues to be addressed is the development of active reaction phases, the bonding and mass transport of species at the catalyst surface and subsurface, the modifications or degradation of the catalysts under operation, etc. Since catalysts are dynamic systems and their performance depends on coupling between different electronic, structural, and mass transport events, occurring at timescales from fs to days, and spacescales from less than nm to mm, we are still far from fully understanding how to design and control the catalytic performance. In this respect, the complementary capabilities of synchrotron-based microscopy methods in terms of imaging, spectroscopy, spatial resolution, and variable probing depths have opened unique opportunities to explore the structure and chemical composition of morphologically complex catalytic systems at relevant length and timescales and correlate them to the fabrication or operating conditions [2–6].

Considering the fact that most of the processes in heterogeneous catalysis occur at reactant/surface interfaces, surface and chemically sensitive techniques are highly required in order to shed light on the catalytic reactions. X-ray photoelectron spectroscopy (XPS), or as will be frequently called in this chapter Photoelectron Spectroscopy (PES), has endorsed remarkable achievements in the characterization of all types of matter. As described in detail in the previous chapters, this is due to the fact that XPS provides multiple information for the elemental composition, chemical state, and electronic structure of the system under investigation [7]. The sample probing depth, determined by the kinetic energy of the emitted photoelectrons, varies from less than 1 to a few nm, reaching ~10 nm only for slow secondary electrons or photoelectrons excited by hard X-rays. This makes XPS a surface and near subsurface sensitive technique, appropriate for the investigation of surface and interfacial phenomena that govern the properties of many functional materials relevant to catalysis, energy harvesting and storage, electronic devices, corrosion, biomedical appliances, etc.

For all surface-interfacial systems, essential factors are the atomic arrangement and the compositional profile across the interface. In the case of catalysis, it is also important to know how the interfacial structure, composition, and functionality respond to external stimuli, such as temperature, electric field, light, or exposure to various gas environments. However, it has long been recognized that surfaces and interfaces of almost all natural and man-made materials are not uniform, comprising lateral heterogeneities at different microscopic and sub-microscopic length scales. This has pushed advances in electron analyzers of classical XPS instruments (i.e., non-synchrotron-based) that allow imaging, implemented first in 1990 [8], while recently reaching a resolution of ~5 nm using monochromatic Mg Kα or Al Kα X-rays [9]. Submicron resolution with laboratory instruments was first achieved with Photoelectron Emission Microscopes (PEEM) but the provided information was limited to the lateral variation of the surface work function using UV light [10]. In these imaging instruments, the emitted electrons are collected by an objective lens, and by means of electrostatic and/or magnetic lenses, a magnified image of the probed area is projected onto a screen, converting the electron signal into visible light. For implementing band pass electron energy filtering in PEEM instruments, the X-ray beam illuminating the sample should be very intense and small in size, properties that cannot be provided by the conventional X-ray sources [10,11]. Another approach for getting high spatial resolution is to use a micro/nano photon beam combined with a sample scanning stage, as in

Scanning Probe Microscopes [12,13]. As for PEEM, also for this approach, the major obstacle was the low brightness of the conventional X-ray sources.

The “photon source” limits have been overcome by the light generated at the 3rd generation synchrotron facilities, which combines ultrahigh brightness with tunability and variable polarization [14]. Consequently, the tunability provides a great advantage to select the photon energy, so as to use conditions with a high cross-section for the atomic core levels under consideration. Another opportunity opened using the ultrabright synchrotron light is that along with XPS one can also perform another very powerful spectroscopy — X-ray Absorption Spectroscopy (XAS) and, in particular, X-ray Absorption Fine Structure (XAFS). Absorption spectroscopies are based on the monitoring of the total electron yield while scanning the energy of the incident photons [15]. As in XPS, the chemical sensitivity of XAFS is due to the characteristic binding energies of core electrons, giving rise to X-ray absorption thresholds and resonant electronic transitions from core levels into unfilled valence states, governed by well-established selection rules. Using circularly or linearly polarized synchrotron beams, XAFS can also provide information on the magnetic moment and symmetry of the chemical bonds, called X-ray Magnetic Circular or Linear Dichroism (XMCD or XMLD) [16]. It should be noted that since in photoelectron-based instruments XAFS is measured in total electron yield, the signal is dominated by secondary electrons, the probing depth of which is up to few nm. Figure 11.1 illustrates the principle setups of SPEM and X-ray photoemission electron microscopy (XPEEM) instruments.

After the first experiments demonstrating the imaging and microspectroscopy abilities of Scanning PhotoElelectron Microscopy (SPEM) using synchrotron light [17] and X-ray PhotoElectron Emission Microscopy (XPEEM) [18], these microscopes have undergone many developments and presently are among the highest requested for exploiting the properties and the evolution of specimen. Please note that the spatial resolution of these techniques is between light and electron microscopy [19–22].

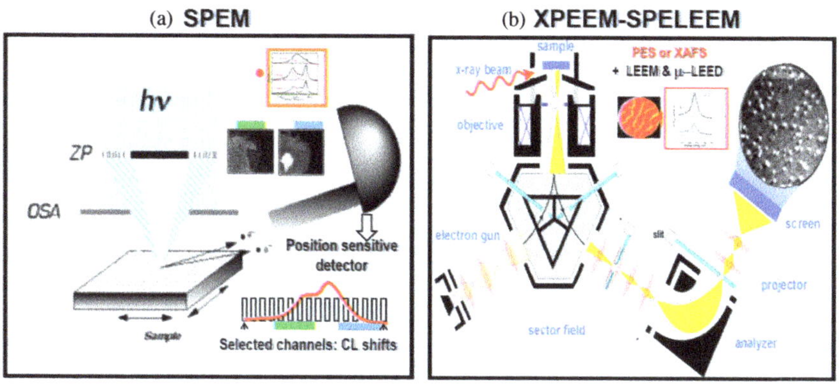

Fig. 11.1. (a) SPEM with zone plate/order sorting aperture photon focusing system where images are obtained by scanning the sample with respect to the microspot of 50–100 nm. Using multichannel detection systems, the number of simultaneously collected images is equal to the number of the channels, each of them monitoring specific electron kinetic energy within the selected energy window. The multichannel imaging allows not only mapping different chemical states with a single sample scan but also spectra-imaging, i.e., to get the spectrum corresponding to the energy window from selected image regions. (b) XPEEM, where the synchrotron beam is focused into a few micron spot on the sample and the electron energy analyzer provides PES-XAFS spectroscopy. In SPELEEM instruments, XPEEM is combined with an electron beam for LEEM-LEED measurements.

The lateral resolution of XPEEM is determined only by lens aberration, so theoretically, it may reach ~2 nm, but the high fields at the sample (up to 20 keV), necessary for high collection efficiency and lateral resolution, limit the XPEEM use only for flat and conductive samples. In addition, the spectral resolution cannot compete with that of the best non-microscopy photoelectron spectroscopy instruments. XPEEM being a direct imaging instrument is very suitable for exploiting dynamic processes and has been one of the most desired instruments at synchrotron facilities for exploiting surface chemical waves, phase separation, and magnetization dynamics with spatial resolution in imaging mode down to 20 nm [21–23]. For comparison, in the case of microspot PES measurements (μ-PES), the spatial resolution is restricted to a spot of 1–2 microns.

Some of the XPEEM instruments are combined with an electron beam probe and are called Spectroscopic Photoemission and Low Energy Electron Microscope (SPELEEM) [11,21]. The combination of LEEM together with micro Low Energy Electron Diffraction technique (micro LEED) makes SPELEEM a multitechnique instrument. The backscattering electron signal provides complementary structural imaging for monitoring surface events, such as structural changes and phase formation and phase transition processes.

SPEM, being a scanning instrument, cannot compete with XPEEM in *time* resolution, while at present, its routine *spatial* resolution is 50–100 nm. The latter is determined by the focusing optics and the photoemitted electrons and collection angle that influence the signal. However, since SPEM spatial resolution does not depend on the kinetic energy of the emitted electrons, it preserves the spectral performance of classical XPS instruments. Please note that the imaging and microspot spectroscopy are independent, therefore the instrument can be optimized in terms of spectral resolution and acquisition time of each mode, accordingly [19,20].

Another very important advantage of SPEM, as compared to XPEEM, is that it can be used to study rough, low-conductive samples and free-standing individual nanostructures, such as nanotubes, nanowires, and nanobelts [13]. Last, but not least, since the specimen under investigation by SPEM is not part of the detection system, it is much easier to develop options for different sample environments, such as, for example, implementing near-ambient pressure setups and applying electrical fields [13,24–27]. It should be noted that the first approach for overcoming the pressure gap in non-microscopic XPS has been suggested by Siegbahn in 1969, and today, (near-)ambient pressure XPS (or NAP-XPS) is very commonly used at synchrotron facilities where it can be combined with XAS measurements as well [28] (see Chapters 10, 12, and 13).

This chapter focuses on the illustration of the unique capabilities of the two types of synchrotron-based PES microscopes, in terms of chemical imaging and microspectroscopy. The chapter explores chemical processes occurring at surfaces and interfaces, where issues, such as complexity at the microscopic length scale, should be faced and understood. The most recent achievements in this respect will be illustrated by presenting selected results, spanning over different catalytic and electrochemical systems.

11.2 Metal Catalysts: From Single Crystals to Model-Supported Catalysts

It is well known that the composition of metal catalysts developed under reaction conditions encompasses desired information about the catalysts' active state. For shedding light on this subject, a great number of studies have been done starting with model single-crystal or polycrystalline metal samples. In a later stage, supported model catalysts were prepared, an important step towards overcoming the so-called materials gap (see Chapter 7). Here, we would like to recall that the model catalytic systems were first explored in the frame of surface science approach, under well-defined vacuum conditions using XPS, STM, LEED, LEEM, and PEEM. These surface science studies have provided important information about (i) molecular and dissociative adsorption of reagents and the interaction of the species with metal surfaces; (ii) possible structural and compositional changes of metal surfaces and the role of surface structural imperfections as steps, defects, or grains of polycrystalline samples; (iii) effects of additive atoms, some playing role as promoters or poisons by affecting the adsorption bond strength, dissociative probability, and surface reactivity; (iv) dynamic processes, as development and propagation of reaction fronts, phase separation resulting in the formation of stationary structures, etc. In the following sections, we will report how the advantage of SPEM for detailed characterization of the local chemical state of the substrate and the adsorbed species under variable ambient, and the advantage of XPEEM-LEEM to follow *in situ* the dynamic processes, have shed light on the complex process occurring under reaction conditions.

11.2.1 *Compositional heterogeneity of metal catalyst surfaces developed under oxidation–reduction reaction conditions*

The first important step to understand the mechanism of catalytic action is to verify the microscopic morphology of the catalyst surface developed under reaction conditions, which is related to the active phase. In the last two decades, both experimental and theoretical studies have confirmed that the formation of non-stoichiometric transient "surface" oxides is a common phenomenon under reaction conditions for metals relevant to catalysis, such as Pd, Ru, Rh, Pt, and Ag [29–36]. The "surface" oxide structures differ from that of the bulk oxides, and their formation after an increasing atomic oxygen coverage is preceded by atomic oxygen adsorption phases. Indeed, the multicomponent metal and O core-level spectra from NAP-XPS studies clearly showed the compositional complexity of the catalyst surfaces under dynamic reaction conditions, indicating that small differences in the oxygen content in the so-called transient oxides may lead to sensible selectivity variations [37]. Following the STM, LEED, and PEEM results, supported by NAP-XPS, the SPEM studies shed more light on the development, the spatial distribution, and the actual chemical state of the coexisting phases on transition metal surfaces, including stoichiometric and non-stoichiometric oxides and "dissolved" and adsorbed oxygen.

The first SPEM experiments were performed using Ru(0001) [38] and Rh(110) [39] single-crystal samples that were exposed to reaction conditions in a separate high-pressure cell or using an atomic oxygen source. These studies clearly demonstrated that even a single-crystal metal surface develops heterogeneity in the oxidation state under usual reaction conditions. The chemical imaging is based on exploiting the chemical shifts undergone by the Ru 3d and Rh 3d core levels corresponding to different oxidation states. Since the core-level energy shifts are determined by the oxygen-metal bonding and coordination number, this makes it possible to discriminate between stoichiometric oxide, intermediate suboxides, and O chemisorption phases. As reported in Refs. [28–33], in the initial oxidation state, both Ru 3d and Rh 3d spectra still contain metallic components, but their intensity decreases continuously with the growth of an oxide layer on top. The signal attenuation can be used to estimate the oxide layer thickness. It should be noted that the formation of stoichiometric oxide usually requires high temperatures; it is therefore kinetically hindered at low-temperature treatments. The incorporation of oxygen into the metal lattice (subsurface), at the initial step of the oxide growth, necessitates to reach a critical coverage of about 4–5 monolayers (ML). Detailed studies of Ru(0001) catalyst at various temperatures have shown that the RuO_2 growth starts only at temperatures above 600 K when the subsurface oxygen reaches 4–5 ML.

Figure 11.2(a) illustrates chemical images and their spectroscopic verification for the local composition of the phases developing on a Ru(0001) surface under oxidation conditions. The two

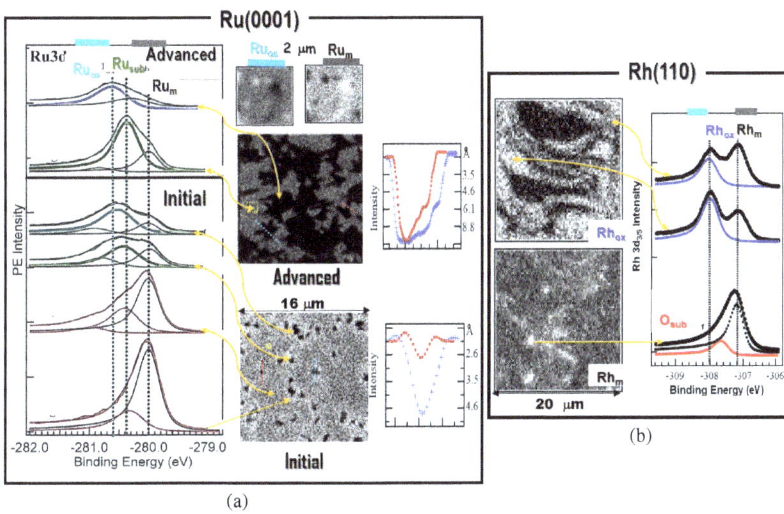

Fig. 11.2. **(a) Ru(0001) single crystals:** (left) Microspot Ru $3d_{5/2}$ spectra, measured in selected dark and bright areas in the images, indicated by the arrows. (Middle) 16x16 μm^2 Ru 3d chemical images tuned to the binding energy range of the Ru_m component, indicated with the gray line on top of the spectra panel, representing the initial (bottom) and advanced (center) oxidation states of Ru(0001) surface. The high-resolution 2×2 μm^2 images on top are taken from the area denoted with a yellow square in the advanced oxidation image and represent the energy range of Ru_{ox} component, indicated with a blue line in the spectra panel, and Ru_m component. (Right) Intensity profiles measured across the features indicated with red and blue dashes in the images. The left axis of the intensity profiles shows the calculated thickness of the "oxide" layer based on the attenuation of Ru_m photoelectron emission. **(b) Rh(110) single crystals:** (left) Rh chemical images representing the binding energy range of the Rh_{ox} and Rh_m components, indicated with blue and gray lines ion top of the spectra panel. (Right) Rh $3d_{5/2}$ spectra, taken in the indicated by arrows spots, representing the lateral inhomogeneity developed during the growth of Rh oxide. The Rh_{ox} and Rh_m images represent the binding energy range of the Rh_{ox} and Rh_m components, indicated with blue and gray lines on top of the spectra panel.

16x16 μm^2 Ru images "visualize" the spatial anisotropy of the initial and advanced stages of oxide growth. They are obtained by measuring only the intensity variations of the Ru 3d component, Ru_m (m for metal) corresponding to the initial metallic state. The dark features visible in the image are due to the local growth of transient and stoichiometric oxide phases represented by the Ru_{sub} and Ru_{ox} components in the Ru $3d_{5/2}$ spectra (the subscripts *sub* and *ox* denote the suboxide and the stoichiometric oxides, respectively). At the initial oxidation stage, they appear only as small nuclei. The Ru_m intensity profiles taken across these dark patches (see line profile graphs in Figure 11.2(a)) can be used as a measure of the local "oxide" thickness attained under actual oxidation conditions. For the two cases, illustrated in Figure 11.2(a), the initial "oxide-like" phase does not exceed two layers, whereas in advanced oxidation, reached at higher temperatures, it exceeds three layers exhibiting a directional oxide growth [38]. The breaks in the slope in the profiles across around 5.5 and 9 Å correspond to ~2 and 3 oxide layers indicating layer-by-layer RuO_2 growth, whereas this is not the case in the initial state where the profile slopes are alike.

Another finding is contained in the small area-high resolution (2x2 μm^2) chemical images on top, taken in the indicated brighter area of the advanced stage. They outline the presence of oxide nuclei appearing bright in the image taken within the energy range of the Ru_{ox} component and dark in the image taken within the energy range of the Ru_m component. This finding suggests that the bright areas in the advanced stage represent somewhat an intermediate oxidation state with oxide nuclei. A notable feature of the initial oxidation stage is that the oxide nuclei are decorated by "brighter" O-depleted areas, probably due to limited mobility of the O adatoms or RuO_x clusters. The different local O content in the dominant bright areas and the brighter decoration around the nucleus are confirmed by the corresponding microspot Ru 3d spectra. The latter contains a smaller Ru_{sub} component, attributed to the transient state with subsurface oxygen. Also, the dark nucleus spectra exhibit different oxidation advancements, which is confirmed by the intensity profiles as well. The two spectra representing dark and bright areas of the advanced stage confirm that Ru oxide is dominant in the dark areas, whereas the bright areas resemble somewhat the dark areas of the initial state. The measured corresponding O 1s spectra confirm the information provided by the Ru 3d spectra [31,38].

As shown in Figure 11.2(b), very similar lateral inhomogeneity in the oxidation state was also observed for Rh(110) surfaces where oxide islands can coexist with different transient oxide phases. Similar to the case of Ru, the incorporation of oxygen and the following oxide formation in O_2 atmosphere requires higher temperatures [39].

The next step towards overcoming the material gap between single-crystal metal surfaces and real catalysts is studies on polycrystalline samples. A polycrystalline material consists of small grains (crystallites) with different crystallographic orientations; the interfaces between two or more grains are called the grain boundaries. The grain boundaries are extended to the surface and are considered defects of the crystal structure that undoubtedly affect the reactivity. The first relevant SPEM experiments tackled the reactions that occurred when Pd is deposited on Ni polycrystalline samples, which revealed that the amount of adsorbed Pd and the probability that a Pd-Ni alloy is formed depend on the crystal structure of the actual surface grain [40,41].

Following the oxidation of Ru (0001) single-crystal surface, demonstrated in Figure 11.2(a), the oxidation of polycrystalline Ru samples was also explored with SPEM. It has been found that oxygen accumulation via penetration into the subsurface is much faster on a polycrystalline Ru sample as compared to a closed-packed Ru (0001) single crystal. The image in Figure 11.3(a) represents two grains of the Ru polycrystalline sample at which the oxidation is monitored. They are taken on an atomically clean surface, before exposure to oxygen, and represent the whole binding energy range of metallic Ru, which comprises both the bulk and the surface Ru 3d components. Supposing that the intensity of the emitted signal is proportional to the number of emitting atoms, we assigned the grains appearing darker (1) and brighter (2) to Ru (1010) and

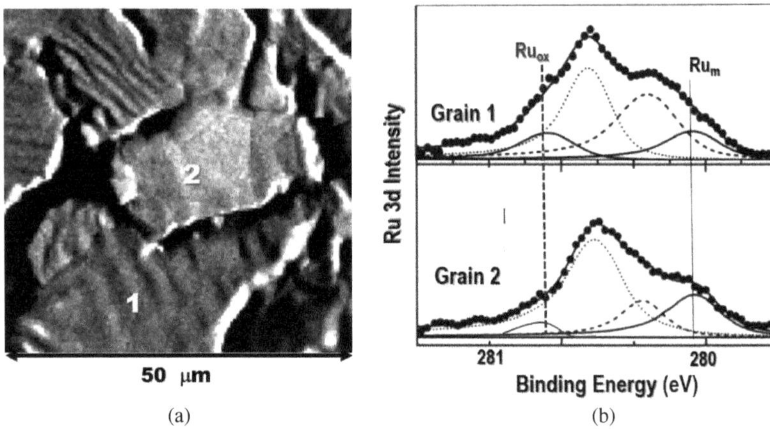

Fig. 11.3. (a) Ru $3d_{5/2}$ images of Ru polycrystalline sample before exposure to oxygen; (b) Rh 3d spectra taken in the indicated grains, showing their different Ru oxidation states after exposure to oxygen.

Ru(0001) surfaces, respectively. It should be noted that the topography due to grains' edges and structural imperfections of the grains are well visible, and these imperfections are the reason for the poorer spectral resolution, as compared to the well-ordered single-crystal surfaces.

The dependence of the oxidation rate on the local grain crystallographic structure is demonstrated by comparing the spectra from the two grains, shown in the right part of Figure 11.3(b). Judging from the intensity of the Ru_m and Ru_{ox} components, it is clear that the Ru (1010) (Grain 1) has attained a more advanced oxidation stage under the same reaction conditions than the Ru (0001) surface. The other components are tentatively assigned to subsurface oxygen and transient surface oxide.

Very recent SPEM studies complemented with direct PEEM imaging of Rh polycrystalline samples have revealed not only the anisotropy of surface oxidation but also its impact on the catalytic oxidation of hydrogen [42]. In particular, the latest SPEM results added also new information about the effect of irregularities as steps and kinks on the rate of surface oxide formation that has been correlated to the different local reactivity in hydrogen oxidation.

11.2.2 *Reactivity differences in supported micro and nanoparticles and compositional heterogeneity within microparticles*

The next step in overcoming the material gap in SPEM studies of model catalysts is to explore the catalytic behavior of Rh nanoparticles, microparticles, and nanocrystalline films deposited (supported) on oxide supports. The samples were prepared by means of pulsed laser deposition of Rh on a MgO planar support. The SPEM studies addressed the impact of particles' size and surface structure of microparticles. To do that we compare the reactivity of Rh nanoparticles, microparticles, and nanocrystalline films grown on the MgO supports under exposure to identical oxidizing and reducing conditions [43,44]. By selecting the deposition conditions, it was possible to have the coexistence of microparticles with nanoparticles or with nanofilms, as illustrated in Figure 11.4(a).

Before SPEM characterization, the morphology of the prepared samples was confirmed by Scanning Electron Microscopy (SEM) and Atomic Force Microscopy (AFM). SEM revealed that the round microparticles, easily resolved with SPEM, expose more defined grains, whereas the oval-shaped particles appear more facetted. These significant structural differences have a strong impact on the reactivity. This was demonstrated by SPEM results showing that after exposure to the same oxidative conditions, the oval-shaped particles attain a higher oxidation state. SEM images also showed that, for very similar sizes and shapes of Rh particles, the deposited microparticles exhibit different grain structures. The different reactivity of these areas towards oxygen was confirmed by SPEM (see Figure 11.4(b)).

This study has also revealed marked variations in the reactivity of supported microparticles with sizes from a few hundred nm to a few micrometers. The size of the randomly distributed nanoparticles was 5–20 nm and the resolution of SPEM allowed comparing, on an average scale, the reactivity of a single microparticle with the nanoparticles in the surrounding area. The spectra in Figure 11.4(a) clearly indicate the different reactivities of the nanostructured film,

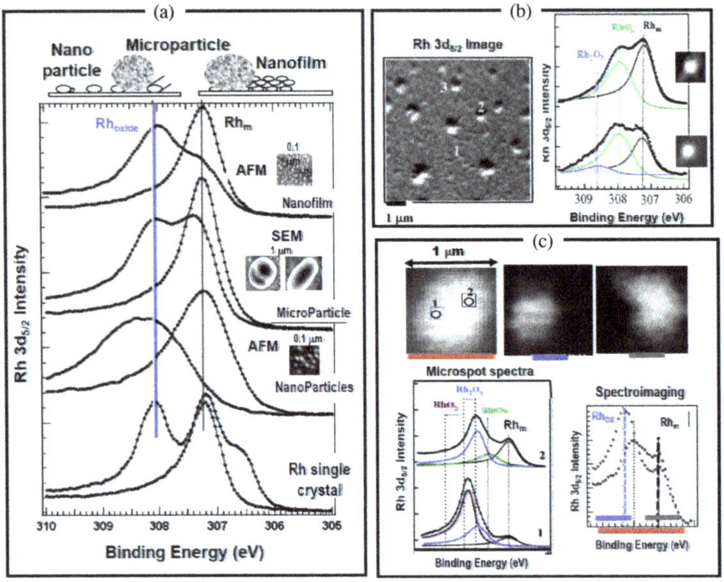

Fig. 11.4. (a) Rh $3d_{5/2}$ spectra taken after exposure to oxygen ambient at 550 K showing the reactivity difference of the exposed species. The inserted SEM and AFM images represent typical microparticles, nanoparticles, and nanofilm morphology. The top sketches represent the two types of samples where microparticles are surrounded by nanofilm or nanoparticles. (b) Rh $3d_{5/2}$ image where 1 indicates the area with nanoparticles (not resolvable with SPEM). The microparticles are the bright features, while 2 and 3 indicate the two microparticles with identical sizes (the corresponding higher resolution SPEM images are shown in the spectra panel). The Rh $3d_{5/2}$ spectra taken from the two microparticles exposed to the same oxidizing conditions demonstrate their different reactivity. (c) Top: Rh particle images taken using different parts of the Rh $3d_{5/2}$ energy window, indicated with red, blue, and gray lines in the spectroimaging panel. The left image taken within the whole Rh energy window (red line) shows the whole particle. The central image corresponds to the energy window, indicated by the blue line, outlining RhO_2. The right image corresponds to the energy window, indicated by the gray line, outlining Rh_m. The circles and squares in the left Rh $3d_{5/2}$ image indicate the places of microspot and spectroimaging measurements, respectively.

a microparticle, and an area of nanoparticles. The latter appears on average to be the most reactive of the three. In this case, the Rh(110) single-crystal spectrum was used as a reference to compare the position of Rh 3d metallic and oxide components.

Apparently, due to structural differences, the width of the Rh_m and Rh_{ox} components increases and becomes rather broad in the case of nanoparticles. This is a typical characteristic of spectra recorded on various-sized nanoparticles due to spectra superposition. It should be also notified that the microparticle spectrum represents only one selected microparticle, since, as shown in Figure 11.4(b), even microparticles with identical sizes and shapes have different reactivity. As mentioned above, the results in Figure 11.4(a) show that on average, nanoparticles reach a more advanced oxidation state as compared to the microparticles (i.e., easier to be oxidized). However, this reactivity trend is reversed when they are exposed to reducing ambient (H_2), where the microparticles undergo a faster reduction than the nanoparticles.

Comparison of the oxidation–reduction state of different microparticles of the same size after exposure to an oxidative environment unraveled the difference in their relative reactivity. In particular, the Rh $3d_{5/2}$ spectra shown in Figure 11.4(b) indicate differences in the contribution of oxidized Rh components (RhO_x and Rh_2O_3) for the two selected particles. This confirms the utmost importance of particle morphology over size, i.e., each particle acts as a microreactor in a sense that it has an individual reactivity. One can extrapolate this behavior in the case of nanoparticles on a real-technical catalyst and explain why it is not possible to derive simple trends and correlate the catalyst particle sizes to the reactivity.

Further on, the sub-micrometer resolution of SPEM has allowed us to explore lateral variations in the oxidation state within the same microparticle, which has to be expected considering the variable grain structures and structural irregularities evidenced by SEM. Figure 11.4(c) shows Rh chemical images and Rh $3d_{5/2}$ spectra taken in different zones of a single Rh microparticle using microspectroscopy and specro-imaging modes. The chemical imaging results clearly illustrated the chemical heterogeneity attained after oxidation. Mapping a narrow Rh $3d_{5/2}$ energy window, dominated by the emission from the RhO_2 phase, revealed that the left-hand side of the particle has attained the most advanced oxidation state closer to RhO_2. The emission from the lower oxidation states, Rh_2O_3

and RhO_x, cannot be well separated, but apparently, these two phases dominate the remaining portion of the particle.

This heterogeneity in the oxidation state is preserved upon following partial reduction with H_2 at 500 K. Under our reaction conditions ($PH_2 < 10^{-5}$ mbar), complete reduction of the formed RhO_2 required a higher temperature (~600 K). These findings are clear evidence of the utmost importance of the nanoscopic particle structure, which can also be dramatically altered under reaction conditions. In fact, complementary LEEM-LEED studies have evidenced major structural evolution of the microparticles with the advancement of the oxidation that ends with the formation of a disordered oxide phase.

Overall, the presented results confirm that there is still a long way to go in order to gain structural information supported by detailed chemical analysis for single catalytic nanoparticles. However, this should be the critical information which will lead to full control of the catalyst's properties.

11.3 Complex Surface Morphology Induced by Propagation of Reaction Fronts

11.3.1 *Reactant adsorbate structures induced by reaction fronts*

One of the difficulties in the quantitative description of catalytic reactions on solid surfaces is the fact that they are dynamic, involving mass transport processes coupled to chemical interactions, that may change the surface structure. This topic is tackled by Gerhard Ertl, who was awarded the Nobel Prize in Chemistry in 2007 for his "groundbreaking studies in surface chemistry". Dynamically created reaction-diffusion patterns, and the lateral propagation of reaction front on metal catalyst surfaces, were observed more than 30 years ago using laboratory PEEM [45,46]. These adsorbate structures with dimensions ranging in the sub-μm to a few μm are the result of the interplay between energetics and the diffusion of adspecies. They are strongly dependent on the adsorption bonding strength and the interactions between the adsorbed species and/or reorganization of the metal catalyst surface structure under reaction conditions. Often, the interactions between the constituents, in the mixed adsorbate layer, favor phase separation processes and even "freezing" of the phase separation under certain reaction conditions. Changes in the reaction conditions, i.e., temperature and reactant pressure, can lead to dramatic reorganizations of the adlayers. A revolutionary step ahead in understanding the pattern formation processes was made thanks to the chemical sensitivity of XPEEM and SPEM instruments providing direct information about the concentration and chemical state of the involved species. Complementary LEEM imaging and micro-LEED in SPEM-LEEM instruments have added better time resolution plus structural information about the catalyst surface and adsorbed layers.

Among the first applications of SPEM and XPEEM-LEEM in this field was the study of the local composition of the spatial patterns during $NO+H_2$ reaction on Rh (110) [47,48]. This work added crucial information for understanding in detail the excitation and the wave propagation mechanisms. In the $NO+H_2$ reaction, NO and H_2 adsorb dissociatively and the interaction between H, O, and N adatoms leads to H_2O and N_2 gas-phase products. The key factor for the reaction mechanism is the destabilization of chemisorbed nitrogen by coadsorbed oxygen. PEEM-LEED studies have shown that under reactions conditions, both adsorbed O and N form two-dimensional c(2×6)-O, (3×1)-N, (2×1)-N, c(2×4)-O+N, and c(2×6) patterns, involving (1xn) and (nx1) substrate reconstructions [49]. The latter was also clearly verified by LEEM-micro-LEED studies [18,50]. The coexisting of N and O induced Rh(110) (nx1) and (1xn) reconstructions, involving both adatom diffusion and mass transport of Rh atoms. This results in a state-dependent anisotropy reflected by the observed different pattern shapes [51]. Quantification of the lateral distribution of N and O adspecies was possible only by chemical imaging, first measured with SPEM [47] and later with XPEEM as well [50].

Figure 11.5 illustrates the first SPEM results, complemented later by LEEM-XPEEM measurements, where the surface chemical composition was evidenced by XPEEM and the corresponding local

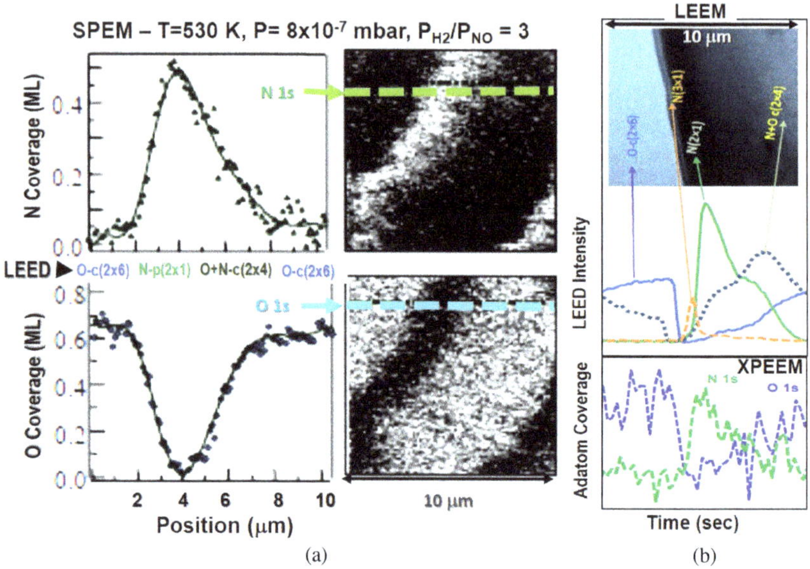

Fig. 11.5. Local chemical composition and structure induced by the propagating fronts during NO+H_2 reaction, measured with SPEM, XPEEM, and LEEM. (a): N 1s and O1s SPEM images and concentration profiles, indicated by the dashed lines, showing the lateral distribution of adsorbed O and N. The LEED patterns are based on the LEEM measurements. (b): (top) LEEM-MEM image of WF variations during wave propagation; (middle) intensity variation of the LEED diffraction spots of the different structures developed during wave propagation; (bottom) XPEEM showing the corresponding variations of the N and O coverage during wave propagation.

structures were revealed by LEEM. In SPEM imaging, that appears somewhat more precise, the reaction conditions were tuned to ensure a slower velocity of the propagating waves. The pulses under the chosen reaction conditions are ~5 μm wide and propagate with a velocity of 0.5 μm/s. The N 1s and O 1s core-level intensities present the dynamics of O and N accumulation/depletion during wave propagation. This is clearly demonstrated by the N 1s and O 1s concentration profiles in SPEM results and the variations in the N 1s and O 1s intensities measured by XPEEM during wave-front propagation. The Work Function (WF) contrast in LEEM-MEM images also corresponds to the different local compositions, and, as indicated by the arrows, there is a correlation with micro-LEED structures developed as a result of the changes in the local N and O coverage.

11.3.2 *Reaction-induced spatial redistribution of metals' submonolayer deposited on catalyst surface*

A great number of adsorption and reaction studies employing surface-sensitive techniques have demonstrated that a small quantity of foreign adatoms, e.g., another metal on the catalyst surface, can change significantly its reactivity [52]. Along with the modification of the surface properties, a distinct feature of the metal adatoms is the dependence of their mobility and bonding on the presence of other adsorbed species (reactants), which in the case of propagating reaction fronts can lead to the formation of microareas with different local compositions, first monitored by PEEM and LEEM. These can be considered as the creation of independent chemical microreactors of different reactivity and selectivity, and only XPEEM and SPEM provided chemical information about the local compositions of the developed stationary patterns.

The first observation of induced transport of a surface modifier by propagating reaction fronts during H_2+O_2 reaction was for Rh(110) surface covered with submonolayer of K. Alkali metals play a prominent role in heterogeneous catalysis due to their function as promoters. Without exhibiting substantial catalytic activity by themselves, they can enhance strongly the efficiency of certain transition metal catalysts. Alkali metal adatoms are very mobile on metal surfaces and exert high chemical affinity to the so-called electronegative adatoms, such as O, N, and OH. However, in order to understand and correctly predict the role of alkali promoters in surface reactions, we should first clarify their spatial organization under dynamic reaction conditions.

Direct evidences on this issue have been obtained with SPEM showing that self-organization processes associated with alkali mass transport take place under reaction conditions, leading to a heterogeneous surface with large-scale structures in the micrometer range.

SPEM images showed how, depending on the reaction conditions, K, which is initially homogeneously distributed, can reversibly condense into microscopic assemblies where it is coadsorbed with oxygen and OH [53]. The mass transport of K, governed by its different mobility and bonding strength on the reduced and on the O-covered Rh surface, proceeds via propagating reduction fronts. By selecting the proper reaction conditions (T and O_2/H_2 pressure), stable stationary concentration patterns can be formed, as illustrated by the K 2p map shown in Figure 11.6(a). The spatial distribution of K across the K-rich feature, indicated by the arrow in Figure 11.6(a), clearly illustrates how the reduction front reacting with oxygen mobilizes K to migrate to the shrinking oxygen-covered parts, where also O concentration reaches very high levels. Under the applied reaction conditions, with advancing K condensation, the velocity of the reduction fronts is slowing down until stationary state is attained. The lifetime of adsorbed H is negligible at the applied reaction temperatures, so only a small amount of coadsorbed OH, stabilized by K, might coexist. This is indicated in the O 1s spectra in Figure 11.6(c) and reported with higher resolution spectra in Ref. [53]. The O 1s and Rh 3d spectra measured at different spots

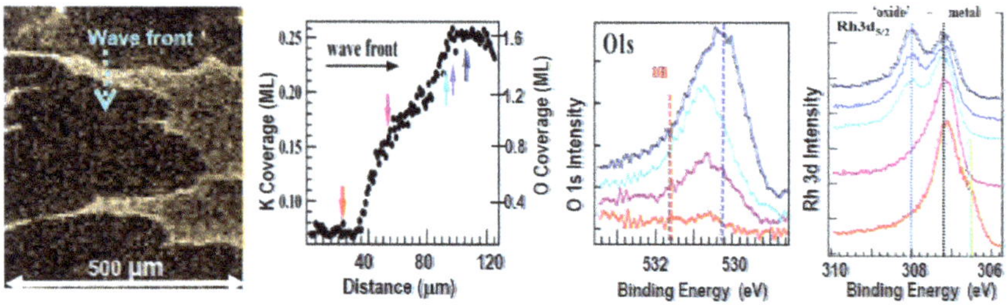

(a) (b) (c) (d)

Fig. 11.6. (a) K 2p SPEM image of stationary state illustrating the inhomogeneous distribution of K induced by the propagating reaction fronts. (b) K concentration profile measured along the line indicated by arrows in K 2p map. The corresponding oxygen coverage is shown on the right axis. (c) and (d) O 1s and Rh $3d_{5/2}$ spectra measured in the spots indicated in the profile plots shown in (b). Reaction conditions: 580 K, PO_2 and PH_2 ~2×10^{-7} mbar.

outside and inside the condensed phase (Figures 11.6(c) and 11.6(d)) clearly show how the increasing oxygen coverage exceeding a monolayer leads to the local formation of Rh surface oxide. The LEEM-LEED measurements under identical reaction conditions also showed that the increasing O coverage induces locally the well-known (1xn) reconstructions of the Rh surface. A characteristic feature of the formed condensed K+O islands and O-free Rh(110) surface is that they are stable only under water formation reaction conditions, i.e., the condensation process is reversible and the patterns are completely dissolved under oxidation or reduction conditions [53,54]. In reducing ambient, the stationary patterns dissolve much faster than in oxidizing ambient because the mobility of K on the O-free surface is much higher.

Combined XPEEM and SPEM studies also confirmed that the reaction fronts trigger modifiers, such as Au, and that it is much less reactive than alkali metals or Rh, to redistribute and condense in separate adsorption phases. In addition, depending on the reaction conditions' regular patterns of different wavelengths can be formed [50,55–57]. The development of Au-modified Rh and the formation of stationary patterns during the H_2+O_2 reaction are similar to that of K-modified Rh. However, in contrast to the case of K, the phase separation results in separate O-rich and Au-rich islands. Figure 11.7(a) shows the Au 4f maps measured with XPEEM in the latest stage of the same H_2+O_2 reaction on an Rh(110) surface covered by ~0.5 ML of Au. Please note that before the reaction, Au was homogeneously distributed over Rh. The map clearly manifests the formation of Au condensation patterns, similar to the case of K shown in Figure 11.6. However, in this case, Au has condensed in the O-depleted areas, reaching a maximum coverage of ~0.9 ML.

The local structure measured by LEED (in the right part of Figure 11.7(a) shows (1×2) reconstructed Rh(110) in the Au-depleted O-rich areas, where the streaks are indicative of O disorder, due to the presence of Au residues. In contrast to the case of K condensation, the patterned at the Au/Rh interface is preserved in oxygen ambient and is dissolved quickly only under reduction conditions restoring the initial homogeneous distribution of Au (not shown).

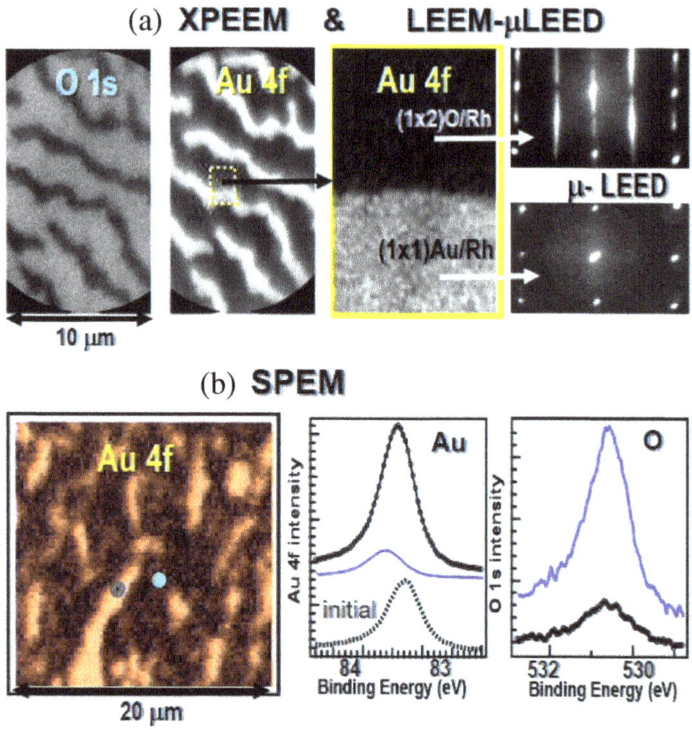

Fig. 11.7. Au deposited on Rh substrate: (a) XPEEM O 1s and two large area and small area Au 4f images and the corresponding local structure measured by LEEM. (b) SPEM Au 4f image and Au $4f_{7/2}$ and O 1s spectra measured in the microspots indicated in the Au 4f, representing the local composition of Au-rich and Au-depleted areas of the stationary pattern. T = 570 K, $H_2 + O_2$ pressure ~6×10^{-7} mbar.

The advantage of using the combination XPEEM-LEEM-LEED for the characterization of dynamic systems is that along with chemical contrast, it adds correlated structural information. For example, under pure oxygen ambient, the Au-depleted areas undergo more advanced (1xn≥4) reconstruction that indicates surface oxide formation.

Higher-resolution spectroscopic measurements with SPEM have revealed in detail the response of the Rh 3d and Au 4f core levels to the local Au and O coverage [57]. The selected Au $4f_{7/2}$ and O 1s microspot spectra in Figure 11.7(b) clearly show the chemical shift of residual Au in the O-rich areas mainly as a result of the substantial O-induced local structural changes. Au-enriched islands on Rh have reduced activity towards dissociative H_2 adsorption as compared to the O-rich ones, Rh. The result is that, under reduction conditions, the reduction fronts that gradually homogenize the Au distribution are ignited inside the O-rich islands [56].

To describe the mechanism of these condensation processes, we should take into account several aspects. The first is the energy factor; since the phase separation is favored by the total energy gain determined by the strength of O-Rh, M (K or Au)-Rh, and the M-O bonding, in the present cases' interactions, the total energy would be $E_{(M\text{-}O)}[\Theta_{M+O}] - (E_M[\Theta_M] + E_O[\Theta_O]) >$ or < 0 (E is the energy and Θ is the coverage of each type of bond). Apparently, in the case of K, the $E_{(K\text{-}O)}$ prevails, whereas for Au system is the $E_{Au} + E_O$ term, since Au is chemically inert to oxygen. This means that in the case of K-addition on Rh, the O-covered surface is energetically favored, whereas in the case of Au-addition, it is the reduced one.

Another factor affecting the reaction fronts is the Au and K mobility, which in both cases is lower in the presence of oxygen. Therefore, the reorganization is determined exclusively by the energy gain as a result of phase separation ignited by the propagating reaction fronts. The local reactivity of the condensed phases is very different: hydrogen dissociative adsorption propensity, and reactivity of oxygen to hydrogen, the condensed K+O phase is strongly reduced, whereas chemically "inert" Au exerts blocking effect decreasing both hydrogen and oxygen dissociative adsorption rates.

Further complexity arises in cases when two or more elements are used as modifiers, since each of them exerts different reactivity with respect to the substrate and the reactants and, in addition, their mutual interactions should be considered as well. In the next example, the Rh surface was modified by two metals: Au and Pd. We recall here that AuPd alloys might find applications in catalytic hydrogenation reactions and hydrogen fuel cells since the presence of Au improves the stability and selectivity of Pd catalysts preserving its high affinity to hydrogen. Therefore in this case, along with Au and Pd bonding with Rh, we need to take into account the Au-Pd miscibility. In fact, Au-Pd alloying is energetically more favored as compared to Rh-Au and Rh-Pd. In addition, the Au+Pd/Rh system behaves differently than Au/Rh since Pd has a very high affinity towards hydrogen as compared to Au and Rh.

Figure 11.8 shows XPEEM images and spectra representing the local composition of the developed patterns during H_2+O_2 reaction on Rh (110), modified with a Pd+Au submonolayer. Similar to the case of Au, O-rich and mixed Au+Pd-rich islands are developed.

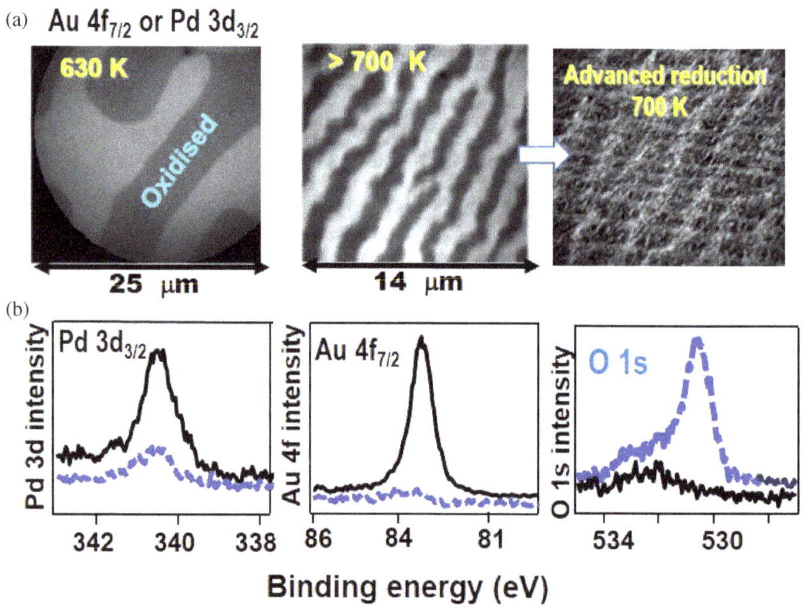

Fig. 11.8. (a) XPEEM Au 4f images of stationary patterns developed on Rh(110) covered with 0.3 ML Au and 0.3 ML Pd during H_2+O_2 reaction (P = $5.3 \cdot 10^{-7}$ mbar; $P[H_2]/P[O_2] = 0.45$) at temperatures lower and higher than 700 K and the following dissolution of the lamellae pattern during transition to reduction conditions. The bright regions are those rich in AuPd. The same contrast was observed also in Pd 3d images. (b) Pd $3d_{3/2}$, Au $4f_{7/2}$, and O 1s spectra representing the local composition of the O-rich (oxidized) and AuPd regions of the stationary patterns obtained at 630 K.

The images taken under different reaction conditions illustrate that the morphology and dimensions of the developed stationary structures are strongly dependent on the reaction temperature. In particular, the pattern size changes from over 10 μm at 630 K to about 1 μm at temperatures higher than 700 K. It should be noted that it is not only the pattern dimensions that change but high temperature also favors the formation of oriented striped structures. Interestignly, some variations in the local composition were observed. The high-temperature enhances the mobility of Pd and Au, which start to penetrate into the O-rich areas. However, since the affinity of the two metals towards oxygen is different, the Pd concentration in O-rich areas grows faster [50,58].

This observation confirms the delicate interplay between thermodynamics and kinetics governing surface reactions and phase separation. As reported in Ref. [59], the wavelength of the formed lamellae structures obeys power scaling low with respect to the reaction rate, in agreement with the theoretical predictions. Since the AuPd islands are more active for hydrogen adsorption, the reduction fronts under reduction conditions are ignited from the AuPd boundaries and propagate inside the O-rich areas, as illustrated in Figure 11.8. The reaction induced the re-organization of the bimetallic Au+Pd adlayer on Rh clearly demonstrating the complexity governing the surface state when approaching realistic multicomponent catalytic systems.

The results collected on the "model modified system" confirmed that energetically and kinetically driven re-organisations of adspecies under catalytic reaction conditions are rather common events. Once such processes and the conditions for the stabilization of the stationary patterns will be fully understood, we will attain the basis for producing substrate-supported microreactors of desired local and total chemical activities.

11.4 Electrochemical Systems

SPEM has been extensively used in the last decade for exploiting the properties of electrochemical systems. Some of the obtained results for solid oxide-based electrochemical systems are already summarized in a recent review [60]. Here, we will give two characteristic examples of SPEM applications related to the evolution of the electrode components typically used in solid oxide fuel cells and the aging of Mn-based electrocatalysts.

11.4.1 *Solid oxide fuel cells*

The operation principle of solid oxide fuel cells (SOFC) involves the reduction of O_2 at the cathode to O^{2-} ions and diffusion of the O^{2-} ions through the solid electrolyte (YSZ) to the anode side of the cell where the oxidation of the fuel (usually H_2 or hydrocarbons) takes place (see also Chapter 13). In the absence of an applied potential, the SOFC anode and cathode electrodes react with the gases (H_2, O_2, hydrocarbons, H_2O, etc.) up to the point that their surface state is stabilized reaching an equilibrium with the gas. This equilibrium breaks down when an external potential is applied between the anode and the cathode. In this case, the electrochemical reaction takes place at the electrode/electrolyte interface and an overpotential is risen. The overpotential has several contributions, namely, ohmic, charge-transfer, and mass transport of electroactive species, and depends on the exact location. The highest charge-transfer contribution is at the electrode/electrolyte interface where actually the electrochemical reaction occurs.

The overpotential distribution affects the local chemical state of the electrode and can also promote the diffusion of electrode species across the electrode/electrolyte interface that it is one of the reasons for the deterioration of the SOFC performance. Thanks to the lateral resolution of SPEM, it has been possible to follow both local chemical changes and atomic migration across the interface resulting in significant structural alterations [61]. Using μ-PES along with the chemical state encoded in the core level shifts, one can also monitor the rigid energy shifts of the photoelectron spectra due to potential drop on the electrode moving from the interface with the electrolyte to areas further inside the electrode (i.e., far from the electrolyte). The latter provides a measure of the local overpotential resulting from the current flow generated by the oxidation reaction at the anode/electrolyte interface.

Figure 11.9 illustrates Ni 3p SPEM maps and Ni 2p spectra summarizing the evolution of the anode/electrolyte interface of SOFC operated at 900 K in hydrogen or hydrocarbon ambient under an applied potential of 1 V. It should be stressed that in such working ambient, the Ni electrode is reduced and the electrochemical oxidation occurs only after applying the voltage. The map of the same interfacial region taken before the operation of the SOFC (i.e., voltage application) evidences the Ni electrode morphology, which has developed during the initial pretreatment in O_2 atmosphere at 900 K. The second map of the same region was taken after many working cycles alternated with cycles of pure chemical oxidation and reduction by changing the gas ambient. By comparing the maps

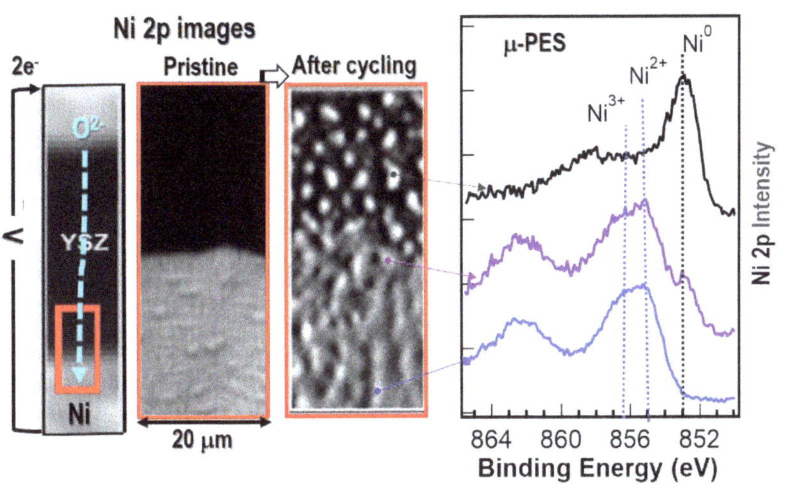

Fig. 11.9. (Left) Ni 3p images showing the morphology changes of the model SOFC as a result of cycling. The dark parts correspond to the YSZ electrolyte, where in the interfacial region, the migrated Ni species start to aggregate on island. (Right) Ni 2p spectra taken in the spots indicated by the arrows, showing the dependence of the Ni oxidation state on the overvoltage distribution moving inside the anode and the lack of activity in the electrically isolated island on the YSZ electrolyte.

taken before and after cycling, it is apparent that during functioning some Ni has diffused inside the YSZ electrolyte aggregating in isolated islands and also that the Ni electrode develops more island-like morphology. It should be noted that the imaging "after cycling" follows a reduction cycle of the cell in C_2H_2+H_2O ambient and the Ni anode was in a reduced metallic state before applying a potential.

The representative Ni 2p spectra (as shown in the right-hand side of Figure 11.9) taken at the points indicated on the map evidence that the working anode underdoes electrochemical oxidation. The degree of nickel oxidation becomes less (nickel is partially reduced) close to the Ni/YSZ interface, in full coherence with the potential drop of ~0.6 eV, measured from the Ni 2p spectra shifts to the higher binding energy. The isolated Ni islands at the YSZ area cannot be polarized (the electrons cannot be transferred in this area due to the luck of electrical contact). Therefore, their chemical state remains entirely controlled by the reducing gas environment and the nickel metallic state remains dominant.

In brief, such SPEM studies not only confirm the correlation between electrochemical activity and local potential but also reveal dynamically evolving structural-compositional changes of the electrode and the electrolyte as a result of the cell operation. These changes can be correlated with cell degradation mechanism taking place under real operation conditions.

Using one of the near-ambient pressure setups of SPEM, we were also able to monitor both the ongoing reaction on the electrode and the generated electric current, as described in Refs. [62,63]. This is possible because the spectral transients recorded under operating conditions, resulting in the generation of electric current, encode both the time-dependent electrochemical kinetics through spectral energy shifts and the electrode oxidation states resulting from the electrochemical and chemical processes. The results obtained exploiting the changes at the interfacial region of Ni anode in CH_4+O_2 or H_2+O_2 ambient have shown that there is no correlation between the oxidation state of the anode resulting from the chemical reaction and the generated potential from the electrochemical reaction. By comparing the reduction rate and potential changes, monitored via SPEM imaging and spectroimaging, it was found that the potential reaches saturation within 5–10 minutes, whereas the chemical steady state is attained after 25–30 minutes. This indicates that there is a competition between chemical reduction rate and anodic oxidation rate to reach the chemical steady state of the electrode. The best result in this SPEM study is that the potential, experimentally measured from the Ni 3p "charge" shifts, is in agreement with thermodynamic predictions for the potential generated by this electrochemical reaction.

11.4.2 *Aging of electrocatalysts*

The last example is relevant to the aging of non-precious metal-based electrocatalysts used at the cathodes of energy converting systems as alkaline fuel cells and metal/air batteries. At these cathodes, the oxygen reduction reaction (ORR) takes place in alkaline solutions, and the physico-chemical aspects of the electrocatalyst stability are still an issue. A number of SPEM studies were dedicated to exploiting the aging of metal-based electrocatalysts [64–66]. Among the metals that have attracted interest due to their relatively good activity to oxygen reduction reaction (ORR) in alkaline solution and their low cost is pyrolyzed Mn/PPy (polypyrrole). Using an electrochemical cell connected with SPEM, we were able to follow, under quasi *in situ operando* conditions, the morphological and chemical modifications of the Mn/PPy electrocatalyst surface as a result of voltammetric ORR cycling in an O_2 saturated alkaline electrolyte [65].

The chemical images and μ-PES in Figure 11.10 show how the aging of Mn/PPy electrocatalyst as a result of ORR cycles leads to irreversible changes in both morphology dimensions of the catalyst particles and their composition. The Mn 2p images clearly evidence the aggregation trend by increasing the number of ORR cycles that was also confirmed by higher resolution SEM images placed in the Mn 3p spectra panel. Along with this morphology evolution, the Mn 3p, Mn 2p, and O 1s microspot spectra taken on the catalyst particles show substantial changes in the chemical state due to advanced

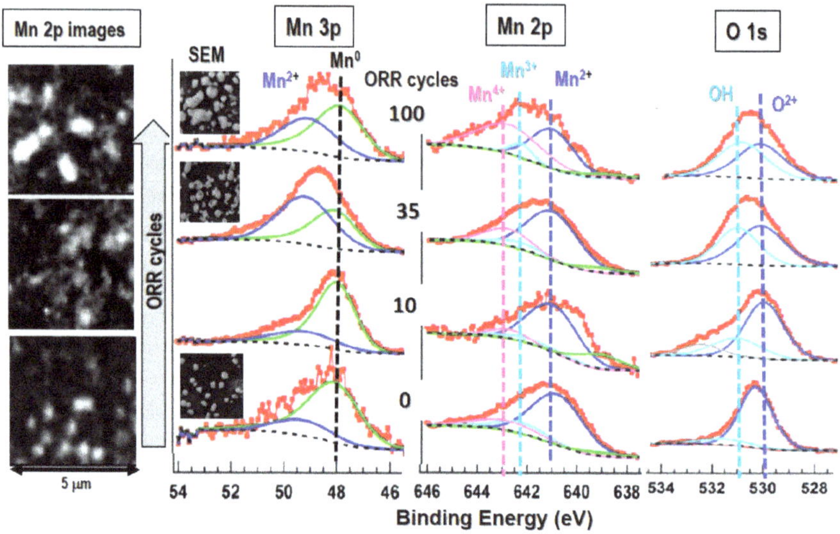

Fig. 11.10. (Left) Mn 2p SPEM images showing the aggregation tendency by increasing the ORR cycles. (Right) Mn 3p, Mn 2p, and O 1s spectra taken on pristine and evolved samples after the indicated number of ORR cycles. The components assigned to the different Mn oxidation states and the O-oxide and OH are indicated in the panels. The SEM images inside the Mn 3p panel confirm the aggregation monitored by SPEM.

oxidation and formation of high valence oxides. Due to the increase in the photoelectron mean free path by increasing their kinetic energy, it was possible, by comparing the Mn 2p and Mn 3p spectra, to reveal that whereas the oxide skin of the pristine particles is predominantly MnO, cycling leads to the formation of a thicker skin containing higher valence Mn oxides.

The emerging OH-related component in the O 1s spectra supposes the co-existence of some intermediate Mn–O–OH state as well. In brief, the SPEM results suggest that the degradation of the Mn/PPy electrocatalyst can be attributed to the potential induced solubility of Mn oxide and hydroxide species in the alkaline electrolyte and the parallel dissolution/re-deposition of these species. The effect of this process is the expansion of the oxide particle skin, enriched with higher valence oxides, which favors particle agglomeration. The observed substantial changes due to the ORR aging of Mn/PPy electrocatalysts can be effectively mitigated by adding Co. The SPEM studies showed that although the tendency to formation of high valence Mn oxides is preserved, the Co suppresses the dissolution/redeposition rate and the particle agglomeration [66].

11.5 Concluding Remarks

The complexity of catalytic systems is multilevel and needs profound knowledge of their properties and their chemical-structural evolution under reaction conditions. In this respect, the complementary capabilities of synchrotron-based photoelectron microscopy methods in terms of imaging, spectroscopy, and spatial resolution have opened unique opportunities to explore the structure and chemical composition of catalyst systems at relevant length and timescales and correlate them to the fabrication or operating conditions. Using selected results, this chapter illustrates the potential of scanning and full-field imaging photoelectron microscopes operated at synchrotron facilities in the characterization of metal catalyst and electrocatalyst systems. These studies help to better understand phenomena such as the evolution in lateral inhomogeneity in the composition and chemical state at sub-micrometer length scales. We believe that with the examples selected in this chapter, we have demonstrated the compositional complexity of metal catalysts developed under variable reaction conditions, where the studies with SPEM and SPELEEM have allowed us to correlate the spatial variations in the chemical composition of the catalyst surfaces, with the surface morphology and the mobility of the reactant species. Using SPEM, we have also made significant advances in monitoring *in situ* the lateral evolution of electrochemical device constituents, gaining deeper insight into electrochemical and chemical processes and structural changes, and correlating them to the actual operating conditions.

Ongoing efforts in improving lateral resolution, time resolution, and performing real *operando* studies will open new opportunities to fully overcome the material and pressure gaps. These are viable routes in our attempts to reveal all possible desired and undesired events that determine the performance of the complex functional materials based on catalysis.

References

[1] Schlögl R. Heterogeneous catalysis. Angew Chem Int Ed. 2015;54:3465–3520.
[2] Catlow CRA, Wells P, Gianolio D. Synchrotron radiation techniques in catalytic science. Phys Chem Chem Phys. 2020;22(34):18725–19378.
[3] Zhong L, Chen D, Zafeiratos S. A mini review of *in situ* near-ambient pressure XPS studies on non-noble, late transition metal catalysts. Catal Sci Technol. 2019;9:3851–3867.
[4] Roy K, Artiglia L, van Bokhoven JA. Ambient pressure photoelectron spectroscopy: Opportunities in catalysis from solids to liquids and introducing time resolution. ChemCatChem. 2018;10:666–682.
[5] Meirer F, Weckhuysen BM. Spatial and temporal exploration of heterogeneous catalysts with synchrotron radiation. Nature Rev Mater. 2018;3:324–340.
[6] Sheng B, Cao D, Liu C, Chen S, Song L. Support effects in electrocatalysis and their synchrotron radiation- based characterizations. J Phys Chem Lett. 2021;12:11543–11554.
[7] Siegbahn K. From X-ray to electron spectroscopy and new trends. J Electr Spectr Rel Phenom. 1990;51:11.
[8] Coxon P, Krizek J, Humpherson M, Wardell IRM. Escascope — A new imaging photoelectron spectrometer. J Electr Spectr Rel Phenom. 1990;52:821–836.
[9] Morgan DJ. Imaging XPS for industrial applications. J Electr Spectr Rel Phenom. 2019;231:109–117 and references therein.
[10] Bauer E. A brief history of PEEM. J Electr Spectr Rel Phenom. 2012;185:314–322 and references therein.
[11] Bauer E. LEEM, SPLEEM and SPELEEM and references therein, *Springer Handbook of Microscopy*, Editors: Peter W. Hawkes and John C. H. Spence. 2019;487–535.
[12] Kirz J, Jacobsen C. The history and future of X-ray microscopy. J Phys Conf Ser. 2009;186:12001.

[13] Amati M, Barinov A, Gregoratti L, Sezen H, Kiskinova M. Chapter 14: Scanning photoelectron microscopy: Past, present and future. In M Rocca, TS Tahman, L Vattuone (Eds.), *Springer Handbook of Surface Science* (pp. 427–448). Springer Handbooks, 2020.

[14] Kunz C. Synchrotron radiation: Third generation sources. J Phys Condens Matter. 2001;13:7499.

[15] Stöhr J. *NEXAFS Spectroscopy*. Berlin: Springer Verlag, 1992.

[16] Stöhr J, Wu Y, Hermsmeier D, Samant G, Harp R, Koranda D, Dunhamand B, Tonner P. Element specific magnetic microscopy with circularly polarized light. Science. 1993;259:658.

[17] Ade H, Kirz J, Hulbert S, Johnson E, Anderson E, Kern D. Scanning photoelectron microscope with a zone plate generated microprobe. Nucl Instr Meth A. 1990;291:126.

[18] Schmidt Th., Heun S, Slezak J, Diaz J, Prince KC, Lilienkamp G, Bauer E. SPELEEM: Combining LEEM and spectroscopic imaging. Surf Rev Lett. 1998;5:1287.

[19] Günther S, Kaulich B, Gregoratti L, Kiskinova M. Photoelectron microscopy and applications in surface and material science. Prog Surf Sci. 2002;70:187 and references therein.

[20] Kaulich B, Thibault P, Gianoncelli A, Kiskinova M. Transmission and emission X-ray microscopy: Operation modes, contrast mechanisms and applications. J Phys Condens Matter. 2011;23:83002.

[21] Locatelli A, Bauer E. Recent advances in chemical and magnetic imaging of surfaces and interfaces by XPEEM. J Phys Condens Matter. 2008;20:93002.

[22] Schneider CM, Wiemann C, Patt M, Feyer V, Plucinski L, Krug IP, Escher M, Weber N, Merkel M, Renault O, Barrett N. Expanding the view into complex material systems: From micro-ARPES to nanoscale HAXPES. J Electron Spectrosc Relat Phenom. 2012;185: 330–339.

[23] Schneider CM, Krasyuk A, Nepijko SA, Schönhense G. Accessing fast magnetization dynamics by XPEEM: Status and perspectives. J Magn Magn Mat. 2006;304:6.

[24] Amati M, Aleman B, Bozzini B, Gregoratti L, Sezen H, Kiskinova M. Characterization of catalytic materials with scanning photoelectron microscopy: Present and future. Surf Sci. 2016;652:20–25.

[25] Kolmakov A, Gregoratti L, Kiskinova M, Günther S. Recent approaches for bridging the pressure gap in photoelectron microspectroscopy. Top Catal. 2016;59:448–468.
[26] Gregoratti L, Al-Hada M, Amati M, Brescia R, Roccella D, Sezen HH, Zeller P. Spatially resolved photoelectron spectroscopy from ultra-high vacuum to near ambient pressure sample environments. Top Catal. 2018;61(12-13):1274–1282.
[27] Amati M, Gregoratti L, Zeller P, Greiner M, Scardamaglia M, Junker B, Ruß T, Weimar U, Barsan N, Favaro M. Near ambient pressure photoelectron spectro-microscopy: From gas–solid interface to operando devices. J Phys D Appl Phys. 2021;54:204004. Special Issue on *In Situ*/Operando Investigation of Catalytic and Electrocatalytic Interfaces.
[28] Bluhm H, Hävecker M, Knop-Gericke A, Kiskinova M, Schlögl R, Salmeron M. MRS Bull. 2007;32:1022–1030 and references therein. https://doi.org/10.1557/mrs2007.211.
[29] Lundren E, Mikkelsen A, Andersen JN, Varga P. Surface oxides on close-packed surfaces of late transition metals. J Phys Condens Matter. 2006;18:R481–499.
[30] Böttcher A, Krenzer B, Conrad H, Niehus H. Mesoscopic-scale growth of oxygen-rich films on Ru (0001) investigated by photoemission electron microscopy. Surf Sci. 2002;504:42.
[31] Blume R, Niehus H, Conrad H, Böttcher A, Aballe L, Gregoriatti L, Barinov A, Kiskinova M. Identification of subsurface oxygen species created during oxidation of Ru(0001). J Phys Chem B. 2005; 109:14052.
[32] Dudin P, Barinov A, Gregoratti L, Kiskinova M, Esch F, Dri C, Africh C, Comelli G. Initial oxidation of a Rh(110) surface using atomic or molecular oxygen and reduction of the surface oxide by hydrogen. J Phys Chem B. 2005;109:13649.
[33] Gustafson J, Mikkelsen A, Borg M, Andersen JN, Lundgren E, Klein C, Hofer W, Schmid M, Varga P, Köhler L, Kresse G, Kasper N, Stierle A, Dosch H. Structure of a thin oxide film on Rh(100). Phys Rev B. 2005;71:115442.
[34] Diebold U, Li S-C, Schmid M. Oxide surface science. Annu Rev Phys Chem. 2010;61:129–148.

[35] Reuter K, Scheffler M. Oxide formation at the surface of late 4d transition metals: Insights from first-principles atomistic thermodynamics. Appl Phys A Mater Sci Process. 2004;78:793 and references therein.

[36] Gong X-Q, Liu Z-P, Raval R, Hu P. A systematic study of CO oxidation on metals and metal oxides: Density functional theory calculation. J Am Chem Soc. 2004;126:8 and references therein.

[37] Knop-Gericke A, Kleimenov E, Hävecker M, Blume R, Teschner D, Zafeiratos S, Schlögl R, Bukhtiyarov VI, Kaichev V, Prosvirin IP, Nizovskii AI, Bluhm H, Barinov A, Dudin P, Kiskinova M. X-ray photoelectron spectroscopy for investigation of heterogeneous catalytic processes. Adv Catal. 2009;52: 213–272, Chapter 5.

[38] Böttcher A, Starke U, Conrad H, Blume R, Niehus H, Gregoratti L, Kaulich B, Barinov A, Kiskinova M. Spectral and spatial anisotropy of the oxide growth on Ru(0001). J Chem Phys. 2002;117:8104.

[39] Dudin P, Barinov A, Gregoratti L, Kiskinova M, Esch F, Dri C, Africh C, Comelli G. Initial oxidation of a Rh(110) surface using atomic or molecular oxygen and reduction of the surface oxide by hydrogen. J Phys Chem B. 2005;109:13649.

[40] Potts W, Morrison GR, Gregoratti L, Kiskinova M, Locatelli A. Spectromicroscopy study of the dynamics of submonolayer Pd on a polycrystalline Ni surface. Phys Rev B. 2003;67:45411.

[41] Potts AW, Morrison GR, Barinov A, Gregoratti L, Kiskinova M. Photoemission microscopy study of the temperature evolution of a Pd film deposited on a polycrystalline Ni substrate. Phys Rev B. 2005;72:193403.

[42] Winkler P, Zeininger J, Suchorski Y, Stöger-Pollach M, Zeller P, Amati M, Gregoratti L, Rupprechter G. How the anisotropy of surface oxide formation influences the transient activity of a surface reaction. Nat Commun. 2021;12:69.

[43] Dudin P, Barinov A, Gregoratti L, Scaini D, He YB, Over H, Kiskinova M. MgO-supported rhodium particles and films: Size, morphology, and reactivity. J Phys Chem C. 2008;112:9040.

[44] Dalmiglio M, Amati M, Gregoratti L, Mentes TO, Niño MA, Felisari L, Kiskinova M. Oxidation of supported PtRh particles: Size and morphology effects. J Phys Chem C. 2010;114:16885.

[45] Imbihl R, Ertl G. Oscillatory kinetics in heterogeneous catalysis. Chem Rev. 1995;697:95.

[46] Ertl G. Reactions at surfaces: From atoms to complexity (Nobel Lecture). Angew Chem Int Ed. 2008;47:3524 and references therein.

[47] Schaak A, Günther S, Esch F, Schutz A, Marsi M, Kiskinova M, Imbihl R. Excitation mechanism for pulses in the system Rh(110)/NO+H_2. Phys Rev Lett. 1999;83:1882.

[48] Schmidt Th., Schaak A, Günther S, Ressel B, Bauer E, Imbihl R. *In situ* imaging of structural changes in a chemical wave with low-energy electron microscopy: The system Rh(110)/NO+H_2. Chem Phys Lett. 2000;318:549–554.

[49] Kiskinova M. Surface structure and reactivity. Chem Rev. 1996;96: 1431.

[50] Locatelli A, Kiskinova M. Imaging with chemical analysis: Adsorbed structures formed during surface chemical reactions. Eur J Chem. 2006;12:8890 and references therein.

[51] Makeev A, Hinz M, Imbihl R. Modeling anisotropic chemical wave patterns in the reaction on a Rh(110) surface. J Chem Phys. 2001; 114:9083.

[52] Kiskinova M. Poisoning and promotion in catalysis based on surface science concepts and experiments. In B Delmon, JT Yates (Eds.), *Studies in Surface Science and Catalysis* (Vol. 70). Amsterdam: Elsevier, 1992.

[53] Marbach H, Günther S, Lürssen B, Gregoratti L, Kiskinova M, Imbihl R. Selforganization of alkali metal on a catalytic metal surface. Cat Lett. 2002;83:161.

[54] Marbach H, Hinz M, Günther S, Gregoratti L, Kiskinova M, Imbihl R. Mass transport of alkali metal in reaction fronts on a catalytic metal surface. Chem Phys Lett. 2002;364:207.

[55] Locatelli A, Heun S, Kiskinova M. Direct observation of reaction-induced lateral distribution of submonolayers of Au deposited ona Rh(110) surface. Surf Sci. 2004;566:1130.

[56] Locatelli A, Sbraccia C, Baroni S, Heun S, Kiskinova M. Energetically driven reorganization of a modified catalytic surface under reaction conditions. J Am Chem Soc. 2005;127:2351.

[57] Locatelli A, Barinov A, Gregoratti L, Aballe L, Heun S, Kiskinova M. Spectroscopic identification and imaging of surface processes occurring at microscopic and mesoscopic scales. J Electr Spec Rel Phenom. 2005;144:361.

[58] Locatelli A, Aballe L, Mentes TO, Guo FZ, Kiskinova M. A spectromicroscopic study of the reactive phase separation of Au+Pd and O on Rh(110). Surf Sci. 2007;601:4663.

[59] Locatelli A, Mentes TO. Aballe L, Mikhailov A, Kiskinova M. Formation of regular surface-supported mesostructures with periodicity controlled by chemical reaction rate. J Phys Chem B. 2006;110: 19108.

[60] Bozzini B, Amati M, Gregoratti L, Rossi F, Kiskinova M. Chapter 4: *In situ* photoelectron spectromicroscopy for the investigation of solid oxide–based electrochemical systems — Book: Solid oxide-based electrochemical devices: Advances, smart materials and future energy applications. Elsevier; 2020. ISBN 978-0-12-818285-7. Doi:10.1016/B978-0-12-818285-7.00003-4.

[61] Bozzini B, Amati M, Gregoratti L, Abyaneh M, Prasciolu M, Trugup AL, Kiskinova M. Microscale evolution of surface chemistry and morphology of the key components in operating hydrocarbon-fuelled SOFCs. J Phys Chem C. 2012;116:23188–23193.

[62] Bozzini B, Amati M, Gregoratti L, Abyaneh M, Prasciolu M, Kiskinova M. *In situ* photoelectron microspectroscopy during the operation of a single-chamber SOFC. Electrochem Commun. 2012;24:104.

[63] Bozzini B, Amati M, Gregoratti L, Kiskinova M. *In situ* photoelectron microspectroscopy and imaging of electrochemical processes at the electrodes of a self-driven cell. Sci Rep. 2013;3:2848.

[64] Bocchetta P, Amati M, Bozzini B, Catalano M, Gianoncelli A, Gregoratti L, Taurino A, Kiskinova M. Quasi *in situ* single-grain photoelectron microspectroscopy of Co/PPy nanocomposites under oxygen reduction reaction. ACS Appl Mater Interfaces. 2014;6:19621–19629.

[65] Bocchetta P, Amati M, Gregoratti L, Kiskinova M, Sezen H, Taurino A, Bozzini B. Morphochemical evolution during ageing of pyrolysed Mn/polypyrrole nanocomposite oxygen reduction electrocatalysts: A study based on quasi *in situ* photoelectron spectromicroscopy. J Electroanal Chem. 2015;758:191–200.

[66] Bocchetta P, Alemán B, Amati M, Fanetti M, Goldoni A, Gregoratti L, Kiskinova M, Mele C, Sezen H, Bozzini B. ORR stability of Mn–Co/polypyrrole nanocomposite electrocatalysts studied by quasi *in situ* identical-location photoelectron microspectroscopy. Electrochem Commun. 2016;69:50.

https://doi.org/10.1142/9781800613294_0012

Chapter 12

Applying XPS to Study Solid/Liquid Interfaces

Pinar Aydogan Gokturk[*,§], Yifan Ye[*,†,‡,¶] and Ethan J. Crumlin[*,‡,||]

[*]*Advanced Light Source, Lawrence Berkeley National Laboratory, Berkeley, CA 94720, USA*

[†]*Joint Center for Artificial Photosynthesis, Lawrence Berkeley National Laboratory, Berkeley, CA 94720, USA*

[‡]*Chemical Sciences Division, Lawrence Berkeley National Laboratory, Berkeley, CA 94720, USA*

[§]*paydogan@lbl.gov*

[¶]*e92@ustc.edu.cn*

[||]*ejcrumlin@lbl.gov*

Abstract

Understanding the fundamental processes that take place at the electrode/electrolyte interfaces of electrochemical systems is crucial for the development of new technologies with higher efficiency and improved performance. Unfortunately, *ex situ* analyses do not always reflect the same properties under operating conditions. That is why *in situ* and *operando* characterization tools are required for precise understanding. It is nonetheless challenging to isolate and study a particular process at the solid/liquid interfaces where and while they occur. Herein, we review four different types of strategies to probe electrode/electrolyte interfaces with X-ray photoelectron spectroscopies. These include the static liquid approach using low-volatile liquids, the vapor condensation approach, the dip and pull method, and liquid cell approaches. Although each method has its own specific limitations at present, all of them enable *in situ/operando* characterization and provide important mechanistic insights into many electrochemical systems. We show some exemplary cases that address different aspects of the interface, including (i) potential distribution at the electrified solid/liquid interfaces and (ii) redox reactions at the interface driven by applied potential. Given the highly specialized nature of current experimental tools, answering a specific question will necessitate understanding the advantages and limitations of each and every technique and then designing the experiment accordingly. We believe such efforts are crucial for moving the field forward from a collection of disconnected *ex situ* findings to a complete understanding of electrode/electrolyte interfaces.

12.1 Introduction: The Solid/Liquid Interface

Solid/liquid interfaces play a fundamental role in many scientific areas, including electrochemical reactions, catalysis, corrosion, biology, microfluidics, water purification, and environmental chemistry, as many of the most important molecular processes take place at such interfaces [1–6]. Figure 12.1 is a pictorial representation of some such processes. In the electrolyte, most ions are solvated by a certain number of solvent molecules, depending on their size and charge. On the other hand, some ions of opposite charge come together by electrostatic forces in order to form a neutral and distinct chemical entity, called an ion pair. Depending on the strength of interactions between those ions, ion pairs can be fully or under-coordinated by the surrounding solvent molecules. Ion association in the electrolyte becomes more pronounced at high concentrations and in solutions of polyvalent salts.

Similarly, ions and/or neutral molecules can also interact with the electrode and be adsorbed onto the surface. Knowledge of the adsorption behavior of these ions or neutral molecules is also essential for elucidating the reaction kinetics on the electrode as adsorption can block sites for the reaction, poison electrode surfaces, and affect the potential distribution across the electrode/electrolyte interface. Depending on the structure of the electrode materials, some ions may also intercalate or be adsorbed into the electrode. For instance, lithium-ion batteries store energy by the insertion or intercalation of lithium ions into the electrode material. During such adsorption and intercalation processes, the solvation shells of ions alter greatly.

Once potential is applied to the electrode, ions and dipole molecules (e.g., water) in the electrolyte migrate to the electrode/electrolyte interfaces, rearrange, and change their orientation to maintain the local electroneutrality. The structure formed at the interface is called the electrical double layer (EDL). As a result, a concentration and potential profile are formed at the electrode/electrolyte interface. Numerous theories and models have attempted to describe the structure of ions at electrode/electrolyte interfaces. According to the earliest model, known as the Helmholtz model, the interface between the electrical conductor and ion-conducting liquid electrolyte behaves like a plate capacitor, and the electrical double layer forms within the nanometer-thin electrolyte. Gouy and Chapman extended this model by introducing a diffusion layer. Later, Stern combined the Helmholtz and diffuse layer models and also divided the compact layer into two planes: the inner Helmholtz, which passes through the center of the adsorbed and under-coordinated ions on the electrode surface, and the outer Helmholtz, which passes through the center of the solvated ion, Figure 12.1 [1].

In addition to non-Faradaic processes, there is a wide range of charge transfer and conversion reactions, as well as dissolution and deposition processes that occur at electrode/electrolyte interfaces. At these interfaces, the electrode serves as either the donor or the acceptor of electrons. Specifically, electrochemical reactions of hydrogen

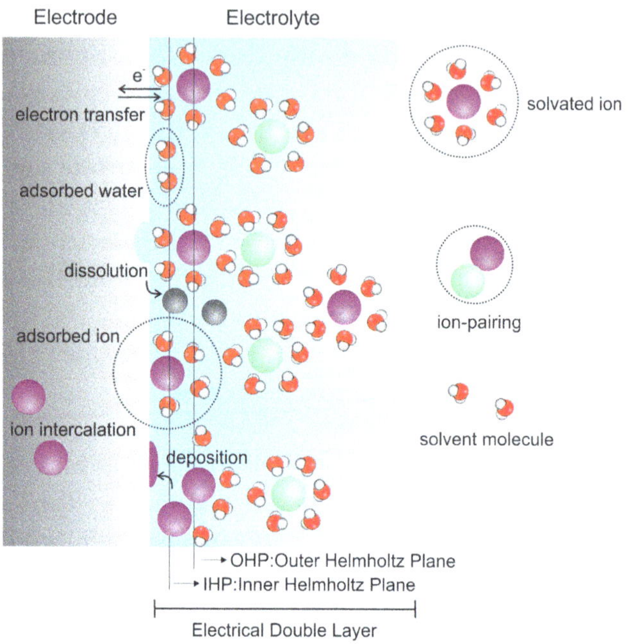

Fig. 12.1. Schematics of electrode/electrolyte interface showing some of the important processes taking place. Purple and green spheres are counter-ions and co-ions in the solution, respectively. Both types of ions are solvated by water molecules, which is the solvent in our representation. At close proximity to the solid/liquid interface, some of the ions interact with the electrode surface. Upon polarization, ions and solvent molecules at the interface rearrange and change their orientation to screen the charge on the electrode. In addition to the capacitive processes, electrochemical redox reactions, which involves a charge transfer between the electrode and electroactive species in electrolyte, dissolution and deposition processes also take place at electrode/electrolyte interfaces.

and oxygen evolution (water splitting) and their reverse reactions have been extensively investigated over the past decades [7–13].

In order to develop new technologies with high efficiency and improved performance, it is crucial to obtain a direct molecular-level understanding of each of the chemical and electrical processes that take place at solid/liquid interfaces. The ability to probe such processes at the times and places they occur has become the next great challenge.

Over the past decades, considerable effort has been made to adapt UHV-based surface sensitive techniques for investigations of solid/liquid interfaces. Surface-sensitive and surface-enhanced spectroscopies that can operate under (near) ambient conditions include spectroscopies, such as infrared (IR) spectroscopy [14–17], Raman spectroscopy [18], vibrational sum-frequency generation (VSFG) [19,20], X-ray photoelectron (XPS) [21], and X-ray absorption spectroscopies (XAS) [22]. In addition, X-ray scattering and diffraction techniques at grazing angles allow the characterization of the surface at hydrated conditions and provide useful structural information. Among the imaging techniques, scanning tunneling microscopy (STM), scanning electron microscopy (SEM) [23], Kelvin probe force microscopy [24,25], and transmission electron microscopy (TEM) [26] have also enabled observation of solid/liquid interfaces.

Within the scope of this chapter, we will focus only on the *in situ* and *operando* characterization of solid/liquid interfaces by XPS and Ambient Pressure XPS (APXPS), with primary emphasis on the different methodologies, and will discuss current capabilities and new developments. We begin with some technical discussions and then extend our understanding to current approaches for preparing solid/liquid interfaces. Next, we highlight several examples of APXPS measurements of solid/liquid interfaces to illustrate the possible applications, as well as to show the utility of different approaches. Lastly, we discuss the limitations of current instrumentations and strategies for addressing them.

12.2 General Considerations

X-ray photoelectron spectroscopy is one of the most powerful methods for investigating electrode/electrolyte interfaces, due to its elemental and chemical sensitivity as well as its facile capability for contactless probing of local electrical potentials. This capability makes XPS suitable for determining chemical compositions and electrical potentials across the interface.

Although XPS is a well-known technique for investigating solid samples, the main challenge of the solid/liquid study with XPS is associated with the short inelastic mean free paths (IMFPs) of photoelectrons in condensed matter, the very same characteristic that makes the technique surface sensitive. That is, emitted photoelectrons can travel only a few nanometers, and in order to lessen the further collision of electrons with gas molecules and a loss of signal, use of the technique is traditionally limited to ultra-high vacuum (UHV) conditions. This obstacle can be overcome with the so-called ambient pressure XPS (APXPS). APXPS allows studies at elevated pressures of gases in the torr range. Details of the technique's working principles can be found in Chapter 10. In general, a sample is placed in an analysis chamber with elevated pressure. This high-pressure chamber (or cell) is separated from the analyzer at UHV through a differentially pumped electrostatic lens system that reduces the travel length of emitted photoelectrons through the gas. Specially designed electron optics elements are typically incorporated in the differential pumping section to guide photoelectrons through apertures and increase their transmission. Today, APXPS instruments exist at synchrotron facilities around the world, and the availability of lab-based systems is increasing as well [27–29]. This second generation of setups has been developed largely to probe solids in a gaseous environment; many areas of research have benefited from these advances, in particular heterogeneous catalysis [30]. Yet the study of buried interfaces, particularly between solids and liquids, requires newly developed approaches or modifications of existing APXPS instruments.

Another challenge for APXPS investigation of solid/liquid electrolyte interfaces is the preparation of solid or liquid films that are sufficiently stable and robust to be the representative of a realistic solid/liquid interface but thin enough to allow signal detection from the interfacial region. For that reason, it is particularly difficult to use soft X-ray photoemission spectroscopy to probe solid/liquid interfaces. For instance, a photoelectron traveling in water with a KE of 500 eV is characterized by an IMFP of about 2 nm. Defining the probing depth as three times the IMFP (i.e., the depth in the material at which 95% of the ejected photoelectrons are inelastically scattered), we find that the maximum achievable probing depth of water with soft X-rays is about 6 nm. However, utilizing higher energy photons creates electrons with larger IMFPs, such that the probing depth of a photoelectron traveling in water with a KE of 3500 eV is equal to 29 nm. Hence, the use of high-energy photons further helps ease the thickness restrictions of liquid or solid film. In this way, attenuation of the XP signal from the gas molecules also becomes less pronounced, and APXPS equipped with tender or hard X-rays can operate at pressures up to 110 Torr throughout the entire chamber [21] and has been recently shown to go up to 5 bar with a specialized local gas delivery cone [31]. The ability to operate at these pressures opens up an opportunity to study such real ambient liquids as water, which has a saturation pressure of around 20 Torr at room temperature.

12.3 Existing XPS Methods to Study Solid/Liquid Interfaces

As we stated earlier, a true solid/liquid interface study requires the detection of signals mainly from the interface layer, without extensive interference from the bulk solid or liquid. To achieve this, we must create either a liquid or a solid film thin enough to allow the detection of photoelectrons from the interfacial region, depending on which side to illuminate from. At present, both approaches are being employed and several different strategies are being developed, Figure 12.2. Each method has its own pros and cons, but all of them exhibit great potential for revealing fundamental aspects of solid/liquid interfaces. We will discuss some of those approaches in detail, including the static liquid approach using low volatile liquids, the vapor condensation approach, the dip and pull method, and the liquid cell approach. Another promising strategy, which will not be addressed in this chapter, is utilizing X-ray standing waves for depth selective photoemission from the interface. In this technique, a substrate with a buried multilayer mirror generates a standing wave by the interference of the incoming and reflected X-rays. Sub-nm resolution

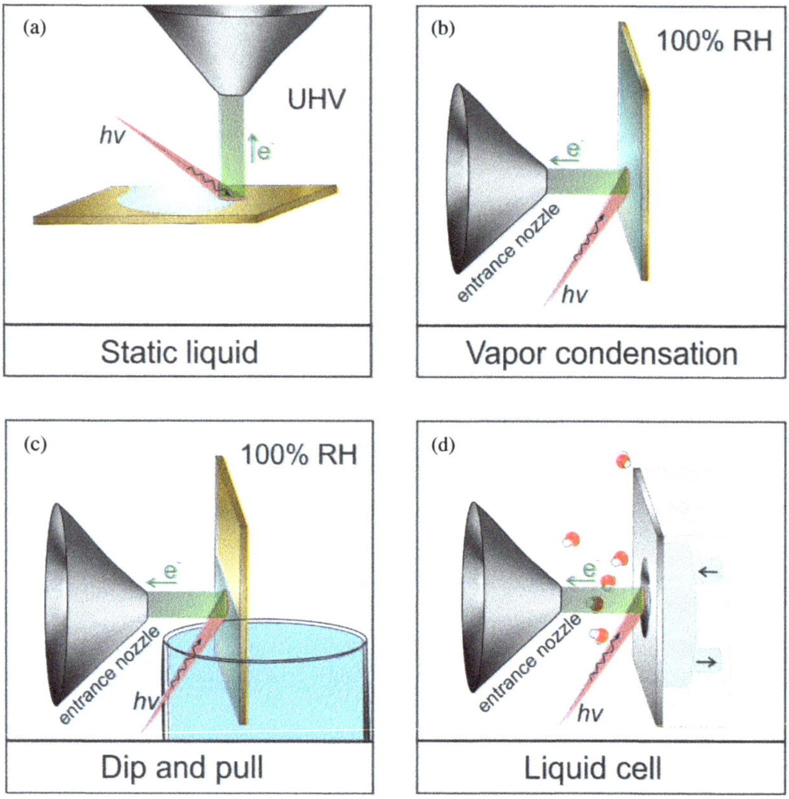

Fig. 12.2. Previously demonstrated methods of photoelectron spectroscopy studies at the solid/liquid interface, namely (a) static liquid method by low volatile liquids, (b) vapor condensation method, (c) dip and pull method, and (d) liquid cell method.

of chemical species at the solid/liquid interface has recently been obtained by combining standing-wave photoemission spectroscopy with APXPS. Nemsak *et al.* combined Standing Wave Ambient Pressure Photoelectron Spectroscopy with the vapor condensation approach (Figure 12.2(b)) to study a solid/liquid interface [32]. Karslıoglu *et al.* further implemented this approach with the dip and pull method (Figure 12.2(c)) and studied the interface of a polycrystalline Ni/KOH solution under potential control [33]. Examples and further details of this technique can be found in other review articles [30,34].

12.3.1 *Static liquid method by low volatile liquids*

Perhaps the most direct way to probe a solid/liquid interface without the necessity of intensive technical modifications is using liquids with a sufficiently low vapor pressure that sample loss through evaporation is insignificant. First studies of liquids were performed in the 1970s on pure low volatile liquids, such as formamide and glycols, as well as their mixtures [35,36]. Later, the first O 1s spectrum for liquid water was reported from a very high concentration of LiCl solution [37].

Over the past decade or so, similar routes have been pursued, mostly utilizing laboratory-based UHV instruments and using ionic liquids (ILs) and other viscous polymeric electrolytes [38–42]. This new line of research has opened up a way to probe electrochemical properties and reactions in various electrochemical systems. Especially after the recognition of ILs as new generation electrolytes for many electrochemical processes, considerable attention has been given to understanding and characterizing IL electrolyte/electrode interfaces under operating conditions.

In one of the first studies, a drop of ionic liquid was placed on an angled electrode. This experimental setup allowed the formation of a liquid film thin enough to probe the IL/electrode interface. Investigators were able to detect simultaneously the core level photoelectron spectra of the elements from the imidazolium-based ionic liquid as well as the Pt substrate underneath. They applied an external potential to the electrode and probed the electrochemical processes by tracing the shifts in binding energies [43]. In another paper, researchers reported the effect of electrical double layer (EDL) and charge screening of ionic liquid electrolytes in a simple electrical double layer capacitor geometry between two coplanar gold electrodes [44]. The same geometry has been used to probe the electrochemical corrosion of gold electrodes in ionic liquids [45]. In one of the more recent studies, a drop of an ionic liquid and polyethylene glycol were placed on an electrode coated with a hydrophobic and dielectric layer. A wire electrode was then immersed into the droplet so that the external bias could be introduced between the liquid and the insulated electrode. This so-called electrowetting on dielectric (EWOD) geometry resulted in an extension of the liquid on the substrate to form a very thin layer of the liquid creep that is thin enough to probe the photoelectron spectra of the hydrophobic layer-related peak but still in electrical contact with the droplet. The authors were able to extract the time (frequency)-dependent electrical potential developments at or around the solid/liquid interface during the electrowetting processes [46,47].

In addition to probing the potential developments, various ionic liquid mediated reactions have been probed *in situ* with lab-based XPS systems equipped with Al K alpha X-ray sources. For instance, License and coworkers reported on the electrochemical reduction of Fe(III) to Fe(II) in an ionic liquid mixture [48]. A similar experimental route has been followed to identify the electrochemical generation and diffusion of Cu^+ as a function of time and position [49]. More recently, by using a simple two-electrode electrochemical system, Aydogan Gokturk *et al.* demonstrated an *in situ* electrochemical reduction of imidazolium-based ionic liquids to N-heterocyclic carbenes within the XPS chamber [50,51].

Although this experimental route is simple to implement standard XPS systems and the information obtained from *in situ* studies has enriched the knowledge of various phenomena that take place at ionic liquid/electrode interfaces, it is limited to only low vapor pressure liquids. This method cannot be implemented to study higher vapor pressure liquids, such as water and aqueous solutions.

12.3.2 *Vapor condensation method*

Another straightforward way to create a thin liquid layer on the top of a solid of choice is through condensation from vapor. This process requires raising the vapor pressure to values close to saturation pressure (at relative humidity (RH) close to 100%). The RH can be controlled by varying the chamber pressure and/or sample temperature [52]. While water is the most commonly used liquid in this method, it is possible to use others, depending on the saturation pressure of the vapor. The condensation of water vapor becomes more pronounced for hydrophilic and ionic species, such as salts and polyelectrolytes, due to their wetting properties. Previous studies showed that multilayer thick liquid films can be grown at relative humidity (RH) close to 100% [52]. Considering the thickness of the liquid layer, even soft X-rays can probe the solid/liquid interface.

So far, this approach has been pursued mainly to study thin films of highly concentrated solutions prepared by the deliquescence of various alkali halide salts. For instance, using synchrotron-based soft X-ray photoelectron spectroscopy, Ghosal *et al.* showed the enhanced halide anion concentration at the surface compared with the bulk of the solution of KBr. The enhancement of anion concentration became more dramatic for the larger, more polarizable anions upon comparison with KI [53].

In a more recent study, APXPS equipped with tender X-rays has been used to investigate the water/polymer interface at saturation vapor pressure and room temperature. This study demonstrated the critical role of side-chain chemistry (hydrophilicity) and the type of counterion contained in the water [54].

Additionally, the vapor condensation approach has been used to create an electrochemical cell made with a Pt (111) working electrode and an Ir counter electrode isolated by a non-conductive SiO_2 film [55]. The electrolyte on the top of the electrodes was formed by rehydration of H_2SO_4 deposited *ex situ* by adjusting the relative humidity close to 100%. The thickness of the concentrated acid layer was estimated to be 4–10 nm. This electrochemical cell was utilized to follow chemical changes and dissolution processes on a Pt (111) electrode during potential cycling under extremely low pH conditions. However, this method's main challenges are its lack of ability to precisely control the liquid layer thickness, the concentration, and the chemical composition of the solution during the experiments. Additionally, increasing carbon contamination at high relative humidity introduces yet another experimental challenge, and solid/liquid interface preparation by this method is therefore limited to a few special cases.

12.3.3 *Dip and pull method*

Another method for the generation of thin electrolyte films on a solid material of interest is called the dip and pull method. A solid of interest is dipped into a liquid reservoir in the analysis chamber and pulled out slowly to form a nanometer thin liquid layer that extends from the liquid layer, see Figure 12.3. This method requires hard or tender X-rays to probe the interface because the thickness of the liquid film produced exceeds the probing depth of soft X-rays [21]. Although the film thickness achieved by the dip and pull approach is around 10–30 nm, the thickness depends on various factors, including the hydrophilicity of the solid surface, the height of the measurement spot above the bulk liquid, and the evaporation rate of the electrolyte under the experimental conditions.

Due to the pressure gradient in the vicinity of the differentially pumped nozzle of the electron analyzer, water from the liquid layer evaporates. Excessive water evaporation from the liquid layer can introduce challenges in controlling the thickness of the liquid layer and the concentration. To avoid this problem, either water is dosed into the analysis chamber through a leak valve connected to a heated water reservoir or a secondary and larger water reservoir is placed inside the analysis chamber to create a pressure close to that of water vapor at room temperature. That is, an equilibrium is achieved where the evaporation and condensation rates are the same.

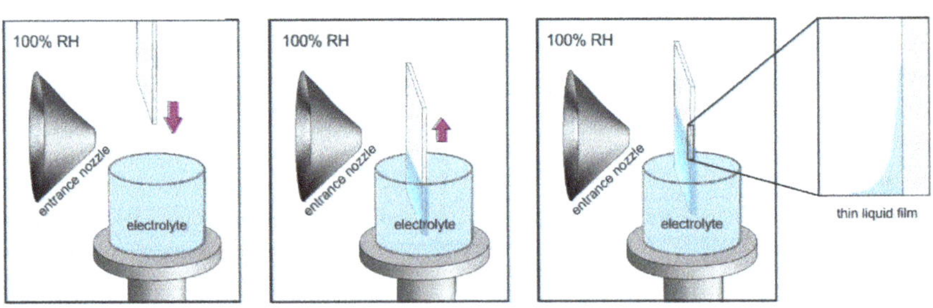

Fig. 12.3. Schematic representation of the dip and pull method coupled with Tender APXPS experiments. See text for the detailed explanation of the experimental procedure.

Placing counter and reference electrodes in the liquid reservoir offers electrochemical control and makes this approach particularly suitable for *operando/in situ* studies of various electrochemical systems. During the measurement, all three electrodes can be kept in contact with the bulk electrolyte such that the electrolyte film on the probed electrode stays in contact with the bulk electrolyte.

Continuity of the thin film from the bulk electrolyte is critical for electrochemical systems and can be easily assessed from its conductivity. It is expected that electrolyte-related XPS core-level peaks show exact binding energy shifts with the applied potentials if the thin film is indeed conductive and connected to the bulk electrolyte. On the other hand, a film that is discontinuous or composed of isolated droplet islands will not be in electrical contact with the bulk electrolyte and will exhibit the potential of the solid surface [21]. Loss of potential control or a deviation of the BE shift of liquid layer-related peaks due to increasing IR drop with time can indicate mass transport limitations in the electrolyte layer, which is the main drawback of this method. This mass transport limitation due to the restricted thickness of the film can, in principle, affect the ion concentrations and the pH value of the solution.

The dip and pull method is relatively new, and so far, it has been used to understand various electrochemical systems and phenomena including local potential developments at the metal/electrolyte and semiconductor/electrolyte interface, chemical and structural changes in the electrode under reaction conditions, and chemical changes/degradation of electrolyte solutions at the electrode surface [9–11,13,56–59].

12.3.4 *Liquid cell method*

An alternative approach to investigating electrode/electrolyte interfaces is to illuminate them from the solid side instead of the liquid side. With such geometry, the thickness of the solid film, often called the electron transparent membrane, becomes limiting and should be lower than the probing depth of the instrument to allow signal detection from the interface. This membrane-based approach has already been widely applied for environmental cell transmission electron microscopy (EC-TEM) and X-ray absorption spectroscopy (XAS) using thin silicon nitride membranes that are transparent to high-energy electrons and X-rays [60–63].

In the typical liquid cell geometry, the membrane of choice separates the liquid phase from the analysis chamber, while the photons and photoelectrons can still travel through the membrane layer to probe the interface. That is why the membrane that encloses the reaction cell should be impermeable and mechanically robust.

Previously, graphene-capped micro-channels and electrochemical cells with a graphene or graphene oxide window were used to enable soft X-ray spectroscopy of a solid–liquid interface [64,65]. In theory, a leak-tight liquid cell can even be used in standard UHV instruments. However, the majority of previous studies utilize APXPS because fragile windows pose high risks to the experimental setup. Figure 12.4 illustrates several liquid cell designs in the literature utilizing graphene windows. The first demonstrations were performed using a single cavity static liquid cell having a micron size

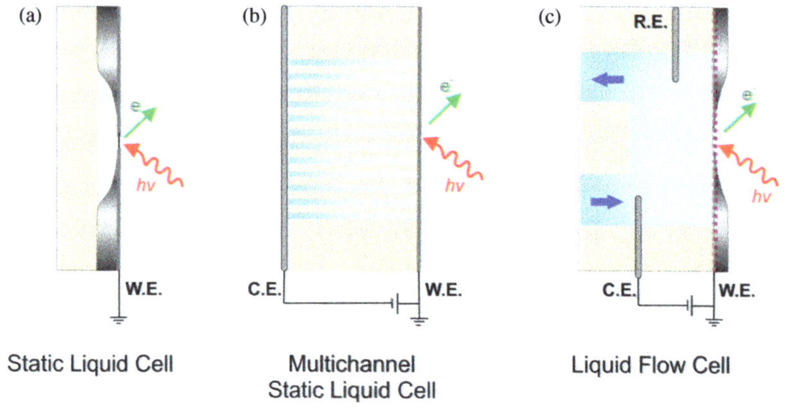

Fig. 12.4. Pictorial representation of different liquid cell designs used in the literature utilizing graphene or graphene oxide windows and soft X-rays: (a) static liquid cell, (b) graphene capped multichannel arrays of liquid cell, and (c) liquid flow cell. WE: working electrode, CE: counter electrode, RE: reference electrode.

opening, which limits the use of this approach to micro-focused X-rays (Figure 12.4(a)) [64]. Later on, Guo *et al.* introduced an alternative approach using graphene capped micro-channel arrays to increase the active area for experiments (Figure 12.4(b)) [23,66]. Similarly, some designs have utilized Si_3N_4 grids coated with bilayer graphene to enclose a single and relatively large volume of liquid [65,67]. Liquid cell designs have been further advanced to incorporate electrodes and flow liquids or gases.

In addition, thicker silicon nitride and polymer electrolyte membrane (PEM) windows require hard or tender X-ray spectroscopy [68]. Reference and counter electrodes can be implemented into the design to achieve a potential control across the interface. The major shortcoming of this approach is that the solid materials must be in the form of either thin films or particles decorated on a membrane. PEM cell designs have been used to mimic and investigate fuel cell relevant interfaces under operating conditions where a particle sputtered polymer membrane allows protons and water to transport between two sides while blocking the passage of electrons. However, the main limitation of this PEM cell design was the rapid consumption of electrolytes by evaporation and the resulting loss of membrane ion conductivity [12]. This effect can be minimized using hard or tender X-ray spectroscopies, which enable higher water vapor pressures inside the analysis chamber, or by using a flow cell, which guarantees a water supply. More recently, Mom *et al.* upgraded this approach by covering the polymer membrane with graphene. In such a geometry, graphene also serves as an evaporation barrier and allows higher pressure water below it [69]. This membrane-based approach has been adopted and developed into a variety of electrochemical cell designs. Each reported cell design, with its own unique attributes, is desirable for a certain system.

The liquid cell method offers several benefits over other approaches. For example, in a flow cell setup (Figure 12.4(c)), the solution concentration and pH can be changed during the course of the experiment. Additionally, gases or liquids flowed through the system provides facile mass transport. Table 12.1 summarizes the advantages and drawbacks of the existing XPS methods applied to the study of solid/liquid interfaces.

Table 12.1. Advantages and drawbacks of existing approaches in comparison.

	X-ray energy range	Advantages	Drawbacks
Static liquid	Soft X-rays	• Electrochemical control • Easy to implement	• Only applicable to low volatility liquids • Liquid open to chamber • Thick liquid film
Vapor condensation	Soft or tender X-rays	• Easy to implement	• Electrochemical control is not easy • Hard to control pH and concentration
Dip and pull	Tender X-rays	• Real electrolyte • Electrochemical control • Minimal charging	• Limited mass transport • Liquid open to chamber
Liquid cell	Soft or tender X-rays	• Real electrolyte • Electrochemical control • Applicable to high vapor pressure liquids	• Beam-sensitive polymer • Fragile window

12.4 Exemplary Cases

In the following section, we will discuss several exemplary XPS studies of solid/liquid interfaces. We have organized these examples into two categories to demonstrate the breadth of two possible applications, probing (i) electrical potentials and (ii) electrochemical reactions at electrified interfaces. These were also chosen to show the utility of different approaches, discussed in the previous section, to studying solid/liquid interfaces.

12.4.1 *Probing electrical potentials*

As already shown in the previous chapters, XPS enables quantitative analysis with chemical specificity and surface sensitivity. However, another feature of XPS to carry information on local potentials is relatively unknown. Upon the application of electrical potential that effectively shifts all energy levels up (with negative applied potential) or down (positive) relative to the reference state of the grounded analyzer, the binding energy of core-level photoelectrons measured by XPS shows a shift that is related to the local potential (V) on the surface by $\Delta BE = V$ eV. For this reason, with the application of external electrical bias during data acquisition, XPS can be useful for investigating various electrochemical systems. Figure 12.5 shows the energy-level diagrams of a solid/liquid/gas system for biased electrode and biased solution. All the kinetic energies of the emitted photoelectrons are given with respect to the Fermi energy of the spectrometer. Many research groups, depending on the experimental conditions and setups, have utilized both the biased electrode and the biased solution approach. For surface potential measurements or a two-electrode electrochemical system, the potential is commonly applied directly to the electrode [70] where a positive bias applied to the conductive electrode shifts all of the sample levels down with respect to those of the analyzer, having the effect of decreasing the kinetic energy of the outgoing photoelectron and vice versa (Figure 12.5(a)). The shift in kinetic energy follows precisely the applied potential. In the three-electrode electrochemical system, the electrode and analyzer are grounded and the Fermi energy is shifted by V with respect to the reference electrode. Hence, upon the potential excitation, electrode-related core-level binding energies show no shift, while the binding energy positions of electrolyte-related core levels shift with the potential (Figure 12.5(b)). Both cell designs, with their own unique attributes, are preferable for a certain system, but it is of utmost importance to discriminate the chemical shifts from the potential-induced shifts in order to correctly interpret the spectra.

12.4.1.1 *Band bending at semiconductor/liquid interface*

In the semiconductor/liquid junction, in addition to the screening effects of ions in solution phase (e.g., EDL), semiconductor

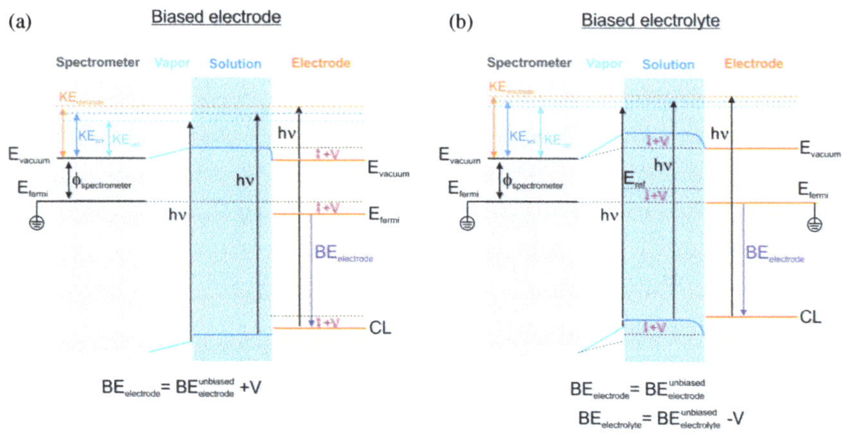

Fig. 12.5. The energy-level diagram of the conductive electrode/electrolyte solution/vapor system for (a) biased electrode and (b) biased electrolyte configurations. E_{fermi}: Fermi level, E_{vacuum}: vacuum level, E_{ref}: Reference Level, Φ: work function, KE: photoelectron kinetic energy, BE: binding energy, CL: core level, V: applied potential.

electrodes experience an additional potential drop within the space-charge region due to immobilized charge carriers in the lattice. This so-called band bending in the semiconductor drives the separation of electron–hole pairs, and it is very important particularly for photoelectrochemical (PEC) energy conversion systems.

Tender APXPS combined with the dip and pull method has been used to probe the semiconductor/liquid junction of PEC cells composed of a layer of TiO_2 on a p-type silicon working electrode in a three-electrode setup, while applying an electrical potential ranging from −1.2 V to +0.4 V [71]. The energy-level diagrams of each cell component are illustrated in Figure 12.6(a). The resulting electrical potential developed in the space charge region of the semiconductor, as well as in the electrolyte of 1M KOH, causes binding energy shifts in the related core levels. Such a high concentration of electrolyte causes an electrochemical double layer with a very small Debye length so that a potential drop at EDL does not contribute to the electrolyte spectra.

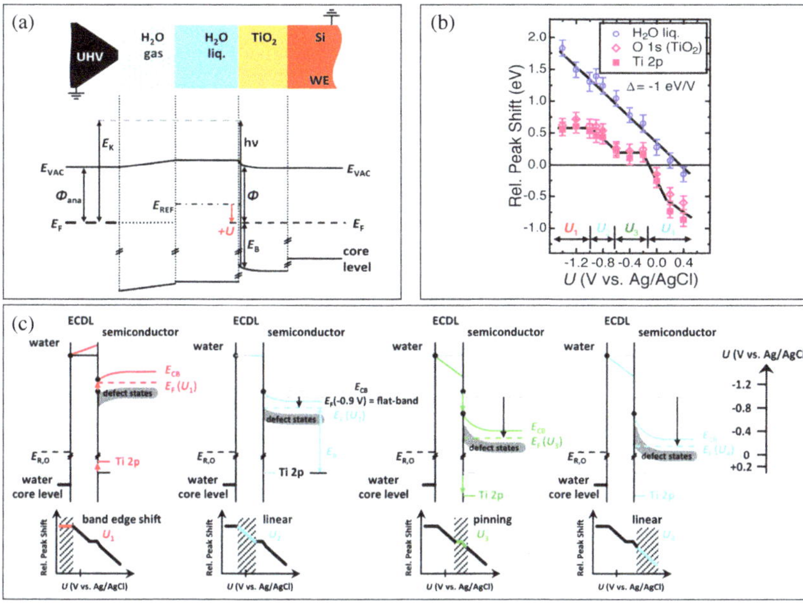

Fig. 12.6. (a) The energy-level diagram of the SiO_2/TiO_2 electrode/liquid solution/gas system with the aqueous electrolyte layer (1M KOH) in equilibrium with water vapor under applied potential U. Both the aperture and the electrode are grounded. (b) Binding energy shifts of O 1s of the water (blue circles), O 1s of the TiO_2 (pink diamonds), and the Ti 2p (pink squares) with the applied potential and (c) corresponding energy-level diagram of the TiO_2/electrolyte junction for highly negative bias (U1 region, red lines), the ideal semiconductor region U_2, from −0.9 V to −0.6 V (blue lines), the increased positive biased (U_3 region, green lines), and positive potentials from −0.2 V (region U_4). The Ti 2p binding energy shifts linearly for band bending regimes (U_2 and U_4) and remains constant for the band shifting regimes (U_1 and U_3).

Detected binding energy shifts of the O 1s peak from the electrolyte and the O 1s and Ti 2p peaks from the TiO_2 are plotted in Figure 12.6(b). As expected, the O 1s peak originates from the liquid phase water follows the applied external potential with a slope of −1. The binding energies of the core levels associated with TiO_2 exhibit more complex behavior that can be explained with four different potential intervals, as depicted in Figures 12.6(b) and 12.6(c). Starting from the most negative potentials, in region 1 (U_1),

TiO_2-related core levels do not shift because semiconductor surface states of conduction band are filled with electrons, resulting in a negative surface charge, and the Fermi level is pinned to the conduction band minimum. For potentials between −1.0 and −0.6 V vs. Ag/AgCl (region U_2), standard semiconductor behavior occurs. Here, potential drops at the space charge region of the semiconductor instead of the electrochemical double layer. For increased positive bias in region U_3, the peak positions again remain constant due to Fermi level pinning of the conduction band. At potentials higher than −0.2 V (region U_4), ideal behavior is once again observed.

12.4.1.2 *Electrical double layer*

The EDL plays a key role in the electrochemistry of electrified interfaces. When an electrolyte solution comes into contact with a polarized/charged surface, ions at the interface rearrange to screen this charge. As a result, a concentration and a potential profile form at the electrode/electrolyte interface, which governs charge transfer processes, and influence the kinetics and thermodynamics of the electrochemical systems. Numerous theories and models have been developed to describe the structure of ions and the shape of potential profiles at the electrode/electrolyte interface. However, it has been challenging to probe the EDL experimentally and verify these models.

Using the combined Tender APXPS and dip and pull method, Favaro *et al.* have succeeded in directly probing the potential drop at the electrified solid/liquid interface by following the changes in spectral broadening and the magnitude of full width half maximum (FWHM) of electrolyte-related core-level peaks as a function of potential (Figure 12.7). Their experiment used the three-electrode electrochemical cell (Figure 12.4(c)) containing an Au working electrode grounded to the analyzer, a polycrystalline Pt foil counter electrode, and an Ag/AgCl reference electrode. Aqueous potassium hydroxide was used as the electrolyte solution. In addition, a neutral probing molecule, pyrazine, was added to the solution to make the spectral analysis easier. In order to follow only the EDL-related processes, the

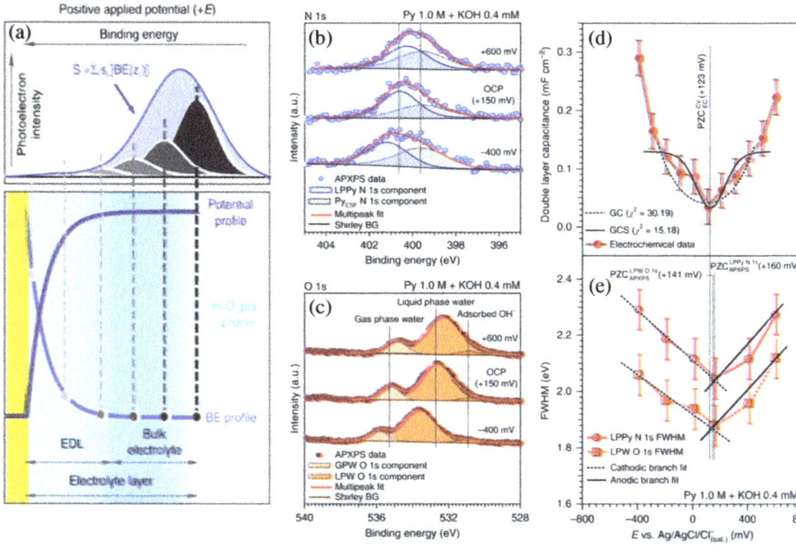

Fig. 12.7. (a) Schematic representation of spectral broadening of the liquid phase related elements' core level due to the potential profile formed at the interface for KOH 0.4 mM aqueous solution containing 1.0 M pyrazine; (b) and (c) Intensity-normalized N 1s and O 1s core-level peaks, respectively, acquired at different applied potentials to the WE; (d) double-layer capacitance as a function of the applied potential from the electrochemical characterization. The double-layer capacitance trend has been fitted by using both Gouy–Chapman and Gouy–Chapman–Stern models; (e) FWHM of N 1s and O 1s peaks coming from the electrolyte layer as a function of the applied potential within the EDL region. Adapted with permission from Ref. [56]. Copyright (2016) Springer Nature.

amplitude of the external potential was chosen such that the Faradaic sources and possible redox reactions would not contribute.

When the net charge is zero at the electrode surface, O 1s and N 1s core-level peaks coming from the electrolyte layer exhibit a narrow and symmetrical peak shape. When the potential is increased or decreased, EDL forms at the interface and molecules experience different electric potentials as a function of its distance to the electrode. Consequently, the O 1s or N 1s photoelectrons from different depths are characterized by different apparent binding energies. The convolution of the single-shifted core-level photoelectron peaks

(Figure 12.7(a)) leads to the experimentally observed spectral broadening and asymmetry in the peak shape.

Expanding the experimental data set to various external electrolyte concentrations and applied potentials enabled the authors to test the observed potential profile of the EDL with various existing electrochemical models (Figure 12.7(d)). This study also suggested a synergistic correlation between photoelectron spectroscopy and electrochemistry, showing good agreement between the potential of zero charge (PZC) values obtained from APXPS and electrochemical analysis.

12.4.2 *Probing reactions at electrified interface*

The electrochemical control during data acquisition, in addition to its chemical sensitivity, makes (AP)XPS very useful for investigating various electrochemical reactions at the electrode/electrolyte interface. With the increasing necessity to transition to clean and sustainable energy, electrochemical half reactions of water splitting (hydrogen evolution reaction (HER) and oxygen evolution reaction (OER)) and fuel cell reactions (oxygen reduction reaction (ORR) and hydrogen oxidation reaction (HOR)) have been extensively investigated with photoelectron spectroscopy over the past decades [7,9–11,13,59,65,72,73].

12.4.2.1 *Pt oxidation in alkaline solution*

Using the Tender APXPS and dip and pull method shown in Figure 12.3, Favaro *et al.* [11] probed the changes in surface chemistry of a Pt electrode under anodic conditions in 1.0 M KOH electrolyte. Their experiment used a three-electrode electrochemical cell (Figure 12.4(c)) containing a polished polycrystalline Pt working electrode grounded to the analyzer, a polycrystalline Pt foil counter electrode, and an Ag/AgCl reference electrode. Both cyclic voltammetry (CV) from –1 V to 1 V vs. Ag/AgCl and chronoamperometry measurements were performed before the data acquisition. Once the CV had stabilized, chemical changes on the

Pt electrode were explored at different potentials along the anodic branch of the CV, from open circuit potential (OCP) up to +900 mV (OER conditions). The continuity of the thin electrolyte film and possible mass transport limitations were assessed by following the binding energy shifts in the electrolyte-related core levels upon the applied voltage and relative areas of ionic species-related core levels, respectively. In addition, the electrochemical reversibility of the reaction was studied by returning back to OCP after OER conditions.

Figure 12.8(a) shows the evolution of the Pt surface chemistry studied as a function of the applied potential. In this study, the Pt 4f spectra deconvoluted into four different components coming from metallic Pt, Pt–OH_{ads}, Pt^{+2}, and Pt^{+4}. Starting from the OCP, three of these components (all except the Pt^{+4}) contributed to the spectra. Interestingly, the relative peak areas of these components, and hence the thickness, remained unchanged with increasing

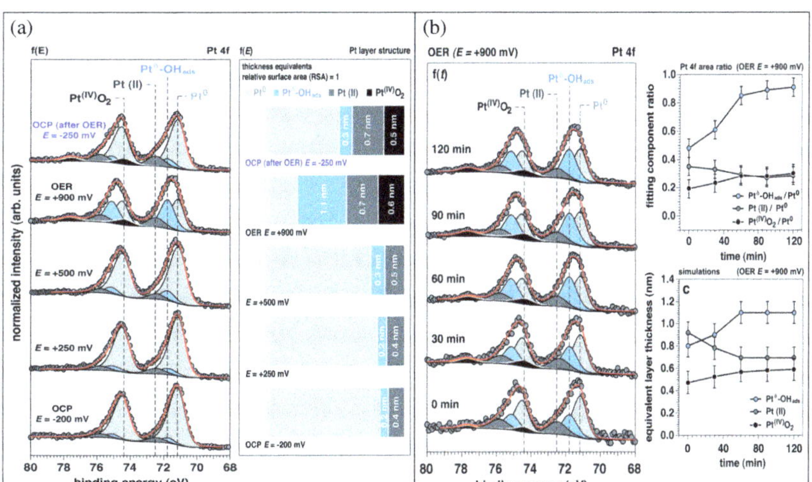

Fig. 12.8. (a) Evolution of the Pt 4f surface chemistry as a function of the applied potential from OCP to OER and corresponding thickness. (b) Chemical evolution of platinum surface under oxygen evolution conditions, 900 mV, as a function of time together with the experimental trends and simulated layer thicknesses.

potentials below the OER, but they changed drastically under OER conditions. The amount of Pt–OH_{ads} and Pt^{+2} species increased once the potential was raised to +900 mV (OER conditions). Under this condition, the formation of Pt^{+4} from platinum oxide became observable as well. At OCP after OER, the intensity of the Pt-OH component, which provides particular insights into the OER mechanism, became comparable to the initial values, indicating the reversibility while Pt^{+2} and Pt^{+4} remain similar.

In addition, the time evolution of the Pt surface chemistry under OER conditions was followed for 120 minutes, Figure 12.8(b). Spectral data showed the initial transitions of Pt species, resulting in the overall increase in $Pt+^{2}$ and Pt^{+4} and the small decrease in Pt-OH before reaching the steady state after 60 minutes. This observation leads to the conclusion that the Pt species observed at oxygen evolution reaction potentials are a mixture of different Pt states, which is different from the pure divalent state suggested by the Pourbaix diagram at this pH range.

12.4.2.2 *Electrochemical reactions using polymer electrolyte membrane cells*

To date, the hydrogen and oxygen cycles for electrochemical energy storage and conversion systems have also been investigated using a variety of electrochemical cell designs that utilize a choice of polymer electrolyte membranes. Most recently, a graphene-capped polymer membrane liquid cell has been used to study the reversible oxidation of Pt at potentials from 0.05 V to 1.85 V vs RHE by soft APXPS under wet conditions [69]. Figure 12.9(a) illustrates the electrochemical liquid flow cell used in this study, where electroactive Pt nanoparticles (NPs) are sandwiched between the polymer membrane and a graphene layer. In this setup, the graphene layer functions as an electron transparent window as well as a contact for the Pt NPs, but it also helps minimize the loss of the liquid water layer via evaporation. A choice of commercial polymer electrolyte membrane, either the Nafion or the FAD, will function as a solid electrolyte connecting the working electrode with the

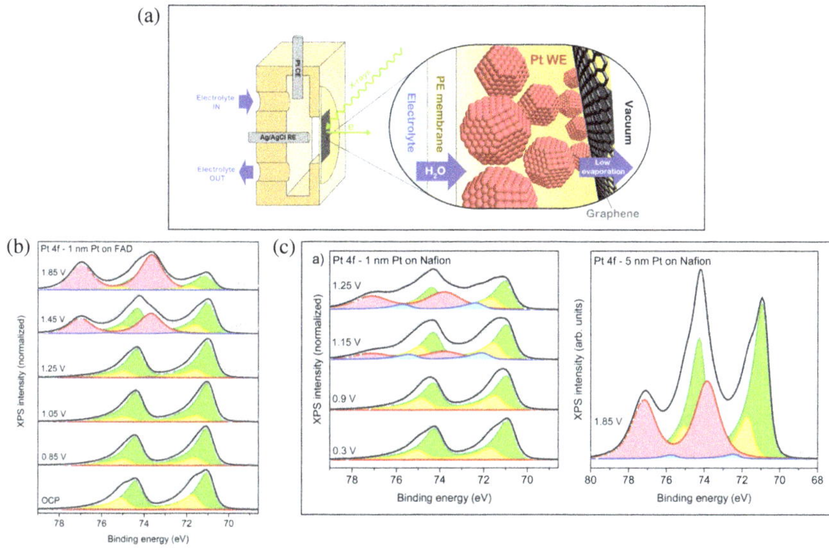

Fig. 12.9. (a) Schematic representation of the electrochemical cell used in the study. Potential dependent stepwise oxidation of a single layer of Pt nanoparticles on (b) FAD and (c) Nafion in 0.1 M H_2SO_4. Pt 4f region is deconvoluted into Pt^0 (green), $Pt^{\delta+}$ (yellow), Pt^{2+} (blue), and Pt^{4+} (red) doublets. Copyright (2019) American Chemical Society. Further permissions related to the material excerpted should be directed to the ACS.

planar Pt counter and Ag/AgCl reference electrode immersed in the liquid reservoir. Nafion is a well-known proton exchange membrane that only water and protons can pass through, while FAD is an anion exchange membrane with Br counterions that allows the passage of anions. The authors evaluated the potential dependent chemical changes in Pt NPs under operating conditions for electrochemical cells made of these two different polymer membranes.

Figure 12.9(b) shows the potential-induced changes in the Pt on the FAD membrane. At OCP, the $Pt^{\delta+}$ chemistry is observed in addition to the metallic peak. With increasing potential up to 1.05 V, the $Pt^{\delta+}$ contribution decreases. Upon further increase of potential to 1.25 V, the Br 3d peak around 69 eV rises, and its interaction with Pt is believed to suppress surface oxide formation and Pt^{2+} chemistry. However, the initial oxidation of Pt NPs on Nafion shows a surface oxide with mixed oxidation states of $Pt^{\delta+}$, Pt^{2+}, and Pt^{4+}, see Figure 12.9(c). For both polymer membranes, anodization to higher potentials, where bulk oxidation can occur, yields pure Pt^{4+}.

In addition to this study utilizing soft X-rays, Takagi *et al.* performed *operando* hard-APXPS experiments on a Pt/C electrode of a Nafion-based polymer electrolyte membrane fuel cell (PEMFC) at a photon energy of 7.94 keV [74]. However, this cell setup was different from the previously designed APXPS liquid cells. Instead of sealing the cell, they used a Nafion membrane to separate the two chambers. The cathode side, where the analysis was performed, was exposed to H_2O pressure to 21 Torr, which corresponds to 100% RH around room temperature, while the mixture of H_2/N_2 gas at atmospheric pressure flowed on the anode side. Due to the low cross-section of Pt 4f at high photon energies, they examined the changes in the Pt $3d_{5/2}$ core level to understand the potential dependent reversible oxidation of the Pt nanoparticles at the cathode. The peak originating from the Pt^{+4} appeared at an applied voltage of 1.3 V and disappeared at 0.1 V.

12.5 Concluding Remarks

As we discussed in this chapter, the advancements of APXPS in the last couple of decades have enabled us to explore various scientific phenomena, including chemistry and potential distribution at electrode/electrolyte interfaces. We have discussed some of the main technical challenges for the use of XPS in *operando* studies of solid/liquid interfaces. The fundamental challenge of this type of experiment is to expand the pressure limit to more realistic conditions, close to atmosphere, and to study real ambient liquids. In this regard, operating with high photon energies is particularly promising. We have also examined several existing approaches to tackle these challenges. Although each approach has its own advantages and limitations, all of them exhibit great significance for enabling *in situ/operando* characterization and revealing the fundamental aspects of solid/liquid interfaces. Yet, all of these approaches are still new, and many new challenges as well as opportunities lie ahead for us to address.

One of the challenges of (AP)XPS is that current studies provide vital steady state or equilibrium information, but real-time observation of fast processes, occurring in a time span of a few milliseconds, is difficult to capture. Novel pump–probe techniques coupling photoelectron spectroscopy with laser systems can provide unique time-resolved information [75]. Such studies with temporal resolution can help unravel the reaction intermediates and kinetics and provide further insight into electrocatalysis under dynamic reaction conditions.

Another aspect, which is less discussed but relatively important, is the thorough understanding of beam-induced chemical changes in the sample. The degree of beam-induced damage on the sample depends significantly on the brightness and the flux of the photon source. With increasing gas pressure inside the analysis chamber, this effect is enhanced due to the formation of reactive radicals from ionized gas species. Thorough evaluation of chemical changes induced by the beam is particularly important for electrochemical cells involving soft materials, such as polymer electrolyte membranes, because changes in the structure or chemistry in such membranes greatly influence the ion/charge transport and hence the entire performance of the device. To mitigate the beam-induced damage, the X-ray photon flux can be lowered or the time beam exposure can be minimized by constantly changing the analysis position. But most importantly, a detailed mechanistic study of beam damage in the field of APXPS does not yet exist and is urgently needed.

Furthermore, increasing spatial resolution and employing mapping capabilities are also important for a better understanding of realistic systems, which greatly suffer from non-homogeneity. Today, micrometer spatial resolution is achievable but nanometer resolution is desirable, as in the case of recent advancements in AP scanning photoelectron spectromicroscopy [76,77].

References

[1] Bard AJ, Faulkner LR. *Electrochemical Methods: Fundamentals and Applications.* New York: Wiley, 2000.

[2] Jerkiewicz G. Chapter 1: From electrochemistry to molecular-level research on the solid–liquid electrochemical interface. In *Solid-Liquid Electrochemical Interfaces* (pp. 1–12). Washington, DC: Oxford University Press, 1997.

[3] Bach K. *Electrical Double Layers in Biology.* New York, USA: Springer, 2012.

[4] Schoch RB, Han J, Renaud P. Transport phenomena in nanofluidics. Rev Mod Phys. 2008;80:839–883.

[5] Bormashenko E. Physics of solid–liquid interfaces: From the Young equation to the superhydrophobicity (Review Article). Low Temp Phys. 2016;42:622–635.

[6] Björneholm O, Hansen MH, Hodgson A, Liu L-M, Limmer DT, Michaelides A, Pedevilla P, Rossmeisl J, Shen H, Tocci G, Tyrode E, Walz M-M, Werner J, Bluhm H. Water at interfaces. Chem Rev. 116:7698–7726.

[7] Casalongue HS, Kaya S, Viswanathan V, Miller DJ, Friebel D, Hansen HA, Nørskov JK, Nilsson A, Ogasawara H. Direct observation of the oxygenated species during oxygen reduction on a platinum fuel cell cathode. Nat Commun. 2016;4:2817.

[8] Sanchez Casalongue HG, Ng ML, Kaya S, Friebel D, Ogasawara H, Nilsson A. *In Situ* observation of surface species on iridium oxide nanoparticles during the oxygen evolution reaction. Angewandte Chemie, International Edition. 2014;53:7169–7172.

[9] Ali-Löytty H, Louie MW, Singh MR, Li L, Sanchez Casalongue HG, Ogasawara H, Crumlin EJ, Liu Z, Bell AT, Nilsson A, Friebel D. Ambient-pressure XPS study of a Ni–Fe electrocatalyst for the oxygen evolution reaction. J Phys Chem C. 2016;120:2247–2253.

[10] Favaro M, Yang J, Nappini S, Magnano E, Toma FM, Crumlin EJ, Yano J, Sharp ID. Understanding the oxygen evolution reaction mechanism on CoOx using operando ambient-pressure X-ray photoelectron spectroscopy. J Am Chem Soc. 2017;139:8960–8970.

[11] Favaro M, Valero-Vidal C, Eichhorn J, Toma FM, Ross PN, Yano J, Liu Z, Crumlin EJ. Elucidating the alkaline oxygen evolution reaction mechanism on platinum. J Mater Chem A. 2017;5:11634–11643.

[12] Streibel V, Hävecker M, Yi Y, Velasco Vélez JJ, Skorupska K, Stotz E, Knop-Gericke A, Schlögl R, Arrigo R. *In Situ* electrochemical cells to study the oxygen evolution reaction by near ambient pressure X-ray photoelectron spectroscopy. Top Catal. 2018;61:2064–2084.

[13] Stoerzinger KA, Favaro M, Ross PN, Yano J, Liu Z, Hussain Z, Crumlin EJ. Probing the surface of platinum during the hydrogen evolution reaction in alkaline electrolyte. J Phys Chem B. 2018;122: 864–870.

[14] Hind AR, Bhargava SK, McKinnon A. At the solid/liquid interface: FTIR/ATR — The tool of choice. Adv Colloid Interface Sci. 2001; 93:91–114.

[15] Andanson J-M, Baiker A. Exploring catalytic solid/liquid interfaces by *in situ* attenuated total reflection infrared spectroscopy. Chem Soc Rev. 2010;39:4571–4584.

[16] Meemken F, Müller P, Hungerbühler K, Baiker A. Simultaneous probing of bulk liquid phase and catalytic gas-liquid-solid interface under working conditions using attenuated total reflection infrared spectroscopy. Review of Scientific Instruments. 2014;85:084101.

[17] Ataka K-i, Yotsuyanagi T, Osawa M. Potential-dependent reorientation of water molecules at an electrode/electrolyte interface studied by surface-enhanced infrared absorption spectroscopy. J Phys Chem. 1996;100:10664–10672.

[18] Tian Z-Q, Ren B. Adsorption and reaction at electrochemical interfaces as probed by surface-enhanced raman spectroscopy. Annual Review of Physical Chemistry. 2004;55:197–229.

[19] Eisenthal KB. Liquid interfaces probed by second-harmonic and sum-frequency spectroscopy. Chem Rev. 1996;96:1343–1360.

[20] Miranda PB, Shen YR. Liquid interfaces: A study by sum-frequency vibrational spectroscopy. J Phys Chem B. 1999;103:3292–3307.

[21] Axnanda S, Crumlin EJ, Mao B, Rani S, Chang R, Karlsson PG, Edwards MOM, Lundqvist M, Moberg R, Ross P, Hussain Z, Liu Z. Using "Tender" X-ray ambient pressure X-ray photoelectron

spectroscopy as a direct probe of solid-liquid interface. Sci Rep. 2015;5:9788.

[22] Zaera F. Surface chemistry at the liquid/solid interface. Surf Sci. 2011;605:1141–1145.

[23] Yulaev A, Guo H, Strelcov E, Chen L, Vlassiouk I, Kolmakov A. Graphene microcapsule arrays for combinatorial electron microscopy and spectroscopy in liquids. ACS Appl Mater Interfaces. 2017;9:26492–26502.

[24] Collins L, Jesse S, Kilpatrick JI, Tselev A, Varenyk O, Okatan MB, Weber SAL, Kumar A, Balke N, Kalinin SV, Rodriguez BJ. Probing charge screening dynamics and electrochemical processes at the solid–liquid interface with electrochemical force microscopy. Nat Commun. 2014;5:3871.

[25] Collins L, Kilpatrick JI, Kalinin SV, Rodriguez BJ. Towards nanoscale electrical measurements in liquid by advanced KPFM techniques: A review. Rep Prog Phys. 2018;81:086101.

[26] Zeng Z, Liang W-I, Liao H-G, Xin HL, Chu Y-H, Zheng H. Visualization of electrode–electrolyte interfaces in LiPF6/EC/DEC electrolyte for lithium ion batteries via *in Situ* TEM. Nano Lett. 2014; 14:1745–1750.

[27] Newberg JT, Åhlund J, Arble C, Goodwin C, Khalifa Y, Broderick A. A lab-based ambient pressure X-ray photoelectron spectrometer with exchangeable analysis chambers. Rev Sci Instrum. 2015;86: 085113.

[28] Arble C, Jia M, Newberg JT. Lab-based ambient pressure X-ray photoelectron spectroscopy from past to present. Surf Sci Rep. 2018;73: 37–57.

[29] Lim H, Yu Y, Kim D, Koh YE, Mun BS, Lehane V. Performance test of a laboratory-based ambient pressure X-ray photoelectron spectroscopy system at the Gwangju Institute of Science and Technology. J Korean Phys Soc. 2019;75:541–546.

[30] Han Y, Zhang H, Yu Y, Liu Z. *In situ* characterization of catalysis and electrocatalysis using APXPS. ACS Catal. 2021;11:1464–1484.

[31] Amann P, Degerman D, Lee M-T, Alexander JD, Shipilin M, Wang H-Y, Cavalca F, Weston M, Gladh J, Blom M, Björkhage M, Löfgren P, Schlueter C, Loemker P, Ederer K, Drube W, Noei H, Zehetner J, Wentzel H, Åhlund J, Nilsson A. A high-pressure X-ray

photoelectron spectroscopy instrument for studies of industrially relevant catalytic reactions at pressures of several bars. Rev Sci Instrum. 2019;90:103102.

[32] Nemšák S, Shavorskiy A, Karslioglu O, Zegkinoglou I, Rattanachata A, Conlon CS, Keqi A, Greene PK, Burks EC, Salmassi F, Gullikson EM, Yang, S-H, Liu K, Bluhm H, Fadley CS. Concentration and chemical-state profiles at heterogeneous interfaces with sub-nm accuracy from standing-wave ambient-pressure photoemission. Nat Commun. 2014;5:5441.

[33] Karslıoğlu O, Nemšák S, Zegkinoglou I, Shavorskiy A, Hartl M, Salmassi F, Gullikson EM, Ng ML, Rameshan C, Rude B, Bianculli D, Cordones AA, Axnanda S, Crumlin EJ, Ross PN, Schneider CM, Hussain Z, Liu Z, Fadley CS, Bluhm H. Aqueous solution/metal interfaces investigated in operando by photoelectron spectroscopy. Faraday Discuss. 2015;180:35–53.

[34] Trotochaud L, Head AR, Karslıoğlu O, Kyhl L, Bluhm H. Ambient pressure photoelectron spectroscopy: Practical considerations and experimental frontiers. J Phys Condens Matter. 2016;29:053002.

[35] Siegbahn H, Siegbahn K. ESCA applied to liquids. J Electron Spectrosc Relat Phenom. 1973;2:319–325.

[36] Siegbahn H. Electron spectroscopy for chemical analysis of liquids and solutions. J Phys Chem. 1985;89:897–909.

[37] Lundholm M, Siegbahn H, Holmberg S, Arbman M. Core electron spectroscopy of water solutions. J Electron Spectrosc Relat Phenom. 1986;40:163–180.

[38] Aydogan Gokturk P, Taner Camci M, Suzer S. Lab-based operando X-ray photoelectron spectroscopy for probing low-volatile liquids and their interfaces across a variety of electrosystems. J Vac Sci Technol A. 2020;38:040805.

[39] Uzundal CB, Sahin O, Aydogan Gokturk P, Wu H, Mugele F, Ulgut B, Suzer S. X-ray photoelectron spectroscopy with electrical modulation can be used to probe electrical properties of liquids and their interfaces at different stages. Langmuir. 2019;35:16989–16999.

[40] Lovelock KRJ, Villar-Garcia IJ, Maier F, Steinrück H-P, Licence P. Photoelectron spectroscopy of ionic liquid-based interfaces. Chem Rev. 2010;110:5158–5190.

[41] Steinrück H-P. Recent developments in the study of ionic liquid interfaces using X-ray photoelectron spectroscopy and potential future directions. PCCP. 2012;14:5010–5029.

[42] Smith EF, Rutten FJM, Villar-Garcia IJ, Briggs D, Licence P. Ionic liquids in vacuo: Analysis of liquid surfaces using ultra-high-vacuum techniques. Langmuir. 2006;22:9386–9392.

[43] Weingarth D, Foelske-Schmitz A, Wokaun A, Kötz R. *In situ* electrochemical XPS study of the Pt/[EMIM][BF4] system. Electrochem Commun. 2011;13:619–622.

[44] Camci MT, Aydogan P, Ulgut B, Kocabas C, Suzer S. XPS enables visualization of electrode potential screening in an ionic liquid medium with temporal- and lateral-resolution. Phys Chem Chem Phys. 2016;18:28434–28440.

[45] Camci MT, Ulgut B, Kocabas C, Suzer S. *In Situ* XPS monitoring and characterization of electrochemically prepared Au nanoparticles in an ionic liquid. ACS Omega. 2017;2:478–486.

[46] Aydogan Gokturk P, Ulgut B, Suzer S. DC electrowetting of nonaqueous liquid revisited by XPS. Langmuir. 2018;34:7301–7308.

[47] Aydogan Gokturk P, Ulgut B, Suzer S. AC electrowetting modulation of low-volatile liquids probed by XPS: Dipolar vs ionic screening. Langmuir. 2019;35:3319–3326.

[48] Taylor AW, Qiu F, Villar-Garcia IJ, Licence P. Spectroelectrochemistry at ultrahigh vacuum: *In situ* monitoring of electrochemically generated species by X-ray photoelectron spectroscopy. Chem Commun. 2009;39:5817–5819.

[49] Qiu F, Taylor AW, Men S, Villar-Garcia IJ, Licence P. An ultra high vacuum-spectroelectrochemical study of the dissolution of copper in the ionic liquid (N-methylacetate)-4-picolinium bis(trifluoromethylsulfonyl)imide. Phys Chem Chem Phys. 2010;12:1982–1990.

[50] Aydogan Gokturk P, Donmez SE, Ulgut B, Türkmen YE, Suzer S. Optical and XPS evidence for the electrochemical generation of an N-heterocyclic carbene and its CS_2 adduct from the ionic liquid [bmim][PF_6]. New J Chem. 2017;41:10299–10304.

[51] Aydogan Gokturk P, Salzner U, Nyulászi L, Ulgut B, Kocabas C, Suzer S. XPS-evidence for *in situ* electrochemically-generated carbene formation. Electrochim Acta. 2017;234:37–42.

[52] Deng X, Herranz T, Weis C, Bluhm H, Salmeron M. Adsorption of water on Cu_2O and Al_2O_3 thin films. J Phys Chem C. 2008;112: 9668–9672.

[53] Ghosal S, Hemminger JC, Bluhm H, Mun BS, Hebenstreit ELD, Ketteler G, Ogletree DF, Requejo FG, Salmeron M. Electron spectroscopy of aqueous solution interfaces reveals surface enhancement of halides. Science. 2005;307:563–566.

[54] Gokturk PA, Barry M, Segalman R, Crumlin EJ. Directly probing polymer thin film chemistry and counterion influence on water sorption. ACS Appl Polym Mater. 2020;2:4752–4761.

[55] Ali-Löytty H, Hannula M, Valden M, Eilert A, Ogasawara H, Nilsson A. Chemical dissolution of Pt(111) during potential cycling under negative pH conditions studied by operando X-ray photoelectron spectroscopy. J Phys Chem C. 2019;123:25128–25134.

[56] Favaro M, Jeong B, Ross PN, Yano J, Hussain Z, Liu Z, Crumlin EJ. Unravelling the electrochemical double layer by direct probing of the solid/liquid interface. Nat Commun. 2016;7:12695.

[57] Yu Y, Baskin A, Valero-Vidal C, Hahn NT, Liu Q, Zavadil KR, Eichhorn BW, Prendergast D, Crumlin EJ. Instability at the electrode/electrolyte interface induced by hard cation chelation and nucleophilic attack. Chem Mater. 2017;29:8504–8512.

[58] Lichterman MF, Richter MH, Hu S, Crumlin EJ, Axnanda S, Favaro M, Drisdell W, Hussain Z, Brunschwig BS, Lewis NS, Liu Z, Lewerenz H-J. An electrochemical, microtopographical and ambient pressure X-ray photoelectron spectroscopic investigation of Si/TiO_2/Ni/electrolyte interfaces. J Electrochem Soc. 2015;163:H139–H146.

[59] Favaro M, Drisdell WS, Marcus MA, Gregoire JM, Crumlin EJ, Haber JA, Yano J. An operando investigation of (Ni–Fe–Co–Ce)O_x system as highly efficient electrocatalyst for oxygen evolution reaction. ACS Catal. 2017;7:1248–1258.

[60] Pu S, Gong C, Robertson AW. Liquid cell transmission electron microscopy and its applications. R Soc Open Sci. 2020;7:191204.

[61] Velasco-Vélez J-J, Skorupska K, Frei E, Huang Y-C, Dong C-L, Su B-J, Hsu C-J, Chou H-Y, Chen J-M, Strasser P, Schlögl R, Knop-Gericke A, Chuang C-H. The electro-deposition/dissolution of $CuSO_4$ aqueous electrolyte investigated by *in situ* soft X-ray absorption spectroscopy. J Phys Chem B. 2018;122:780–787.

[62] Schön D, Xiao J, Golnak R, Tesch MF, Winter B, Velasco-Velez J-J, Aziz EF. Introducing ionic-current detection for X-ray absorption spectroscopy in liquid cells. J Phys Chem Lett. 2017;8:2087–2092.

[63] Velasco-Velez J-J, Wu CH, Pascal TA, Wan LF, Guo J, Prendergast D, Salmeron M. The structure of interfacial water on gold electrodes studied by X-ray absorption spectroscopy. Science. 2014;346:831.

[64] Kolmakov A, Dikin DA, Cote LJ, Huang J, Abyaneh MK, Amati M, Gregoratti L, Günther S, Kiskinova M. Graphene oxide windows for *in situ* environmental cell photoelectron spectroscopy. Nature Nanotechnol. 2011;6:651–657.

[65] Velasco-Velez JJ, Pfeifer V, Hävecker M, Weatherup RS, Arrigo R, Chuang C-H, Stotz E, Weinberg G, Salmeron M, Schlögl R, Knop-Gericke A. Photoelectron spectroscopy at the graphene–liquid interface reveals the electronic structure of an electrodeposited cobalt/graphene electrocatalyst. Angewandte Chemie, International Edition. 2015;54:14554–14558.

[66] Guo H, Strelcov E, Yulaev A, Wang J, Appathurai N, Urquhart S, Vinson J, Sahu S, Zwolak M, Kolmakov A. Enabling photoemission electron microscopy in liquids via graphene-capped microchannel arrays. Nano Lett. 2017;17:1034–1041.

[67] Weatherup RS, Eren B, Hao Y, Bluhm H, Salmeron MB. Graphene membranes for atmospheric pressure photoelectron spectroscopy. J Phys Chem Lett. 2016;7:1622–1627.

[68] Takagi Y, Wang H, Uemura Y, Nakamura T, Yu L, Sekizawa O, Uruga T, Tada M, Samjeské G, Iwasawa Y, Yokoyama T. *In situ* study of oxidation states of platinum nanoparticles on a polymer electrolyte fuel cell electrode by near ambient pressure hard X-ray photoelectron spectroscopy. PCCP. 2017;19:6013–6021.

[69] Mom R, Frevel L, Velasco-Vélez J-J, Plodinec M, Knop-Gericke A, Schlögl R. The oxidation of platinum under wet conditions observed by electrochemical X-ray photoelectron spectroscopy. J Am Chem Soc. 2019;141:6537–6544.

[70] Brown MA, Abbas Z, Kleibert A, Green RG, Goel A, May S, Squires T. M. Determination of surface potential and electrical double-layer structure at the aqueous electrolyte-nanoparticle interface. Phys Rev X. 2016;6:011007.

[71] Lichterman MF, Hu S, Richter MH, Crumlin EJ, Axnanda S, Favaro M, Drisdell W, Hussain Z, Mayer T, Brunschwig BS, Lewis NS, Liu Z, Lewerenz H-J. Direct observation of the energetics at a semiconductor/liquid junction by operando X-ray photoelectron spectroscopy. Energ Environ Sci. 2015;8:2409–2416.

[72] Pfeifer V, Jones TE, Velasco Vélez JJ, Arrigo R, Piccinin S, Hävecker M, Knop-Gericke A, Schlögl R. *In situ* observation of reactive oxygen species forming on oxygen-evolving iridium surfaces. Chem Sci. 2017;8:2143–2149.

[73] Arrigo R, Hävecker M, Schuster ME, Ranjan C, Stotz E, Knop-Gericke A, Schlögl R. *In situ* study of the gas-phase electrolysis of water on platinum by NAP-XPS. 2013;52:11660–11664.

[74] Takagi Y, Wang H, Uemura Y, Ikenaga E, Sekizawa O, Uruga T, Ohashi H, Senba Y, Yumoto H, Yamazaki H, Goto S, Tada M, Iwasawa Y, Yokoyama T. *In situ* study of an oxidation reaction on a Pt/C electrode by ambient pressure hard X-ray photoelectron spectroscopy. Appl Phys Lett. 2014;105:131602.

[75] Borgwardt M, Mahl J, Roth F, Wenthaus L, Brauße F, Blum M, Schwarzburg K, Liu G, Toma FM, Gessner O. Photoinduced charge carrier dynamics and electron injection efficiencies in Au nanoparticle-sensitized TiO_2 determined with picosecond time-resolved X-ray photoelectron spectroscopy. J Phys Chem Lett. 2020;11:5476–5481.

[76] Sezen H, Alemán B, Amati M, Dalmiglio M, Gregoratti L. Spatially resolved chemical characterization with scanning photoemission spectromicroscopy: Towards near-ambient-pressure experiments. 2015;7:3665–3673.

[77] Zeller P, Amati M, Sezen H, Scardamaglia M, Struzzi C, Bittencourt C, Lantz G, Hajlaoui M, Papalazarou E, Marino M, Fanetti M, Ambrosini S, Rubini S, Gregoratti L. Scanning photoelectron spectro-microscopy: A modern tool for the study of materials at the nanoscale. 2018;215:1800308.

https://doi.org/10.1142/9781800613294_0013

Chapter 13

NAP-XPS Studies of Mixed Conducting Electrodes During High-Temperature Electrochemical Reactions

David N. Mueller

Peter Grünberg Institute,
Forschungszentrum Jülich GmbH, Germany
dav.mueller@fz-juelich.de

Abstract

In this chapter, we will give an overview of how NAP-XPS can be used to investigate processes in and properties of mixed ionic electronic conducting materials at elevated temperatures, with a special focus on *operando* studies on working electrochemical devices. After a brief introduction of fundamental concepts of solid-state electro- and defect chemistry, we will focus on the reactions of prototypical materials exposed to thermal, chemical, and electric stimuli and how those can be elucidated with NAP-XPS. After discussing the experimental challenges and general setups used, we use examples to showcase the capabilities of NAP-XPS. Starting with simple oxides, such as CeO_2, where NAP-XPS is used to clarify mechanisms of electrocatalytic reactions with H_2O and O_2, we give several examples of how the technique was used to significantly improve the understanding of mixed conductors, the results themselves being used to further drive technique development. NAP-XPS shines brightly here, as surface-sensitive assessment of both electrostatic as well as chemical properties is unrivaled by any other technique. We extend the discussion to more complex multicomponent transition metal perovskites used in both cathodes and anodes of high-temperature electrochemical devices, where the element and chemical state sensitivity of NAP-XPS allows us to track atomic rearrangements on the surface induced by changes in electrochemical potentials, finishing with a short description of NAP-XPS experiments on cermets, a special case of a mixed conductor.

This chapter is both intended for the solid-state (electro) chemist to showcase how NAP-XPS might help them to tackle physicochemical questions, as well as to introduce the basic concepts of solid-state ionics and electrochemistry to the spectroscopist, aiming to further the understanding between those two fields.

13.1 Introduction

The advent and increased availability of *in situ* variants of X-ray spectroscopic tools had tremendous impact on the field of solid-state ionics, in general, and on solid-state electro- and defect chemistry, in particular. Here, the class of mixed ionic electronic conductors (MIEC) is of great interest, as they often show lithe redox activity of (some of) their constituents, and electrochemical processes involve redox reactions driven by electrochemical potentials. Changes in formal oxidation states and surface chemistry during those reactions are ubiquitous, and the use of electron spectroscopy to assess them is a natural consequence. Especially due to its relative surface sensitivity, X-ray photoelectron spectroscopy (XPS) can complement the understanding of these processes gained by electrochemical methods, which are mostly only sensitive to the bulk. At this point, it has to be made clear, though, that the equivalence of the chemical concepts of oxidation state, ionic charge, and their "direct" observance through spectroscopic means has to be scrutinized heavily [1]. Nevertheless, the importance of *in situ* and *operando* variants of XPS is hard to overstate as it offers surface-sensitive understanding of ionic and electronic processes during electrochemical reactions on MIEC and potentially elucidates mechanisms of solid/gas reactions.

In contrast to general heterogeneous catalysis, high-temperature solid-state electrochemical reactions with MIEC usually concern the interaction of a solid with a gas, where the solid is actively participating in the reaction (undergoing chemical or electronic changes) and not merely providing a reaction site. In oxides, for example, it is often seen that oxygen from the gas phase is exchanged with the solid, concomitantly transferring electrons. This has the consequence that the defect chemistry and thus the physicochemical properties can be profoundly influenced by the concentration of the reactants and electrochemical biasing.

One of the major breakthroughs championed by NAP-XPS is the understanding that surface (defect) chemistry may vary vastly from bulk (defect) chemistry depending on the applied biases, be they of thermodynamic (changes in temperature or oxygen partial pressure) or electrochemical (voltage) origin. Mobility of ions at the high temperatures of applications (usually $T \geq 500°C$) also makes it virtually impossible to quench states of interest, making *in situ* and *operando* spectroscopic studies of high-temperature electrochemical processes an indispensable tool for the field.

In this chapter, we will focus on mixed ionic electronic conducting materials for solid-state electrochemical applications and how they are studied with NAP-XPS, with a short detour to cermets and lithium battery cathodes as those present special cases of MIEC.

Naturally, this will put the focus on non-stoichiometric oxides as those are found most abundantly in the processes discussed; the concepts are however applicable universally. We will start with a brief primer on fundamental aspects of solid-state electrochemical cells (SSEC) and non-stoichiometric materials, followed by the description of experimental techniques showcasing certain aspects of MIEC materials and how NAP-XPS can help us to understand them. We will finish with more detailed examples on different materials where the use of NAP-XPS elevated the understanding in the field and challenged paradigms. In these examples, experimental techniques and analysis concepts will be explained on the go, to give the reader an understanding of the peculiarities MIEC might introduce to the NAP-XPS technique.

13.2 Solid-State Electrochemical Cells

One of the main applications of MIEC is as electrodes in SSEC. All components of the SSEC (anode, electrolyte, and cathode) are comprised of solids (see Figure 13.1) that conduct either ionic or electronic charge carriers in MIEC components. The working principle for all cells is the presence of an electrochemical potential gradient across the cell; the electrochemical potential $\bar{\mu}_i$ is defined as [2]

$$\bar{\mu}_i = \mu_i + z_i e_0 \phi, \tag{13.1}$$

with μ_i the chemical potential, z_i the formal charge of species i, e_0 the elemental charge, and ϕ the electrostatic potential.

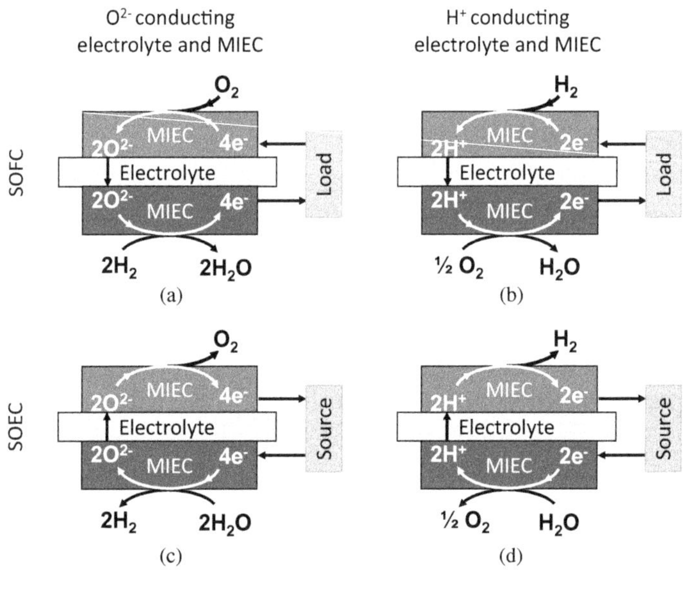

Fig. 13.1. Working principle of solid-state electrochemical cells. (a) Solid oxide fuel cell with oxygen ion conducting electrolyte and MIEC. (b) Solid oxide fuel cell with proton conducting electrolyte and MIEC. (c) Solid oxide electrolysis cell with oxygen ion conducting electrolyte and MIEC. (d) Solid oxide electrolysis cell with proton conducting electrolyte and MIEC.

Since the vast majority of SSEC are comprised of oxides, we will focus on those here; the concepts are however applicable to other cell types, e.g., Li ion batteries.

The cells can be roughly divided into two categories, the determining factor being the driving force. In solid oxide fuel cells (SOFC), the gradient is governed by the chemical potential μ_i of the species that is conducted through the electrolyte, having different magnitudes on either side of the cell. In its counterpart, solid oxide electrolyzer cells (SOEC), the gradient is applied as an external voltage ϕ.

As conduction of ions through the electrolyte is a prerequisite, the species the electrochemical gradient is applied on is limited to mobile species, such as O^{2-} or H^+, and operation is facilitated by high temperatures. The reactants on either side of the SSEC hence are mostly gaseous, and the redox reactions they undergo at the surface of the electrodes of the cell are the subject to be studied by high-temperature NAP-XPS.

At both electrodes of the SSEC, electron and ion transfer across the solid/gas interface occurs. We will use the electrolysis of steam to hydrogen and oxygen and its respective back reaction as examples for SOEC and SOFC, respectively. Formulated as an equilibrium, we get

$$O_2(g) + 2H_2(g) \rightleftharpoons 2H_2O(g). \tag{13.2}$$

The respective half-cell reactions, here with O^{2-} as mobile species in the electrolyte, are for the oxidant side:

$$O_2(g) + 4e^-(\text{cathode}) \rightleftharpoons 2O^{2-}(\text{cathode}). \tag{13.3a}$$

And for the fuel side:

$$3H_2(g) + 2O^{2-} \rightleftharpoons 4H_2O(\text{anode}) + 4e^-(\text{anode}). \tag{13.3b}$$

The cell potential E is given by the Nernst equation

$$E = E_0 - \frac{RT}{zF}\ln\frac{p(H_2O)^2}{p(O_2)\,p(H_2)^2}, \tag{13.4}$$

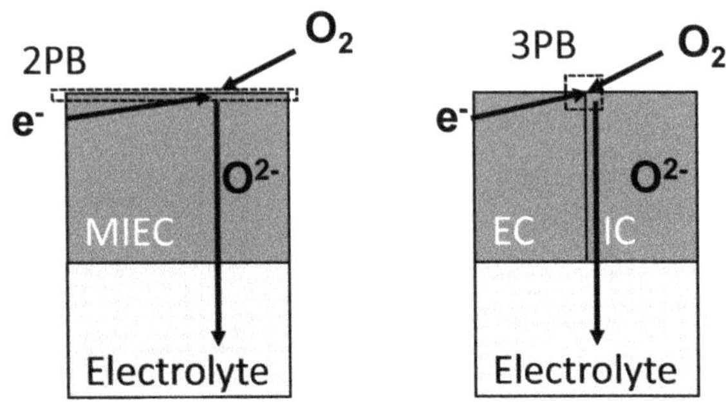

Fig. 13.2. Reaction pathways of the oxygen incorporation reaction on an MIEC via the dual-phase boundary (2PB, left) and the triple-phase boundary (3PB) of gas phase, electronic (EC), and ionic conductor (IC) (right). The 2PB extends over the whole surface of the MIEC, whereas the 3PB is only accessible where gas phase, EC, and IC meet.

with E_0 the standard potential of the cell, R the gas constant, T the temperature, z the number of electrons transferred (four in this case), F Faraday's constant, and $p(i)$ the partial pressures of the reactant i.

Depending on whether protons or oxide ions are transported through the electrolyte, recombination of both to molecular water occurs at the surface of the cathode or anode, respectively. To exemplify the important concepts to understand the NAP-XPS studies on these materials and devices, we will focus on oxygen (mixed) conductors in this chapter.

Cathode and anode materials thus also need to conduct both O^{2-} and electrons (or holes). There are two principles of how this can be achieved: First, MIEC where both species are mobile in the same material, and the cathode can be a monolithic ceramic. Second, composite materials, where each component offers mobility to either electronic or ionic charge carriers. The most used variant here is the combination of an ionically conducting ceramic, and a metallic electronic conductor, hence those materials are referred to as “cermets”.

An important distinction between the two is the way the half reactions take place. While in MIEC only two phases (gas and solid) need to be in contact at the two-phase boundary (2PB) in order to facilitate the transfer of ions and electrons, a composite material requires a triple-phase boundary (3PB) where both solids and the gas phase are in contact (Figure 13.2). This provides challenges for the morphological control, especially since electronic and ionic charge carriers are going in different directions. Since many monolithic mixed conductors have limited stability towards reducing conditions, cermets are the choice material for the fuel side.

13.3 Fundamental Concepts of MIEC

MIEC are one way to enable the transport of electronic, as well as ionic charge carriers simultaneously through a material. The total conductivity σ_t is the sum of that of the ionic and electronic charge carriers

$$\sigma_t = \sigma_e + \sigma_h + \sigma_i = e_0 \left(nu_e + pu_h \right) + z_i e_0 N_i u_i \, , \tag{13.5}$$

where e_0 is the elemental charge, n and p the concentrations of electrons and holes with their mobilities u_e and u_h, respectively, z_i the formal charge, N_i the concentration, and u_i the mobility of the ion i.

Mixed conduction in a monolithic material can be achieved by introducing point defects in oxides, either intrinsically through temperature-dependent non-stoichiometry or by extrinsic doping [3].

13.3.1 *Non-stoichiometric oxides*

Since MIEC require ionic as well as electronic conductivity, the compounds offering this functionality in a single phase are recruited to a large extent from the family of non-stoichiometric oxides. Oxygen vacancies can be accommodated in the lattice, allowing for the mobility of oxide ions. These oxygen vacancies can be either compensated electronically or ionically, the former mostly facilitated through redox of the cations. In its simplest form, the formation of oxygen vacancies is achieved through oxygen exchange with the atmosphere according to the equilibrium

$$MO(\mathrm{s}) \rightleftharpoons MO_{1-\delta}(\mathrm{s}) + \frac{\delta}{2}\mathrm{O}_2(\mathrm{g}). \tag{13.6}$$

For electronic compensation, the metal M thus needs to change its formal oxidation state during this reaction, which in turn changes the amount of available electronic charge carriers, resulting in both ionic and electronic conductivity provided the mobility of the species is reasonable. One way to express these defect chemistry equilibria to include charge carriers, as well as point defects, is the Kröger–Vink notation [4], which turns the above equation into

$$2M_M^{\mathrm{x}} + \mathrm{O}_\mathrm{O}^{\mathrm{x}} \rightleftharpoons 2M_M' + \mathrm{v}_\mathrm{O}^{\bullet\bullet} + \frac{1}{2}\mathrm{O}_2. \tag{13.7}$$

Here, the subscripts refer to the lattice site the species occupies. The superscripts indicate the charge *relative* to the ideal lattice sites with "·" denoting a positive and "′" a negative charge. Please note that we are adopting the notation for lower case "v" and "i" for vacancies and interstitials, respectively, to avoid confusion with Vanadium and Iodine [5].

Oxygen vacancies can also be created deliberately by aliovalent cation substitution, essentially forcing ionic compensation, as, for example, Y_2O_3 in ZrO_2 to form yttria-stabilized zirconia (YSZ), a standard electrolyte material in SSEC.

$$\mathrm{Y_2O_3} + 2\mathrm{v}_{\mathrm{Zr}}^{''''} + 4\mathrm{v}_\mathrm{O}^{\circ\circ} \rightarrow 2\mathrm{Y}_{\mathrm{Zr}}' + 3\mathrm{O}_\mathrm{o}^{\mathrm{x}} + \mathrm{v}_\mathrm{O}^{\circ\circ}. \tag{13.8}$$

In the bulk, defect concentrations are governed by the thermodynamics of the aforementioned equilibria, meaning that the concentration of charge carriers can be tuned both by deliberate aliovalent substitution of cations and choice of thermodynamic conditions.

13.3.2 *Surface reactions at MIEC under operating conditions*

What distinguishes reactions at the surface of most mixed conductors at high temperatures from other processes, such as heterogeneous catalysis or liquid electrochemistry, is the fact that both mass and charge can be transferred across the solid/gas interface. At high enough temperatures, this can include cation species, leading to phenomena quite unique for these conditions. This presents an ideal subject to be studied by NAP-XPS, as quenching the surface chemical states is virtually impossible.

13.3.2.1 *Oxygen exchange*

The most ubiquitous reaction occurring at the surface of a non-stoichiometric oxide is described by the oxygen intercalation equilibrium (here for a material where electrons are the majority of electronic charge carriers).

$$\frac{1}{2}O_2 + v_O^{\bullet\bullet} \rightleftharpoons O_O^{x} + 2e'. \tag{13.9}$$

It is important to note the distinction towards what is often referred to as "oxygen evolution reaction" (OER) in electrocatalysis: in that case, oxygen gas is produced by reducing water (or hydroxide) from the liquid electrolyte, whereas in the case of MIEC, the oxygen provided here is from the catalyst material, meaning that the surface chemistry (and with that, electronic structure) is dynamic.

Intercalation is not limited to oxygen. In proton-conducting mixed conductors, hydroxyl anions can occupy the oxygen site, which can be achieved either in a purely ionic mechanism (no electron transfer)

$$H_2O + v_O^{\bullet\bullet} + O_O^{x} \rightleftharpoons 2OH_O^{\bullet}, \tag{13.10}$$

or concomitant with a redox reaction between gas and solid.

$$H_2 + 2O_O^x \rightleftharpoons 2OH_O^{\bullet} + 2e'. \quad (13.11)$$

One should note that intercalation reactions are not limited to anions. For example, Li intercalation into layered transition metal oxides is the working principle of state-of-the-art Li-ion battery cathodes.

13.3.2.2 *Cation segregation/precipitation*

At high enough temperatures, the cations become mobile, and thus rearrangement of the chemistry as a response to thermodynamic bias becomes possible. $SrTiO_3$, for example, shows macroscopic SrO islands after annealing at 1273 K in oxygen [6]. This is due to the (partial) Schottky equilibrium [7]:

$$Sr_{Sr}^x + O_O^x \rightleftharpoons v_{Sr}'' + v_O^{\bullet\bullet} + SrO(surf.), \quad (13.12)$$

which may also occur at much lower temperatures, though in a less obvious (and thus not as easy to detect) manner, calling for surface-sensitive *in situ* analysis.

Together with the oxygen exchange (Eq. 13.9), one arrives at the oxygen partial pressure-dependent precipitation of SrO:

$$Sr_{Sr}^x + \frac{1}{2}O_2 + 2e' \rightleftharpoons v_{Sr}'' + SrO(surf.). \quad (13.13)$$

Both the precipitated SrO phase and the accumulation of electronic charge carriers can be detected and quantified with XPS[6]. With NAP-XPS, it then has been shown that even at temperatures as low as 300°C (still some several hundred degrees lower than your typical SSEC operation), Sr cations in Nb-doped $SrTiO_3$ become mobile enough to alter the surface chemistry [8,9]. Here, surface SrO precipitates could be evidenced by the Sr3d core levels, which compensated the buildup of a surface space charge layer both discovered through electrochemical means and corroborated by the apparent binding energy shifts of the core levels. The use of synchrotron radiation with variable incident energy allowed for a semi-quantitative depth profiling as the extent of the surface space charge layer (tens of nm) was greater than the IMFP of the resulting kinetic energies of detected photoelectrons.

Increasing the chemical complexity by having mixed A-site as in the late 3d transition metal perovskite oxides exacerbates chemical surface restructuring, though the driving force can be different. In $(La,Sr)CoO_{3-\delta}$ (LSC), a state-of-the-art SOFC cathode material, not only precipitation of SrO but also variation of La to Sr (or the respective alkaline earth substituent for La) ratio as a reaction to exposure to temperature and oxygen atmosphere were found [10]. NAP-XPS studies later revealed the reversibility and dynamics of these rearrangements, which we will elaborate on later (see Section 13.5.2.2).

13.4 Operando Methods

13.4.1 *General geometry*

Since mixed conductors show their most used utility in solid-state electrochemical cells, we will give special focus on the approaches used in full *operando* spectroscopy, that is XPS on an operating electrochemical cell. Naturally, these setups can also be used for "mere" *in situ* studies, i.e., spectroscopy at elevated temperature in reactive gas environments. An interesting feature arises from the fact that, thermodynamically speaking, electrochemical biasing can emulate partial pressures, potentially giving access to partial pressure regimes far beyond the experimental limitations of NAP-XPS.

There are different approaches with respect to the cell geometry and sample morphology, each with its own drawbacks and benefits. Figure 13.3 summarizes cell geometries used most frequently for *operando* NAP-XPS studies on solid-state electrochemical cells. The top row represents the single chamber approach where the whole cell is exposed to a single pressure so that the electrochemical potential gradient can only be affected by applying an electric bias *E*. One advantage of those is that the cell can be moved freely under the analyzer cone, enabling spatial mapping of the surface. The bottom center shows a sketch of a dual-chamber setup, where two gas

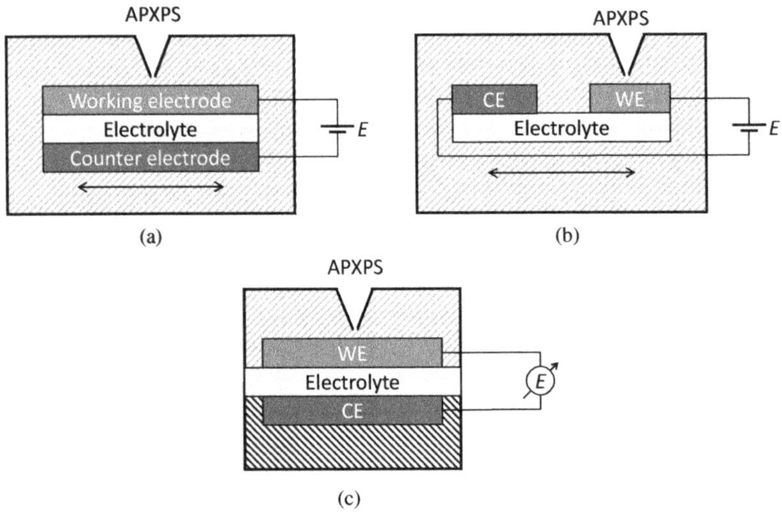

(c)

Fig. 13.3. Overview of different cell geometries used in *operando* NAP-XPS. Top row shows the single chamber approach, where all cell components (working electrode WE, electrolyte, and counter electrode CE) are exposed to the gas in the analysis chamber. Operation of the cell can be only driven by an electrostatic potential gradient caused by electrical biasing with voltage *E*. The cell can be moved laterally (indicated by the double arrow), allowing for spectroscopic access to different sites on the surface. Bottom center shows the dual-chamber approach where the gas volume of the counter electrode is sealed to the working electrode/analysis chamber. Choosing gases of different compositions, a chemical potential gradient is imposed across the cell, and the cell potential *E* can be measured simultaneously with the spectroscopy.

volumes are separated (here exemplarily by the electrolyte). Since only the working electrode's gas side is subject to limitations in pressure by the differential pumping of the NAP-XPS analyzer, the counter electrode's side can be subjected to any gas and pressure. In this setup, a chemical potential gradient is thus applied, and the resulting electrical potential can be measured.

13.4.2 *Dual-chamber setup: Chemical potential gradient*

Fuel cells need two separate gas volumes of different chemical potentials in order to drive the electrochemical reactions at the electrodes. Classically, this is realized by separating the fuel and oxidizer sides by the electrolyte material. In purely electrochemical setups, the whole stack is heated (e.g., in a furnace) to ensure homogeneity of the temperature in all components. This approach is unfeasible in a spectroscopic experiment, where components cannot be kept at such high temperatures. Additionally, in a synchrotron environment, the incident X-ray beam and analyzer positions are fixed, requiring the whole cell assembly to be able to be moved when the need to change the analysis locale arises. Local heating is thus a necessary compromise.

Kooser *et al.* developed a dual-chamber solid oxide fuel cell assembly (Figure 13.4) where the cathode side is encapsulated and thereby can be supplied with gases of atmospheric pressure. The cathode side is sealed against the analysis chamber by the electrolyte [11]. The NAP-XPS analysis chamber can then be operated in the usual manner with pressures being limited to the capabilities of the differential pumping (usually up to tens of mbars). The advantage

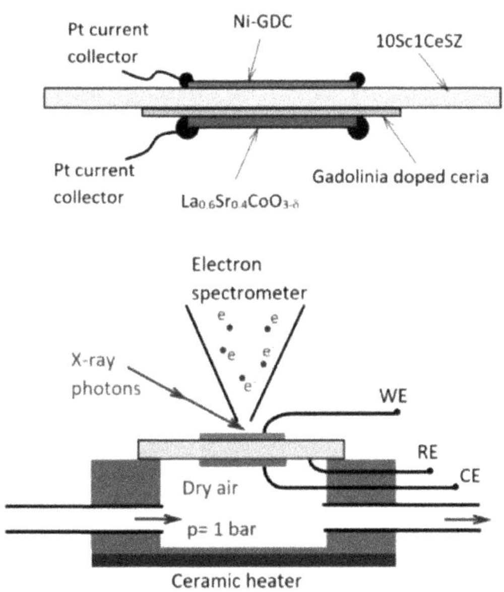

Fig. 13.4. Schematic of the dual-chamber setup used by Kooser *et al.* for investigations on solid oxide fuel cells. Adapted with permission from Ref. [11].

over the single-chamber setup is the ability to use different gases on the two sides in order to decouple effects related to the gas chemistry from those of electrochemical polarization, e.g., differentiating whether low oxygen partial pressures achieved by polarization or H_2O/H_2 mixtures show the same behavior of the catalyst material.

13.4.3 *Single-chamber setups: Electric potential gradient*

In general, single-chamber setups provide the advantage that the electrochemical cell stack can be assembled outside the chamber and then placed onto a versatile heater with electrical contacting capabilities such as described by Whaley *et al.* [12]. No gas-tight sealing of the components is necessary, and the risk of catastrophic failure (exposure of the NAP-XPS side to atmospheric pressure) is minimized. The drawback is that both the electrodes are subjected to the same gas atmosphere, limiting the study of chemical effects on the electrodes as would be possible in a dual-chamber setup.

Electrochemically, this is a workaround, as the cell can now only be run in electrolyzer mode as the external driving force has to be electrical, i.e., no chemical potential gradient is present. That this approach is still valid to assess electrochemical reactions can be demonstrated by a simple experiment using equivalent chemical and electrical driving forces and checking the spectroscopic answer.

The potential E across an oxygen concentration cell is given by the Nernst equation in the form of

$$E = \frac{RT}{nF} \ln\left(\frac{pO_2{}^{A}}{pO_2{}^{B}}\right)^{‰}, \tag{13.14}$$

with n the number of electrons transferred in (1.9). If we define B as the analysis chamber pressure and A as the effective oxygen partial pressure we want to exert on the material, we can do so by applying bias E:

$$pO_2{}^{A} = \exp\left(\frac{EnF}{RT}\right)^{2} pO_2{}^{B}. \tag{13.15}$$

For an electrochemical cell consisting of an (La,Sr)$FeO_{3-\delta}$ (LSF) cathode and an (Ba,Sr)(Co,Fe)$O_{3-\delta}$ (BSCF) anode separated by an YSZ electrolyte, this has been tested with X-ray absorption spectroscopy, where the O-K-edge can be used to quantify the electron–hole concentration [13], and both electrical and chemical polarization gave the same result (see Figure 13.5).

This concept has later been adopted and extended by Nenning *et al.* to allow *operando* XPS studies *in vacuo* by providing the effective oxygen partial pressure for exsolution experiments by means of electrical polarization [14], removing the necessity of ambient pressure and the associated experimental complexity for some investigations on MIEC.

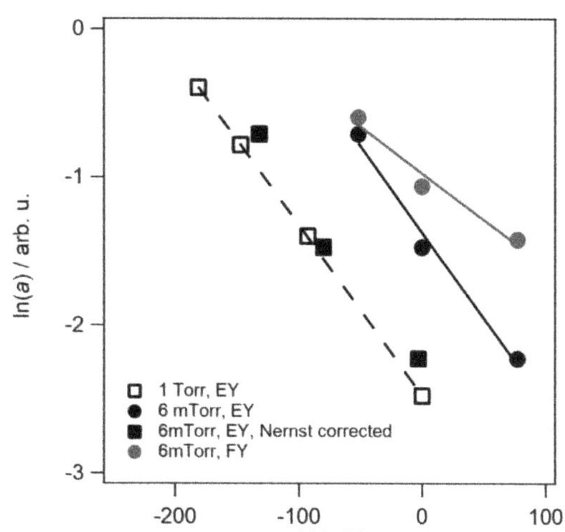

Fig. 13.5. Quantification of the electron–hole concentration (a) by O-K-Edge XAS [13] as a function of overpotential on an LSF cathode at 450°C. Open squares show measurements taken at 1 Torr of oxygen partial pressure and closed circles are taken at pO_2 = 6 mTorr. Values for closed squares were calculated from the 6 mTorr data corrected for the Nernst potential due to the difference in pO_2 to the 1 Torr data. The fact that the Nernst corrected 6 mTorr data is quite close to the 1 Torr data shows the interchangeability of electrostatic and chemical potential as driving forces in SSEC. All data were collected in partial electron yield (EY); for comparison, the bulk sensitive fluorescence yield (FY) data are shown as gray solid circles, emphasizing that the surface defect chemistry differs from that of the bulk.

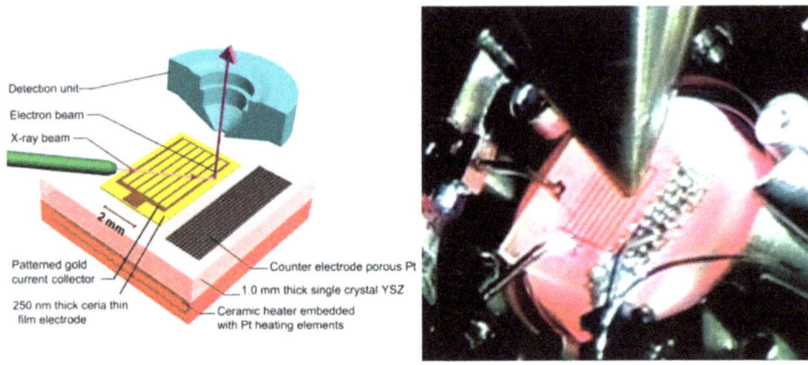

(a) (b)

Fig. 13.6. Schematic (a) and photograph during operation (b) of a coplanar SSEC as used by DeCaluwe *et al.* Adapted with permission from Ref. [17]. Copyright (2010) American Chemical Society.

Single-chamber experiments can be divided once more into two geometries, which differ in the arrangement of the cell components, both again with their advantages and disadvantages.

The coplanar geometry shown in Figure 13.6 offers the advantage that all parts of the cell are accessible for spectroscopic analysis. This is of great use if both electrodes are to be investigated without (often times consuming) the removal of the sample. One other advantage is that, if heated from the back side as usually done, no temperature gradient parallel to the electrochemical gradient is present so that both electrodes and electrolyte are at the same conditions.

Accessibility of the whole cell cross-section was used by El Gabaly *et al.* who could map the potential landscape with rigid shifts of the kinetic energy of the Zr 3d core levels [16] and could differentiate the various overpotentials while accurately reproducing the total cell voltage measured by the potentiostat. With a similar geometry, DeCaluwe *et al.* and Zhang *et al.* quantified the Ce oxidation states of mixed conducting CeO_2 in operating SSEC [15,17]. Besides showing that the Ce^{3+} concentration is dynamic during electrochemical reactions depending on the applied bias, the spatial mapping

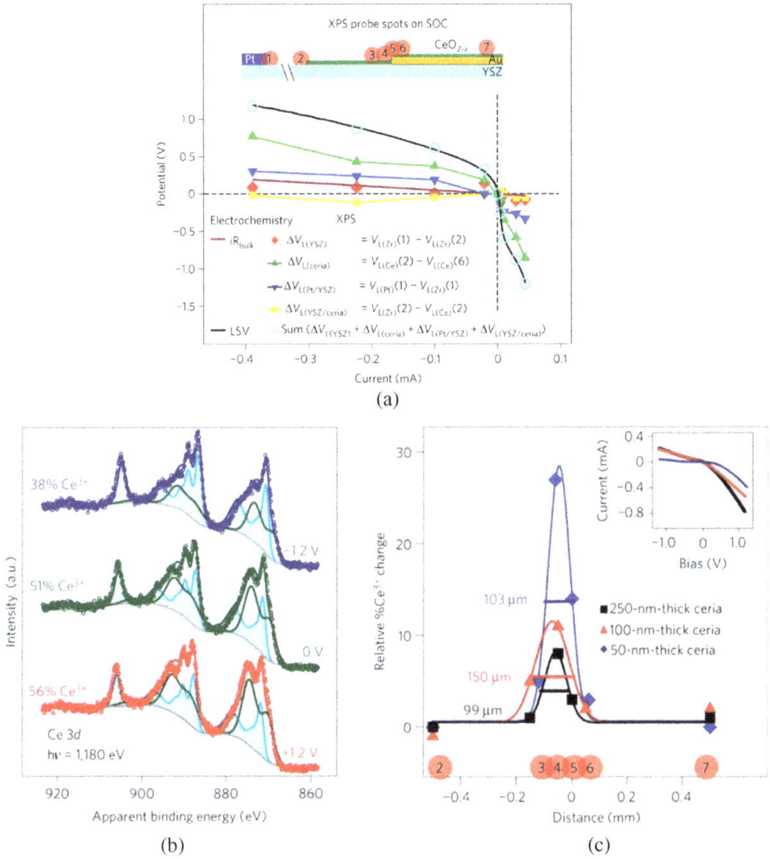

Fig. 13.7. Potential landscape of a coplanar SSEC as measured *in situ* from binding energy shifts of the respective core levels (a) [17]. Evolution of the Ce^{3+} concentration as a function of bias (b) and spatial distribution across the cell (c). Adapted with permission from Ref. [17].

also revealed that electrochemical activity in the electrode extends to a distance of more than 100 μm from the current collectors (see Figure 13.7).

In the single-chamber through plane geometry (Figure 13.3(a) and Figure 13.8), the electrochemical cell is executed as a standard fuel cell stack, the electrode of interest facing the analyzer cone of the spectrometer.

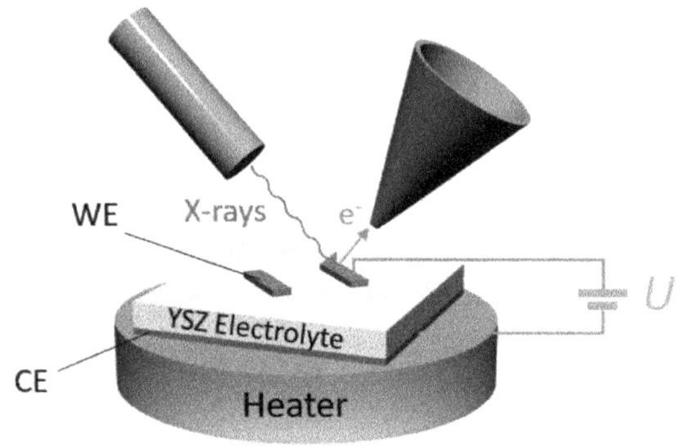

Fig. 13.8. Artistic rendition of a through plane single-chamber electrochemical NAP-XPS setup. The counter electrode is placed directly on the heater (with current collector), and the working electrode as well as the electrolyte are accessible by the analyzer cone. Electrochemical reactions can be driven by an applied potential *U*.

One advantage of using a through plane geometry is that the electrochemical setup is much closer to what is used in standard (non-spectroscopic) experiments. If the working electrode is much smaller (laterally) than the electrolyte and counter electrode, as depicted in Figure 13.8, the electrochemical potential gradient is quite uniform across the whole area. This makes electrochemical analyses such as impedance spectroscopy, which access the whole electrochemical cell, much more comparable to the spectroscopic analysis which can access only a small portion of the electrode. Any effect of inhomogeneous distribution of the electrochemical potential gradient through edge effects is minimized. The drawback of this geometry is the necessity to heat from the back, which causes a temperature gradient parallel to the electrochemical potential gradient. If the electrolyte's electrochemical behavior is highly dependent on the temperature, and not easily separated from that of the working electrode, difficulties in connecting the spectroscopic to the electrochemical data might arise.

13.4.4 *Sample types: Model vs. real systems*

If the aim of *operando* spectroscopic studies is the elucidation of mechanisms on an atomistic scale, one commonly taken approach is the use of electrocatalyst surfaces that are flat on an atomic scale. To avoid influences of grain boundaries or step edges, the materials should furthermore be single crystalline or at least highly oriented. Physical vapor deposition techniques can provide samples that satisfy these prerequisites quite well, and especially pulsed laser deposition (PLD) has become the main synthesis technique for NAP-XPS studies on MIEC. One important issue that needs to be accounted for, though, is that due to the limited thickness of tens to maybe a few hundreds of nm (in order to retain single crystalline growth), the sheet conductance may be too low to efficiently carry the current so that the surface reaction one aims to observe is not the (only) limiting step anymore. The drawback of using such thin-film geometries is their potential disconnect from technologically relevant morphologies that are polycrystalline porous ceramics providing a vastly increased surface area. Comparison of electrochemical performance and the atomistic insights from NAP-XPS studies on model systems thus can be difficult, leading to another school of thought that opts to study SSEC as they are used in electrochemical experiments. Both approaches have their merits and drawbacks, and thus should be chosen carefully with respect to the scientific challenge addressed.

13.5 Examples

13.5.1 *Fluorites*

The fluorites are (chemically speaking) simple AX_2 compounds, of which the oxides $AO_{2-\delta}$ have the ability to accommodate oxygen vacancies. The nonstoichiometry δ is either achieved ionically by substitution of some of the tetravalent cations with tri- or divalent ones (see Eq. (13.8)) or electronically by reducing the cation A. Yttrium-stabilized zirconia (YSZ), for example, is a good ionic, but poor electronic conductor, and is used as either electrolyte or ceramic component in cermets. One of the advantages of being electronically inactive with respect to any electrochemical reaction is that the electrostatic potential can be measured directly from rigid core-level shifts, as discussed earlier (Section 13.4.3 and Refs. [15,16]) to quantify polarization.

When the A-site cation is reducible under certain conditions, such as Ce, mixed conductivity can occur due to the presence of electronic (polarons) and ionic (oxygen vacancies) charge carriers. $CeO_{2-\delta}$ (ceria) and its doped variants show considerable redox activity in reducing conditions, rendering it a mixed conductor at, e.g., the fuel side of an SSEC. Substitution with aliovalent cations, such as Sm, Gd, and Pr, further increases conductivity [18]. One advantage of NAP-XPS studies on ceria is that the defect chemistry is comparably simple and well understood, at least for the bulk: Compensation of vacancies created by the release of oxygen is achieved electronically [18]:

$$O_O^x + 2Ce_{Ce}^x \rightleftharpoons v_O^{\cdot\cdot} + \frac{1}{2}O_2 + 2Ce'_{Ce}, \tag{13.16}$$

resulting in the simple charge neutrality condition

$$2\left[v_O^{\infty}\right] = \left[Ce'_{Ce}\right]. \tag{13.17}$$

As the polarons can be detected and quantified by both Ce 3d core-level and Ce 4f valence band XPS [19,20], a single quantitative XPS experiment can describe the defect chemistry exhaustively. With NAP-XPS, this had become possible as a function of pO_2 and T without the ambiguity any quenching process introduces. Generally, space charge effects need to be considered (see Section 13.3.2.2), which for ceria fortunately are negligible for certain experiments [21] so we can use it as a simple example here.

The first groundbreaking revelation facilitated by high-temperature NAPXPS was that the Ce^{3+} polaron concentration was orders of magnitude higher than that of the bulk value, and also only diminutively dependent on the oxygen partial pressure [16,22]. Figure 13.9 shows the valence band (a) and Ce 3d core-level spectra, the different colors indicating bulk oxidation states as expected from defect chemical equilibrium (open symbols in Figure 13.9, bottom). The quantification of *in situ* spectra at different temperatures and oxygen partial pressures yielded an increase of two orders of magnitude of Ce^{3+} on the surface.

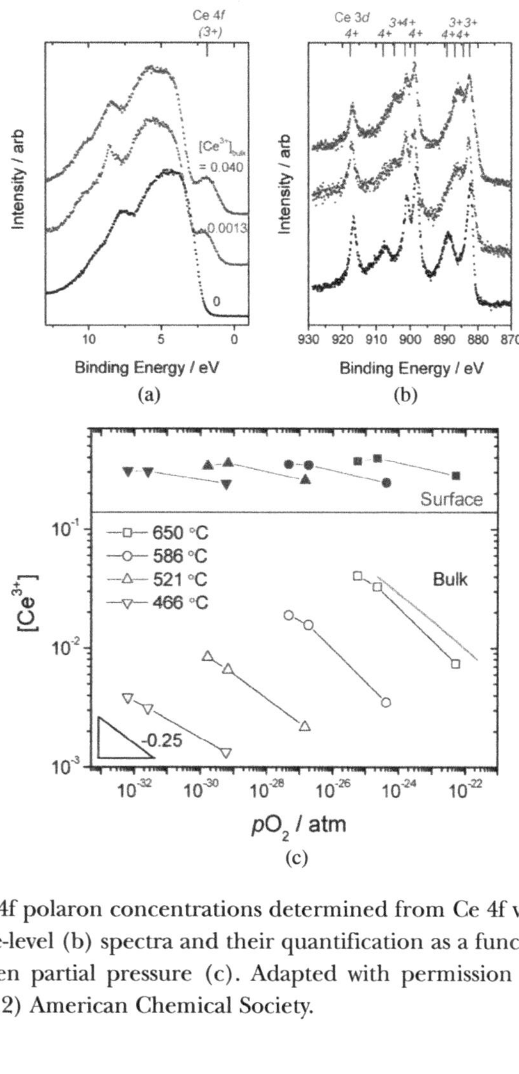

Fig. 13.9. Ce 4f polaron concentrations determined from Ce 4f valence band (a) and Ce 3d core-level (b) spectra and their quantification as a function of temperature and oxygen partial pressure (c). Adapted with permission from Ref. [22]. Copyright (2012) American Chemical Society.

With such quantification capabilities and the use of *operando* electrochemical cells, mechanistic studies have become feasible. On a mixed conductor, the water splitting equilibrium can be formulated in the Kröger–Vink notation as follows:

$$H_2O + 2Ce'_{Ce} + v_O^{\bullet\bullet} \rightleftharpoons H_2 + O_O^x + 2Ce_{Ce}^x. \tag{13.18}$$

In addition to the Ce'_{Ce} species that can be quantified by NAP-XPS, the different possible intermediate oxygen species (O_O^x, $OH_O^{\bullet}$ $H_2O_O^{\bullet\bullet}$) can also be identified and quantified from the O 1s core-level XPS. Please note that, for the sake of consistency, here we choose the Kröger–Vink notation for those species that are sometimes also referred to as adsorbates, e.g., OH^-(ad). Since in a defect chemical sense those are not distinguishable, and this chapter focuses on the mixed conducting properties of the materials, this didactic reduction is used.

Zhang *et al.* used the cell geometry shown in Figure 13.7(a) to track concentrations of O and Ce species and their binding energies with NAPXPS in a working electrochemical cell (Figure 13.10), driving oxygen exchange with positive and negative overpotentials [23]. The spatial resolution of the O 1s peak allowed them to identify the local potential difference between oxide and hydroxide species that was attributed to the potential drop across the electrochemical double-layer ($\Delta\chi$). The enrichment and depletion of $OH_O^{\bullet}$ (OH^-) and O_O^x (O^{2-}) species at positive and negative biasing, respectively, were used as evidence to identify the rate-determining step as the dissociative charge transfer step from Ce'_{Ce} (Ce^{3+}) to the adsorbed water molecule:

$$H_2O + Ce'_{Ce} \rightleftharpoons Ce^x_{Ce} + OH^- + H^+. \tag{13.19}$$

Later, Feng *et al.* quantified Ce polarons and O-species as a function of overpotential and found a crucial participation of oxygen vacancies in the mechanism [24]. In addition to the measured Ce 4f polaron concentration on the surface (see Figures 13.11(a)–(c)), species containing oxygen were quantified from the O 1s core levels (Figures 13.11(d) and (e)). What proved to be serendipitously useful here was a SiO_2 contamination, not participating in any (electro-)chemical reaction, providing an internal quantification standard in the O 1s spectra.

Surprising was the fact that the actual vacancy concentration did not change with overpotential (the total amount of oxygen as the sum of O^{2-} and OH^-) which was interpreted such that, in contrast to

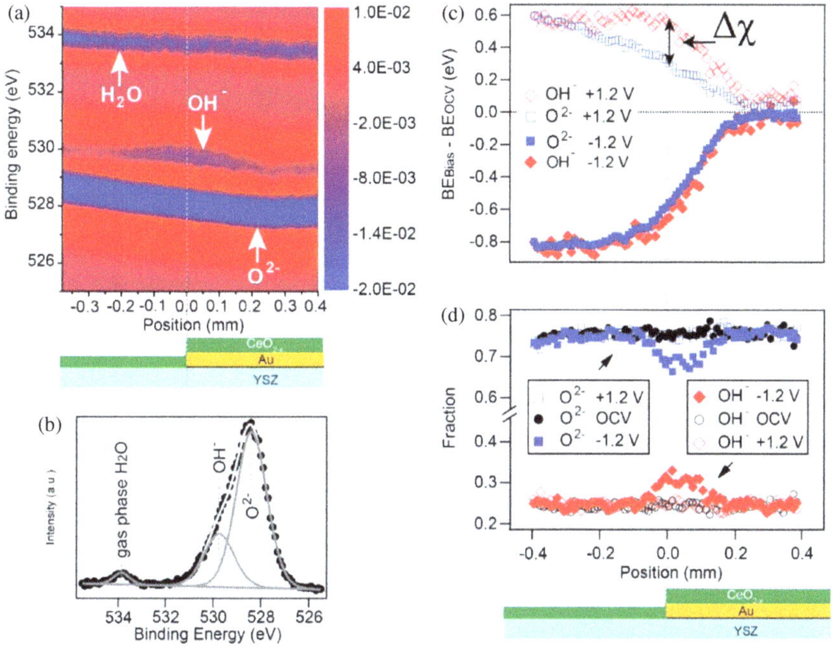

Fig. 13.10. (a) Second derivative of the O 1s XPS as a function of position along the cell (schematic below the graph) in order to track the binding energies of the different oxygen species. (b) O 1s core levels at $x = -0.155$ mm to show contributions of the oxide, hydroxide, and water species, respectively. (c) Binding energy shifts as measured under applied bias for OH^- and O^{2-} core levels. $\Delta\chi$ denotes the surface potential step. (d) Quantification of OH^- and O^{2-} species at open-cell conditions (no bias) and ±1.2 V applied biases showing depletion and enrichment of the O^{2-} and OH^- species, respectively. Adapted with permission from Ref. [23]. Copyright (2013) American Chemical Society.

Zhang *et al.* [23], the rate-determining step of the water-splitting reaction on Ceria is the charge transfer from hydroxyl to Ce:

$$OH_O^{\circ} + Ce'_{Ce} \rightarrow OH_O^{x} + Ce_{Ce}^{x}. \tag{13.20}$$

While the aforementioned studies focused on spectroscopy of the solid phase, NAP-XPS also offers the opportunity to collect spectra of the gas phase and — with the proper geometry — direct and accurate work function measurements can be undertaken

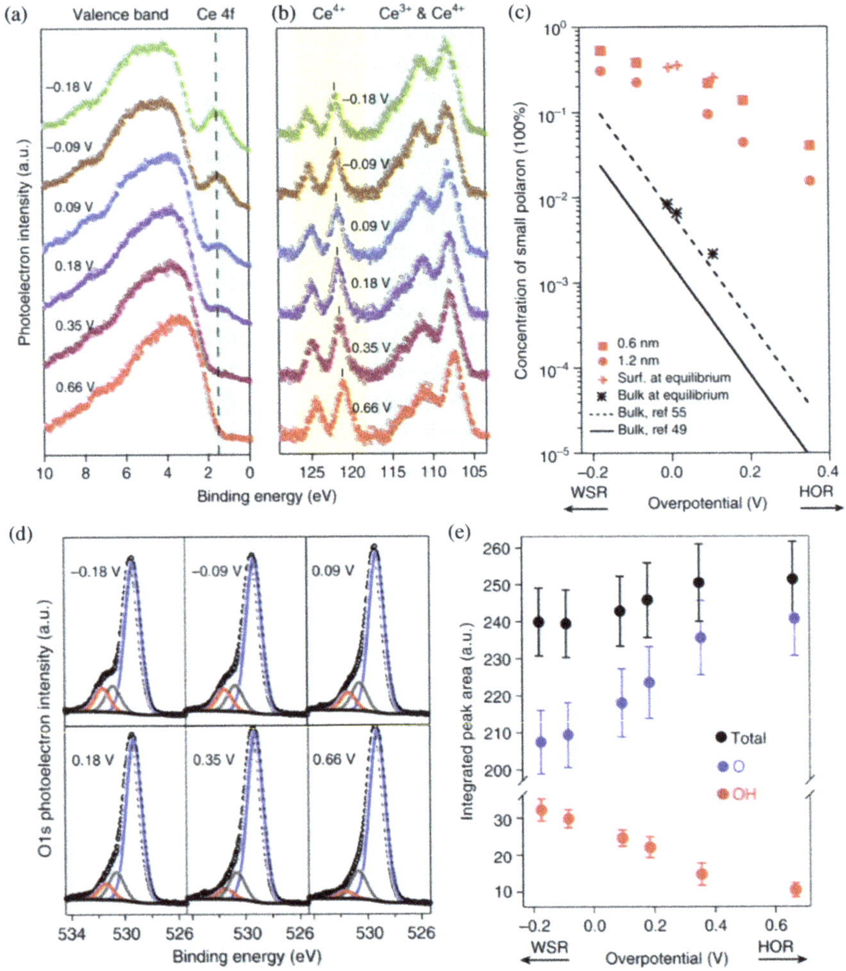

Fig. 13.11. Quantitative analysis of Ce^{3+} polarons from Ce 4f valence band (a) and Ce 3d core-level XPS (b) as a function of applied bias. Results are compared to bulk values (c). Quantification of oxygen species from O 1s XPS (d), showing that the oxygen content (vacancy concentration) does not change with electrochemical polarization (e) [24]. Adapted with permission from Ref. [24].

[25]. By using a grazing incidence X-ray beam (see Figure 13.12), only gas molecules very close to the surface are illuminated, yielding an apparent binding energy shift of the gas molecule core levels (e.g., Ar 2p) that accurately represents the change in work function.

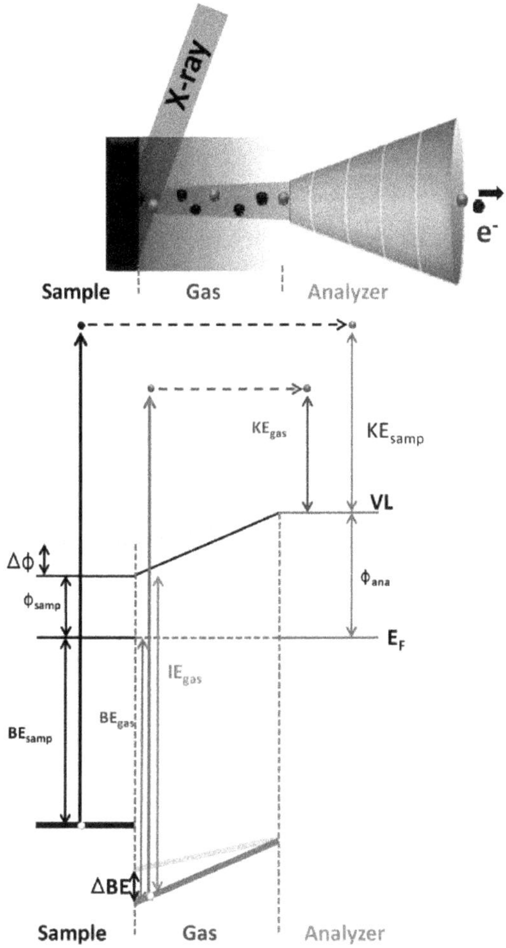

Fig. 13.12. Energetics of core-level XPS on a solid in contact with the gas phase. The binding energy shift of the gas phase core levels describes changes in the work function. Adapted with permission from Ref. [25]. Copyright (2013) American Chemical Society.

Feng *et al.* used this method with electrified Sm doped ceria [26] and thus could deconvolve the contribution of the electron chemical potential $\Delta\mu_e$ and the surface dipole potential $e\Delta\chi$ to the change in work function $\Delta\Phi_S$ as given by

$$-\Delta\Phi_S = \Delta\mu_e + e\Delta\chi + e\Delta\psi_{SC}. \qquad (13.21)$$

Here, as typical for mixed conductors with high defect concentrations, the high concentration of oxygen vacancies induced by the extrinsic Sm doping screened any space charge effectively so that the contribution of the space charge potential $e\Delta\psi_{SC}$ could be neglected. Since the change in the electron chemical potential is affected by electrochemical biasing, any deviation from the 1:1 ratio of $\Delta\Phi_S$ and $\Delta\mu_e$ has to be attributed to a surface dipole of adsorbate molecules. And indeed, the polar OH^- adsorbates present in an H_2/H_2O mixture showed a contribution of 0.6 eV/V, whereas the CO_3^{2-} (with apolar tridentate adsorption geometry) and adsorbate free cases showed no such contribution.

13.5.2 *Late 3d transition metal perovskites*

The ABO_3 perovskites add one level of complexity as they have two non-equivalent cation sites A and B. This offers more flexibility in functionalization but also adds to the complexity, both chemically and spectroscopically. Fine tuning of the chemistry by co-substitution of either site with two (or more) constituents allows for the creation of state-of-the-art SSEC materials, such as BSCF and (La,Sr)(Co,Fe)$O_{3-\delta}$ (LSCF), but comes at the expense of opening pathways for decomposition reactions and phase transitions. We will first focus on studies concerning the electronic charge carriers and properties in these materials as modulated by electrochemical processes and thus governing their performance and then discuss any (usually unwanted) reaction that alters the surface chemistry.

13.5.2.1 *Detecting charge carriers and electrochemical behavior*

In LSF, the oxygen vacancy concentration can be tuned by choice of the "dopant" concentration, the term "dopant" being used liberally here since the LSF system is an almost complete solid solution [27]. Substitution of La^{3+} with Sr^{2+} can either be compensated ionically (with oxygen vacancies) or electronically (with electron holes) according to the charge neutrality condition:

$$[Sr'_{La}] = 2[V_O^{\circ\circ}] + [h^{\circ}]. \quad (13.22)$$

Oxygen vacancies and electron holes are then once again connected through the oxygen exchange with the surrounding atmosphere (Eq. 13.9). What makes this specific material a good subject to tackle with NAPXPS is the fact that it has already been studied extensively both spectroscopically and physicochemically — just individually. A comprehensive understanding of the bulk defect chemistry and the underlying thermodynamics has been summarized by Kuhn *et al.* [28], whereas the electronic structure elucidated by X-ray absorption as well as photoelectron spectroscopy has been extensively described by Abbate *et al.* [29].

In contrast to the ceria examples, where the electronic charge carriers could be measured and quantified directly from both valence band and Ce 3d core-level XPS, identification of those in LSF (and late 3d transition metals in general) is much less straightforward. Since those compounds are negative charge transfer materials and show high covalency [30], the electron hole can be localized predominantly on the ligand and not the cation, whose core-level XPS might be much less informative regarding the oxidation state of the material [29]. Electrochemical oxidation and reduction of LSF, for example, did not show any changes in the Fe 2p core levels or Fe L_{32} absorption edges, and only assessment of the O–K edge allowed identification and quantification of electron holes [13] which makes the Fe^{4+} formalism used in electrochemistry questionable [1].

In reducing conditions, however, Nenning *et al.* were able to find signatures of Fe in its formal +2 oxidation state in a reducing H_2/H_2O atmosphere [31], though in a much higher concentration than predicted from bulk thermodynamics where decomposition of the perovskite phase begins at already small (<10%) Fe^{2+} concentrations [32]. Further (electrochemical) reduction induced the formation of metallic Fe^0, exsolved as particles from the perovskite host lattice [14], which was accompanied by the emergence of electronic states close to the Fermi level visible in the valence band XPS.

Nenning *et al.* could demonstrate with *operando* XPS that these metallic Fe particles are responsible for an increased water splitting activity [31]. In the same work, Nenning *et al.* presented an elegant formal deduction of the equivalency of change in oxygen chemical potential $\Delta\mu_{O_2}$ and shift of the Fermi level E_f (with F: Faraday's constant)

$$\Delta\mu_{O_2} = -4F\Delta E_f, \tag{13.23}$$

that has already been introduced empirically in Section 13.4.3 (Figure 13.5). In conjunction with the generalized Nernst equation (1.14) for an oxygen concentration cell

$$E = \frac{\Delta\mu_{O_2}}{4F}, \tag{13.24}$$

it explains the 1:1 relation of apparent binding energy shifts in eV and electrode potential in V (Figure 13.13), provided that a) the species observed by XPS are "redox inactive", *i.e.*, the Fermi level shift within the band gap is the only changing contribution to the binding energy [33], and b) oxygen vacancies are not the minority charge carriers. Nenning *et al.* could consequently show that in (La,Sr)$CoO_{3-\delta}$ (LSC), this simple 1:1 relation does not hold anymore, especially for the Co core levels, since LSC behaves like a metallic conductor [28,31], which is corroborated by a valence band edge intersecting the Fermi level.

13.5.2.2 *Decomposition, segregation, precipitation, and phase transitions*

In general, the Schottky equilibrium (13.12) as introduced earlier for the $SrTiO_3$ case is also present in the late 3d transition metal oxides, especially facile at operating temperatures. Defining Sr as a dopant on the La site in $La_{0.5}Sr_{0.5}FeO_{3-\delta}$ (LSF55), we get

$$\mathrm{Sr}_{\mathrm{La}}' + \frac{1}{2}\mathrm{O}_2 + 2\mathrm{e}' \rightleftharpoons \mathrm{v}_{\mathrm{La}}''' + \mathrm{SrO}. \tag{13.25}$$

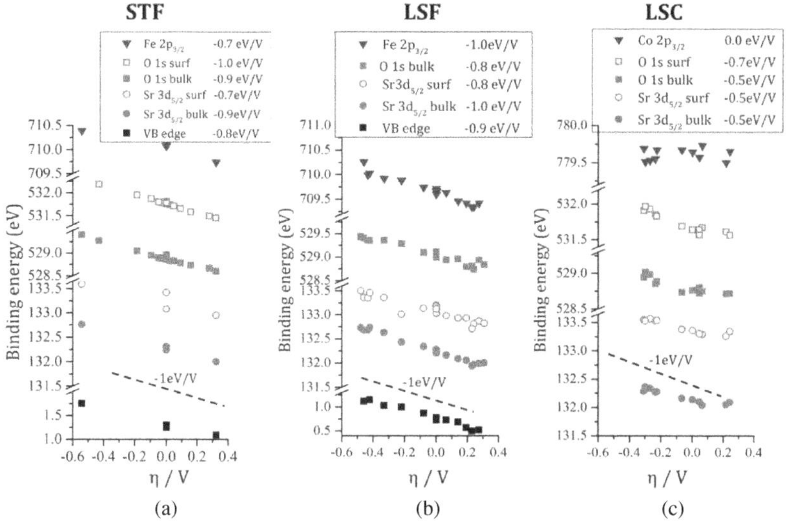

Fig. 13.13. Binding energy shifts as a function of overpotential on $Sr(Ti,Fe)O_{3-\delta}$ (a), $(La,Sr)FeO_{3-\delta}$ (b), and $(La,Sr)CoO_{3-\delta}$ (c) cathodes measured *in situ*. Adapted with permission from Ref. [31]. Copyright (2016) American Chemical Society. Further permission related to this material excerpted should be directed to the ACS.

Since the Sr is a charged defect, it will react to electrochemical biasing, driving Sr transport that can result in a concentration gradient:

$$Sr'_{La,\alpha} + La^{x}_{La,\beta} \rightleftharpoons La^{x}_{La,\alpha} + Sr'_{La,\beta}, \qquad (13.26)$$

where α and β denote different layers in the perovskite. Since synchrotron-based XPS offers depth-dependent chemical analysis through tunability of incident photon energy, the distribution of both precipitated and segregated constituents can be tracked. Since these processes are pO_2 dependent, NAP-XPS once again is a mandatory tool to study and quantify these phenomena *in situ*. In the extreme case, these processes can lead to decomposition or phase transition phenomena.

In the $(La,Sr)CoO_{3-\delta}$ system, Muturo *et al.* found that heating of the sample in an oxygen environment produced hints of the SrO precipitation as evidenced in the Sr 3d core levels, concomitant with a transformation of the perovskite (dubbed LSC_{113}) into a Ruddlesden–Popper-type phase (LSC_{214}) with additional CoO_x precipitation [34]:

$$2(\mathrm{La,Sr})\mathrm{CoO}_{3-\delta_\mathrm{P}} \rightarrow (\mathrm{La,Sr})_2\mathrm{CoO}_{4+\delta_\mathrm{R}} + \mathrm{CoO}_\mathrm{x} + (2-2\delta_\mathrm{P}-\delta_\mathrm{R}-x)\mathrm{O}_2. \tag{13.27}$$

The authors deliberately kept the assignment of the emerging secondary Sr 3d doublet (cyan in Figure 13.14(a)) ambiguous, as a higher BE component compared to Sr in the perovskite lattice can have several origins. The separation in BE from the perovskite lattice Sr signal (green in Figure 13.14(a)) is compatible with SrO, $Sr(OH)_2$ [35], or LSC_{214} [36] and has also been attributed to Sr in the termination layer of the LSC_{113} perovskite phase [37].

This study is noteworthy as it shows that if one defines "ambient" pressure liberally, down to the base pressure of standard ultra-high vacuum (UHV) equipment (10^{-6} mbar in this study), NAP-XPS can be undertaken without the additional experimental expenditure of differentially pumped analyzers or closed cells. This approach should be made use of as an intermediate step from UHV to NAP-XPS, but only when the scientific question of interest falls into the low-pressure range. In this particular study, however, it is arguable whether these are surface reactions of catalytic importance, as the reaction (13.27) can be regarded as a halfway point to the total reduction of the perovskite to binary La- and Sr-oxides and metallic Co. This would be expected since the conditions (650°C at 10^{-6} mbar) used are beyond the bulk stability limit of LSC [28]. Nevertheless, this study also showed that surface chemistry is reversibly dependent on electrochemical biasing. Whether this is because of the emulation of high pO_2 as described before or a direct effect of the electrical potential was not quantified.

Later, the same group employed NAP-XPS at a pO_2 of 10^{-3} at (~1 mbar) on the same material system and showed a temperature dependence on surface chemistry, once again mostly visible in the Sr 3d core level [38]. Importantly, they also found that the

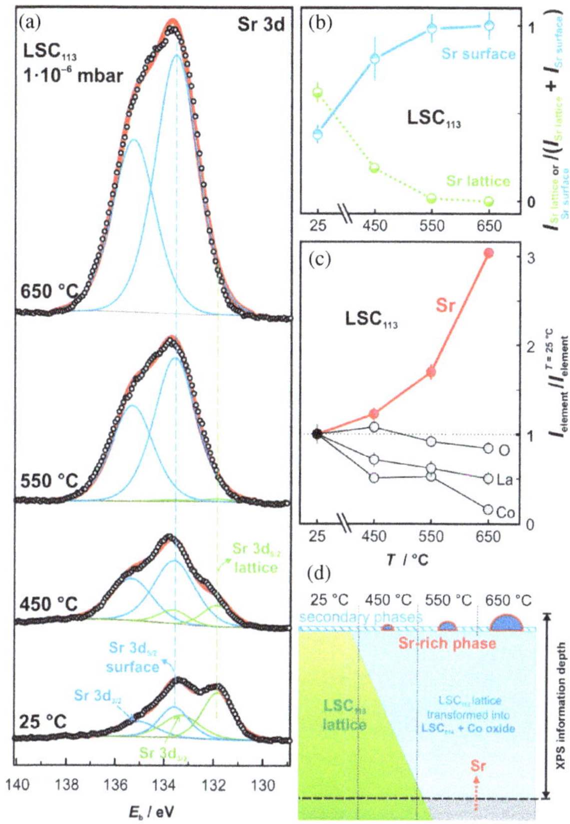

Fig. 13.14. Sr 3d core-level XPS of (La,Sr)$CoO_{3-\delta}$ demonstrating the evolution of a secondary surface phase with increasing temperature in reducing conditions (a). Quantification of the evolving surface phase (b) and chemical composition of the surface (c). Schematic depiction of the phase relations in the LSC_{113} system as found with NAP-XPS (d). Adapted with permission from Ref. [34]. Copyright (2012) American Chemical Society.

enrichment of secondary Sr phases that were observed on PLD-prepared thin films was completely absent on sintered ceramic pellets. Several effects, such as strain [39] and increased defect concentrations [40], can be discussed as the reason for this discrepancy. It should serve as a warning and invitation to heavily scrutinize whether PLD thin films are suitable as a model system for the scientific question at hand — and not only for NAP-XPS experiments — since the technologically relevant morphologies are usually sintered ceramics or powders. Though academically interesting, it has to be made clear whether any effects observed stem from a material's property and not rather solely be affected by the unusual length scales found in thin films [41,42].

In contrast to experiments on (La,AE)$FeO_{3-\delta}$ where no precipitation for AE = Ca was found [13], the segregation and precipitation of alkaline earth dopants have been found to be universal for AE = Ca, Sr, and Ba in the (La,AE)$MnO_{3-\delta}$ system [43]. It was quantified as a function of electrochemical bias in an elegant experiment (shown schematically in Figure 13.15), employing a cell geometry

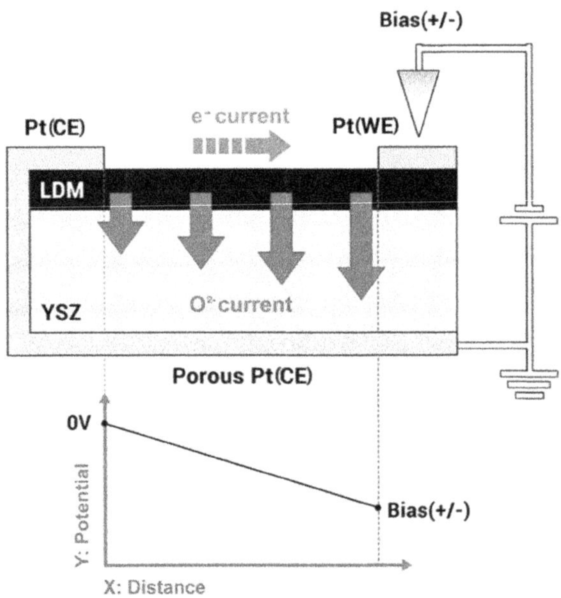

Fig. 13.15. Schematic of a specialized electrochemical cell to give access to varying oxygen chemical potential gradients on one sample. LDM denotes the cathode (La,D)$MnO_{3-\delta}$ with D = Ca, Sr, Ba and YSZ the zirconia electrolyte. The graph on the bottom shows the corresponding lateral gradient in oxygen chemical potential gradient. Adapted with permission from Ref. [43]. Copyright (2020) American Chemical Society.

that allowed accessing a polarization gradient across a single thin-film electrode to obtain a range of oxygen chemical potentials [44]. Later, the same group found they could prevent the precipitation by the addition of Hf, again evidenced through NAP-XPS, preserving cell performance for a much longer time [45].

A strong dependence of the extent of Sr precipitation in LSF on the presence of water vapor and sulfur impurities — ubiquitous contaminants in SOFC oxidant gases in technological applications — has been found in addition to the now generally accepted pO_2 dependence [46]. Here, Zhang *et al.* used the versatility of the O 1s core levels to identify hydroxyl and SO_x species as a function of pH_2O and pO_2 with NAP-XPS.

That the Sr segregation and precipitation phenomenon can be described as universal in mixed conductors is corroborated by studies of Paloukis *et al.* [47], where exposure of $La_{0.75}Sr_{0.25}Cr_{0.9}Fe_{0.1}O_{3-\delta}$ (LSCrF) to reducing gas atmospheres (H_2, H_2O, and mixtures) at 300°C (after annealing in the same atmosphere at 500°C) lead to a complete vanishing of the Sr3d signal of the subsurface perovskite phase, indicating that the surface was almost completely covered with the precipitated phase. Since the experiments were carried out on screen-printed samples resulting in a more powder-like morphology as opposed to the PLD thin films discussed earlier, it suggests that the precipitation phenomenon is not solely a function of morphology, at least at severely reducing atmospheres that were not tested in the LSC case. It does seem that in less reducing conditions (O_2 and UHV, i.e., ~10^{-6} mbar) the morphology stabilizes the materials against segregation and precipitation as the same group found much smaller effects in these cases [48].

All the above-discussed processes are in principle not limited to (mixed) A-site cations of the ABO_3 perovskite structure but have not been found for B = Mn, Fe, and Co. For Cr, however, Chen *et al.* could show that in oxidizing atmospheres, Cr^{3+} in LSCrF is oxidized to Cr^{6+} and subsequently precipitates [49]. In contrast to the A-site precipitation which hampers performance, in this instance, performance is not impacted, possibly due to inhibition of the detrimental SrO precipitation similar to the Hf decoration in LSM [45].

The Sr-rich end member of LSC, $SrCoO_3$, is a good example of a mixed conductor undergoing a phase transition which is governed by the oxygen stoichiometry. In contrast to some complex mixed conducting perovskite which can accommodate oxygen vacancy concentrations of up to $\delta = 0.8$ through entropic stabilization [50], many of the late 3d transition metal perovskites have a natural limit of $\delta = 0.5$: at this degree of deficiency, a brownmillerite $A_2B_2O_5$ phase with ordered oxygen vacancies forms [51,52]. Controlling the oxygen content electrochemically, *Lu et al.* found a reversible topotactic phase transition from $Sr_2Co_2O_5$ to $SrCoO_3$ [53] and tracked changes in the electronic structure by NAP-XPS [54]. They concluded that the Fermi level shifts downwards in the brownmillerite phase upon oxidation, whereas after the phase transition to the perovskite, the Fermi level stays approximately constant, even during increased oxygen intercalation into the lattice. This also corroborates the findings of Nenning *et al.*, where in metallic LSC no shift of the Fermi level during electrochemical oxidation and reduction was observed [31].

13.5.3 *Cationic mixed conductors*

Ionic conduction in MIEC is not limited to anions, as the cationic rearrangements resulting in segregation reactions and phase transitions discussed earlier suggest. A magnitude of cationic conductivity is useful for SSEC devices, however, either requires extremely high temperatures as the diffusivity of cations is orders of magnitude lower than that of anions [55] or is limited to light elements, such as Li and Na. The intercalation of Li

$$Li^+ + e^- + MO_x \rightleftharpoons LiMO_x, \qquad (13.28)$$

for example, is the reaction during the charge and discharge of a Li battery with a metal oxide MO_x as the cathode [56]. Here, we refrain from using the Kröger–Vink notation, as the use of point defect chemistry formalisms — though tremendously useful [57] — is much more sparsely encountered than in the examples discussed earlier. Most of the concepts introduced earlier are applicable here, too. NAP-XPS is also used as a tool to elucidate processes in Li battery cathodes, though used less frequently [58]. The reason for this might be that the "standard" use case of NAP-XPS — investigating gas–solid interactions — does not align with the pressing scientific question intrinsic to Li battery materials. For those, other, more bulk-sensitive spectroscopic techniques [59], such as RIXS[60] and STXM [61,62], are more useful. NAP-XPS has been used by Lu *et al.* to reveal reversible oxygen reduction and evolution at the $Li_xV_2O_5$ cathode surface in a Li-air battery [63]. The group of Hahlin *et al.* demonstrated recently that the charge and discharge of a battery with Li_xCoO_2 cathodes could be tracked by the Co 2p BE shifts [64].

13.5.4 *Cermets*

The material class of cermets (a portmanteau of "Ceramic" and "Metal") is a special case of a mixed electronic ionic conductor as it is a composite of an ionically and an electronically conducting phase and hence will only be discussed briefly for completeness' sake.

One state-of-the-art material is Ni/YSZ cermet, where metallic Ni is the electron and YSZ is the ion-conducting phase. When used as an electrode, the oxygen intercalation reaction has to take place at the triple-phase boundary (see Figure 13.2, right). One primary feat of this material when used as an SSEC anode is that the metallic phase changes reversibly from metallic Ni to NiO, depending on oxygen partial pressure and electrochemical polarization [65]. This redox reaction can be observed and quantified elegantly *operando* with NAP-XPS, as demonstrated, for example, by Paloukis *et al.* [47]. Here, the authors could show that the amount of Ni in the first few nm is significantly depleted in reducing conditions, which is reversible by exposure to oxidizing conditions.

Exchanging the purely ionic conductor YSZ with a ceramic component that can be reduced to become a mixed conductor, such as Gd-doped ceria (GDC) [18], the overall performance of these cermets can be increased. Mewafy *et al.* quantified the ratio of reduced metallic Ni to NiO by NAP-XPS as a function of applied overpotential [66] and showed that, in contrast to Ni/YSZ, the electrode

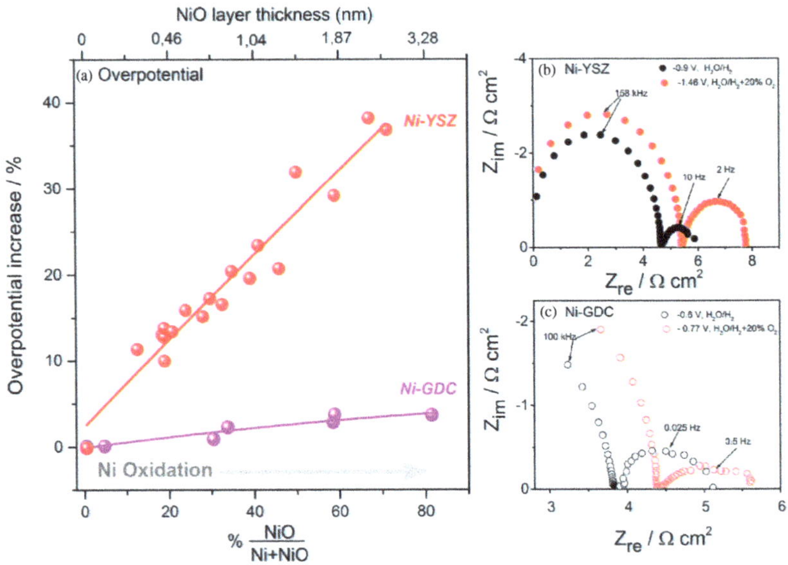

Fig. 13.16. Comparison of cermets with Ni as metallic phase and GDC or YSZ as the ceramic phase. The overpotential increase at the same NiO content is much larger for the YSZ variant, indicating the participation of the $Ce^{3+/4+}$ redox couple in the case of GDC. Adapted with permission from Ref. [66]. Copyright (2019) American Chemical Society.

polarization resistance remained constant (Figure 13.16). This was ascribed to the fact that the $Ce^{3+/4+}$ redox couple could compensate for the oxidation of Ni. Furthermore, this signified that the active region of the catalyst now expanded beyond the triple-phase boundary, as reduced ceria is electrochemically active as well. The emergence of such Ce^{3+} states in Ni/GDC has also been shown by Kooser *et al.* [11] in a dual-chamber SOFC setup.

13.6 Concluding Remarks

By no means should this chapter be taken as an exhaustive review of all materials and applications related to MIEC but rather as a guide of how the fundamental and scientifically important concepts governing processes in these materials can be investigated with NAP-XPS.

NAP-XPS has become a mainstay in solid-state electrochemistry research, as evidenced by the examples given here. Though there are still limitations to the technique with respect to pressure and temperatures, the insights gained from studies on the model system in idealized environments could provide mechanistic understanding not attainable through other means.

It is important to emphasize, though, that, as in any research field, the choice of technique should follow the necessity defined by the scientific question at hand and not vice versa. Consequently, the most enlightening data are gathered from the combination of feasible experiments, each contributing to the understanding in a different way. The use of *operando* NAP-XPS on working electrochemical cells is an example where this is achieved in a single experiment to a certain extent, though it always has to be heavily scrutinized whether two (or more) individual experiments can yield similar results, often with higher accuracy for each dataset than would be possible from a combinatorial approach and with less experimental challenges.

Likewise, it is of utmost importance that it is understood how electro- and physico-chemical concepts that are used to describe MIEC can be related to data obtained from spectroscopic tools. The concept of oxidation states, formal charges, and point defects used frequently in the former does not have a straightforward correspondence in spectroscopic signatures [1]. Keeping this in mind, NAP-XPS did already have a tremendous impact on the understating of mixed conductors under operating conditions, as the materials at hand behave quite dynamic on length scales (in terms of distance from the solid/gas interface) that are hardly accessible by conventional techniques.

Acknowledgments

The author gratefully acknowledges financial support from the German Science Foundation (DFG) in the frame of the priority program "Manipulation of matter controlled by electric and magnetic field: Towards novel synthesis and processing routes of inorganic materials" (SPP 1959) under project MU 4591/1-2.

Furthermore, the author likes to thank Roger A. De Souza, Tomáš Duchoň, and Stefan Cramm for helpful discussions.

References

[1] De Souza RA, Mueller DN. Electrochemical methods for determining ionic charge in solids. Nat Mater. 2020. Doi:10.1038/s41563-020-0790-9.
[2] Chueh WC. Haile SM. Electrochemistry of mixed oxygen ion and electron conducting electrodes in solid electrolyte cells. Annu Rev Chem Biomol Eng. 2012;3:313–341.
[3] West AR. Solid electrolytes and mixed ionic-electronic conductors: An applications overview. J Mater Chem. 1991;1:157–162.
[4] Kröger FA, Vink HJ. Relations between the concentrations of imperfections in crystalline solids. Solid State Phys. 1956;3:307–435.
[5] Norby TA. Kröger-Vink compatible notation for defects in inherently defective sublattices. J Korean Ceram Soc. 2010;47:19–25.
[6] Szot K *et al.* Formation of micro-crystals on the (100) surface of $SrTiO_3$ at elevated temperatures. Surf Sci. 2000;460:112–128.
[7] Moos R, Hardtl KH. Defect chemistry of donor-doped and undoped strontium titanate ceramics between 1000°C and 1400°C. J Am Ceram Soc. 1997;80:2549–2562.
[8] Andrä M *et al.* Oxygen partial pressure dependence of surface space charge formation in donor-doped $SrTiO_3$. APL Mater. 2017;5.
[9] Andrä M *et al.* Chemical control of the electrical surface properties in donor-doped transition metal oxides. Phys Rev Mater. 2019;3:1–9.
[10] Lee W, Han JW, Chen Y, Cai Z, Yildiz B. Cation size mismatch and charge interactions drive dopant segregation at the surfaces of manganite perovskites. J Am Chem Soc. 2013;135:7909–7925.
[11] Kooser K *et al.* Operando high-temperature near-ambient pressure X-ray photoelectron spectroscopy and impedance spectroscopy study of Ni–$Ce_{0.9}Gd_{0.1}O_{2-\delta}$ solid oxide fuel cell anode. Int J Hydrogen Energy 2020;45:25286–25298.
[12] Whaley JA *et al.* Note: Fixture for characterizing electrochemical devices in-operando in traditional vacuum systems. Rev Sci Instrum. 2010;81:086104.

[13] Mueller DN, Machala ML, Bluhm H, Chueh WC. Redox activity of surface oxygen anions in oxygen-deficient perovskite oxides during electrochemical reactions. Nat Commun. 2015;6:6097.

[14] Nenning A, Fleig J. Electrochemical XPS investigation of metal exsolution on SOFC electrodes: Controlling the electrode oxygen partial pressure in ultra-high-vacuum. Surf Sci. 2019;680:43–51.

[15] Decaluwe SC *et al.* In situ characterization of ceria oxidation states in high-temperature electrochemical cells with ambient pressure XPS. J Phys Chem C. 2010;114:19853–19861.

[16] El Gabaly F *et al.* Measuring individual overpotentials in an operating solid-oxide electrochemical cell. Phys Chem Chem Phys. 2010;12: 12138–12145.

[17] Zhang C *et al.* Measuring fundamental properties in operating solid oxide electrochemical cells by using in situ X-ray photoelectron spectroscopy. Nat Mater. 2010;9:944–949.

[18] Mogensen M, Sammes NM, Tompsett GA. Physical, chemical and electrochemical properties of pure and doped ceria. Solid State Ionics. 2000;129:63–94.

[19] Pfau A, Schierbaum KD. Pfau, Schierbaum — The electronic structure of stoichiometric and reduced CeO, surfaces an XPS, UPS and HREELS study. Surf Sci. 1994;321:71–80.

[20] Duchoň T *et al.* Ordered phases of reduced ceria as epitaxial films on Cu(111). J Phys Chem C. 2014;118:5058–5059.

[21] Balaji Gopal C *et al.* Equilibrium oxygen storage capacity of ultrathin $CeO_{2-\delta}$ depends non-monotonically on large biaxial strain. Nat Commun. 2017;8:1–12.

[22] Chueh WC *et al.* Highly enhanced concentration and stability of reactive Ce^{3+} on doped CeO_2 surface revealed in Operando. Chem Mater. 2012;24:1876–1882.

[23] Zhang C *et al.* Mechanistic studies of water electrolysis and hydrogen electro-oxidation on high temperature ceria-based solid oxide electrochemical cells. J Am Chem Soc. 2013;135:11572–11579.

[24] Feng ZA, El Gabaly F, Ye X, Shen ZX, Chueh WC. Fast vacancy-mediated oxygen ion incorporation across the ceria-gas electrochemical interface. Nat Commun. 2014;5:4374.

[25] Axnanda S *et al.* Direct work function measurement by gas phase photoelectron spectroscopy and its application on PbS nanoparticles. Nano Lett. 2013;13:6176–6182.

[26] Feng ZA *et al.* Origin of overpotential-dependent surface dipole at CeO_{2-x}/gas interface during electrochemical oxygen insertion reactions. Chem Mater. 2016;28:6233–6242.

[27] Mizusaki J, Yoshihiro M, Yamauchi S, Fueki K. Nonstoichiometry and defect structure of the perovskite-type oxides $La_{1-x}Sr_xFeO_{3-\delta}$. J Solid State. 1985;58:257–266.

[28] Kuhn M *et al.* Oxygen nonstoichiometry and thermo-chemical stability of Perovskite-type $La_{0.6}Sr_{0.4}Co_{1-y}Fe_yO_{3-\delta}$ (y = 0, 0.2, 0.4, 0.5, 0.6, 0.8, 1) Materials. J Electrochem Soc. 2012;160:F34–F42.

[29] Abbate M *et al.* Controlled-valence of $La_{1-x}Sr_xFeO_3$ and $La_{1-x}Sr_xMnO_3$ studied by soft X-ray absorption spectroscopy. Phys Rev B. 1992;46: 4511–4519.

[30] Bocquet A E *et al.* Electronic structure of early 3d-transition-metal oxides by analysis of the 2 p core-level photoemission spectra. Phys Rev B. 1996;53:1161–1170.

[31] Nenning A *et al.* Ambient pressure XPS study of mixed conducting perovskite-type SOFC cathode and anode materials under well-defined electrochemical polarization. J Phys Chem C. 2016;120: 1461–1471.

[32] Kuhn M, Hashimoto S, Sato K, Yashiro K, Mizusaki J. Oxygen nonstoichiometry, thermo-chemical stability and lattice expansion of $La_{0.6}Sr_{0.4}FeO_{3-\delta}$. Solid State Ionics. 2011;195:7–15.

[33] Fujimori A *et al.* Core-level photoemission measurements of the chemical potential shift as a probe of correlated electron systems. J Electron Spectros. Relat. Phenomena. 2002;124:127–138.

[34] Mutoro E *et al.* Reversible compositional control of oxide surfaces by electrochemical potentials. J Phys Chem Lett. 2012;3:40–44.

[35] Vasquez RP. X-ray photoelectron spectroscopy study of Sr and Ba compounds. J Electron Spectros Relat Phenomena. 1991;56: 217–240.

[36] Yang X, Park DW, Kim M-I. Selective oxidation of hydrogen sulfide over $LaCoO_3$ and $LaSrCoO_4$ mixed oxides. Korean J Chem Eng. 2007;24:592–595.

[37] Van Der Heide PAW. Systematic X-ray photoelectron spectroscopic study of $La_{1-x}Sr_x$-based perovskite-type oxides. Surf Interface Anal. 2002;33:414–425.

[38] Crumlin EJ *et al.* Surface strontium enrichment on highly active perovskites for oxygen electrocatalysis in solid oxide fuel cells. Energy Environ Sci. 2012;5:6081.

[39] Cai Z, Kuru Y, Han JW, Chen Y, Yildiz B. Surface electronic structure transitions at high temperature on perovskite oxides: The case of strained $La_{0.8}Sr_{0.2}CoO_3$ thin films. J Am Chem Soc. 2011;133: 17696–17704.

[40] Keeble DJ *et al.* Identification of A- and B-site cation vacancy defects in perovskite oxide thin films. Phys Rev Lett. 2010;105:3–6.

[41] Christensen A, Stoltze P, Norskov JK. Size dependence of phase separation in small bimetallic clusters. J Phys Condens Matter. 1995;7:1047–1057.

[42] Mueller DN, *et al.* Nanoscopic surface decomposition of $Pr_{0.5}Ba_{0.5}CoO_{3-\delta}$ perovskites turns performance descriptors ambiguous. J Phys Chem C. 2021. Doi:10.1021/acs.jpcc.1c00976.

[43] Kim D, Bliem R, Hess F, Gallet JJ, Yildiz B. Electrochemical polarization dependence of the elastic and electrostatic driving forces to aliovalent dopant segregation on $LaMnO_3$. J Am Chem Soc. 2020; 142:3548–3563.

[44] Huber TM *et al.* Experimental design for voltage driven tracer incorporation and diffusion studies on oxide thin film electrodes. J Electrochem Soc. 2017;164:F809–F814.

[45] Bliem R, Kim D, Wang J, Crumlin EJ, Yildiz B. Hf deposition stabilizes the surface chemistry of perovskite manganite oxide. J Phys Chem C. 2021;125:3346–3354.

[46] Zhang D *et al.* Hydroxylation and cation segregation in $(La_{0.5}Sr_{0.5})FeO_{3-\delta}$ electrodes. Chem Mater. 2020;32:2926–2934.

[47] Paloukis F *et al.* Insights into the surface reactivity of cermet and perovskite electrodes in oxidizing, reducing, and humid environments. ACS Appl Mater Interfaces. 2017;9:25265–25277.

[48] Paloukis F *et al.* In situ X-ray photoelectron spectroscopy study of complex oxides under gas and vacuum environments. Appl Surf Sci. 2017;423:1176–1181.

[49] Chen D *et al.* Revising the role of chromium on the surface of perovskite electrodes: Poison or promoter for the solid oxide electrolysis cell performance? J Catal. 2020;381:520–529.
[50] Mueller DN, De Souza RA, Yoo H-I, Martin M. Phase stability and oxygen nonstoichiometry of highly oxygen-deficient perovskite-type oxides: A case study of (Ba,Sr)(Co,Fe)$O_{3-\delta}$. Chem Mater. 2012; 24:269–274.
[51] McIntosh S, Vente JF, Haije WG, Blank DHA, Bouwmeester HJM. Structure and oxygen stoichiometry of $SrCo_{0.8}Fe_{0.2}O_{3-\delta}$ and $Ba_{0.5}Sr_{0.5}Co_{0.8}Fe_{0.2}O_{3-\delta}$. Solid State Ionics. 2006;177:1737–1742.
[52] Vente JF, McIntosh S, Haije WG, Bouwmeester HJM. Properties and performance of $Ba_xSr_{1-x}Co_{0.8}Fe_{0.2}O_{3-\delta}$ materials for oxygen transport membranes. J Solid State Electrochem. 2006;10:581–588.
[53] Lu Q, Yildiz B. Voltage-controlled topotactic phase transition in thin-film $SrCoO_x$ monitored by *in situ* X-ray diffraction. Nano Lett. 2016;16:1186–1193.
[54] Lu Q, Chen Y, Bluhm H, Yildiz B. Electronic structure evolution of $SrCoO_x$ during electrochemically driven phase transition probed by *in situ* X-ray spectroscopy. J Phys Chem C. 2016;120:24148–24157.
[55] Harvey SP, De Souza RA, Martin M. Diffusion of La and Mn in $Ba_{0.5}Sr_{0.5}Co_{0.8}Fe_{0.2}O_{3-\delta}$ polycrystalline ceramics. Energy Environ Sci. 2012;5:5803–5813.
[56] Manthiram A. A reflection on lithium-ion battery cathode chemistry. Nat Commun. 2020;11:1–9.
[57] Maier J. Review — Battery materials: Why defect chemistry? J Electrochem Soc. 2015;162:A2380–A2386.
[58] Han Y, Zhang H, Yu Y, Liu Z. *In Situ* characterization of catalysis and electrocatalysis using APXPS. ACS Catal. 2021;11:1464–1484.
[59] Miller EC, Toney MF. X-ray studies of energy materials. In Jaeschke EJ, Khan S, Schneider JR, Hastings JB (Eds.), *Synchrotron Light Sources and Free-Electron Lasers: Accelerator Physics, Instrumentation and Science Applications* (pp. 1803–1824). Springer International Publishing, 2020. Doi:10.1007/978-3-030-23201-6_65.
[60] Gent WE *et al.* Coupling between oxygen redox and cation migration explains unusual electrochemistry in lithium-rich layered oxides. Nat Commun. 2017;8:2091.

[61] Li Y *et al.* Current-induced transition from particle-by-particle to concurrent intercalation in phase-separating battery electrodes. Nat Mater. 2014;13:1149–1156.
[62] Lim J *et al.* Origin and hysteresis of lithium compositional spatiodynamics within battery primary particles. Science (80-). 2016;353: 566–571.
[63] Lu YC *et al.* In situ ambient pressure X-ray photoelectron spectroscopy studies of lithium-oxygen redox reactions. Sci Rep. 2012;2:1–6.
[64] Zhu S *et al.* HIPPIE: A new platform for ambient-pressure X-ray photoelectron spectroscopy at the MAX IV laboratory. J Synchrotron Radiat. 2021;28:624–636.
[65] Graves C, Ebbesen SD, Jensen SH, Simonsen SB, Mogensen MB. Eliminating degradation in solid oxide electrochemical cells by reversible operation. Nat Mater. 2015;14:239–244.
[66] Mewafy B *et al.* Influence of surface state on the electrochemical performance of nickel-based cermet electrodes during steam electrolysis. ACS Appl Energy Mater. 2019;2:7045–7055.

https://doi.org/10.1142/9781800613294_bmatter

Index

CATALYTIC SCIENCE SERIES

(Continued from page ii)

Vol. 11 *Supported Metals in Catalysis (Second Edition)*
by J. A. Anderson

Vol. 10 *Concepts in Syngas Manufacture*
by J. Rostrup-Nielsen and L. J. Christiansen

Vol. 9 *Deactivation and Regeneration of Zeolite Catalysts*
edited by M. Guisnet and F. R. Ribeiro

Vol. 8 *Petrochemical Economics: Technology Selection in a Carbon Constrained World*
by D. Seddon

Vol. 7 *Combinatorial Development of Solid Catalytic Materials: Design of High-Throughput Experiments, Data Analysis, Data Mining*
edited by M. Baerns and M. Holeňa

Vol. 6 *Catalysis by Gold*
edited by G. C. Bond, C. Louis and D. T. Thompson

Vol. 5 *Supported Metals in Catalysis*
edited by J. A. Anderson and M. F. García

Vol. 4 *Isotopes in Heterogeneous Catalysis*
edited by Justin S. J. Hargreaves, S. David Jackson and Geoff Webb

Vol. 3 *Zeolites for Cleaner Technologies*
edited by Michel Guisnet and Jean-Pierre Gilson

Vol. 2 *Catalysis by Ceria and Related Materials*
edited by Alessandro Trovarelli

Vol. 1 *Environmental Catalysis*
edited by F. J. J. G. Janssen and R. A. van Santen

www.ingramcontent.com/pod-product-compliance
Ingram Content Group UK Ltd.
Pitfield, Milton Keynes, MK11 3LW, UK
UKHW022106170926
8624IPUK00024B/169

9 781800 613287